W9-CPF-946

PUBLICATIONS OF THE NEWTON INSTITUTE

Mathematics of Derivative Securities

Bank of America, N.A.
233 SOUTH WACKER DRIVE
CHICAGO, IL 60606
IL 1-003-27-35

233 SOUTH WACKER DRIVE
CHICAGO, ILLINOIS 60606

Publications of the Newton Institute

Edited by H.P.F. Swinnerton-Dyer

Executive Director, Isaac Newton Institute for Mathematical Sciences

The Isaac Newton Institute of Mathematical Sciences of the University of Cambridge exists to stimulate research in all branches of the mathematical sciences, including pure mathematics, statistics, applied mathematics, theoretical physics, theoretical computer science, mathematical biology and economics. The four six-month long research programmes it runs each year bring together leading mathematical scientists from all over the world to exchange ideas through seminars, teaching and informal interaction.

Associated with the programmes are two types of publication. The first contains lecture courses, aimed at making the latest developments accessible to a wider audience and providing an entry to the area. The second contains proceedings of workshops and conferences focusing on the most topical aspects of the subjects.

MATHEMATICS OF DERIVATIVE SECURITIES

edited by

M.A.H. Dempster

University of Cambridge

and

S.R. Pliska

University of Illinois, Chicago

CAMBRIDGE
UNIVERSITY PRESS

PUBLISHED BY THE PRESS SYNDICATE OF THE UNIVERSITY OF CAMBRIDGE
The Pitt Building, Trumpington Street, Cambridge CB2 1RP, United Kingdom

CAMBRIDGE UNIVERSITY PRESS
The Edinburgh Building, Cambridge CB2 2RU, United Kingdom
40 West 20th Street, New York, NY 10011-4211, USA
10 Stamford Road, Oakleigh, Melbourne 3166, Australia

© Cambridge University Press 1997

This book is in copyright. Subject to statutory exception
and to the provisions of relevant collective licensing agreements,
no reproduction of any part may take place without
the written permission of Cambridge University Press.

First published 1997

Printed in the United Kingdom at the University Press, Cambridge

Typeset in 12pt Computer Modern

A catalogue record for this book is available from the British Library

Library of Congress cataloguing in Publication data

ISBN 0 521 58424 8 hardback

CONTENTS

PART IV. TERM STRUCTURE AND INTEREST RATE DERIVATIVES

PART V. NUMERICAL METHODS

CONTRIBUTORS

Phillippe Artzner, Institut de Recherche Mathématique Avancée, Université Louis Pasteur, 7 rue Rene Descartes, F 67084 Strasbourg Cedex, France. artzner@math.u-strasbg.fr

Simon Babbs, First Vice President and Head of Quantitative Research, First National Bank of Chicago, First National House, 90 Long Acre, London WC2E 9RB, UK. sbabbs@fnbc.co.uk

Martin W. Baxter, Statistical Laboratory, University of Cambridge, 16 Mill Lane, Cambridge CB2 1SB, UK. M.W.Baxter@statslab.cam.ac.uk

Jean-Philippe Bouchaud, Service de Physique de l'Etat Condensé, CEA-Saclay, 91191 Gif sur Yvette, CEDEX, France *and* Science et Finance, 109–111 rue V. Hugo, 92 532 Levallois, France.

Alan Brace, FMMA, PO Box 731, Grosvenor Place, Sydney 2000, Australia.

M.J. Brennan, Irwin and Goldyne Hearsh Professor of Banking and Finance, University of California, Los Angeles CA 90024, USA *and* Professor of Finance, London Business School.

M. Chesney, Département Finance et Economie, 1, rue de la libération, 78351 Jouy en Josas Cedex, France. chesney@gwsmtp.hec.fr

N.I. Crew, University of California, Los Angeles CA 90024, USA.

N.J. Cutland, School of Mathematics, University of Hull, Hull HU6 7RX, UK. n.j.cutland@maths.hull.ac.uk

Jakša Cvitanić, Department of Statistics, Columbia University, New York, NY 10027, USA. cj@stat.columbia.edu

Mark H.A. Davis, Tokyo-Mitsubishi International plc, 6 Broadgate, London EC2M 2AA, UK. mark.davis@t-mi.com

Michael A.H. Dempster, Judge Institute of Management Studies, University of Cambridge, Trumpington St., Cambridge CB2 1AG, UK. mahd2@cam.ac.uk

Bruno Dupire, Paribas Capital Markets, Swaps and Options Research Team, 33 Wigmore Street, London W1H 0BN, UK.

Philip H. Dybvig, Olin School of Business, Washington University, Campus Box 1133, St. Louis, MO 63130, USA. dybvig@dybfin.wustl.edu

Lina El-Jahel, Department of Economics, Birkbeck College, 7-15, Gresse Street, London, W1P 2LL, UK.

Robert J. Elliott, Department of Mathematical Sciences, University of Alberta, C.A.B. 632, University of Alberta, Edmonton, Alberta, Canada T6G 2G1.
relliott@gpu.srv.ualberta.ca

Bjorn Flesaker, Bear Stearns, 245 Park Avenue, New York, NY 10167, USA.
wdec98a@prodigy.com

Marco Frittelli, Facoltà di Economia, Università degli Studi di Milano, Via Sigieri 6, 20135 Milano, Italy.
frittema@imiucca.csi.unimi.it

S.K. Gandhi, NatWest Markets, 135 Bishopsgate, London EC2N 3UR, UK.

H. Geman, Université Paris IX Dauphine *and* ESSEC, Département Finance, Avenue Bernard Hirsch, BP 105, 95021 Cergy-Pontoise Cedex, France.
p_geman@ped.essec.fr

Stewart D. Hodges, Financial Options Research Centre, Warwick Business School, University of Warwick, Coventry CV4 7AL, UK.
forcsh@razor.wbs.warwick.ac.uk

Lane P. Hughston, Merrill Lynch International, 25 Ropemaker Street, London EC2Y 9LY, UK.
lane@lonnds.ml.com

P.J. Hunt, ABN AMRO Bank, 6th floor, Broadgate Court, 199 Broadgate, London EC2M 3TY, UK.

M. Jeanblanc-Picqué, Equipe d'analyse et probabilités, Université d'Evry Val d'Essonne, Boulevard des coquibus, 91025 Evry Cedex, France.
jeanbl@lami.univ-evry.fr

P.E. Kopp, School of Mathematics, University of Hull, Hull HU6 7RX, UK.
P.E.Kopp@maths.hull.ac.uk

Harold J. Kushner, Division of Applied Mathematics, Brown University, Providence, RI 02912, USA.
hjk@dam.brown.edu

David Lando, University of Copenhagen, Institute of Mathematical Statistics, Universitetsparken 5, DK-2100 Copenhagen Ø, Denmark.
dlando@math.ku.dk

Charles H. Lahaie, College of Business and Management, University of Maryland, College Park, MD 20742, USA.
clahaie@bmgtmail.umd.edu

John P. Lehoczky, Department of Statistics, Carnegie Mellon University, Pittsburgh, PA 15213-3890, USA.
jpl@stat.cmu.edu

Hans Lindberg, Economics Department, Sveriges Riksbank, Brunkebergstorg, 11, S-103, 37 Stockholm, Sweden.

Dilip B. Madan, College of Business and Management, University of Maryland, College Park, MD 20742, USA.
dilip@bmgtmail.umd.edu

A. Mbanefo, Credit Suisse Financial Products, First Boston Corporation, 55 E 52nd Street, 34th Floor, New York, NY 10055, USA.
ambanefo@csfpny.fbc.com

F. Mercurio, Tinbergen Institute, Erasmus University Rotterdam, Burg. Oudlaan 50, 3062 PA Rotterdam, Netherlands.
mercurio@tir.few.eur.nl

Marek Musiela, Department of Statistics, School of Mathematics, The University of New South Wales, Sydney 2052, NSW, Australia.
musiela@maths.unsw.edu.au

Spassimir H. Paskov Union Bank of Switzerland, 299 Park Avenue, New York, NY 10171, USA.
nypav@ubss.com
and Department of Computer Science, Columbia University, New York, NY 10027, USA.
paskov@cs.columbia.edu

William Perraudin, Department of Economics, Birkbeck College, 7-15, Gresse Street, London, W1P 2LL, UK.
100757.130@CompuServe.COM

Marc Potters, Science et Finance, 109–111 rue V. Hugo, 92 532 Levallois, France.

L.C.G. Rogers, School of Mathematical Sciences, University of Bath, Bath BA2 7AY, UK.
L.C.G.Rogers@maths.bath.ac.uk

Michael J.P. Selby, 73, Grove End Gardens, Abbey Road, St John's Wood, London NW8 9LN, UK.

Didier Sornette, Laboratoire de Physique de la matière Condensée, Université de Nice-Sophia Antipolis, B.P. 70, Parc Valrose, 06108 Nice CEDEX 2, France *and* Science et Finance, 109–111 rue V. Hugo, 92 532 Levallois, France.

T.C.F. Vorst, Department of Finance, Erasmus University Rotterdam *and* Erasmus Center for Financial Research, P.O. Box 1738, 3000 DR Rotterdam, Netherlands.
vorst@opres.few.eur.nl

Nick Webber, Warwick Business School, University of Warwick, Coventry CV4 7AL, UK.
bsrdq@csv.warwick.ac.uk

W. Willinger, Bellcore, Morristown NJ 07962, USA.

M.C. Wyman, Scarborough College, Scarbourough, North Yorkshire, UK.
mattheww@ucscarb.ac.uk

M. Yor, Laboratoire de probabilités, Tour 56, 3-ième étage, Université PMC, 4, Place Jussieu, 75252 Paris Cedex 05, France.
secret@proba.jussieu.fr

Foreword

Robert C. Merton

The two-day discussion meeting on mathematical models in finance at the Royal Society of London in November 1993 and its subsequent publication in *Philosophical Transactions* provided signal recognition to the rapidly advancing but still comparatively new discipline which relates mathematical finance theory and finance practice. But this strong signal was outdone by the extraordinary commitment of time, breadth of scholarship, and economic resources devoted to the Mathematical Finance Programme of six-month duration at the Isaac Newton Institute for Mathematical Sciences. That the contents of this volume represent only a small portion of the work produced testifies to the scale of the Programme's contributions. The fine introductory chapter by the two co-editors happily removes any need for me to comment here on the content of the volume. It is enough to say that the chapters focused for the most part on derivative securities are representative of the high quality and mathematical sophistication of research in the field. Instead I try my hand at locating the relation between mathematical research in finance theory and finance practice, as it has evolved and in prospect.

The essence of finance theory is the study of allocation and deployment of economic resources, both spatially and across time, in an uncertain environment. To capture the influence and interaction of time and uncertainty effectively requires sophisticated analytical tools. Indeed, mathematical models of modern finance contain some truly elegant applications of probability and optimization theory. But, of course, all that is elegant in science need not also be practical; and surely, not all that is practical in science is elegant. Here we have both. Over the past two decades, the mathematically complex models of finance theory have had a direct and important influence on finance practice. This conjoining of intrinsic intellectual interest with extrinsic application is central to research in modern finance.

The origins of much of the mathematics in modern finance can be traced to Louis Bachelier's 1900 dissertation on the theory of speculation. This work marks the twin births of both the continuous-time mathematics of stochastic processes and the continuous-time economics of derivative-security pricing. Although largely lost to financial economists for more than a half century, this same mathematics rediscovered by Kiyoshi Ito and Paul Samuelson, has been an essential tool in the development of modern mathematical finance theory.

Through the resulting theory's influence on finance practice, this mathematics has played a fundamental role in supporting the creation of new financial products and markets around the globe. In the present and the impending future, that role is expanding to support the design of entirely new financial institutions, decision-making by senior management , and public policy on the financial system. To underscore that point, I begin with a few remarks about financial innovation of the past, this to be followed by some observations on the directions for change in the future.

New financial product and market designs, improved computer and tele-communications technology and advances in the theory of finance during the past quarter-century have led to dramatic and rapid changes in the structure of global financial markets and institutions. The scientific breakthroughs in financial modeling in this period both shaped and were shaped by the extraor-dinary flow of financial innovation which coincided with those changes. To put this into perspective, we have only to recall the fall of Bretton Woods lead-ing to floating-exchange rates for currencies, the development of the national mortgage market in the United States, the first oil shock, and the creation of the first listed options exchange which accompanied publication of the famous Black–Scholes option-pricing model, all in 1973; ERISA and the subsequent development of the U.S. pension fund industry and the first money-markets fund with check writing that took place in 1974; and the fact that 25 years ago total assets in all U.S. mutual funds amounted to $48 billion. Today those funds are some 50 times larger, with one institution, Fidelity, accounting for approximately $450 billion. In this same period, average daily trading volume on the New York Stock Exchange grew from 12 million shares to more than 300 million. Even more dramatic were the changes in Europe and in Asia. The cumulative impact has significantly affected all of us—as users, producers, or overseers of the financial system.

Nowhere has this been more the case than in the development, refinement and broad-based implementation of contracting technology. Derivative secu-rities such as futures, options, swaps and other contractual agreements, the underlying substantive subject of this mathematical volume, provide a prime example. Those innovations in financial-contracting technology have improved efficiency by expanding opportunities for risk sharing, lowering transactions costs and reducing information and agency costs. Some observers see the ex-traordinary growth in the use of derivatives as fad-like, but a more likely explanation is the vast saving in transactions costs derived from their use. The cost of implementing financial strategies for institutions using deriva-tives can be one tenth to one twentieth of the cost of executing them in the underlying cash-market securities. The significance of reducing spread costs in financing can be quite dramatic for corporations and for sovereigns: for instance, a 1 percent (i.e., 100 basis point) reduction in debt-spread cost in Italian government debt would reduce the deficit by 1.25 percent of the Gross Domestic Product of Italy.

Further improved technology, together with growing breadth and experience in the applications of derivatives, should reduce transactions costs secularly as both users and producers of derivatives move down the learning curve. Like retail depositors with automatic teller machines in banks, initial resistance by institutional clients to contractual agreements can be high, but once used, customers tend not to return to the traditional alternatives for implementing financial strategies.

A central process in the past two decades has been the remarkable rate of globalization of the financial system. Inspection of the diverse financial systems of individual nation-states would lead one to question how much effective integration across geopolitical borders could have taken place since those systems are rarely compatible in institutional forms, regulations, laws, tax structures, and business practices. Still, significant integration did take place. This was made possible in large part by derivative securities functioning as adapters. In general, the flexibility created by the widespread use of contractual agreements, other derivatives, and specialized institutional designs provides an offset to dysfunctional institutional rigidities. More specifically, derivative-security technologies provide efficient means for creating cross-border interfaces among otherwise incompatible domestic systems, without making widespread or radical changes within each system. For that reason, future development of derivative-security technology and markets within smaller and emerging-market countries could help form important gateways of access to world capital markets and global risk-sharing. Furthermore, derivatives and other contracting technologies are likely to play a significant role in the financial engineering of the major transitions required for European Monetary Union and for restructuring financial institutions in Japan.

As the preceding remarks are intended to indicate, innovation is a central force driving the financial system toward greater economic efficiency with considerable economic benefit having accrued from the changes over the past several decades. Moreover, both finance research and practitioner experience over that period has lead to vast improvements in our understanding of how to use the new financial technologies to manage risk. Despite all this, we still find today an intense uneasiness among managers, regulators, politicians, the press, and the public over these new derivative-security activities and the associated risks to financial institutions, relative to their more traditional risks such as commercial, real-estate, and less-developed-country lending and loan guarantees. Why? One conjecture about the sources of this collective anxiety holds that their implementation has required major changes in the basic institutional hierarchy and in the infrastructure to support it. As a consequence, the knowledge base required to manage and oversee financial institutions differs considerably from the traditional training and experience of many financial managers and government regulators. Changes of this sort are threatening. It is difficult to deal with change that is exogenous with respect to one's traditional knowledge base and framework and therefore comes to seem beyond our control. Less understanding of the new environment can create a sense of greater risk even when the objective level of risk in the system is unchanged or actually reduced. That knowledge gap may widen since the current pace of financial innovation is anticipated to accelerate into the 21st Century. Moreover, greater complexity of products and the need for more rapid decision-making will likely increase the reliance on models, which implies a growing place for elements of mathematical maturity in the managerial knowledge base. Managing this knowledge gap offers considerable challenge to private institutions and government as well as considerable opportunity to schools of management and engineering and to university departments of mathematics.

The successful financial-service providers and governmental overseers in the impending future will be those that can address the dysfunctional aspects of innovation in financial technology while still fully exploiting their functional benefits. What types of research and training will be needed to manage financial institutions? With this in mind, consider only a few thoughts on the direction of product and service demands of users of the financial system.

The household sector of users in the more developed financial systems have experienced a secular trend of disaggregation in financial services. There are those who see this trend continuing with existing products such as mutual funds being transported into technologically less developed systems. Perhaps so, especially in the more immediate future. However, deep and wide-ranging disaggregation has left households with the responsibility for making important and technically complex, micro financial decisions involving risk (such as detailed asset allocation and estimates of the optimal level of life-cycle saving for retirement) that they had not had to make in the past, are not trained to make in the present, and are unlikely to execute efficiently even with attempts at education in the future. I therefore believe that the trend will shift toward more aggregated financial products and services, which are easier to understand and more tailored toward individual profiles. Those products and services will include integration of human capital considerations, hedging, and income and estate tax planning into the asset-allocation decisions and the creation of financial instruments that eliminate 'short-fall' or 'basis' risk for the households with respect to targeted financial goals such as tuition for children's higher education and desired consumption-smoothing throughout the life-cycle (e.g., preserving the household's standard of living in retirement). Paradoxically, making the products more user-friendly and simpler to understand for customers will create considerably more complexity for the producers of those products. Hence, financial-engineering creativity and the technological and transactional bases to implement that creativity, reliably and cost-effectively, are likely to become a central competitive element in the industry. The resulting complexity will require more elaborate and highly quantitative risk-management systems within financial-service firms and a parallel need for more sophisticated approaches to government oversight. Neither of these can be achieved without greater reliance on mathematical financial modeling, which in turn will be feasible only with continued improvements in the sophistication and accuracy of financial models.

Nonfinancial firms currently use derivative securities and other contractual agreements to hedge interest rate, currency, commodity, and even equity price risks. With improved lower-cost technology and learning-curve experience, this practice is likely to expand. Eventually, this alternative to equity capital as a cushion for risk could have a major change on corporate structures as more firms use hedging to substitute for equity capital; thereby moving from publicly traded shares to private closely-held shares. The big potential shift in the future, however, is from tactical applications of derivatives to strategic ones. For example, a hypothetical oil company with crude oil reserves and gasoline and heating oil distribution but no refining could complete the vertical integration of the firm by using contractual agreements instead of physical acquisition. Thus, by entering into contracts that call for the delivery of crude oil by the firm on one date in return for receiving a mix of refined petroleum products at a prespecified later date, the firm in effect creates a synthetic refinery. Real-world strategic examples in natural gas and electricity are described in Harvard Business School case studies, 'Enron Gas Services' (1994) and 'Tennessee Valley Authority: Option Purchase Agreements' (1996), by Peter Tufano. It is no coincidence that the early applications are in energy- and power- generation industries that need long-term planning horizons and have major fixed-cost components on a large scale with considerable uncertainty. Since energy and power generation are fundamental in every economy, this use for derivatives may become mainline applications in both developed and developing countries. Eventually, such use of derivatives will become standard tools for implementing strategic objectives.

A major requirement for the efficient broad-based application of these contracting technologies in both the household and nonfinancial-firm sectors will be to find effective organizational structures for ensuring contract performance, including global clarification and revisions of the treatment of such contractual agreements in bankruptcy. The need for assurances on contract performance is likely to stimulate further development of the financial-guarantee business for financial institutions. Such institutions will have to improve further the efficiency of collateral management as assurance for performance. As has been known for some two decades, the mathematical models for pricing and hedging derivative securities can be applied directly to the valuation and risk-exposure measurement of financial guarantees.

A consequence of all this prospective technological change will be the need for greater analytical understanding of valuation and risk management by users, producers, and regulators of derivative securities. Furthermore, improvements in efficiency from derivative products will not be effectively realized without concurrent changes in the financial 'infrastructure'—the institutional interfaces between intermediaries and financial markets, regulatory practices, organization of trading, clearing, settlement, other back-office facilities, and management-information systems. To perform its functions as both user and overseer of the financial system, government will need to innovate and make use of derivative-security technology in the provision of risk-accounting standards, designing monetary and fiscal policies, implementing stabilization programs, and financial-system regulation.

In summary, in the distant past, mathematical models had only limited and side-stream effects on finance practice. But in the last quarter century, such models have become mainstream to practitioners in financial institutions and markets around the world. In the future, mathematical models will surely play an indispensable role in the functioning of the global financial system.

Even this brief discourse on the virtues of the application of mathematical models in finance practice would be negligently incomplete without an added word of caution about their use. At times we can lose sight of the ultimate purpose of the models when their mathematics become too interesting. The mathematics of financial models can be applied precisely, but the models are not at all precise in their application to the complex real world. Their accuracy as a useful approximation to that world varies significantly across time and place. The models should be applied in practice only tentatively, with careful assessment of their limitations in each application. This stated, the reader should not allow this cautionary note on practical limitations to get in the way of enjoying the elegance of the mathematics still to come.

Robert C. Merton

Harvard University

June 1997

Foreword

Michael Atiyah

The Isaac Newton Institute for Mathematical Sciences was set up in 1992 as a high-level international centre which would, from year to year, concentrate on exciting new developments in mathematics, interpreted in its broadest sense. The physical sciences figured prominently in its activities as did more recent work in the biological sciences. However, the decision to run a programme on 'financial mathematics' was an unexpected novelty and it reflected the great upsurge of interest in this field. The meeting opened very shortly after the Barings crash, a dramatic indication of the high-profile and extremely topical nature of Derivatives.

A few years later it is clear that there is going to be long-term interest in the application of sophisticated mathematics to the financial markets. This book, which brings together many of the contributions made at the Newton Institute, gives an indication of the scope and significance of the field. The Newton Institute is glad to have played a part in stimulating this activity.

Michael Atiyah

April 1997

Part I
Introduction

Editors' Introduction

This book is one of several volumes of proceedings of the Mathematical Finance Programme held from January through June 1995 at the Isaac Newton Institute for Mathematical Sciences, Cambridge, England. Its chapters represent a wide ranging collection of contributions to theory and methods for financial derivative instrument valuation and hedging. They are based largely – but not exclusively – on papers presented to the Bank of England Conference entitled 'Mathematics of Finance: Models, Theories and Computation' which was held at the Institute from 22 May to 2 June 1995. All chapters reflect presentations made at the Newton Institute at some time during the Programme in Mathematical Finance.

In addition to this summary, the first section of this book deals with several introductory matters – treating preliminaries or background assumptions for the valuation and hedging of derivative financial products. 'Stochastic Calculus and Markov Methods' by Rogers is a review of the main tools of mathematical probability which are in daily use in mathematical finance. Although it contains no substantial proofs, the level of mathematics introduced is roughly that needed to comprehend the remainder of the volume and a number of references are given for further study. Key to the idea of the risk-neutral pricing of derivative securities is the concept of a martingale. The first section of this chapter introduces the basic ideas of martingale theory in a discrete time setting, including convergence theorems and the decomposition of sub- and super-martingales. Next, continuous time martingales are treated, including Brownian motions, stopping times and local martingales and ending with the result that up to a change of time-scale every continuous martingale is a Brownian motion. Semimartingales are introduced next, and Ito's formula and stochastic integrals are developed in this general setting. Markov processes are treated in a similar manner, beginning with discrete time processes and the martingale representation of the fundamental theorem of integral calculus for a nonlinear functional of a Markov chain and moving on to stochastic differential equations (SDEs) – including examples of the principal SDEs currently used in mathematical finance. Also introduced is the Cameron–Martin–Girsanov theorem on change of measure which is basic to risk neutral valuation and the fundamental theorem of asset pricing stating the equivalence of lack of arbitrage opportunities and the existence of a unique risk neutral (equivalent martingale) measure in a complete market.

The 'Risk Premium in Trading Equilibria Supporting Black–Scholes Option Pricing' by Hodges and Selby treats an interesting question concerning the behaviour of the risk premium on the market portfolio of risky assets under Black–Scholes–Merton assumptions. It turns out that under these conditions the market risk premium satisfies a nonlinear partial differential equation (PDE) called Burgers' equation and the chapter provides some new insights into its analysis. At first glance it might seem that such a time homogeneous market would allow the risk premium to vary inversely with the level of the market, so that some kind of mean reversion could take place. The authors show however that under these conditions the risk premium must be either increasing in the market level or constant, providing arbitrage opportunities are ruled out. Interesting conclusions are drawn regarding standard turnpike results for asymptotic portfolio selection which imply in practice that re-investing individuals with distant horizons should follow a power utility investment policy providing their horizon is sufficiently distant. The authors observe that a limitation in the applicability of their results is that they are based on a single representative investor model rather than a full multi-agent equilibrium. As such they allow one solution in which significant arbitrage opportunities are possible.

The final introductory chapter, 'On the numeraire portfolio' by Artzner, discusses the concept of numeraire which is basic to all valuation in economics. This note focusses on the question of when the physical – as opposed to the risk-neutral – probabilities may be used to value securities in terms of expected discounted cash flows, analogous to the situation under the risk-neutral equivalent martingale process. The author concludes that the physical probabilities can often be used for martingale pricing and are associated with an interesting numeraire, but that this does not deprive the standard Black–Scholes money market account numeraire – and its associated martingale measure – of their exceptional rôle.

Part II of this book is concerned with option pricing and hedging techniques under essentially Black–Scholes–Merton assumptions, although a number of the contributions treat certain departures from one or more of these standard postulates. The first paper, 'Some combinations of Asian, Parisian and Barrier Options' by Chesney, Geman, Jeanblanc-Piqué and Yor, treats European options whose terminal payoff is related to a path dependent term, such as the average of the underlying asset price, and involves stopping times related to its excursions. The relatively difficult mathematics involved are aimed at providing at least the Laplace transform of the option price in a form whose complexity varies with exotic features. A typical example of an instrument studied is an up-and-out Parisian call option – the owner of this option loses it if the underlying asset price reaches a specified level before maturity and remains constantly above this level for a time longer than a fixed number termed the option window; otherwise the owner will receive the standard European call payoff. Thus it is necessary to examine the excursions of the underlying geometric Brownian motion. In resolving such questions, use is made of Brownian meanders, Lamperti's representation of geometric Brownian motion as a time-changed Bessel process and the theory of Hartman densities. Numerical methods based on the formulae derived in this paper depend on numerical inversion of the Laplace transforms obtained. Such methods have recently been developed for Asian double barrier options and Parisian options, to which references are given.

'Co-movement Term Structure and the Valuation of Energy Spread Options' by Mbanefo sets out a framework which may be applied to valuing an option on the difference between two underlying prices. A specific concern is with energy markets, but the problem of valuing a derivative security based on the difference between two price processes modelled as geometric Brownian motions applies in a wide variety of contexts. The author argues that a satisfactory valuation model must involve three types of term structure in order to price an energy spread option correctly: that of forward prices of the underlying assets, of their volatility and of their correlation. He then discusses the shortcomings of the best of the current models – Shimko's model – which is based on a normal approximation to the difference of two geometric Brownian motions. Shimko's model specifies a constant correlation which is easily shown by the author to be violated in some relatively simple experiments. Mbanefo concludes with the introduction of a new model which explicitly takes into account the time varying nature of the correlation of the underlying processes and appears to be related to squared Bessel processes.

The next paper 'Pricing and Hedging with Smiles' by Dupire, is a detailed version of his seminal work on incorporating Black–Scholes implied volatilities into option valuation. Black–Scholes–Merton theory assumes a constant volatility of the underlying price process. However, when market prices are assumed to be generated by the Black–Scholes formula and the resulting implicit function is solved for the implied volatility, options with a common maturity but different strike prices exhibit a pattern termed a smile. A similar pattern results from options with a common strike price but different maturities. This chapter addresses the problem of existence, uniqueness and construction of a diffusion process compatible with all the option prices observed in the market. A partial differential equation is derived in a continuous time setting which is similar to the Black–Scholes PDE, but for a given option involves derivatives of its value with respect to time and the strike price. This leads to an expression for the instantaneous – as opposed to the implied – volatility at a given time and underlying price. Upon discretization, a trinomial tree representation of the diffusion is obtained which may be fitted to vanilla option data of different strikes and maturities and then used to price American exotic (path dependent) options. Moreover, derivatives of the instantaneous volatility manifold at a given point can be used to construct robust hedging schemes.

'Option Pricing in the Presence of Extreme Fluctuations' by Bouchaud and Sornette, discusses a different variation from the Black–Scholes–Merton world – precisely, the geometric Brownian motion nature of the underlying stochastic price process. The authors distinguish the form of the underlying stationary price distribution at different time scales, noting that for short enough time scales a Lévy stable process – as originally suggested by Mandelbrot – is a better fit than geometric Brownian motion. Although such price distributions possess extreme fluctuations, they show nevertheless how Black–Scholes arguments can be extended to price and hedge options in their presence. After summarizing the properties of power law distributions, including the Lévy stable laws, the authors examine empirical evidence for a cross-over time between the Lévy and diffusion regimes for the returns. They note that the leptokurtosis – i.e. heavy tail weights – evidenced by price distributions on a short time scale tends to disappear as the time-step increases. They treat option pricing in the presence of non-Gaussian distributions of both infinite and finite variance, leading to a rational way of fixing the bid-ask spread around the resulting price of the option.

'Convergence of Snell Envelopes and Critical Prices in the American Put' by Cutland, Kopp, Willinger and Wyman uses the methods of non-standard analysis to study convergence results in the theory of optimal stopping applied to the Black–Scholes model. In particular, the authors show that the optimal stopping times for an American put option in a sequence of Cox–Ross–Rubenstein tree pricing models converges to the unique optimal stopping time in the Black–Scholes model in a sense which is stronger than the weak convergence of probability theory. This leads to a new convergence result for the associated Snell envelopes of the discounted return processes and in the limit determines the critical prices exactly. The methods of non-standard analysis are implemented through the hyperfinite Cox–Ross–Rubenstein market model which intuitively is a lifting of the standard Cox–Ross–Rubenstein discrete model to the continuous by the basic model-theoretic techniques of non-standard analysis. The paper serves as a brief introduction to these techniques in the context of mathematical finance.

The final chapter in this part, 'Filtering Derivative Security Valuations from Market Prices' by Elliott, Lahaie and Madan, is a novel alternative approach to explaining the volatility smile effect discussed above. The authors attribute the deviations between market and model prices to three sources: errors of observation, model errors – due to the presence of rational bubbles, errors in parameter estimation or the consequences of approximation strategies inherent in econometric modelling – and market mispricing – due to the slow adjustment of certain markets to changes in market fundamentals. Due to these errors the true model, derived from theoretical valuation in terms of expected discounted cash flow under the risk neutral measure, is never observed. The authors therefore treat such a theoretical option price as a predetermined nonlinear function of the unobservable theoretical underlying price which together constitute a hidden, or unobserved, Markov model. They construct the conditional expectation of a function of such a model by recursive updating coupled with the simultaneous updating of any parameters in the pricing function. This is a problem in nonlinear filtering whose techniques – both theoretical and numerical – are well understood. The chapter thus illustrates the application of hidden Markov model filtering and estimation techniques to the simultaneous theoretical valuation of primary and derivative assets under the classical Black–Scholes–Merton assumptions with application to stock index options using data on Standard and Poor 500 index future options for the year 1992. The authors observe that the unfiltered Black-Scholes prices exhibit the well known strike-related volatility smile effect, but that this effect tends to be corrected by the use of filtered option prices.

Part III of the book is concerned with valuation and hedging issues for markets with imperfections. By this we mean extensions of the classical frictionless Black–Scholes–Merton model to include complications such as transaction costs, stochastic volatility, and constraints on trading strategies like prohibitions on short sales. The first chapter here, 'Hedging Long Maturity Commodity Commitments with Short-Dated Futures Contracts' by Brennan and Crew, studies an example where the market is incomplete, that is, there are not enough tradable securities to replicate the relevant contingent claims. Adding to the literature that was prompted by the Metallgesellschaft episode of December 1993, this paper evaluates four strategies for hedging a long-term commitment to deliver a fixed quantity of a commodity by trading short term futures contracts. These strategies are the stack and roll, the Edwards–Canter minimum variance, the Brennan, and the Gibson–Schwartz, with the last two involving a stochastic model for the dynamics of the term structure of futures prices versus time to expiration. These strategies were tested against monthly futures prices for NYMEX light crude oil for the period from March 1983 to December 1994. The authors concluded that the Brennan and the Gibson–Schwartz strategies were significantly better than the other two.

'Option Pricing and Hedging in Discrete Time with Transaction Costs and Incomplete Markets' by Mercurio and Vorst is concerned with the optimal hedging of an option when the underlying is governed by the classical discrete time binomial model, there are proportional transaction costs, and the hedge is adjusted only at a specified subset of the trading periods. Two approaches are considered. One includes super-replication and leads to an initial cost that is independent of investors' preferences. The other involves the minimization of the squared local deviation from perfect replication and consequently does depend on investors' preferences and predictions. Simulations suggest this latter method can be effective for practical purposes.

'Option Pricing in Incomplete Markets with an Application to Transaction Cost Models' by Davis gives a summary of recent research by the author on the pricing and hedging of options in incomplete markets. Specifically, the focus is on contingent claims which cannot be replicated by a portfolio of traded securities and thus cannot be priced by arbitrage. The approach here is to assume the price is consistent with a representative agent who wishes to maximize the expected utility of his wealth at the time when the option expires. Under a standard d–dimensional Brownian motion model, the theory of multiplicative functionals is used to derive a formula for what the price of this option must be. This formula is in terms of the risk neutral probability measure that corresponds to the agent's marginal utility. The paper concludes with the extension to a securities market where there are proportional transaction costs.

'Nonlinear Financial Markets: Hedging and Portfolio Optimisation' by Cvitanić is an up-to-date survey of the techniques and results of the theory of optimal trading for a single agent with a nonlinear wealth process in a continuous time framework. The theory presented includes situations in which the agents' strategy can influence asset prices, interest rates are different for borrowing and lending, and cases with portfolio constraints. The martingale duality techniques discussed here have recently been developed for the complete markets case in a number of papers and the framework presented includes these works while offering simplified proofs. In all the results encompassed only the drift term of the wealth process becomes nonlinear. Valuation and hedging problems under Markovian and smoothness assumptions are solved in a framework which is based on forward-backward stochastic differential equations – forward time equations related to pricing and backward to hedging. Interesting results unknown in classical models are derived for European options sold by a large investor. For example, selling an option for its fair Black–Scholes price insures that perfect hedging is possible, but selling it for more than that price does not guarantee the existence of a hedge.

Finally, 'Semimartingales and Asset Pricing under Constraints' by Frittelli extends the well known fundamental theorem of asset pricing cited above to the case of security markets with frictions. In particular, the frictions here are in the form of a requirement that the trading strategies be subject to a convex constraint. The key result is that there exists an equivalent quasi-martingale measure if and only if there do not exist what are called 'extreme arbitrage opportunities'. It follows that an adapted stochastic process is a semimartingale if and only if, when it is viewed as a price process, there are no extreme arbitrage opportunities. This reinforces the view that one should be wary about price process models which are not semimartingales, because extreme arbitrage opportunities should not appear in a reasonable market.

Part IV of this volume is concerned with term structure models and interest rate derivatives. 'Bond and Bond Option Pricing Based on the Current Term Structure' by Dybvig examines the Ho and Lee model from a modern perspective and then extends that model to a wider class where the model's term structure can still be made to match the observed term structure. This is accomplished by a proper choice of a deterministic function which, when added to the sample path of a specified process, gives the sample path of the spot interest rate. The paper makes the important point that the resulting models, as well as most other models in the literature, are *ad hoc* in the sense that each period the whole model (i.e. its parameters in the case of spot rate models or its state space in the case of HJM models) must be adjusted in order to fit the newly observed term structure. The paper goes on to conduct an exploratory data analysis, the results of which suggest one-factor models with slight mean reversion explain nearly all the variability of interest rates. This has implications for how the choice of the model should depend on the maturity of the derivative being priced or hedged.

The next two papers are concerned with the fundamental theory of term structure models. In particular, both address the link between the spot rate approach and the Heath–Jarrow–Morton style forward rate curve approach. 'Dynamic Models of Yield Curve Evolution' by Flesaker and Hughston view term structure models in terms of 'financial observables', notably the initial yield curve and the zero coupon bond price volatility term structure, which is given by a one-parameter family of vector processes adapted to the filtration of a d-dimensional Brownian motion. Simple exact representations for the money market account process and the zero coupon bond price processes are derived by the use of martingale arguments. A new framework is also presented for interest rate dynamics based on a general characterization for the condition that interest rates should be positive.

'General Interest Rate Models and the Universality of HJM' by Baxter explicitly examines the link between the two term structure model approaches. Subject to some technical restrictions, this paper shows how to transform, at least in principle, a spot rate model into a Heath–Jarrow–Morton style forward rate curve model, and vice versa. Explicit links with the zero coupon bond prices are also provided. The paper concludes with two interesting counter-examples. One is an example of a perfectly reasonable market of zero coupon bonds, but this market cannot be cast as either a spot rate model or an HJM style model. The other is a market which can be cast as either a spot rate model or an HJM model, but some of the bond prices have non-vanishing volatility as time approaches maturity.

'Swap Derivatives in a Gaussian HJM Framework' by Brace and Musiela is noteworthy for two reasons. In the first place, by using the change of numeraire technique this paper derives price formula for a number of interest rate derivatives under the Gaussian (i.e. the volatility structure is deterministic) Heath–Jarrow–Morton style framework. In the second place, and just as importantly, this paper collects in one place the precise mathematical definitions of a wide variety of interest rate derivatives. For a domestic economy the derivatives include forward swaps and swap rates, forward caps and floors, swaptions, options on swap rate spreads, captions and floortions, compound swaptions, and exotic caps and swaptions. In the case of foreign economies, the derivatives include differential forward swaps, basket caps, and cross-currency swaptions.

The next chapter in this section, 'Modelling Bonds and Derivatives with Default Risk' by Lando, changes direction by introducing the notion of credit risk. This paper provides a comprehensive and very comprehensible survey of models of bonds subject to the risk of default. There are two main approaches. In one, default occurs when the value of the firm, modelled as a stochastic process, hits a barrier (e.g. the book value hits zero). The bond's price will thus depend on the probability distribution of a first passage time by a process (the firm's value) which may or may not be independent of the process describing the Treasury yield curve. In the other approach, the default time resembles the first jump time of a Poisson process, in that it is a totally inaccessible stopping time whose probability distribution is described by a specified intensity process. Again, the default time may or may not be independent of the Treasury yield curve. In the dependent case, the bond's price may be the solution of a backwards stochastic differential equation, the solution of which can be nonlinear with respect to the promised payoff.

The final two papers in Part IV are concerned with term structure models where the interest rate processes do not have continuous sample paths. Extending earlier work by the same authors, 'Term Structure Modelling under Alternative Official Regimes' by Babbs and Webber develops a class of term structure models where there are N state variables modelled as diffusion processes and an additional M state variables modelled as pure jump processes. In general, the processes are not independent of each other, and the spot interest rate is a specified function of the $N + M$ state variables. An example is provided where one jump process is a floor and another is a ceiling for the spot rate. The authors argue that their approach is realistic, because government monetary authorities exercise control – by setting discount rates, Lombard rates, and so forth – over short term interest rates.

Finally, 'Interest Rate Distributions, Yield Curve Modelling and Monetary Policy' by El-Jahel, Lindberg and Perraudin treats a similar problem in terms of a special case of the Babbs-Webber model which has closed form solutions. The authors concentrate on explaining two phenomena which influence the distribution of short term interest rates; namely, the practice of many monetary authorities of pegging an interest rate at the short end of the yield curve and periodically adjusting it in discrete jumps, and the attitude of the monetary authorities in their reaction to inflationary shocks – either stringent or relaxed. The first of these phenomena affects the leptokurtosis of short rate state distributions and the second affects the mean reverting nature of the short rate process. The chapter first examines the distributional properties of short term interest rates in Germany, the UK and the US, employing a variety of econometric techniques including nonparametric kernel estimates, unit root tests and simple autoregressions. The authors then estimate two commonly applied single state variable yield curve models – those of Vasicek and Cox–Ingersoll–Ross – to find significant evidence in the data for misspecification of both models. In particular, mean reversion rates of the short term rate process tend to be overestimated. Armed with this empirical evidence, a Babbs–Webber model is proposed in which short term interest rates are specified by a pure jump process whose jump rate is a function of a diffusion process. Assuming an Ornstein–Uhlenbeck process for the diffusion variable and a quadratic jump rate function, the authors are able to apply the Karhunen–Loeve eigenfunction expansion techniques of physics to obtain power series representations of the conditional distribution of the diffusion state variable given its past and bond yields, which latter may be applied to give similar representations of interest rate based derivative values.

The last part of this volume gives a comprehensive overview of numerical methods used in current derivative valuation and hedging practice. Each chapter treats specific instruments as instances to which the numerical techniques described are applied. Examples considered are Bermudan swaptions in a single currency, cross-currency Bermudan swaps and swaptions, American options – including consideration of transactions costs and stochastic volatilities – and collatoralized mortgage obligations. The authors of the first paper of Part V make the point that it is vital for a trader to have a detailed understanding of all the trades in his or her book and of the risk to which he or she is exposed as a consequence. Such information requires not only the rapid calculation of a value of the trade, but also the ability to evaluate for risk management purposes the sensitivity of these values to various input variables in changing market conditions. Ideally this should be in real time, for precisely the time when the information is most needed is when the market is changing rapidly. The 'real time' goal for most complicated instruments is presently far from achievement. However, advances in techniques – some of which are presented here – are rapid towards this goal and are of course enhanced by accelerating performance of general computer technology.

'Numerical Option Pricing Using Conditional Diffusions' by Gandhi and Hunt shows how to use problem structure to speed-up standard numerical techniques. The main idea is to apply probability theory to perform explicit calculations regarding the behaviour of the underlying price process between the discrete grid points considered in the pricing algorithms. Specifically, the authors consider tree valuation of a Bermudan swaption in a single currency in which the short rate is generated by a mean reverting Gaussian extended-Vasicek model with an explicit solution in terms of a time-changed Wiener process. They exploit results concerning the Brownian bridge to explicitly calculate values on the nodes of the tree at exercise times of the swap. The authors observe that the methods can be extended to log Gaussian models, such as those of Black–Karasinski and Black–Derman–Toy, as well as to higher dimensional problems – which extensions are topics of current research. Applications of these ideas to exotic instruments such as barrier options with American features pose more difficult problems, but are still amenable to similar techniques.

'Numerical Valuation of Cross-currency Swaps and Swaptions' by Dempster and Hutton represents the state-of-the-art for the numerical partial differential equation approach to derivative instrument valuation. The authors treat a range of cross-currency swaps and swaptions which require a three-dimensional state variable parabolic PDE representation of the deal value.

This chapter also uses extended-Vasicek Gaussian models for the term structure in both currencies. The deal value is represented for numerical solution in terms of three correlated driftless Gaussian processes to give, by standard Ito's Lemma calculations, the associated three state variable parabolic PDE. While PDE approaches are more generally applicable than tree-type methods, they pose significant computational challenges in two or three state variables. In the one-dimensional Black–Scholes–Merton world, a Crank–Nicolson second order accurate approach is the method of choice. However, for the 3-D PDE of this chapter, implicit or mixed approaches require at best $O(n^7)$ floating point operations – versus $O(n^5)$ for a second order accurate explicit scheme – where n is the (maximum) number of spatial grid points in a spatial dimension. Equipped with this discrete approximation to the valuation problem in a single discretionary period, terminable differential swaps and other deals are valued numerically by backward recursion through the payment dates. The solutions found are investigated graphically – visualisation being a critical tool for the development of numerical methods in multiple dimensions. The authors also investigate the sensitivity of the deal values found to the specification of a variety of boundary conditions. Further developments are likely to incorporate multi-grid methods for numerical PDE solution at great savings in computation time.

The third chapter in Part V, 'Numerical Methods for Stochastic Control Problems in Finance' by Kushner, surveys a numerical technique applicable under weaker assumptions than those required for the partial differential equation approach. The technique described – termed Markov chain approximation – has been pioneered by the author for the numerical solution of a wide variety of stochastic control problems. It approximates the underlying continuous time and space stochastic processes involved by discrete time and space approximations which are intuitive in that they are physically close to the original model. Convergence in the discretization steps can be demonstrated by the weak convergence methods of probability theory under very general conditions. After introducing several examples for which the technique is applicable – including American options with stochastic volatility, path dependent options and the consideration of transaction costs – the author studies the infinite time optimal stopping problem in detail. By way of example, the valuation of an American option under the physical probabilities with a given exogenous interest rate process over an infinite horizon is studied and the results obtained modified to yield the valuation of an ordinary American option with finite maturity.

The last two chapters of this book treat the most general approach to derivative valuation and hedging calculations, namely simulation techniques. Although easy to code, these techniques converge to the required accuracy extremely slowly relative to those of earlier chapters due to their sub-arithmetic convergence in sample size – even with known acceleration enhancements. 'Simulation Methods for Option Pricing' by Lehoczky provides a survey of current ideas and methods associated with Monte Carlo simulation as a method for computing derivative prices and hedging parameters. After a discussion of the convergence of sample averages due to the strong law of large numbers and the provision of their error estimates using the central limit theorem, the author describes several variance reduction techniques – i.e. acceleration methods – such as antithetic and control variables, importance sampling and stratified sampling. Techniques for simulating the values of derivative instruments are discussed in terms of both sample paths and multiple integrals, as are techniques for simulating the derivatives of deal values required to evaluate hedging parameters such as the Greeks and sensitivities to other market parameters. The chapter concludes by summarizing methods recently introduced by Broadie and Glasserman, in the context of American option valuation with a stochastic price volatility model, in terms of both antithetic and control variables.

The final contribution, 'New Methodologies for Valuing Derivatives' by Paskov, discusses the application of low discrepancy sequence algorithms to the valuation of high dimensional integrals. The idea of these deterministic methods is that, rather than finding evaluation points of the integrand by pseudo random number generation techniques, number theory is used to produce a uniform coverage by valuation points of the d-dimensional space involved. As a result, the resulting convergence in sample sizes is typically $O((\log n)^d/n))$ versus the $O(1/\sqrt{n})$ of the central limit theorem. The author describes first the representation of the value of a collateralized mortgage obligation (CMO) instrument as an evaluation of several high-dimensional integrals. He then tests two low discrepancy algorithms – those of Sobel and Halton – versus two standard Monte Carlo algorithms – classical and antithetic variables – to conclude that for this CMO the Sobel algorithm is always superior to the other algorithms. Other authors have reported even faster practical convergence rates for the Faure low discrepancy algorithm. Parallel computation may in general be used to speed up simulation techniques in a straightforward manner and the author describes a software system which runs on a network of heterogeneous work stations such as may be found on many dealing floors. Taken all-in-all, Monte Carlo and related simulation techniques remain the most general methods for evaluation of complex derivative instruments and the generation of their hedging parameter information. Overcoming the slow convergence rates of these techniques remains a major computational challenge which in specific cases will no doubt require the skilled use of vector parallel computing technology.

Before closing this section the editors would like to acknowledge the contributions of a number of colleagues. First, all the authors contributing to this volume have responded to our call with enthusiasm, patience and good humour – both regarding their own papers and those papers many were asked to referee. We are very grateful to them and we hope they find their efforts justified. Secondly, we would like to thank Darren Richards and Chris Jones of the Finance Research Group at The Judge Institute of Management in the University of Cambridge for skillful, diligent and cheerful help in typesetting and proofreading the volume. Thirdly, we take this opportunity to thank the staff of the Isaac Newton Institute who made the stays of all of us in Cambridge during the Mathematical Finance Programme so pleasant. One of us (SRP) expresses his gratitude to the Master and Fellows of Emmanuel College, Cambridge, for the award of a Visiting Fellowship tenable during the Programme and for their kind and stimulating hospitality during its term. Finally, this book – and so many others like it – would not have been possible without the judgement and high professionalism of David Tranah of Cambridge University Press. We owe him our sincerest gratitude.

M.A.H.D. and S.R.P.

Cambridge and Chicago

March 1997

Stochastic Calculus and Markov Methods

L. C. G. Rogers

Abstract

The aim of these notes is to provide the briefest possible *resumé* of the main probabilistic tools in daily use in mathematical finance. Ideally, the reader needs to have a good understanding of probability, up to about the level of conditional expectation. So far as is possible, technical matters are kept in the background, but it is as well to remember that there *is* a background, and inevitably the more one knows of that background the more comprehensible these notes become.

1 Introduction

In mathematical finance, it is commonly assumed that the log of a share price is modelled by a Brownian motion (with drift), and models of other processes (for example, the spot interest rate) which are solutions of stochastic differential equations driven by Brownian motion are also in widespread use. Such models are still just about tractable, and are not too unrealistic. The main goal of these notes is to explain a little of what Brownian motion and stochastic differential equations are, and how one can work and calculate with these things. The secondary goal of these notes is to describe some basic ideas of Markov processes. While it is entirely possible to discuss Brownian motion and stochastic differential equations without mentioning Markov processes, this would be artificial; Brownian motion and the solutions of stochastic differential equations *are* Markov processes, and Markovian ideas are essential to the most effective techniques of calculation, as well as being the source of many interesting examples of martingales.

So let us begin with a little orientation. We are going to consider some real-valued continuous random process $(x_t)_{t \geq 0}$ which evolves according to the differential equation

$$\dot{x}_t = \sigma(x_t)\dot{W}_t + \mu(x_t). \tag{1.1}$$

Here, σ and μ are some suitably nice (for example, Lipschitz continuous) functions, and $(W_t)_{t \geq 0}$ is some continuous random function. As is common in applied mathematics, the dot over the symbols x, W denotes differentiation with respect to time.

Example. If $\mu \equiv 0$, $\sigma \equiv 1$, then $x_t = a + W_t$ for some $a \in \mathbb{R}$.

Example. If σ is constant, and $\mu(x) = -\beta(x - c)$, with $\beta > 0$, then we have a model for a random process which is drawn back to its 'equilibrium' position c by the restoring term $\mu(\cdot)$, but is disturbed from equilibrium by the random process W.

Example. If $\sigma(x) = \sigma|x|^{\frac{1}{2}}$, $\mu(x) = -\beta(x - c)$ for constants $\alpha, \beta, \sigma > 0$, then we have a model for a process where the effect of the noise is greater the larger the value of $|x|$. Such a process was proposed by Cox, Ingersoll and Ross [2] as a model for the spot-rate of interest.

We often write (1.1) in the equivalent 'differential' form

$$dx_t = \sigma(x_t)dW_t + \mu(x_t)dt \qquad (1.2)$$

or in the 'integrated' form

$$x_t - x_0 = \int_0^t \sigma(x_u)dW_u + \int_0^t \mu(x_u)du. \qquad (1.3)$$

If we wanted to find a stochastic differential equation for the process $y_t \equiv f(x_t)$ (where f is C^1) then the ordinary rules of calculus would tell us that

$$\dot{y}_t = f'(x_t)\dot{x}_t \qquad (1.4)$$

or, in differential notation,

$$\begin{aligned} dy_t &= f'(x_t)dx_t \\ &= f'(x_t)\{\sigma(x_t)dW_t + \mu(x_t)dt\}. \end{aligned} \qquad (1.5)$$

However, all of the above is based on the assumption that the random noise process W is differentiable, and in the case of central importance to us, where W is Brownian motion *this does not happen!* Does this actually matter very much? Indeed it does. What sense do (1.1), (1.2), (1.3) now have if W is Brownian motion? They only have sense via a proper definition of the stochastic integral (the first term appearing on the right-hand side of (1.3)), which is quite a lengthy business to set up. Is it worth the effort? Why don't we simply restrict attention to random disturbances W which *are* differentiable? There are many reasons why it is worth the effort to set up the stochastic integral. Firstly, the solution x to the stochastic differential equation (SDE) will be a Markov process, which is to say that its future behaviour depends on its current value, but not on its history; if W were differentiable, this would not be the case, so to know how x was going to behave we would have to carry around information about its history (perhaps even its entire history!) which would be unacceptable in any real application. The second main reason is that the stochastic integral with respect to Brownian motion is a (local) *martingale*, and this allows one to bring powerful martingale results to bear, particularly the optional sampling theorem (OST). In this context, the OST says that under some mild growth restrictions,

$$E \int_0^T \sigma(x_u)dW_u = 0$$

for any *stopping time* (or *optional time*) T. Informally, the random time T is a stopping time T if you know for sure by time t whether it has happened by time t, whatever $t \geq 0$.

Another reason why setting up stochastic integration properly really matters is that the change-of-variables formula (1.5) is *no longer correct* if W is Brownian motion; the *correct* version of (1.5) is the celebrated *Itô formula*, but to understand this we need to go deep into properties of martingales and Brownian motion.

So what we are going to do is this. We begin with discrete-time martingale theory; the setting is technically much easier than continuous time, but the results are essentially the same. Next we move on to continuous time and the most interesting martingale, Brownian motion; we show how martingale theory already provides powerful computational techniques, and how the (very random) Brownian motion has an astonishing regularity which is the basis of Itô's formula. We then briefly discuss stochastic integration and Itô's formula itself. Next we go back to some very simple Markov chain theory, which gives us all the essential Markovian ideas, and then we synthesise the Markovian and martingale views in our study of stochastic differential equations. As a final topic, we discuss change of measure, a key method in financial mathematics, and in probability more generally.

Notes on the literature. Everything in this course is well known to probabilists, and frequently presented in text books. The big entry fee to probability is measure theory; one can (and, in a British mathematics degree, typically *does*) go a long way without measure theory, and the book [3] of Grimmett and Stirzaker is about as good an account of probability without measure as one could ask for. It also contains a concise account of Markov chains and many applications. For a non-measure-theoretic account of probability, the classic two-volume work of Feller [4] is a joy to read, though its aims are sufficiently different from those of the current discussion that it should be viewed as background material.

And then one comes to measure-theoretic probability. It is not really necessary to swallow large pieces of measure theory; all that is needed is not to choke on small pieces! A reader armed with a clear idea of the definitions and main results and a healthy disregard for 'technical details' will be able to follow the arguments reasonably well, even though he or she may not be capable of building such arguments unassisted. For a clear and masterly account of measure-theoretic probability and how it comes to life in the context of discrete-time martingales, David Williams' book [13] is unsurpassed. This material, and much more on Markov processes and Brownian motion, features in the recent book [10] of Rogers and Williams. The companion volume [11] gives a complete account of stochastic calculus and other topics, as do the books of Ikeda and Watanabe [6], Karatzas and Shreve [7], Revuz and Yor [9], Chung and Williams [1], ... Which one reads depends on questions of taste, and the nature of the applications of interest. A briefer and highly readable account is in Øksendal [8].

2   Discrete-time Martingales

To begin with, let us review some well-known material on conditional expectations. The reader must either be, or become, totally familiar with this.

Probability happens on a probability space $(\Omega, \mathcal{F}, P)$, where Ω is a set, $\mathcal{F}$ is a σ-field of subsets of Ω, called *events*, and P is the probability defined on events, satisfying the familiar rules:

$$0 \leq P(A) \leq 1 \text{ for all } A \in \mathcal{F}; \tag{2.1i}$$
$$P(\Omega) = 1; \tag{2.1ii}$$
$$P(A \cup B) = P(A) + P(B) \text{ if } A \text{ and } B \text{ are disjoint, } A, B \in \mathcal{F}; \tag{2.1iii}$$
$$P(A_n) \uparrow P(\cup_n A_n) \text{ if } A_n \in \mathcal{F}, \ A_1 \subseteq A_2 \subseteq .... \tag{2.1iv}$$

A σ-field on Ω is a collection $\mathcal{F}$ of subsets of Ω which is closed under the formation of countable unions and taking of complements, and which contains Ω. The reason for introducing σ-fields is that it is frequently impossible to define interesting probabilities on *all* subsets of Ω, so one can only make progress by defining probabilities only on *some* subsets of Ω; if we think of $\mathcal{F}$ as being the collection of sets whose probability *is* defined, then the closure properties of a σ-field are forced on us in view of the properties (2.1) of P, which can only make sense if the sets whose probabilities are discussed in (2.1) are actually sets in $\mathcal{F}$.

This is the analyst's view of measure (and is not wrong) – but the probabilist understands that σ-fields are not just a tiresome technical necessity but are needed to define and work with stochastic processes. To do this, we typically consider a *filtration*, a family $(\mathcal{F}_n)_{n \in \mathbb{Z}^+}$ of σ-fields which are increasing ($\mathcal{F}_n \subseteq \mathcal{F}_{n+1} \subseteq \mathcal{F}$ for all n); the interpretation of $\mathcal{F}_n$ is the 'information known at time n'. A stochastic process $(X_n)_{n \in \mathbb{Z}^+}$ is then called *adapted* if X_n is $\mathcal{F}_n$-measurable for all n, and this bears the interpretation that 'X_n is known at time n'. For example, X_n might be the price of a share at the end of trading on day n. If we intended to hold the share during day $n + 1$, we would be interested in $E(X_{n+1}|\mathcal{F}_n)$, the *conditional expectation* of X_{n+1} given $\mathcal{F}_n$ (i.e. given all that was known by the end of day n), because this would tell us how much we expected to gain by holding the share on day $n + 1$. Before discussing this further, let us record what conditional expectation given a σ-field is, and what properties it has.

If $X \in L^1(\Omega, \mathcal{F}, P)$ and $\mathcal{G}$ is a σ-field, $\mathcal{G} \subseteq \mathcal{F}$, then the *conditional expectation of X given $\mathcal{G}$* is the $\mathcal{G}$-measurable random variable $E(X|\mathcal{G})$ with the property that for any $A \in \mathcal{G}$

$$E[E(X|\mathcal{G}); A] \equiv \int_A E(X|\mathcal{G})dP = \int_A X dP \equiv E[X; A].$$

($E(X|\mathcal{G})$ exists, and is unique to within changes on sets of probability 0, in case you were worried.) The conditional expectation of X given $\mathcal{G}$ has the following properties of expectation:

$$E(X|\mathcal{G}) \geq 0 \text{if } X \geq 0; \tag{2.2i}$$
$$E(aX + bY|\mathcal{G}) = aE(X|\mathcal{G}) + bE(Y|\mathcal{G}) \text{for } a, b \in \mathbb{R}, \tag{2.2ii}$$
$$X, Y \in L^1(\Omega, \mathcal{F}, P);$$
$$E(1|\mathcal{G}) = 1; \tag{2.2iii}$$
If $X_n \uparrow X$, $X_n, X \in L^1$, then
$$E(X_n|\mathcal{G}) \uparrow E(X|\mathcal{G}); \tag{2.2iv}$$
If $X_n \to X$, and $|X_n| \leq Y \in L^1$ for all n, then
$$E(X_n|\mathcal{G}) \to E(X|\mathcal{G}); \tag{2.2v}$$
If $f : \mathbb{R} \to \mathbb{R}$ is convex and $X, f(X) \in L^1$, then
$$E[f(X)|\mathcal{G}] \geq f(E(X|\mathcal{G})). \tag{2.2vi}$$

In the special case where $\mathcal{G} = \{\emptyset, \Omega\}$, you can check for yourself that $E(X|\mathcal{G}) = EX$, and in that case the above properties are familiar results concerning expectation. There are also important properties of conditional expectation which have no counterpart for expectation:

('Tower property') If $\mathcal{G}_1 \subseteq \mathcal{G}_2 \subseteq \mathcal{F}$ are σ-fields, then $\tag{2.2vii}$
$$\text{for } X \in L^1, \ E(X|\mathcal{G}_1) = E(E(X|\mathcal{G}_2)|\mathcal{G}_1);$$
('Slot property') If $X, XY \in L^1$, and Y is $\mathcal{G}$-measurable, $\tag{2.2viii}$
$$\text{then } E[XY|\mathcal{G}] = YE[X|\mathcal{G}];$$
If X is independent of $\mathcal{G}$ then $E(X|\mathcal{G}) = EX;$ $\tag{2.2ix}$
If X is $\mathcal{G}$-measurable, then $E(X|\mathcal{G}) = X.$ $\tag{2.2x}$

These are the properties of conditional expectation which will henceforth be used without comment. We need them immediately.

A *martingale* is a sequence $(X_n, \mathcal{F}_n)_{n \in \mathbb{Z}^+}$ where each X_n is a random variable and each $\mathcal{F}_n$ is a σ-field such that

$$\mathcal{F}_n \subseteq \mathcal{F}_{n+1} \subseteq \mathcal{F} \quad \text{for all } n; \tag{2.3i}$$
$$X_n \in L^1(\Omega, \mathcal{F}_n, P) \quad \text{for all } n; \tag{2.3ii}$$
$$X_n = E[X_{n+1}|\mathcal{F}_n] \quad \text{for all } n. \tag{2.3iii}$$

[If we replace the equality in (2.3iii) by $\geq$, then we have the definition of a *supermartingale*, and if we replace by $\leq$, we have the definition of a *submartingale*.]

Examples. (a) If $Y \in L^1(\mathcal{F})$ and $(\mathcal{F}_n)_{n \in \mathbb{Z}^+}$ is a filtration, then

$$(Y_n, \mathcal{F}_n)_{n \in \mathbb{Z}^+} \equiv (E[Y|\mathcal{F}_n], \mathcal{F}_n)_{n \in \mathbb{Z}^+}$$

is a martingale. We say that (Y_n) is a martingale *closed on the right*. (Why is $Y_n = E(Y|\mathcal{F}_n)$ in $L^1(\Omega, \mathcal{F}_n, P)$? Use properties (vi) and (vii), and the fact that $E[X|\mathcal{G}] = EX$ if $\mathcal{G} = \{\emptyset, \Omega\}$.)

When the filtration we are talking about is not ambiguous, we will usually say simply that $(Y_n)_{n \in \mathbb{Z}^+}$ is a martingale.

(b) In the simplest discrete-time model of the movement of share prices, we suppose that the price on day $n+1$ is either uX_n or dX_n, where $0 < d < u$ are fixed constants, with price changes on different days independent. Suppose also that the probability of changing to uX_n is $p \equiv 1 - q \in (0, 1)$. Then

$$E(X_{n+1}|\mathcal{F}_n) = p \cdot uX_n + q \cdot dX_n = (pu + qd)X_n.$$

Thus (X_n) is a martingale if and only if $pu + qd = 1$. Check for yourself that $((pu + qd)^{-n}X_n)_{n \in \mathbb{Z}^+}$ is a martingale.

(c) A *random walk* $(X_n)_{n \in \mathbb{Z}^+}$ is a process such that $X_0 = 0$, and $X_n - X_{n-1}(n = 1, 2, ...)$ are independent and identically distributed. Assuming $X_1 \in L^1$, show that (X_n) is a martingale if and only if $EX_1 = 0$. Is it possible for (X_n) to be a martingale closed on the right?

Our next topic in discrete-time martingale theory is *stochastic integration*. This is too grand a title for what we are about to do, but the continuous-time analogue is significantly more complicated.

Suppose $(X_n)_{n \in \mathbb{Z}^+}$ is a process *adapted* to the filtration $(\mathcal{F}_n)_{n \in \mathbb{Z}^+}$ which means that X_n is $\mathcal{F}_n$-measurable for all n. Write $\Delta X_n \equiv X_n - X_{n-1}$.

Think of X_n as the price at the end of day n of some share. If during day n you hold θ_n shares, then your gain on the day will simply be $\theta_n \Delta X_n$. Thus if you started at time 0 with wealth W_0, your wealth at time n would be

$$W_n = W_0 + \sum_{j=1}^{n} \theta_j \Delta X_j \equiv W_0 + (\theta \cdot X)_n,$$

where the process $\theta \cdot X$ is called the *stochastic integral of θ with respect to X*. In this interpretation of (X_n), it is entirely natural to assume that the process $(\theta_n)_{n \geq 1}$ must be *previsible*, which is to say that θ_n must be $\mathcal{F}_{n-1}$-measurable for all $n \geq 1$. Indeed, it would be absurd to let someone choose the number of shares they are going to hold during day n in the light of the price of the share at the end of that day!

So far, we have made no assumptions about the process X; but if we suppose that X *is a martingale, then W is also a martingale* (if we suppose, say, that each θ_j is bounded). The proof of this is trivial and is a good exercise in the use of conditional expectations:

$$
\begin{aligned}
E[W_{n+1}|\mathcal{F}_n] - W_n &= E[W_{n+1} - W_n|\mathcal{F}_n] \quad \text{(properties (x), (iii))} \\
&= E[\theta_n \Delta X_{n+1}|\mathcal{F}_n] \\
&= \theta_n E[\Delta X_{n+1}|\mathcal{F}_n] \quad \text{(property (viii))} \\
&= \theta_n \{E(X_{n+1}|\mathcal{F}_n) - X_n\} \quad \text{(properties (iii), (x))} \\
&= 0
\end{aligned}
$$

since X is a martingale. This trivial calculation has profound implications for financial mathematics, but it is not our brief to discuss them here. Rather, let us see how we can now trivially deduce the optional sampling theorem from this. This needs the notion of a *stopping time*, which is a random variable $T : \Omega \to \mathbb{Z}^+ \cup \{\infty\}$ with the property that

$$
\{T \leq n\} \in \mathcal{F}_n \quad \text{for all } n \geq 0.
$$

An equivalent statement is that *the process*

$$
\theta_n \equiv I_{\{T \geq n\}} \quad (n \geq 1)
$$

is previsible. So if we form the process $(\theta \cdot X)_n$, we know that this will be a martingale if X is; however,

$$
\begin{aligned}
(\theta \cdot X)_n &= \sum_{j=1}^{n} I_{\{T \geq j\}}(X_j - X_{j-1}) \\
&= X_{T \wedge n} - X_0,
\end{aligned}
$$

where $a \wedge b$ denotes the smaller of a and b. So we have the important optional sampling theorem.

THEOREM 1 *(OST) If $(X_n, \mathcal{F}_n)_{n \in \mathbb{Z}^+}$ is a martingale, and T is a BOUNDED stopping time, then*

$$
E[X_T|\mathcal{F}_0] = X_0,
$$

and hence

$$
EX_T = EX_0.
$$

Proof. If we know that $T \leq N$, then $(\theta \cdot X)_N = X_{T \wedge N} - X_0 = X_T - X_0$ and so we have

$$
\begin{aligned}
0 &= E[(\theta \cdot X)_N|\mathcal{F}_0] \quad \text{since } \theta \cdot X \text{ is a martingale} \\
&= E[X_T - X_0|\mathcal{F}_0] \\
&= E[X_T|\mathcal{F}_0] - X_0.
\end{aligned}
$$

The final statement of the theorem follows from the first by taking expectations. Do notice that the boundedness of T is essential to the result; if we consider the martingale which is symmetric simple random walk on $\mathbb{Z}$ (so ΔX_j are independent random variables with common distribution $P(\Delta X_j = 1) = P(\Delta X_j = -1) = \frac{1}{2}$) and take $T = \inf\{n : X_n = -1\}$, then it can be shown that T is finite with probability 1, but $X_T = -1$, $X_0 = 0$, so the conclusion of the OST is not satisfied. There exist variants of the OST for T which are not bounded, but one needs more regularity on X; for example, if X is closed on the right, then the OST holds for any stopping time T.

[Remark: We have given a cheap version of the OST; see Chapter A14 of [13] for the full story.]

As a little taste of the use of the OST, if we take integers $a < 0 < b$, we can compute $P(X$ hits a before $b)$ when X is the symmetric simple random walk. Indeed, if $T = \inf\{n : X_n = a$ or $X_n = b\}$, then for any N we apply the OST to the bounded stopping time $T \wedge N$:

$$
\begin{aligned}
0 &= EX_0 \\
&= E(X_{T \wedge N}) \\
&= aP(X_T = a, T \leq N) + bP(X_T = b, T \leq N) + E[X_N : T > N].
\end{aligned}
$$

Now $|E[X_N : T > N]| \leq (|a| + b)P(T > N) \to 0$ as $N \to \infty$, since T is almost surely finite. Thus letting $N \to \infty$ we obtain

$$0 = aP(X_T = a) + b\, P(X_T = b)$$

and since $P(X_T = a) = 1 - P(X_T = b)$ we conclude that

$$P(X_T = a) = \frac{b}{b - a}.$$

Exercise. If we took instead $T = \inf\{n : X_n = -1\}$, we could use the OST at the bounded stopping time $T \wedge N$ and then let $N \to \infty$. What happens?

The other great result of martingale theory is the martingale convergence theorem (in reality, a collection of results). Here is what it says.

THEOREM 2 *Let $(X_n, \mathcal{F}_n)_{n \in \mathbb{Z}^+}$ be a martingale.*

(i) If X is bounded in L^1 (that is $\sup_n E|X_n| < \infty$) then there exists $X_\infty \in L^1(\mathcal{F}_\infty)$ such that

$$X_n \stackrel{a.s.}{\to} X_\infty.$$

(ii) If X is bounded in L^1 then the following are equivalent:

(a) $X_n \to X_\infty$ in L^1;
(b) $X_n = E(X_\infty | \mathcal{F}_n)$;
(c) $(X_n)_{n \in \mathbb{Z}^+}$ is uniformly integrable.

(iii) If for some $p > 1$, X is bounded in L^p ($\sup_n E|X_n|^p < \infty$) then there exists some $X_\infty \in L^p(\mathcal{F}_\infty)$ such that

$$X_n \to X_\infty \quad \text{a.s. and in } L^p,$$

and $X_n = E(X_\infty|\mathcal{F}_n)$.

For a proof, see, for example, Williams [13]. Another extremely useful result in this area is the following.

THEOREM 3 *(Positive Supermartingale Convergence Theorem).* If $(Y_n, \mathcal{F}_n)_{n \in \mathbb{Z}^+}$ *is a supermartingale, and $Y_n \geq 0$ for all n, then there exists $Y_\infty \in L^1(\mathcal{F}_\infty)$ such that*

$$Y_n \overset{a.s.}{\to} Y_\infty.$$

Of the two great martingale results, it is the comparatively trivial OST which is the more useful.

Before we move on from discrete-time martingales, we discuss briefly the notion of *compensation*. If $(Z_n)_{n \in \mathbb{Z}^+}$ is a sub-martingale, it tends to rise on average; the idea of compensation is to counteract this tendency in a previsible way. We have the following.

PROPOSITION 4 *There is a unique previsible increasing process A such that $Z_n - A_n$ is a martingale and $A_0 = 0$.*

Proof. If we define $A_0 = 0$,

$$A_n - A_{n-1} = E[Z_n - Z_{n-1}|\mathcal{F}_{n-1}]$$

then plainly A is previsible, increasing (because Z is a sub-martingale) and it is easy to check that $Z - A$ is a martingale. Uniqueness follows because a previsible martingale is constant.

In continuous time, an analogous result holds, and is very deep. One of the most important applications of compensation is to sub-martingales of the form $Z_n = M_n^2$, where M is an L^2 martingale. We see more of this in the next chapter.

3   Brownian Motion and Martingales in Continuous Time

It is now clear what we should take for the definition of a martingale in continuous time, defined on some probability triple $(\Omega, \mathcal{F}, P)$. It should be (and is) an object $(X_t, \mathcal{F}_t)_{t \in \mathbb{R}^+}$ such that

$$\mathcal{F}_t \text{ is a sub-}\sigma\text{-field of } \mathcal{F}, \text{ and } X_t \in L^1(\Omega, \mathcal{F}_t, P) \text{ for all } t \geq 0; \quad (3.1\text{i})$$
$$\mathcal{F}_s \subseteq \mathcal{F}_t \text{ for } s \leq t; \quad (3.1\text{ii})$$
$$X_s = E(X_t|\mathcal{F}_s) \text{ for } 0 \leq s \leq t. \quad (3.1\text{iii})$$

Replacing the equality in (3.1iii) by $\geq$ gives the definition of a supermartingale. Likewise, a stopping time should be (and is) a random variable $T : \Omega \to [0, \infty) \cup \{\infty\}$ such that for all $t \in \mathbb{R}^+$

$$\{T \leq t\} \in \mathcal{F}_t.$$

We expect (and shall have) the analogues of the great discrete-time results, but there is a slight problem. Suppose that we take T to be a random variable uniformly distributed on $[0, 1]$ and define the process $(X_t)_{t \in \mathbb{R}^+}$ by

$$\begin{aligned} X_t(\omega) &= 1 \text{ if } T(\omega) = t \\ &= 0 \text{ if } T(\omega) \neq t. \end{aligned}$$

Then it is easy to see that X is a martingale with respect to its own filtration, because conditional expectation is only unique to within a random variable which is almost surely zero. However, the OST must fail, since $EX_0 = 0$, $EX_T = 1$. Our problem is that the sample paths of X are not nice. So for continuous time, our first need is to regularise the paths.

THEOREM 5 *Let* $(X_t, \mathcal{F}_t)_{t \in \mathbb{R}^+}$ *be a supermartingale on* $(\Omega, \mathcal{F}, P)$ *and suppose that*

> *(i)* $t \mapsto EX_t$ *is right-continuous;*
> *(ii)* $(\Omega, \mathcal{F}, P)$ *is complete (that is, if* $A \in \mathcal{F}, P(A) = 0$*, then every subset of* A *is in* $\mathcal{F}$*);*
> *(iii)* $\mathcal{F}_t = \bigcap_{u > t} \mathcal{F}_u \equiv \mathcal{F}_{t+}$ *for all* $t \geq 0$*;*
> *(iv) each* $\mathcal{F}_t$ *contains all the null sets in* $\mathcal{F}$*.*

Then there exists a supermartingale $\tilde{X}_t$ *such that for all* ω *the function* $t \mapsto \tilde{X}_t(\omega)$ *is right continuous with left limits, and for each* $t \geq 0$

$$P(X_t = \tilde{X}_t) = 1.$$

The conditions (iii)–(iv) are what are commonly referred to as the *usual conditions*; we shall always assume then, and may (and shall) assume that all supermartingales (and, in particular, martingales) in continuous time are right-continuous with left limits. Under *these* conditions, the analogues of the OST and martingale-convergence theorem hold. A function $t \mapsto f(t)$ defined for $t \in \mathbb{R}^+$ is called an *R-function* if it is right continuous in $[0, \infty)$ with left limits in $(0, \infty)$.

THEOREM 6 *(OST)* *If* $(X_t, \mathcal{F})_{t \in \mathbb{R}^+}$ *is a martingale (with R-paths) and* T *is a bounded stopping time, then*

$$X_0 = E(X_T | \mathcal{F}_0),$$

and $EX_T = EX_0$.

The martingale convergence theorem also goes across immediately.

So let us now look at the most interesting continuous-time martingale, *Brownian motion* (also called the *Wiener process*). Brownian motion is a process $(B_t)_{t \in \mathbb{R}^+}$ such that

$$B_0 = 0; \tag{3.2i}$$

$$t \mapsto B_t(\omega) \text{ is continuous for all } \omega \in \Omega; \tag{3.2ii}$$

$$\text{for } 0 \leq s \leq t, \ B_t - B_s \text{ is independent of } \{B_u : u \leq s\}$$

$$\text{and has a } N(0, t - s) \text{ distribution.} \tag{3.2iii}$$

Only properties (ii) and (iii) are essential; we will often say that $(a + B_t)_{t \in \mathbb{R}^+}$ is a Brownian motion, with starting point a. The existence of such a process is not obvious, but the martingale property is quite easy; if $\mathcal{F}_t = \sigma(\{B_u : u \leq t\})$ then we use the third property to show that for $0 \leq s \leq t$

$$E(B_t - B_s | \mathcal{F}_s) = E(B_t - B_s), \text{ by property (ix) of conditional}$$
$$\text{expectation;}$$
$$= 0, \text{ by property (iii) of Brownian motion.}$$

It is also not hard to prove that $(B_t^2 - t)_{t \in \mathbb{R}^+}$ *is a martingale.* Indeed

$$E[B_t^2 - t - (B_s^2 - s)|\mathcal{F}_s]$$
$$= E[(B_s + B_t - B_s)^2 - B_s^2 | \mathcal{F}_s] - (t - s)$$
$$= E[(B_t - B_s)^2 + 2B_s(B_t - B_s)|\mathcal{F}_s] - (t - s)$$
$$= E[(B_t - B_s)^2 | \mathcal{F}_s] + 2B_s E[B_t - B_s | \mathcal{F}_s] - (t - s)$$
$$= E(B_t - B_s)^2 - (t - s), \text{ by property (ix) of conditional}$$
$$\text{expectation and the martingale property of } B;$$
$$= 0 \quad \text{by property (iii) of } B.$$

A remarkable converse to the above is that if $(M_t)_{t \in \mathbb{R}^+}$ is a continuous martingale such that $M_t^2 - t$ is also a martingale, *then M is a Brownian motion!*

As an easy exercise in the use of conditional expectation, you should prove that for any $\alpha \in \mathbb{R}$ the process $(\exp(\alpha B_t - \frac{1}{2}\alpha^2 t))_{t \geq 0}$ is a martingale. [*Hint:* If $Y \sim N(\mu, \sigma^2)$ then $E \exp(\alpha Y) = \exp(\alpha\mu + \frac{1}{2}\alpha^2\sigma^2)$.]

We can now use this in various ways; here is just one application. If we fix $a < 0 < b$ and let $T = \inf\{t : B_t = a \text{ or } b\}$, then the process

$$M_t \equiv e^{-\frac{1}{2}\alpha^2 t} \cosh(\alpha(B_t - \mu))$$

is a martingale ($\mu \equiv (b + a)/2$). The key point is that $\cosh\alpha(b - \mu) = \cosh\alpha(a - \mu) = \cosh\alpha(b - a)/2$ and therefore by the optional sampling theorem, for any $N > 0$

$$EM_0 = EM(T \wedge N)$$

$$\begin{aligned}
&= E[e^{-\alpha^2 T/2}; T \le N] \, \cosh \alpha (b-a)/2 \\
&\quad + e^{-\alpha^2 N/2} E[\cosh \alpha (B_N - \mu); T > N] \\
&\to E(e^{-\alpha^2 T/2}) \cosh \alpha (b-a)/2
\end{aligned}$$

as $N \to \infty$. We therefore conclude that

$$E(e^{-\alpha^2 T/2}) = \frac{\cosh \alpha (b+a)/2}{\cosh \alpha (b-a)/2}. \tag{3.3}$$

This gives us the (Laplace transform of the) distribution of T, which determines the distribution of T. There is in general no closed-form expression for this distribution, though. However, if we let $a \to -\infty$, we get in the limit

$$E e^{-\alpha^2 T/2} = e^{-\alpha b} = \int_0^\infty e^{-\alpha^2 t/2} \frac{b e^{-b^2/2t}}{\sqrt{2\pi t^3}} dt, \tag{3.4}$$

which identifies the density of the first-passage time to level $b > 0$.

Before we go on to discuss Itô's formula in the next section, we need to know the following. *If* $(M_t)_{t \in \mathbb{R}^+}$ *is a continuous martingale, then there exists a continuous increasing adapted process* $\langle M \rangle$ *such that* $\langle M \rangle_0 = 0$ *and for each* $T \ge 0$

$$\sup_{0 \le t \le T} \left| \sum_{i=1}^\infty \{ M(t \wedge i 2^{-n}) - M(t \wedge (i-1)2^{-n}) \}^2 - \langle M \rangle_t \right| \xrightarrow{p} 0 \tag{3.5}$$

as $n \to \infty$. *This process has the property that*

$$M_t^2 - \langle M \rangle_t \text{ is a (local) martingale.} \tag{3.6}$$

In the special case where M is Brownian motion, the increasing process $\langle M \rangle_t$ is, of course, just t! In view of the fact that if M is a continuous martingale and $M_t^2 - t$ is a continuous martingale then M is Brownian motion, it is not hard to accept the extremely useful fact that *if* M *is a continuous martingale then on (some enrichment of)* $(\Omega, \mathcal{F}, P)$ *there exists a Brownian motion* $(B_t)_{t \ge 0}$ *such that*

$$M_t = B(\langle M \rangle_t). \tag{3.7}$$

This is one reason why Brownian motion is so important; *up to change of time scale, every continuous martingale is a Brownian motion!*

4   Itô's Formula and Stochastic Integrals

We present here a basic version of Itô's formula (also called the change-of-variables formula); there are variants of the formula for processes with discontinuous paths (see, for example, Chapter VI of [11]), but for our purposes we shall consider only continuous adapted processes.

In the light of what we know already, it is not hard to understand why the change-of-variables formula (1.5) is incorrect, and what the correction ought to be (but beware – the ordinary Newtonian calculus *is* correct if one works with Stratonovich stochastic integrals, but these are not appropriate except in very restricted circumstances).

So let us suppose that $(X_t)_{t \in \mathbb{R}^+}$ is a continuous adapted semimartingale, which means that we have an expression

$$X_t = M_t + A_t$$

where A is a continuous adapted process all of whose paths have finite variation, and M is a continuous (local[1]) martingale. Now (following Föllmer [5]) let us suppose that $f : \mathbb{R} \to \mathbb{R}$ is C^2 with bounded derivatives up to order 2 (though this last assumption is not necessary). We then have (with $t_i^n \equiv (i2^{-n}) \wedge t$)

$$f(X_t) - f(X_0)$$
$$= \sum_{i=1}^{\infty} \{ f(X(t_i^n)) - f(X(t_{i-1}^n)) \} \tag{4.1}$$
$$= \sum_{i=1}^{\infty} \{ f'(X(t_{i-1}^n)) \Delta_i^n X_t + \tfrac{1}{2} f''(X(t_{i-1}^n) + \theta_i^n \Delta_i^n X_t)(\Delta_i^n X_t)^2 \} \tag{4.2}$$

where $\Delta_i^n X_t \equiv X(t_i^n) - X(t_{i-1}^n)$ and $0 < \theta_i^n < 1$ for all n, i; this last equality is simply Taylor's formula. Now from (3.5) we know that for each $T > 0$

$$\sup_{0 \le t \le T} \left| \sum_{i \ge 1} (\Delta_i^n X_t)^2 - \langle X \rangle_t \right| \to 0$$

in probability, so, taking a subsequence if necessary, we could assume the convergence is almost sure, and then, with a little care, one can prove the intuitively plausible result that

$$\sum_{i \ge 1} \tfrac{1}{2} f''(X(t_{i-1}^n) + \theta_i^n \Delta_i^n X_t)(\Delta_i^n X_t)^2 \to \int_0^t \tfrac{1}{2} f''(X_s) d\langle X \rangle_s. \tag{4.3}$$

Given that, it must be that the remaining terms in (4.2), namely,

$$\sum_{i=1}^{\infty} f'(X(t_{i-1}^n)) \Delta_i^n X_t,$$

---

[1] $(M_t)_{t \in \mathbb{R}^+}$ is a local martingale if there exist stopping times $T_n \uparrow \infty$ such that $M_t^n \equiv M(t \wedge T_n)$ is a martingale. In this account, we tend to gloss over the distinction between martingales and local martingales, as with many other points of second order; the reader must study one of the texts mentioned in the Introduction to learn how to handle the distinction safely.

must also be converging to something. That something is just the stochastic integral $\int_0^t f'(X_s)dX_s$;

$$\sum_i f'(X(t_{i-1}^n))\Delta_i^n X_t \to \int_0^t f'(X_s)dX_s$$

uniformly on compacts, in probability. Hence we deduce the (basic) Itô formula:

$$f(X_t) - f(X_0) = \int_0^t f'(X_s)dX_s + \frac{1}{2}\int_0^t f''(X_s)d\langle X\rangle_s. \qquad (4.4)$$

You have to view this *either* as amazing and profoundly important (which it is) *or* as merely a definition of what the expression $\int_0^t f'(X_s)dX_s$ is to mean! A careful treatment sets up a proper theory of stochastic integrals (which gives meaning to $\int_0^t f'(X_s)dX_s$) and then the amazing thing is the equality (4.4).

But let us make a few remarks on what these stochastic integrals are. Recalling that $X_t = M_t + A_t$, we have

$$\sum_i f'(X(t_{i-1}^n))\Delta_i^n X_t = \sum_i f'(X(t_{i-1}^n))\Delta_i^n M_t + \sum_i f'(X(t_{i-1}^n))\Delta_i^n A_t \qquad (4.5)$$

and, because the paths of A are of finite variation, the convergence of the second term on the right to $\int_0^t f'(X_s)dA_s$ is ordinary analysis, the Lebesgue-Stieltjes integral being well defined. The interesting part is

$$\sum_i f'(X(t_{i-1}^n))\Delta_i^n M_t. \qquad (4.6)$$

If we take just one term in the sum and write it out more fully, we obtain

$$f'(X(t \wedge (i-1)2^{-n}))\{M(t \wedge i2^{-n}) - M(t \wedge (i-1)2^{-n})\}$$

which will be zero if $t \le (i-1)2^{-n}$, and constant if $t \ge i2^{-n}$, so can be written instead as

$$f'(X((i-1)2^{-n}))\{M(t \wedge i2^{-n}) - M(t \wedge (i-1)2^{-n})\},$$

which is clearly a *martingale*! (If f' were constant, it is trivial that the process is a martingale, and if f' is general, then the random variable $f'(X((i-1)2^{-n}))$ is $\mathcal{F}((i-1)2^{-n})$–measurable, and is known before the martingale term in the curly brackets begins to move. A formal proof is straightforward, and is left as an exercise.)

Thus the expression (4.6) is also a martingale (it is a sum of martingales), so the limit as $n \to \infty$, which is written $\int_0^t f'(X_s)dM_s$, should also be a martingale; this is an example of the general principle that *stochastic integrals with respect to local martingales are local martingales.*

Let us see just a couple of examples of Itô's formula applied to Brownian motion. Firstly, with $f(x) = x^2$, we learn from (4.4) that

$$B_t^2 = \int_0^t 2B_s dB_s + t$$

since $\langle B \rangle_t = t$. Part of this we knew already, namely, that $B_t^2 - t$ is a martingale; the new information is that we can express it as $\int_0^t 2B_s dB_s$.

From the way that the stochastic integral $N_t \equiv \int_0^t f'(X_s) dX_s$ was shown to be a limit of sums of martingales, we would expect that

$$
\begin{aligned}
\langle N \rangle_t &= \lim_{n \to \infty} \sum_{i \geq i} (\Delta_i^n N_t)^2 \\
&= \lim_{n \to \infty} \sum_{i \geq 1} f'(X(t_{i-1}^n))^2 (\Delta_i^n X_t)^2 \\
&= \int_0^t f'(X_s)^2 d\langle X \rangle_s,
\end{aligned}
$$

and this is indeed the case. So we should therefore have that

$$(B_t^2 - t)^2 - 4\int_0^t B_s^2 ds \quad \text{is a martingale,}$$

and this can be verified directly from the definition of Brownian motion.

We also have an extension of Itô's formula to more than one continuous semimartingale; if $X_t = (X_t^1, ..., X_t^n)$ is an n–vector of continuous semimartingales, then

$$f(X_t) - f(X_0) = \int_0^t \sum_{i=1}^n D_i f(X_s) dX_s^i + \frac{1}{2} \sum_{i=1}^n \sum_{j=1}^n \int_0^t D_i D_j f(X_s) d\langle X^i, X^j \rangle_s,$$

(4.7)

where $D_i \equiv \partial/\partial x^i$, and $2\langle X^i, X^j \rangle \equiv \langle X^i + X^j \rangle - \langle X^i \rangle - \langle X^j \rangle$ defines the quadratic covariation of X^i and X^j by polarisation. The argument we gave for the case of one semimartingale could be extended to (a heuristic for) this case also. As an example, suppose that $X_t = (t, B_t)$ so that

$$
\begin{aligned}
f(t, B_t) - f(0, B_0) &= \int_0^t f_1(s, B_s) ds + \int_0^t f_2(s, B_s) dB_s \\
&\quad + \frac{1}{2} \int_0^t f_{22}(s, B_s) ds,
\end{aligned}
$$

where $f_i \equiv D_i f$; taking the specific example $f(t, x) = \exp(\alpha x - \frac{1}{2}\alpha^2 t)$ produces the statement

$$e^{\alpha B_t - \frac{1}{2}\alpha^2 t} - 1 = \int_0^t \alpha e^{\alpha B_s - \frac{1}{2}\alpha^2 s} dB_s,$$

part of which we again knew already.

Our final topic in this section is a result of profound significance.

THEOREM 7 *Let* $(B_t)_{t\geq 0}$ *be a Brownian motion, and let* $\mathcal{F} = \sigma(\{B_t : t \geq 0\})$. *Then any* $Y \in L^2(\mathcal{F})$ *may be expressed as*

$$Y = EY + \int_0^\infty H_s dB_s \qquad (4.8)$$

where H *is a previsible process for which* $E \int_0^\infty H_s^2 ds < \infty$. *The representation is unique, and*

$$\mathrm{var}\ (Y) = E \int_0^\infty H_s^2 ds.$$

For a proof, see any of the texts on stochastic calculus referred to in the Introduction. The previsible integrand H might be some function of t and B_t, but generally can be much more complicated. This result is a key part of the theory of complete markets. In that context, we change measure so as to make discounted prices into martingales, and then we express a general contingent claim as a stochastic integral with respect to B (or, equivalently, with respect to the discounted price process), which then demonstrates a way of replicating the contingent claim by trading on the marketed assets.

5   Markov Chains in Discrete Time

Our goal in this section is to develop some of the intuition of Markov processes in the simplest possible setting – discrete time and a countable state space. Recall that for processes which one can hope to use in practice, we tend to concentrate on Markovian models.

We shall consider a stochastic process $(X_n)_{n\in\mathbb{Z}^+}$ with values in a countable set I (which may be a subset of $\mathbb{R}$, but does not have to be), and which is *Markovian*, in that there exists a matrix $P \equiv (p_{ij})_{i,j\in 0}$ such that

$$P(X_{n+1} = j | X_0 = i_0, ..., X_{n-1} = i_{n-1}, X_n = i) = p_{ij} \qquad (5.1)$$

for all $n \in \mathbb{Z}^+$, $i_0, ..., i_{n-1}, i, j \in I$. Informally, *where the process next jumps depends on where it now is, but not on how it got there.* It is not hard to develop equivalent formulations of the Markov property; for example, one proves easily by induction that

$$P(X_0 = i_0, X_1 = i_1, ..., X_n = i_n) = P(X_0 = i_0) \prod_{j=1}^n p_{i_{r-1}i_r} \qquad (5.2)$$

and from that by summing over $i_1, i_2, ..., i_{n-1}$, one deduces that

$$P(X_n = i_n | X_0 = i_0) = p_{i_0 i_n}^{(n)}, \qquad (5.3)$$

where $(p_{ij}^{(n)})_{i,j\in I} \equiv P^n$ is the matrix of n–step transition probabilities. From the interpretation (5.1) of P, it is clear that $p_{ij} \geq 0$ for all i, j, and $\sum_j p_{ij} = 1$ for all i; such a matrix is called *stochastic*.

The *simple random walk* on $I = \mathbb{Z}$, for which for some $p \in (0,1)$

$$p_{i,i+1} = p, \quad p_{i,i-1} = q \equiv 1 - p \quad \forall\, i \in \mathbb{Z}$$

and $p_{ij} = 0$ if $|i - j| \neq 1$, furnishes an important example of a Markov chain. Many others arise in applications. When $p = q = \frac{1}{2}$, the simple random walk is called *symmetric*, and its behaviour is qualitatively completely different from its behaviour if $p \neq q$. Indeed, if $p = q$, the chain is *recurrent* (which means it visits every state infinitely often) whereas for $p > q$ the chain drifts off to $+\infty$, and will visit each state at most finitely many times. Such a chain is called *transient*.

Let us now consider a problem of a type which arises in finance. We have a Markov chain on a statespace I which will eventually with certainty reach some distinguished state ∂ and remain there for ever. We have the opportunity after each jump of the chain to choose to stop where we are; if the chain is in state i, we receive reward $r(i)$ if we choose to stop, and if we choose to stop, the whole game finishes. If we do not choose to stop, we receive nothing, and the chain takes another step, after which we again must decide whether or not to stop. Assume that $r(i) \geq 0 = r(\partial)$ for all i. Our problem is to determine a stopping rule which maximises the expected reward we receive when we stop.

The problem is solved as follows. Let $f_n(i)$ denote the maximal expected reward if we are compelled to stop after at most n steps, and we start at state i. Then clearly

$$f_0(i) = r(i), \tag{5.4}$$

and we shall have

$$f_{n+1}(i) = \max\left\{ r(i), \sum_j p_{ij} f_n(j) \right\}, \quad n \geq 0. \tag{5.5}$$

The interpretation of (5.5) is transparent; if we are in state i with at most $n + 1$ steps to go, we can *either* stop immediately (yielding reward $r(i)$) *or* take a step, yielding an expected further reward $f_n(j)$ if we happen to jump to state j. But we jump to state j with probability p_{ij}, so our expected reward if we do not stop is $\sum_j p_{ij} f_n(j)$. The decision is to pick the alternative which gives largest expected reward.

In a more compact notation, if we think of f_n as a column vector $(f_n(i))_{i \in I}$, then we have

$$f_0 = r, \quad f_{n+1} = \max\{r, Pf_n\}. \tag{5.6}$$

It is easy to see that $f_0 \leq f_1 \leq \dots$ so the f_n increase to some limit which (if finite) satisfies

$$f = \max\{r, Pf\}. \tag{5.7}$$

The function f is the value of the game, and the stopping rule is to stop when in state i if and only if $f(i) = r(i)$.

NationsBanc-CRT
233 SOUTH WACKER DRIVE
CHICAGO, ILLINOIS 60606

This is not quite all that needs to be said, though; if r was bounded by 1, say, the equation (5.7) is solved by the f we want, but it is also solved by $f(i) = \lambda$ for all i, if $\lambda \geq 1$! That is, the solution to (5.7) is not *unique*. However, the solution we want is the *minimal* solution to (5.7).

Exercise.   Show that if g solves $g = \max\{r, Pg\}$ then $g \geq f$, by firstly observing that $g \geq f_0$.

It is extremely rare to find examples of optional stopping problems which can be solved in closed form, although numerical solution by construction of f_n is always possible provided the statespace is not too big.

Let us use the optimal stopping methodology to solve a very simple example, that of simple random walk on $\mathbb{Z}$ with $p > q$, and with reward function $r(0) = 1$, $r(i) = 0$ for $i \neq 0$. It is of course perfectly clear what the optimal stopping rule is; wait until the chain reaches 0 (if ever) and then stop immediately. Therefore we shall have

$$f(i) = P \text{ (chain reaches } 0|X_0 = i),$$

and this is somewhat more interesting. From (5.7), we shall have for $i \neq 0$

$$f_i = (Pf)_i = pf_{i+1} + qf_{i-1}.$$

This constant-coefficient difference equation admits a general solution of the form

$$f_i = A + B(q/p)^i.$$

For $i \leq 0$, the solution $(q/p)^i$ is unbounded, so it has to be that $f_i = 1$ for all $i \leq 0$. For $i \geq 0$, though, both the solutions are bounded, and the condition $f_0 = 1$ only gives one condition. We cannot determine A and B without appeal to the *minimality* of f; when we use this, we find that the minimal solution is

$$
\begin{aligned}
f_i &= 1 \quad (i \leq 0) \\
&= (q/p)^i \quad (i \geq 0),
\end{aligned}
$$

so for $i > 0$, $P(X \text{ reaches } 0|X_0 = i) = (q/p)^i$.

Now let us consider a variant of the simple random walk which leads us into interesting issues. As before, we fix $p = 1 - q \in (\frac{1}{2}, 1)$, and define a transition matrix

$$
\begin{aligned}
p_{01} &= p_{0,-1} = \frac{1}{2} \\
p_{i,i+1} &= p = 1 - p_{i,i-1} \quad \text{for } i \geq 1, \\
p_{i,i-1} &= p = 1 - p_{i,i+1} \quad \text{for } i \leq -1.
\end{aligned}
$$

This Markov chain looks like a random walk with probability p of an upward step while in the positive integers, and with probability p of a downward step while in the negative integers. It is intuitively clear that the process will ultimately escape either to $+\infty$ or to $-\infty$; *but what is the probability that ultimately the process escapes to $+\infty$?* This obviously depends on the starting point. If we define

$$h(i) = P\ (\text{escape to } +\infty | X_0 = i)$$

then $h(0) = \frac{1}{2}$ by obvious symmetry, and for $i \le 0$, we have

$$h(i) = (q/p)^{-i} h(0) = \frac{1}{2}(q/p)^{-i}$$

since the process must firstly reach 0, and then subsequently escape to $+\infty$. Finally, for $i > 0$, we have

$$h(i) = \{1 - (q/p)^i\} + (q/p)^i \cdot h(0) = 1 - \frac{1}{2}(q/p)^i.$$

[**Exercise.** Why?]. Now if we consider the martingale

$$M_n \equiv P(\text{escape to } +\infty | \mathcal{F}_n),$$

then $M_n = h(X_n)$, because, whatever n, the event that the chain escapes to $+\infty$ is in the future after n, so the probability that the event happens depends only on the current position X_n, not on any of the earlier values of X. Another way to look at this is to notice that

$$E[h(X_{n+1})|\mathcal{F}_n] = \sum_j p_{X_n j} h(j) = h(X_n),$$

since (as is easily checked) $h = Ph$. A function for which $h = Ph \ge 0$ is called *harmonic*, and a function for which $h \ge Ph \ge 0$ is called *superharmonic*. We have here an example where we take an event A concerned with the 'ultimate behaviour' of X, and its probability given the starting point is a harmonic function of the starting point;

$$h(i) = P(A|X_0 = i).$$

Now suppose we *condition the chain on the event A.* Then

$$
\begin{aligned}
P(X_0 = i_0, ..., X_n = i_n | A) &= \frac{P(X_0 = i_0, ..., X_n = i_n, A)}{P(A)} \\
&= \frac{E[h(X_n); X_0 = i_0, ..., X_n = i_n]}{P(A)} \\
&= \frac{P(X_0 = i_0) \prod_{r=1}^n p_{i_{r-1} i_r} h(i_n)}{\sum_i h(i) P(X_0 = i)} \\
&= \tilde{P}(X_0 = i_0) \prod_{r=1}^n \tilde{p}_{i_{r-1} i_r},
\end{aligned}
$$

where $\tilde{P}(X_0 = i_0) \equiv P(X_0 = i_0|A)$ and $\tilde{p}_{ij} = p_{ij}h_j/h_i$. Comparing with (5.2), we see that *by conditioning on A we convert the original Markov chain into a new Markov chain with transition matrix*

$$\tilde{p}_{ij} = \frac{p_{ij}h_j}{h_i} \qquad (5.8)$$

(and the fact that this *is* a transition matrix is just the statement that h is harmonic!) Of course, we could use any (strictly positive) harmonic function h to modify the original Markov chain according to the recipe (5.8); the chain we end up with is called the *Doob h–transform* of the original chain. For a general harmonic h, the interpretation in terms of conditioning is less direct, but underlines what is going on.

By h–transforming, we are in effect *altering the law of the Markov chain*, and the martingale $h(X_n)/h(X_0)$ is the change-of-measure martingale. The method of changing measure is widely used in the semi martingale context; it is worth realising that this methodology is exemplified and inspired by the (Markovian) Doob h–transform.

While we are thinking about the Markov/martingale connection, let us remark that although for a general $f : I \to \mathbb{R}$ the process $f(X_n)$ will not be a martingale, we can make it into a martingale by appropriately compensating it:

$$M_n \equiv f(X_n) - \sum_{r=0}^{n-1}(Pf - f)(X_r)$$

is a martingale, as you can easily check. While this may not be much used in the discrete-time setting, it is well adapted to stochastic calculus in the continuous-time setting, as we shall see.

6 Continuous-time Markov Chains

At the level of generality needed for applications, continuous-time Markov chains can be understood through the theory of discrete-time Markov chains. A continuous-time Markov chain $(X_t)_{t\in\mathbb{R}^+}$ with (countable) state space I is an I–valued process such that for all $t_0 = 0 \le t_1 \le ... \le t_n$ and $i_0, i_1, ..., i_n \in I$

$$P(X(t_r) = i_r, \ r = 0, ..., n) = P(X_0 = i_0) \prod_{r=1}^{n} p_{i_{r-1}i_r}(t_r - t_{r-1}), \qquad (6.1)$$

where for each $t \ge 0, P(t) \equiv (p_{ij}(t))_{i,j\in I}$ is a stochastic matrix. By taking $n = 2$ in (6.1), and summing over i_1, we find that

$$P(s + t) = P(s)P(t) \quad \forall \ s, t \ge 0, \qquad (6.2)$$

which is the semigroup property of $(P(t))_{t\ge 0}$. For various technical reasons, we also assume that the paths of X are right-continuous with left limits in the discrete topology. This means that, wherever the chain starts, it will stay there for a positive time. So if $\tau = \inf\{t > 0; X_t \neq X_0\}$, we have $P(\tau > 0) = 1$. Now

$$
\begin{aligned}
P(\tau > t + s | X_0 = i) &= P(\tau > t + s | \tau > t, X_0 = i) \cdot P(\tau > t | X_0 = i) \\
&= P(\tau > s | X_0 = i) P(\tau > t | X_0 = i)
\end{aligned}
$$

using the Markov property (because $\{\tau > t, X_0 = i\} = \{\tau > t, X_t = i\}$, and if $X_t = i$, the fact that $X_u = i$ for all $u < t$ is irrelevant to the future behaviour of the process). This implies that

$$
P(\tau > t | X_0 = i) = e^{-q_i t} \tag{6.3}
$$

for some $0 \leq q_i < \infty$, so *the time of the first jump is exponentially distributed.*

Under the assumptions we have made so far, it can be shown that

$$
p'_{ij}(0) \equiv q_{ij} \quad \text{exists for all } i, j
$$

and that $q_{ii} = -q_i$, $q_{ij} \geq 0$ for $i \neq j$, and $\sum_j q_{ij} = 0$ for each i (though the last conclusion does not hold in generality). The matrix Q is called the (infinitesimal) *generator* of the chain.

Now the structure of the chain is illuminated by calculating for $i \neq j$

$$
\begin{aligned}
& P[\tau \in dt, X(\tau) = j | X_0 = i]/dt \\
&= \lim_{\delta \downarrow 0} \frac{1}{\delta} P[t < \tau \leq t + \delta, \ X(t + \delta) = j | X_0 = i] \\
&= \lim_{\delta \downarrow 0} \frac{1}{\delta} e^{-q_i t} P_{ij}(\delta) \\
&= q_i e^{-q_i t} (q_{ij}/q_i).
\end{aligned}
$$

Thus *the time of the first jump is* $\exp(q_i)$–*distributed, and the place of the first jump is independent, chosen according to the distribution* $(J_{ij})_{j \in I} \equiv (q_i^{-1} q_{ij})_{j \in I}$. This in fact extends to all subsequent behaviour of the chain, so we have a beautifully simple description of the movement of the continuous-time chain: if $Y_0, Y_1, Y_2, \ldots$ is the sequence of states moved through, then $(Y_n)_{n \in \mathbb{Z}^+}$ is a discrete-time chain with transition matrix J, and, conditional on the sequence $(Y_n)_{n \in \mathbb{Z}^+}$, the times spent in the successive states before jumping to the next state are independent exponential variables, with parameters $q(Y_0), q(Y_1), \ldots$ This sample-path description says all that there is to say about the behaviour of the chain (but be warned that without the simplifying assumptions made here, the sample paths of Markov chains can be indescribably complicated – literally!). However, the analysis of the semigroup $(P(t))_{t \geq 0}$ is worth developing too, because this generalises more readily to other continuous-time Markov processes. From the semigroup relation (5.2) we have that

$$
P(s + t) - P(t) = (P(s) - I) P(t) = P(t)(P(s) - I)
$$

and it is tempting to divide by s and let $s \downarrow 0$ to conclude that

$$P'(t) = QP(t) = P(t)Q \tag{6.4}$$

which is true in the current situation. This suggests that formally

$$P(t) = \exp(tQ), \tag{6.5}$$

which, when suitably interpreted, is true. Moreover, from (6.4) we should be able to conclude that for suitable f

$$P_t f - f = \int_0^t P_s Q f \, ds \tag{6.6}$$

which we can state in probabilistically more meaningful terms by saying that

$$f(X_t) - f(X_0) - \int_0^t Q f(X_s) \, ds \quad \text{is a martingale.} \tag{6.7}$$

It is worth noticing that this conclusion did not depend very crucially on the fact that the statespace of the process X is a countable set; really, only the semigroup property (6.2) and the differentiation of the semigroup (6.4) were involved. We shall proceed to generalise the Markov chain results in the next section.

7   Stochastic Differential Equations

For a Markov chain, we have seen that the (infinitesimal) generator, interpreted as $P'(0)$, plays a big role in determining the behaviour of the process. Is there an analogue for Brownian motion? Indeed there is. If we consider for Brownian motion

$$\begin{aligned}
\frac{1}{t}\{P_t f(x) - f(x)\} &= \frac{1}{t}\{E[f(B_t + x)] - f(x)\} \\
&= \frac{1}{t}E[f(x + \sqrt{t}B_1) - f(x)],
\end{aligned}$$

since $B_t \sim N(0, \sqrt{t})$, we may use Taylor's formula if $f \in C_b^2$ to write

$$\begin{aligned}
\frac{1}{2}\{P_t f(x) - f(x)\} &= \frac{1}{t}E\left[f'(x)\sqrt{t}B_1 + \frac{1}{2}f''(x + \theta\sqrt{t}B_1)tB_1^2\right] \\
&= E\left[\frac{1}{2}f''(x + \theta\sqrt{t}B_1)B_1^2\right] \\
&\rightarrow \frac{1}{2}f''(x) \quad (t \downarrow 0)
\end{aligned}$$

since $|\theta| \le 1$. So if we take the statement (6.7) and rephrase it for Brownian motion, we learn that for $f \in C_b^2$

$$f(B_t) - f(B_0) - \int_0^t \frac{1}{2}f''(B_s)\, ds \quad \text{is a martingale.} \tag{7.1}$$

This statement, arrived at entirely by Markovian arguments, reaches the same conclusion as the semimartingale argument leading to the Itô formula statement

$$f(B_t) - f(B_0) - \frac{1}{2}\int_0^t f''(B_s)ds = \int_0^t f'(B_s)dB_s \qquad (7.2)$$

which is a local martingale, and in fact also a martingale, since f' is assumed bounded.

We now want to consider what happens for a stochastic differential equation (SDE) of the form

$$dX_t = \sigma(X_t)dB_t + \mu(X_t)dt \qquad (7.3)$$

where σ, μ are (say) Lipschitz continuous. This can also be phrased in integrated form as

$$X_t = X_0 + \int_0^t \sigma(X_s)dB_s + \int_0^t \mu(X_s)ds. \qquad (7.4)$$

Does there exist a solution to this, if we are given X_0 and the driving Brownian motion B? Is there at most one solution? Fortunately, the answer to both questions is, 'Yes'. The classical proof by successive approximations appears in all the texts on stochastic calculus listed in the Introduction. Moreover, once we know this, it is intuitively clear that the solution is Markovian, because we actually build the solution process from X_0 and $(B_t)_{t\geq 0}$; thus if we define $\tilde{X}_t \equiv X_{t+c}$, $\tilde{B}_t \equiv B_{t+c} - B_c$, then

$$\tilde{X}_t = X_c + \int_0^t \sigma(\tilde{X}_s)d\tilde{B}_s + \int_0^t \mu(\tilde{X}_s)ds$$

is built from X_c and from $\tilde{B}$ (which is *independent* of $(B_u)_{u\leq c}$). Hence $\tilde{X}$ depends on $(X_0, (B_u)_{u\leq c})$ only through X_c, and this is (informally) the statement that X is Markovian. We can readily discover the generator of X, because if $f \in C^2$ we have by Itô's formula that

$$f(X_t) - f(X_0) - \int_0^t \mathcal{G}f(X_s)ds = \int_0^t \sigma f'(X_s)dB_s$$

is a local martingale (and a martingale if $\sigma f'$ is bounded, for example), where we write

$$\mathcal{G}f(x) = \frac{1}{2}\sigma(x)^2\frac{\partial^2 f}{\partial x^2}(x) + \mu(x)\frac{\partial f}{\partial x}(x). \qquad (7.5)$$

Let us now examine a few examples (which essentially exhaust the supply of SDEs which can be 'solved').

Example: exponential SDE. The SDE

$$dX_t = X_t(\sigma dB_t + \mu dt)$$

is solved explicitly by

$$X_t = X_0 \exp[\sigma B_t + (\mu - \sigma^2/2)t].$$

This log Brownian motion process frequently serves as a model for the movement of a share price. The generator is $\mathcal{G} = \frac{1}{2}\sigma^2 x^2 D^2 + \mu x D$.

Example: Ornstein–Uhlenbeck process. The SDE

$$dX_t = \sigma dB_t + \beta(a - X_t)dt$$

can be solved quite explicitly by considering

$$\begin{aligned} d[e^{\beta t}(X_t - a)] &= e^{\beta t}\{dX_t + \beta(X_t - a)dt\} \\ &= \sigma e^{\beta t} dB_t \end{aligned}$$

so that

$$e^{\beta t}(X_t - a) - (X_0 - a) = \int_0^t \sigma e^{\beta s} dB_s$$

and thus

$$\begin{aligned} X_t &= (1 - e^{-\beta t})a + e^{-\beta t}X_0 + e^{-\beta t}\int_0^t \sigma e^{\beta s} dB_s \\ &= (1 - e^{-\beta t})a + e^{-\beta t}X_0 + e^{-\beta t}\sigma^2 W\left(\frac{e^{2\beta t} - 1}{2\beta}\right) \end{aligned}$$

for some Brownian motion W, in the light of the key result (3.7). The qualitative behaviour for $\beta > 0$ is completely different from the case $\beta < 0$. When $\beta > 0$, X_t converges in distribution to $N(a, \sigma^2/2\beta)$, but X_t does *not* converge almost surely, of course. When $\beta < 0$, we find that $e^{\beta t}X_t$ converges almost surely to $X_0 - a + \sigma^2 W(-1/2\beta)$, so X is growing at exponential rate.

The generator is $\mathcal{G} = \frac{1}{2}\sigma^2 D^2 + \beta(a - x)D$, and the process X is Gaussian; this makes calculations particularly amenable.

Example: squared Bessel processes. If we consider the SDE

$$dX_t = \sigma(X_t^+)^{\frac{1}{2}}dB_t + (\alpha - \beta X_t)dt,$$

it is not clear that there exists a unique solution, since the coefficients are no longer Lipschitz continuous. Nevertheless, there does exist a unique solution. No simple expression exists for the solution, in contrast to the two previous examples, but one can still do quite a lot of explicit calculations. An entire chapter of Revuz and Yor [9] is devoted to the case where $\sigma = 2$, $\beta = 0$ (the true squared Bessel process) – you will find many helpful methods of calculation discussed there.

Remarks: Both of the last two (families of) examples have been proposed as possible models for the term-structure of interest rates, being the Vasicek and Cox-Ingersoll-Ross models, respectively.

8 Changes of Measure and Drift

Our final topic is the topic of change of measure. We have already seen examples of change of measure for Markov chains by h–transforming, and we are going to see a semimartingale version of this idea. Suppose that B is a Brownian motion, and $(\mu_t)_{t \geq 0}$ is a bounded previsible process. If we define $(Z_t)_{t \geq 0}$ by

$$dZ_t = Z_t \mu_t dB_t, \quad Z_0 = 1, \tag{8.1}$$

we have the explicit solution

$$Z_t = \exp\left(\int_0^t \mu_s dB_s - \int_0^t \frac{1}{2}\mu_s^2 ds\right), \tag{8.2}$$

which is a non-negative local martingale (and in fact also a martingale in view of the boundedness of μ). We now define a new probability measure $\tilde{P}$ by

$$\left.\frac{d\tilde{P}}{dP}\right|_{\mathcal{F}_t} = Z_t \tag{8.3}$$

(so, explicitly, for any $A \in \mathcal{F}_t$, we define

$$\tilde{P}(A) = \int_A Z_t dP.$$

Since $A \in \mathcal{F}_u$ for any $u \geq t$, we could therefore have a possibly different definition of $\tilde{P}(A)$ by setting

$$\tilde{P}(A) = \int_A Z_u dP;$$

the fact that the two possible definitions agree is equivalent to the fact that Z is a martingale).

The key fact about $\tilde{P}$ is that

$$under\ \tilde{P},\ \tilde{B}_t \equiv B_t - \int_0^t \mu_s ds\ \ is\ a\ Brownian\ motion. \tag{8.4}$$

Put in other words, $\tilde{P}$ converts the Brownian motion B into a Brownian motion with drift μ. To prove (8.4), it will be enough to show that $\tilde{B}$ is a $\tilde{P}$–local martingale, because the quadratic-variation process $\langle \tilde{B} \rangle_t$ is evidently the same as the quadratic-variation process $\langle B \rangle_t = t$, and so $\tilde{B}$ must be a $\tilde{P}$–Brownian motion. But it is not hard to see that to show that $\tilde{B}$ is a $\tilde{P}$–martingale is equivalent to showing that $Z_t \tilde{B}_t$ is a P–martingale. Now we use Itô's formula on $Z_t \tilde{B}_t$;

$$\begin{aligned}
d(Z_t \tilde{B}_t) &= Z_t d\tilde{B}_t + \tilde{B}_t dZ_t + d\langle \tilde{B}, Z \rangle_t \\
&= Z_t\{d\tilde{B}_t + \mu_t \tilde{B}_t dB_t + \mu_t d\langle \tilde{B}, B \rangle_t\} \\
&= Z_t\{1 + \mu_t \tilde{B}_t\}dB_t
\end{aligned}$$

which is a local martingale.

Thus by changing measure, we can change drift, and this trick is used to great effect in the situation of a complete Brownian market; we change measure to make discounted prices into martingales then use the Brownian integral representation to express the general contingent claim as a stochastic integral with respect to the Brownian motion, equivalently, the discounted price processes. This then demonstrates a trading strategy which exactly replicates the given contingent claim. Simple!

References

[1] K.L. Chung and R.J. Williams, *Introduction to Stochastic Integration*, Birkhäuser, Boston, 1983.

[2] J.C. Cox, J.E. Ingersoll, and S.A. Ross, A theory of the term structure of interest rates, *Econometrica* **53**, 385—408, 1985.

[3] G. Grimmett and D. Stirzaker, *Probability and Random Processes*, Oxford University Press, 1992.

[4] W. Feller,*An Introduction to Probability Theory and its Applications*, Vol 1, 3rd edition, Wiley 1968; Vol. 2, 2nd edition, Wiley, 1971.

[5] H. Föllmer, Calcul d'Itô sans probabilités, *Sem. de Probabilités XV, LNM 850*, Springer, 1981.

[6] N. Ikeda and S. Watanabe, *Stochastic Differential Equations and Diffusion Processes*, North Holland-Kodansha, 1981.

[7] I. Karatzas and S. E. Shreve, *Brownian Motion and Stochastic Calculus*, Springer, 1988.

[8] B. Øksendal, *Stochastic Differential Equations*, Springer, 1985.

[9] D. Revuz and M. Yor, *Continuous Martingales and Brownian Motion*, Springer, 1991.

[10] L.C.G. Rogers and D. Williams, *Diffusion Markov Processes and Martingales, Vol. 1*, 2nd edition, Wiley, 1994.

[11] L.C.G. Rogers and D. Williams, *Diffusion Markov Processes and Martingales, Vol. 2*, Wiley, 1987.

[12] O.A. Vasicek, An equilibrium characterization of the term structure, *J. Fin. Econ.* **5**, 177–188, 1977.

[13] D. Williams, *Probability with Martingales*, Cambridge University Press, 1991.

The Risk Premium in Trading Equilibria which Support Black–Scholes Option Pricing

Stewart D. Hodges and
Michael J. P. Selby

Abstract

This article provides further analysis of the behaviour of the risk premium on the market portfolio of risky assets. Earlier work by Hodges and Carverhill (1993), and by others, has characterised the evolution of the market risk premium in economies where the variance of the return on the market has constant variance and market index options can be priced using the Black–Scholes model. In such economies the risk premium satisfies a non-linear partial differential equation called Burgers' equation. This chapter provides some significant new insights into this analysis. First we describe the nature of the existing results and provide a much simpler and more intuitive derivation. Next, we consider the time homogeneous case. Our original objective was to find a time homogeneous economy which allows the risk premium to vary inversely with the level of the market so that some kind of mean reversion could take place. Sadly, this is impossible. We obtain the interesting, but negative, result that the risk premium must be constant or increasing in the market level for time homogeneous equilibria which rule out arbitrage. Finally, this result is shown to tie in to earlier work on asymptotic portfolio selection. The article also illustrates that caution is required in this kind of modelling to avoid writing down models which admit arbitrage. The analysis also shows the limitations of the representative investor paradigm.

1 Introduction

This article describes properties of the behaviour of the risk premium on the market portfolio of risky assets. It analyses how the market risk premium must evolve in a single representative agent economy where the variance of the equity market return and the risk free interest rate are both constant. These are precisely the assumptions which enable index options to be valued using the Black and Scholes (1973) option valuation model. For the purposes of option pricing – under these assumptions – the behaviour of the risk premium is of no consequence. We transform to a risk neutral measure and obtain prices as expectations under that measure. However, for portfolio management the risk premium is extremely important, and we would like to know how it behaves and, if it varies through time, we will want to know how it varies.

There is some evidence that the required risk premium does vary through time, and in a way which is akin to mean reversion. Thus, studies by Poterba and Summers (1988), Fama and French (1988) and Lo and MacKinlay (1988) have all suggested that equity market returns are to some extent predictable. There is also evidence that the dividend yield may be related to the expected risk premium.

It is quite plausible that if investors have decreasing relative risk aversion then these kinds of relationships will be supported within a market equilibrium. The earlier analysis of Hodges and Carverhill (1993) showed how the risk premium would have to evolve under a non-linear partial differential equation known as Burgers' equation if the economy were characterised by a single representative investor maximising (non-state dependent) expected utility for wealth at terminal date H and the interest rate and volatility of the risky asset were constant. The original motivation for the current paper was to try to find a simple model of an economy of the type just described. A pleasing aesthetic for such a model is the property of time homogeneity: it seems undesirable and unnatural to be able to deduce the date from the form of the structure of the risk premium (as a function of the market price). We therefore investigate economies which have this steady state property. However, it turns out that the only equilibria possible under the time homogeneity assumption either give a risk premium which increases with the market level or else a trivially constant one.

The structure of the chapter is as follows. After this introductory section, we describe the Burgers' equation characterisation of the evolution of the price of risk and some of its key properties. Section three then presents a new and much simpler derivation of the result than has previously appeared. In section four we add the requirement of time homogeneity. The conclusion of this section is that the only three possible steady state alternatives are:

1. that the price of risk becomes completely flat and (boringly) constant as a function of market level;

2. that it increases with market level; or

3. that its relationship contains a singularity of sufficient severity to permit arbitrage.

In a short fifth section this analysis is discussed in the context of work on asymptotic portfolio selection carried out in the mid seventies. It is shown that the two perspectives are consistent with one another. The final section summarises the results contained in the chapter and draws together their various implications.

2 Burgers' Equation and its Properties

We begin by summarising the results of Hodges and Carverhill (1993).

The analysis is of an economy with a single asset whose price S_t follows the process

$$\frac{dS}{S} = [r + \sigma\alpha(.)]dt + \sigma dB^P \tag{1}$$

where r is a constant risk free interest rate, σ is the volatility of the asset which we assume constant, B^P is Brownian Motion under the objective probability measure P, and $\alpha(.)$ is the (adapted) process for the price of risk.

For simplicity we shall also assume that no dividends are paid within the time horizon H. The question posed is how must the price of risk $\alpha(\omega, t)$ behave if we require this economy to correspond to an equilibrium characterised by individuals maximising the expected utility of their wealth at dates greater than or equal to H. This question has also been studied by Bick (1990), and by He and Leland (1993) whose results are very closely related to, and slightly more general than, those described here. Our assumption of the constancy of the variance rate enables us to obtain stronger results.

The kind of equilibrium which we are concerned with here is that characterised by a single representative investor. We assume agents maximise expected utility over utility functions which are not state dependent (and that their initial endowments are not state dependent either). Their demands aggregate to the demands of a single representative agent who holds the market portfolio at all times (see, for example, Huang and Litzenberger (1988)). It is worth noting how this assumption differs from the currently more usual and rather weaker assumption of no-arbitrage. Following Harrison and Kreps (1979), we note that any arbitrage-free price system can be sustained as a competitive equilibrium characterised by a single representative agent. In general, this agent may have a state dependent utility function or possibly just initial endowments which are state dependent. However, in our equilibrium assumption we are ruling out the possibility of this kind of state dependence.

Further, for any economy which does not permit arbitrage, there exists a risk neutral probability measure under which the rate of drift of all assets is the risk free rate, r. We shall define P as the objective probability measure, and Q as the risk neutral one. Thus the process for S under the measure Q

is:

$$\frac{dS}{S} = rdt + \sigma dB^Q \qquad (2)$$

ratio of the risk neutral density to the objective one (i.e., the Radon-Nikodym derivative dQ/dP) gives the state-price density which also defines the marginal utilities of this agent. Later we will use Girsanov's Theorem to understand the nature of this change of measure.

It is convenient first of all to introduce a change of variables. Instead of working with S_t we shall work with the transformed variable

$$x_t = \ln S_t - \left(r - \frac{1}{2}\sigma^2\right)t \qquad (3)$$

We find that the process for x is

$$dx = \alpha(.)\sigma dt + \sigma dB^P$$

under P, and

$$dx = \sigma dB^Q \qquad (4)$$

under Q. The key result (shown by Hodges and Carverhill (1993)) is that given the assumptions we have made, $\alpha = \alpha(x, t)$ is a deterministic function of x and time (so it is path independent) which satisfies a non-linear partial differential equation known as Burgers' equation. The equation is

$$\alpha_\tau = \frac{1}{2}\sigma^2\alpha_{xx} + \sigma\alpha\alpha_x \qquad (5)$$

where $\tau = H - t$ and the subscripts denote partial derivatives.

Note that this equation is very similar to the diffusion equation, but the final term makes it non-linear. As with the diffusion equation, given suitable boundary conditions (e.g. α is a specified function of x at the horizon date H and there are regularity conditions for extreme values of x) we can solve backwards in time from the horizon date to all earlier dates. Equation (5) was proposed by Burgers in 1948 as a model for the one dimensional flow of a viscous fluid. It also occurs in a number of other applications including modelling traffic flows. Closed form solutions are possible depending on the boundary conditions imposed. Properties of this equation and solution methods may be found in Bland (1988), Kevorkian (1990), and Whitham (1974).

In 1950 and 1951, Hopf and Cole showed independently how an analytic solution to Burgers' equation could be derived using a clever transformation which reduces the problem to a conventional diffusion equation. For this transformation we define v by setting

$$\alpha = \sigma\frac{v_x}{v} = \sigma\frac{\partial}{\partial x}[\ln(v(x, t))] \qquad (6)$$

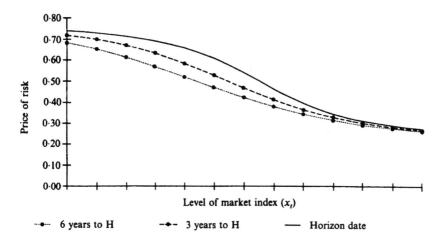

Figure 1. Solution of Burgers' equation

Cole and Hopf showed that if $v(x, \tau)$ satisfies the diffusion equation

$$v_\tau = \frac{1}{2}\sigma^2 v_{xx} \tag{7}$$

then the resulting $\alpha(x, \tau)$ from equation (6) solves Burgers' equation (5). Since, from (6), a terminal date boundary condition on α defines $v(x, 0)$ up to a multiplicative constant, we can then solve (7) for $v(x, \tau)$ and finally use (6) to obtain $\alpha(x, \tau)$.

As an example, the case where $\alpha(x, 0) = a - bx$ leads to the solution $\alpha(x, \tau) = (a - bx)/(1 + b\tau\sigma)$, which, for positive b, leads to an increasingly flat relationship as increases. Figure 1, which is reproduced from Hodges and Carverhill, shows the results of numerical calculations starting from more complicated S-shaped initial conditions. The initial condition for α is shown with a solid line, the dashed line shows α three years earlier, and the dotted one three years before that. Note that the direction in which α moves depends on the slope of α as a function of x. In this example the slope of the risk premium curve is generally smaller the further we are from the horizon date.

3   A Simple Derivation

In Hodges and Carverhill (1993), two derivations are given of the Burgers' equation result. Both approaches directly make use of Dybvig's (1988a,b) insights that wealth must be monotonic decreasing in the state-price density. The first approach involves taking limits of a binomial approximation, while the second is based on algebraic manipulation of the Girsanov change of measure. Neither approach is very simple, and the intuition provided is limited. The derivations in He and Leland (1993) suffer from similar drawbacks. We therefore now provide an alternative derivation which we hope provides better clarity and insight into what is going on. We will make use of both the Girsanov change of measure and of the Hopf and Cole transformation.

We start by assuming the general model (1). We note by Girsanov's Theorem that the change of measure corresponding to

$$\frac{dP}{dQ} = \exp\left\{\int_0^t \alpha_s dB_s^Q - \frac{1}{2}\int_0^t \alpha_s^2 ds\right\} = M(B_t^Q, t) \tag{8}$$

gives a Q-martingale M which satisfies the diffusion equation

$$\frac{\partial M}{\partial t} + \frac{1}{2}\frac{\partial^2 M}{\partial B^2} = 0$$

or, changing to a function of x and using equation (4)

$$\frac{\partial M}{\partial \tau} = \frac{1}{2}\sigma^2\frac{\partial^2 M}{\partial x^2} \tag{9}$$

The economics tells us that the marginal utility (with respect to log wealth) is proportional to the state-price density function dQ/dP. It is therefore inversely proportional to $M(x,t)$ (i.e. $U_x = k/M$ for some positive constant k). Portfolio theory tells us that the market equilibrium condition for holding the market portfolio is that the price of risk α must equate to σ times the Arrow–Pratt coefficient of relative risk aversion[1]

$$\alpha = -\sigma\frac{U_{xx}}{U_x} = -\sigma\frac{d}{dx}[\ln U_x]$$

Substituting for U_x we finally obtain

$$\alpha = -\sigma\frac{d}{dx}[\ln(\frac{1}{M})] = \sigma\frac{M_x}{M} \tag{10}$$

This completes the proof, as we now have shown that $\alpha(x,\tau)$ takes the form of the Cole and Hopf transformation[2]. Equation (10) corresponds to our earlier equation (6), and the diffusion equation (9) corresponds to the earlier (7).

Notice that the v variable, of the Cole and Hopf ratio which gives α, corresponds to the reciprocal of marginal utility (with respect to log wealth). This follows a diffusion, but what we are interested in is the evolution of the measure of (relative) risk aversion given by $-U_{xx}/U_x$.

---

[1]Note that the risk aversion $-SU_{SS}/U_S = -U_{xx}/U_x$ where x is as defined in equation (3).

[2]He and Leland (page 610) also identify $1/U_x$ as following a diffusion and describe the Cole and Hopf transformation for Burgers' equation.

4 The Time Homogeneous Case

In this section we describe our analysis of the time homogeneous case. We wish to find how the risk premium (may depend on the level of the market in such a way that the form of this function does not depend on time. This restriction enables us to reduce the partial differential equation (Burgers' equation) to an ordinary differential equation. We solve this differential equation and consider which solutions are consistent with market equilibrium. Not all solutions to Burgers' equation will be feasible, as this is a necessary condition for equilibrium and not a sufficient one. We conclude that there are only three possible alternatives in the steady state: either the price of risk becomes completely flat and (boringly) constant as a function of market level, or that it increases with market level or its relationship contains a singularity of sufficient severity to permit arbitrage.

We begin with a price of risk function $\alpha(x, \tau)$ which satisfies Burgers' equation (5). We interpret our assumption of homogeneity in time as meaning that α takes the form

$$\alpha(x, \tau) = y(u) \qquad \text{where} u = x + \theta\tau \qquad (11)$$

for some function y and constant θ. Forming the partial derivatives and substituting into Burgers' equation we obtain the ordinary differential equation

$$\theta y' = \frac{1}{2}\sigma^2 y'' + \sigma y y' \qquad (12)$$

Integrating this equation once gives

$$\frac{1}{2}\sigma^2 y' = \theta y - \frac{1}{2}\sigma y^2 + \text{constant} \qquad (13)$$

We may therefore obtain the solutions y by integrating:

$$\sigma \int \frac{dy}{a + (2\theta/\sigma)y - y^2} = \int du \qquad (14)$$

where α is a constant. This integrates to give

$$\frac{\sigma}{2c} \ln\left[\left|\frac{y + c - b}{y - c - b}\right|\right] = u + \quad \text{constant} \qquad (15a)$$

$$\text{where} \quad b = \theta/\sigma \text{ and } c = \sqrt{a + b^2} \qquad (15b)$$

Finally, provided c is real and not complex, we obtain a general solution of the form

$$y = \begin{cases} \dfrac{(c+b)e^{s(u)} + b - c}{e^{s(u)} + 1} & \text{for } y \in (b - c, b + c) \\ \dfrac{(c+b)e^{s(u)} + c - b}{e^{s(u)} - 1} & \text{for } y \notin (b - c, b + c) \end{cases} \qquad (16a)$$

$$\text{where} \quad s(u) = k + \frac{2c}{\sigma}u \qquad (16b)$$

and k is a constant. The first equation provides a stable travelling wave solution with y between $b - c$ and $b + c$. We require $b > 0$ for a positive risk premium. For positive c the solution increases from b_c at $-\infty$ to $b + c$ at $+\infty$. If we take c to be negative it still increases, from $b + c$ to $b - c$. Turning to the second equation, we may easily rule out as having no economic interpretation the other possible mathematical solutions which arise where c is complex. These involve trigonometric functions (instead of hyperbolic ones), and imply that the price of risk as a function of the market level has repeated singularities and oscillates an infinite number of times. No utility function will support these kinds of behaviour.

We now distinguish special cases of equation (16). We first note that the risk premium y has an unacceptable singularity at $u = -\frac{1}{2}k\sigma/c$, except in the case where $c = 0$. For $c = 0$ but $k \neq 0$ we obtain

$$y = b = \frac{\theta}{\sigma} \tag{17}$$

This, of course, is the trivial case of a constant risk premium. It corresponds to the well known assumption of a representative investor with constant relative risk aversion, or, in other words, either power or logarithmic utility (see, for example, Bick (1987)). However, there is one further case which we need to examine, corresponding to $k = 0$ and taking the limit as c tends to 0. This gives a solution

$$y = \frac{\theta}{\sigma} + \frac{\sigma}{u}$$

which implies

$$\alpha(x, \tau) = \frac{\theta}{\sigma} + \frac{\sigma}{x + \theta\tau} \tag{18}$$

Unfortunately, although this solution does indeed satisfy Burgers' equation (as can easily be verified), it also has a singularity, namely, at $x = 0$ in the case where $\theta = 0$. The effect of the singularity is that as x falls close to this point, the drift increases sufficiently to prevent the point of the singularity from ever being reached. Sadly this model cannot represent an economic equilibrium because it admits arbitrage. This is a pity, because the variation of the risk premium, which increases as the market falls, would otherwise have made an interesting model. The hyperbolic form of the drift (under the objective probability measure P) means that the process for x is one which is well known to probability theorists and is called a 3–dimensional Bessel process. Many of its properties are given in Revuz and Yor (1991). The variable x corresponds to the distance of a point P from the origin, where P follows a random walk in 3 dimensions with no drift. Many analytical properties are known for this process, including that this process never reaches the origin. Hence the support for future values under the objective measure P is bounded below by the point of singularity, while that of the risk neutral measure Q is given by the whole of the real line. The two probability measures cannot be equivalent, so there does not exist a risk neutral probability measure. It is therefore clear that this otherwise appealing model admits arbitrage and so cannot correspond to an equilibrium.

We are left with the surprising conclusion that the only possible time homogeneous representative economy supporting a constant variance rate corresponds to an increasing or a constant risk premium. The only way a negative slope in the risk premium function can persist through time is through the existence of a singularity of sufficient severity to allow arbitrage.

5 Relationship to Asymptotic Portfolio Theory

It is interesting to reinterpret the results of the last section in the context of earlier work on asymptotic portfolio theory. Hakansson (1974) showed that in a terminal utility model, under really quite weak assumptions, there are strong convergence results which govern the evolution of the induced utility functions for earlier periods. An abbreviated account of this work is also given in Hakansson (1987). Hakansson finds that, under very general assumptions which are distinct from ours, there is always some coefficient of relative risk aversion such that the induced utility functions converge to the corresponding power utility function as one backs away from the horizon date. Thus reinvesting individuals with distant horizons should follow a power utility investment policy as long as their horizon remains distant. (But, in practice the convergence is slow, so distant may mean very distant!) In the travelling wave solution for our economy, as one backs further and further away from the horizon, one becomes less and less likely to be in the transition region and increasingly likely to be close to either y's lower bound or its upper bound.

Conclusions

This chapter has provided further analysis of the behaviour of the risk premium on the market portfolio of risky assets. Understanding the behaviour of the market risk premium is a problem of fundamental importance for fields such as portfolio management. Earlier work by Hodges and Carverhill (1993) and by others has characterised the evolution of the market risk premium in economies where the variance of the return on the market has constant variance and market index options can be priced using the 1973 Black–Scholes model. Representative agent equilibrium implies that, although the risk premium can vary, it must evolve according to a non linear partial differential equation called Burgers' equation.

We have provided some significant new insights into this analysis. First, we have summarised the existing results and provided a much simpler and more intuitive derivation. Next, we have analysed the time homogeneous case. It would have been nice if we had found a time homogeneous economy which allows the risk premium to vary so that some kind of mean reversion could take place. Sadly, this is impossible. We obtained the interesting but negative result that the risk premium must be increasing in the market level or constant for time homogeneous equilibria which rule out arbitrage. However, this result is one which is consistent with earlier work by Hakansson on asymptotic portfolio selection.

The assumptions we have made appear to be strong but some relaxation would be possible, particularly along the lines of He and Leland (1993) who allow both the risk free rate and volatility to vary. However, stochastic variation of these variables does make the mathematics horribly complex. Burgers' equation holds whether or not there are intermediate dividends, and there should be no real difficulty in extending the formal analysis to include more general situations. In a model with many dividend dates, the risk premium for the claim which pays on a given date will satisfy Burgers' equation subject to a boundary condition imposed at that date. The market portfolio now becomes a portfolio of claims which pay at different dates, and its risk premium is a weighted average of the risk premia for the different dates. For any given payment, the risk premium must behave according to the results presented above. However, for the portfolio, there could be a non-trivial steady state risk premium function formed from the weighted averages. It seems unlikely that such a model could be solved analytically, but numerical solution would not be difficult.

The chapter showed how easy it is to obtain a models which admit arbitrage, such as our Bessel process one. It is surprisingly easy to produce and perform partial analysis of such models without appreciating their inherent contradictions. We wish to reiterate a strong caveat in this regard. The authors are aware of published papers where arbitrage exists in a model: for example a recent paper on valuing options on spreads where the underlying asset price has a process with a reflecting barrier under the (assumed) risk neutral probabilities. We have also seen an option pricing model (derived from a no-arbitrage argument and with no market frictions) where the option prices fail to satisfy put-call parity!

A further implication of our work concerns the limitations of the representative investor paradigm. Under a single representative investor with decreasing relative risk aversion, the price of risk would be likely to become increasingly sensitive to the market level as the market evolves through time. It appears to us that in order to obtain more realistic models of market equilibrium behaviour, and ones which would enable us to study issues such as how to invest within a lifetime cycle, we may need to employ the much richer but more difficult techniques available of overlapping generations models.

Acknowledgements

We are grateful for useful discussions with our colleagues in FORC, particularly Les Clewlow, with Andrew Carverhill, Michael Dempster, Hua He and Chris Rogers, and also to participants at a number of research seminars at which this work has been presented. We are pleased to acknowledge support from FORC Corporate Members and from the Isaac Newton Institute for Mathematical Sciences.

References

Bick, A. (1990) 'On Viable Diffusion Price Processes of the Market Portfolio', *Journal of Finance* **45**, 673–689.

Bick, A. (1987), 'On the Consistency of the Black-Scholes Model with a General Equilibrium Framework', *Journal of Financial and Quantitative Analysis* **22**, 259-275.

Black, F. and M. Scholes (1973), 'The Pricing of Options and Corporate Liabilities', *Journal of Political Economy* **81**, 637–659.

Bland, D.R. (1988), *Wave Theory and Applications*, Oxford University Press.

Dybvig, P.H. (1988a), 'Inefficient Dynamic Portfolio Strategies, or How To Throw Away a Million Dollars', *Review of Financial Studies* **1**, 67-88.

Dybvig, P.H. (1988b), 'Distributional Analysis of Portfolio Choice', *Journal of Business* **61**, 369–393.

Fama, E.F. and K.R. French (1988), 'Dividend Yields and Expected Stock Returns', *Journal of Financial Economics* **22**, 3–25.

Hakansson, N.H. (1974), 'Convergence to Isoelastic Utility and Policy in Multiperiod Portfolio Choice', *Journal of Financial Economics* **1**, 201–224.

Hakansson, N.H. (1987), 'Portfolio Analysis', in *The New Palgrave Finance*, J. Eatwell, J.M. Milgate and P. Newman (eds), MacMillan.

Harrison, J.M. and D.M. Kreps (1979), 'Martingales and Arbitrage in Multiperiod Securities Markets', *Journal of Economic Theory* **20**, 381–408.

He, H. and H. Leland (1993), 'On Equilibrium Asset Price Processes', Finance Working Paper **221**, Haas School of Business, University of California at Berkeley, December 1991, published in *Review of Financial Studies* **6**, 593–617.

Hodges, S.D. and A.P. Carverhill (1993), 'Quasi Mean Reversion in an Efficient Stock Market: The Characterisation of Economic Equilibria Which Support Black–Scholes Option Pricing', *Economic Journal* **103**, 395–405.

Huang, C.-F. and R.H. Litzenberger (1988), *Foundations for Financial Economics*, North-Holland.

Kevorkian, J. (1990), *Partial Differential Equations: Analytical Solution Techniques*, Brooks/Cole.

Lo, A.W. and A.C. MacKinlay (1988), 'Stock Prices Do Not Follow Random Walks: Evidence from a Simple Specification Test', *The Review of Financial Studies* **1**, 41–66.

Poterba, J.M. and L.H. Summers (1988), 'Mean Reversion in Stock Prices: Evidence and Implications', *Journal of Financial Economics* **22**, 27–59.

Revuz, D. and M. Yor (1991), *Continuous Martingales and Brownian Motion*, Springer Verlag.

Whitham, G.B. (1974), *Linear and Nonlinear Waves*, Wiley.

On the Numeraire Portfolio

Philippe Artzner

Abstract

This note gathers properties and uses of the 'numeraire portfolio' – some of them spread in the literature – as they were discussed at the various special emphasis sessions of the program in Mathematical Finance held in Cambridge's Isaac Newton Institute for Mathematical Sciences, Spring 1995.

1 Introduction

The various special emphasis sessions of the program in Mathematical Finance at the Isaac Newton Institute, Spring 1995, have been a splendid opportunity for exchange of ideas among specialists of related fields. It often happened that some questions raised during one session were answered at a later session.

We report on one instance of this, namely, the explanation of the use of the 'physical' or 'historical' probability as a martingale measure for pricing assets by taking conditional expectations. This sounds surprising indeed, since von Neumann-Morgenstern theory leads us to consider expectations of *utility* of asset cash-flows! Similarly, we are used to 'loading factors' applied to 'pure' insurance premiums. Under deterministic interest rates, these premiums are computed by the 'equivalence principle', i.e. by using expectations of discounted future premiums and benefits payments under the physical probability.

The second observation above begins to point towards the explanation to be developed in this paper. The 'discount factors' of actuarial mathematics, under deterministic interest rates, are the reciprocals of future values of a current unit invested either in zero coupon bonds of various maturities, or in a savings account. The values of these instruments will serve as standards for the measurement of future premiums and benefits, that is, future cash flows are measured *relative* to future prices of these particular assets. Under stochastic interest rates, it may also happen that a good choice of asset provides the current prices of all assets as expectations of these relative measurements under the physical probability. It is clear that one has to choose carefully the asset, zero-coupon bond, money market account or some other to be

53

used as benchmark, since the probability measure to be used for expectation computations is imposed *a priori*.

Section 2 presents occurrences in the literature, some as early as 1977, of both the physical probability as a martingale measure and of particular choices of numeraire. The use of the historical probability should be linked to the early work of Kelly (1956). Section 3 explains the special role of the money market account numeraire and Section 4 concludes.

2   Five instances of the numeraire portfolio

Bühlmann (1992, p. 113, 1995, p. 6) presents arbitrage-free (i.e. martingale) pricing *without* changing from the physical probability **P** to a 'risk-neutral' probability. It seems that the drawback of this simplicity is the introduction of a 'stochastic discount function ϕ_k from time k to time 0 '. This function is such that for *any* security dividend process S the 'corn' price process of S, denoted by $\Pi(S)$ which gives the prices of security S at various dates in terms of units at these dates of the unique consumption good of theoretical finance, satisfies in particular:

$$\Pi(S)(0) = \mathbb{E}^{\mathbf{P}}[\Sigma_k \phi_k S(k)]$$

where $S(k)$ is the 'corn' cash-flow of security S at date (time) k.

Natural questions are: What does ϕ_k, or $1/\phi_k$, stand for? How can it be interpreted ? Is there any traded asset corresponding to it ? Answering these questions should be helpful in differentiating topics which must be treated under the physical probability from others to be addressed under a risk-neutral probability.

Jørge Aase Nielsen mentioned in Cambridge that an earlier paper, Vasicek (1977), preceding the new era of change of probability, also contains a martingale under the physical probability (see equation (20), p. 184). The paper by Artzner and Delbaen (1989), while mainly devoted to the use of change of probability for martingale pricing of interest-rate instruments, also mentions (Corollary 1, p. 103) such martingales.

Bajeux and Portait (1992, 1995 a,b) mention the paper by Long (1990) which is devoted to a proof of the existence of a portfolio with the property that 'corn' prices relative to *this* portfolio's 'corn' value K, are martingales under the physical probability. If taken as a numeraire, the 'corn' value of this portfolio is called the numeraire portfolio. It appeared first in the paper by Kelly (1956) (see also Heath (1988, 1990) as well as Cover and Thomas (1991, Ch. 6) for asymptotic properties). In the notation of Bühlmann (1992, 1995), with a finite probability space Ω one immediately sees that $1/\phi_1$ is the 'corn' value $K(1)$ at date 1 of a portfolio chosen by an investor wishing to maximise $\mathbb{E}^{\mathbf{P}}[\log K(1)]$ under the constraint that the initial 'corn' value

$\mathbb{E}^{\mathbf{P}}[\phi_1 K(1)] = 1$. If $p(\omega)$ denotes the physical probability of state ω and if $1/o(\omega)$ is the initial 'corn' price of the Arrow-Debreu security AD_ω paying one unit of 'corn' at date 1 if and only if state ω obtains, then $1/o(\omega) = \phi_1(\omega) \cdot p(\omega)$ and $K(1, \omega) = o(\omega) \cdot p(\omega)$. The investor puts the fraction $p(\omega)$ of his initial wealth into the purchase of $o(\omega) \cdot p(\omega)$ units of security AD_ω. For the continuous case the reader is referred to Long (1990) or Bajeux-Portait (1992, 1995a,b).

Emphasis on *relative* pricing is the key to an understanding of the relation between various possible martingale measures. An early example of relative pricing is Margrabe (1978). given by Jamshidian (1987, formula (12), or 1989, formula (5)) and Huang and Litzenberger (1988, Section 8.5) introduce the ratios of 'corn' prices of some securities to 'corn' prices of pure discount bonds, in order to get martingale formulae under an appropriate probability. Finally, at this point we should note that the terminology 'stochastic *accumulation*' would put more emphasis on the importance of the choice of financial instrument(s) to carry money *forward*, than the terminology 'stochastic *discounting*' does.

3   Which numeraire after all?

Having observed that different continuous numeraires correspond to different equivalent martingale measures, which for a given security provide by conditional expectation different *relative* price processes, i.e. ratios of 'corn' prices to different numeraires, one may ask the following question (see Harrison and Kreps (1979) and Harrison and Pliska (1981)):

> *What is particular to the choice of numeraire, and consequently to the choice of equivalent martingale measure, which appears in general arbitrage pricing?*

The answer is that the change made was the one necessary to get martingales for the 'corn' prices relative to a 'money-market account numeraire' defined, as in Duffie (1992, p. 96), by the assumed existence of a security 'corn' price process B such that:

$$B = \frac{1}{D},$$

with

$$D(t) = \exp\left(-\int_0^t r(u)du\right).$$

The process B describes the 'money market account', or how much 'corn' is being accumulated over time at the stochastic growth rate r, also called the short rate of interest. Under technical no arbitrage conditions, there is a change from the physical probability $\mathbf{P}$ to a probability $\mathbf{Q}^B$ under which 'corn' prices, divided by B, are martingales. What is so convenient about this choice of the money market account as numeraire? The numeraire portfolio idea has already shown that the physical probability could be used as a (relative) martingale pricing measure!

Before answering this question, we should first pause to observe that any continuous finite variation numeraire (FVN) other than B has to be proportional to B since the ratio of FVN to B, a relative price, will be a martingale while being at the same time of finite variation. The discrete time case would similarly involve the proportionality of predictable numeraires.

Björk (1995, Def. 7.1.1, p. 72) introduces an 'heuristic definition' of a martingale measure $\mathbf{Q}$ by requiring that there exists a process r such that under $\mathbf{Q}$ for any security its 'corn' price process $\Pi(S)$ satisfies:

$$d\Pi(S)(t) = r(t)\Pi(S)(t)dt + v(t)\Pi(S)(t)dW(t),$$

with W a Brownian motion and $v(t)$ the volatility of $\Pi(S)(t)$. The process r has to be the one determining the money market account and the probability $\mathbf{Q}$ has to be the probability $\mathbf{Q}^B$ above and then we can write more compactly:

$$d\Pi(S)(t) = r(t)\Pi(S)(t)dt + dM(t),$$

with M a martingale under the probability $\mathbf{Q}^B$.

This is totally in the spirit of the Corollary in Section 2.2 of Artzner–Delbaen (1989, p. 108). From a discussion with Jørge Aase Nielsen about the paper of Vasicek (1977) mentioned above, it appears that caution must be used with this universal form of the drift term of $d\Pi(S)$. One realises that the return process has 0 as universal drift term, return being measured in terms of prices which themselves are actually *relative* to the use of the numeraire given by capitalisation of one initial unit in the money market account.

Another numeraire N would be related to the probability $\mathbf{Q}^N$ defined by the following density with respect to $\mathbf{Q}^B$:

$$d\mathbf{Q}^N/d\mathbf{Q}B = N(T)D(T) = N(T)/B(T).$$

This new probability is a martingale measure for price processes denominated in terms of the numeraire N, and therefore we have for each security 'corn' price process $\Pi(S)$:

$$d(\Pi(S)(t)/N(t)) = dL(t),$$

with L a martingale under $\mathbf{Q}^N$. Girsanov's formula gives:

$$d\Pi(S)(t) = r(t)\Pi(S)(t)dt + (B(t)/N(t))d < \Pi(S)(t),$$
$$N(t)/B(t) > +dL'(t),$$

with L' a martingale under $\mathbf{Q}^N$. This can be rewritten:

$$d\Pi(S)(t) = r(t)\Pi(S)(t)dt + (1/N(t))d < \Pi(t), N(t) > +dL'(t).$$

It appears therefore that, when using the numeraire N and consequently the probability $\mathbf{Q}^N$, unless the numeraire N has zero covariation with a security S, the finite variation part in the security's price differential $d\Pi(S)(t)$ is no longer of the universal form $r(t)\Pi(S)(t)dt$. An extra term has, in general, also to be added, as in Duffie (1992, formula (3), p. 98). This is a strong point in favor of the money market account numeraire B.

4   Conclusion

The physical probability can be used for martingale pricing and is associated with an interesting numeraire. This does not deprive the money market account numeraire and the associated martingale measure(s) of their exceptional rôle.

References

Artzner, P. and F. Delbaen (1989). 'Term structure of interest rates: the martingale approach', *Advances in Applied Mathematics* **10**, 95–129.

Bajeux, I., and R. Portait (1992). 'Dynamic asset allocation in a mean-variance framework', Working Paper, E.S.S.E.C., Cergy-Pontoise, presented at the French Finance Association December meeting, Paris.

Bajeux, I., and R. Portait (1995a). 'The numeraire portfolio: a new approach to continuous time finance', Working Paper, August 1995, George Washington University and E.S.S.E.C.

Bajeux, I., and R. Portait (1995b). 'Pricing contingent claims in incomplete markets using the numeraire portfolio', Working Paper, November 1995, George Washington University and E.S.S.E.C.

Björk, T. (1995). 'Arbitrage Theory in Continuous Time', Lectures at the 12th International Summer School of the Swiss Association of Actuaries, August 29–September 2, Monte Verità TI.

Bühlmann, H. (1992). 'Stochastic discounting', *Insurance: Mathematics and Economics* **11**, 113–127.

Bühlmann, H. (1995). 'Life Insurance with Stochastic Interest Rates', Lectures at the 12th International Summer School of the Swiss Association of Actuaries, August 29–September 2, Monte Verità TI.

Cover, T. and J. Thomas, (1991). *Elements of Information Theory*, Wiley.

Duffie, D. (1992). *Dynamic Asset Pricing Theory*, Princeton University Press.

Harrison, M. and D. Kreps (1979). 'Martingales and arbitrage in multiperiod securities markets', *Journal of Economic Theory* **20**, 381–408.

Harrison, M. and S. Pliska (1981). 'Martingales and stochastic integrals in the theory of continuous trading', *Stochastic Processes and their Applications* **11**, 215–260.

Heath, D. (1988). 'Horserace and arbitrage', paper presented in Strasbourg, January 11.

Heath, D. (1990). 'If you are so smart, why ain't you rich?'. Note, March 3, 1–12.

Huang, C.F. and R. Litzenberger (1988). *Foundations for Financial Economics*, North-Holland.

Jamshidian, F. (1987). 'Pricing of contingent claims in the one-factor term structure model', Working Paper, Merrill Lynch Capital Markets, 1–19.

Jamshidian, F. (1989). 'An exact bond option formula', *Journal of Finance* **44**, 205–209.

Margrabe, W. (1978). 'The value of an option to exchange one asset for another', *Journal of Finance* **33**, 177–186.

Kelly, J. (1956). 'A New interpretation of the information rate', *Bell System Technical Journal* **35**, 917–926.

Long, J. (1990). 'The numeraire portfolio', *Journal of Financial Economics* **26**, 29–69.

Vasicek, O. (1977). 'An equilibrium characterization of the term structure', *Journal of Financial Economics* **5**, 177–188.

Part II
Option Pricing and Hedging

Some Combinations of Asian, Parisian and Barrier Options

M. Yor, M. Chesney, H. Geman, and M. Jeanblanc-Picqué

Abstract

This article addresses some of the valuation problems, in the Black and Scholes setting of a geometric Brownian motion for the underlying asset dynamics, for options whose pay-off is related to the terminal price of the stock and an arithmetic average of fixing and/or involves stopping times related to excursions. In all cases, we are able to provide at least the Laplace transform in time of the option price under a form whose complexity varies with the number of exotic features. We emphasize that we do not give closed form formulas for the general case, but we aim to develop a methodology which may be used in many cases.

1   Introduction

Amongst the large variety of path-dependent options, barrier options enjoy the feature of having been traded and discussed in the literature for quite some time. Back in 1973, Merton offered in his seminal paper a pricing formula for an option whose pay-off is restricted by a floor knock-out boundary and a few years later, Goldman, Sosin and Gatto (1979) provided closed form solutions for all types of single barrier options. Over the last few years, barrier options have become increasingly popular since they may produce, at a lower cost than standard options, the appropriate hedge in a number of risk management strategies (see Reiner and Rubinstein (1991)). Some of them combine several exotic features. For instance, it is well known that when implied volatilities are trading at historically high levels, going to Asian instruments reduces this volatility and the option price, as long as the risk-adjusted drift of the underlying asset is positive or within some interval (see Geman and Yor (1993) for a thorough discussion of this issue). As a way to cut premium even further, the user may choose an instrument which also has a knock-out feature; for instance, consider an American treasurer who is expecting a series of cash-flows denominated in deutschemarks and wants to buy Asian options DM/\$. If the current spot level is DM/\$ 1.4050, the premium can be reduced by the addition of a knock-out barrier at DM/\$

1.4850. This Asian barrier put is still an attractive hedging tool since, if the barrier is reached, it means that the underlying asset is typically moving in favor of the cash position the option buyer wants to protect.

In the same manner, by buying Asian calls on oil, a major airline company may hedge its exposure against an increase in the cost of fuel (both because its needs in resupplying are regularly spread over time and because the nature of oil production, namely, the long time between extraction and delivery, entails that oil indices are typically arithmetic averages). The company may want to reduce the cost of its coverage by asking a down and out specification on the Asian call it will buy on the OTC markets.

French and Australian financial institutions have recently traded so-called 'Parisian' options, whose pay-off is contingent on the fact that the underlying asset remains below or above a given value for a time period longer than a fixed number called the window (see Chesney, Jeanblanc-Picqué and Yor (1997) for a complete description and valuation of these instruments). For a window length equal to zero, the Parisian option reduces to a standard barrier option. When the window is extended until maturity, the Parisian option reduces to a standard European option. In the intermediate case, the option presents its 'Parisian' feature and becomes a flexible financial tool which has some interesting properties: for instance, for some values of the parameters, when the underlying asset price is close to the barrier or when the size of the window is small, its value is a decreasing function of the volatility. Therefore, it allows traders to bet in a simple manner on a decrease of volatility. Last but not least, as far as down-and-out barrier options are concerned, an influential agent in the market who has written such options and sees the price approach the barrier may try to push the price further down, even momentarily and the cost of doing so may be smaller than the option payoff. In the the case of Parisian options, this would be more difficult and more expensive. Therefore, as in the case of Asian versus standard options, the possibility of market manipulations is reduced.

In the 'Asian Parisian' case, the excursion condition is relative to the underlying asset, but the pay-off at maturity corresponds to an Asian option. The price of an 'Asian Parisian' option is, ceteris paribus, lower than the Asian option price. For instance, an up-and-in 'Asian Parisian' option knocks in when the underlying price remains above a given level for a period of time at least equal to the window. In this case, the option only represents a hedge in the worst cases, namely when both the spot price and its average are high.

A discussion of the organisation of this chapter is postponed until the end of section 2.

2 The Setting

We assume in this paper that the underlying asset dynamics are driven by the equation

$$\begin{cases} dS_s &= S_s(\nu\,ds + dB_s) \\ S_0 &= 1, \end{cases}$$

where $(B_s, s \geq 0)$ is a standard Brownian motion under the usual risk-adjusted probability. We have set the volatility equal to 1 in order to obtain simpler formulas, the general drift ν allows for dividend-paying stocks and currencies. We denote by $B_s^{(\nu)} = B_s + \nu s$ the Brownian motion with drift ν. In previous research, we have been interested in options whose prices are, up to the discount factor, defined by the following quantities (t stands for the maturity)

- Asian options:

$$E\left[(A_t^{(\nu)} - k)^+\right]$$

  where $A_t^{(\nu)} = \int_0^t ds\,\exp(2B_s^{(\nu)})$. Asian options are studied in Geman and Yor (1993). See also Kemna and Vorst (1990) and Rogers and Shi (1995) for different approaches.

- Parisian options: Recall the definition of an up-and-out 'Parisian' option: the owner of this option loses it if the underlying asset price $(S_s, s \geq 0)$ reaches a level L before maturity t and remains constantly *above* this level for a time interval longer than a fixed number c, called the option window. If not, the owner will receive the pay-off $(S_t - k)^+$. Therefore, we consider excursions[1] of the process $(B_s^{(\nu)}, s \geq 0)$ above a given barrier. In this paper, we assume that the level L is equal to the spot price at initial time, which implies that the excursions for the process $B_s^{(\nu)}$ are at (more correctly: away from) the level 0. The general case can easily be derived by waiting until the first hitting time of the barrier and then translating that level to 0. See Chesney, Jeanblanc-Picqué and Yor (1997) for more details.

  Let g_s be the left extremity of the excursion which straddles time s, and

$$H_c^+ = \inf\{s : \mathbb{I}_{B_s^{(\nu)}>0}(s - g_s) \geq c\},$$

  the first time when an excursion above 0 is 'older' than the window c. The price of an up-and-out Parisian call is given by

$$E\left[\mathbb{I}_{H_c^+>t}\left(\exp(B_t^{(\nu)}) - k\right)^+\right].$$

  Our methodology applies also to other Parisian options, namely down and knock-in options.

---

[1]See the Appendix for a precise definition.

- Double barrier options:

$$E\left[\mathbb{I}_{T_{-a,b}>t}\left(\exp(B_t^{(\nu)}) - k\right)^+\right],$$

where $T_{-a,b} = \inf\{s : B_s^{(\nu)} \notin [-a,b]\}$. Such options are studied by a number of authors, including Kunimoto and Ikeda (1992), He, Keirstead and Rebholtz (1995) and Geman and Yor (1996).

This has led us to consider the mixed general quantities

$$\begin{aligned} \sigma_+^{(\nu)}(a,b,k;t) &\stackrel{def}{=} E\left[\mathbb{I}_{\Sigma>t}\left(a\exp(B_t^{(\nu)}) + bA_t^{(\nu)} - k\right)^+\right] \\ \sigma_-^{(\nu)}(a,b,k;t) &\stackrel{def}{=} E\left[\mathbb{I}_{\Sigma\leq t}\left(a\exp(B_t^{(\nu)}) + bA_t^{(\nu)} - k\right)^+\right], \end{aligned} \tag{1}$$

where Σ is a stopping time in the filtration generated by $(B_t^{(\nu)})$. We develop below the computations related to the particular cases $\Sigma = H_{c,d}$ and $\Sigma = T_{-a,b}$ where

$$H_{c,d} \stackrel{def}{=} \inf\{s : \mathbb{I}_{B_s^{(\nu)}>0}(s - g_s) \geq c \quad \text{or} \quad \mathbb{I}_{B_s^{(\nu)}<0}(s - g_s) \geq d\}.$$

We use the concise notation

$$\begin{aligned} H_c^+ &\stackrel{def}{=} H_{c,\infty} = \inf\{s : \mathbb{I}_{B_s^{(\nu)}>0}(s - g_s) \geq c\} \\ H_d^- &\stackrel{def}{=} H_{\infty,d}. \end{aligned}$$

In the case $\Sigma = H_c^+$ (resp. H_d^-), only excursions above 0 (resp. below 0) are relevant. To simplify the presentation, we also introduce

$$H_c = \inf\{s : s - g_s \geq c\},$$

since this stopping time allows us to develop our methodology, without taking care of the signs of excursions.

The case $a = 0, \Sigma = t$ is the Asian option case; $b = 0, \Sigma = H_c^+, \sigma_+^{(\nu)}$ (resp. $b = 0, \Sigma = H_c^+, \sigma_-^{(\nu)}$) is the up-and-out (resp. up-and-in) Parisian option case; $b = 0, \Sigma = T_{-a,b}$ is the double barrier case.

The reader will note that the computation of $\sigma_+^{(\nu)}(a,b,k;t) + \sigma_-^{(\nu)}(a,b,k;t)$ reduces to that of $E\left[\left(a\exp(B_t^{(\nu)}) + bA_t^{(\nu)} - k\right)^+\right]$. This last expression involves the law of the pair $(B_t^{(\nu)}, A_t^{(\nu)})$ and is computed in Section 5.

Therefore, we restrict our attention to $\sigma^{(\nu)}(a,b,k;t) \stackrel{def}{=} \sigma_-^{(\nu)}(a,b,k;t)$. We derive in this article the expression of the Laplace transform of $\sigma^{(\nu)}$ with respect to time. This method could even be extended to the situation where the pay-off is a function of $B_t^{(\nu)}$ and $A_t^{(\nu)}$. In the particular case $a = b = 0$, the

Laplace transform of Σ is also obtained, hence the price of the corresponding boost option[2].

In fact, equality (1) covers several types of contingent claims which may be useful in specific situations. For example, our method yields the price of Asian floating options, i.e., with pay-off $(A_t^{(\nu)} - \exp B_t^{(\nu)})^+$, corresponding to the case $k = 0, b = 1, a = -1, \Sigma = t$ and the price of Asian spread options, with pay-off the positive part of the difference between two independent Asian pay-offs.

The chapter is organised as follows: in the next section, we show that, using a Laplace transform in time, the problem can be split into two subproblems. The first one reduces to the computation of the joint law of $(\Sigma, \exp B_\Sigma^{(\nu)}, A_\Sigma^{(\nu)})$ and is studied in Section 4. The second one is to find the law of the pair $(B_t^{(\nu)}, A_t^{(\nu)})$ and is solved in Section 5. An Appendix gathers the main definitions and results about the different stochastic processes and σ-fields which are used in this paper.

3 Laplace Transform in Time

As observed in previous works by the authors, the Markov property and time changes can be used in an essential way for such computations. In what follows, $\overline{B}^{(\nu)}$ denotes a Brownian motion with drift ν, assumed to be independent of the original Brownian motion $(B_s, s \geq 0)$. We want to compute the time Laplace transform of $\sigma^{(\nu)}$ which, thanks to the strong Markov property applied at time Σ, can be written as

$$\int_0^\infty dt\, e^{-\lambda t} \sigma^{(\nu)}(a, b, k; t) = E\Big[\exp(-\lambda\Sigma) \int_0^\infty dt\, e^{-\lambda t}\Big(a \exp(B_\Sigma^{(\nu)} + \overline{B}_t^{(\nu)})$$
$$+ b\,[A_\Sigma^{(\nu)} + \overline{A}_t^{(\nu)} \exp(2B_\Sigma^{(\nu)})] - k\Big)^+\Big].$$

Thus, clearly our original problem decomposes into two subproblems:

- Finding the joint law of $(\Sigma, \exp B_\Sigma^{(\nu)}, A_\Sigma^{(\nu)})$. We call this problem (JL), which itself decomposes into $\begin{cases} (JL)^1, \text{ for } \Sigma = H_c \text{ or } \Sigma = H_{c,d} \\ (JL)^2, \text{ for } \Sigma = T_{-a,b}, \end{cases}$

- Computing the resolvent type quantity

$$v^{(\nu)}(\lambda, \overline{a}, \overline{b}, \overline{k}) \stackrel{\text{def}}{=} E\Big[\int_0^\infty dt\, e^{-\lambda t}\Big(\overline{a} \exp \overline{B}_t^{(\nu)} + \overline{b}\,\overline{A}_t^{(\nu)} - \overline{k}\Big)^+\Big].$$

We use bars to avoid confusion with the original parameters $a, b, k, \ldots$. We refer to this second problem as problem (V). The particular case $\overline{a} = 0$ leads to the Asian option price, whereas the case $\overline{a} = -\overline{b}$ leads to the Asian floating options price.

---

[2]Recall that, for a boost option, the pay-off is proportional to the time spent in a band.

4   A Solution of Problem (JL)

4.1   On problem $(JL)^1$

Here the problem is to compute the law of the triple

$$\left(H_c,\ \exp B_{H_c}^{(\nu)},\ A_{H_c}^{(\nu)}\right).$$

The more general case when $\Sigma = H_{c,d}$ will be studied at the end of this section.

The Cameron–Martin formula allows one to reduce the problem to the case $\nu = 0$. We recall that we are interested in excursions of the Brownian motion away from 0 and that

$$g_t = \sup\{s \le t, B_s = 0\}, \quad H_c = \inf\{t : t - g_t \ge c\}.$$

For this purpose, we use the following[3]

Proposition 1

(i) The σ-field $\mathcal{F}_{g_{H_c}}^-$, the random variable $\eta_c \stackrel{def}{=} \mathrm{sgn}\,(B_{H_c})$, and the process $\{|B_{g_{H_c}} + u|, u \le c\}$ are independent; furthermore, η_c is a symmetric random Bernoulli variable, and $\{|B_{g_{H_c}} + u|, u \le c\}$ is a Brownian meander with length c.

(ii) For any $\mathbb{R}_+$-valued, $(\mathcal{F}_t)$-predictable process $(z_t, t \ge 0)$, the following relationship holds

$$\sqrt{\frac{\pi c}{2}}\, E[z_{g_{H_c}}] = \int_0^\infty ds\, E[z_{\tau_s} \mathbb{I}_{\Delta(\tau_s) \le c}],$$

where $\tau_s = \inf\{u : \ell_u > s\}$, ℓ is the local time of (B_t) at 0, and

$$\Delta(\tau_s) = \sup_{u \le s}(\tau_u - \tau_{u-})$$

is the maximum length of excursions up to time τ_s.

Proof: The first part relies on standard properties of the Brownian meander which are recalled in the Appendix. The second part follows from the 'balayage' formula (see Revuz and Yor (1994), chap. VI, sect. 4) which states that for any bounded, $(\mathcal{F}_t)$-predictable process $(z_t, t \ge 0)$, the process

$$z_{g_t}|B_t| - \int_0^t d\ell_u\, z_u$$

is a $(\mathcal{F}_t)$-martingale; hence, by projection on $(\mathcal{F}_{g_t}^-)$, the process

$$z_{g_t}|\mu_t| - \int_0^t d\ell_u\, z_u$$

---

[3]See the Appendix for the corresponding definitions.

is a $(\mathcal{F}_{g_t}^-)$-martingale, where $\mu_t \overset{def}{=} \text{sgn}(B_t)\sqrt{\dfrac{\pi}{2}(t-g_t)}$ is the so-called Azéma martingale; recall that μ_t and $|\mu_t| - \ell_t$ are $(\mathcal{F}_{g_t}^+)$-martingales. Then, we use the stopping time theorem, at time H_c, to obtain

$$\sqrt{\frac{\pi c}{2}}\, E(z_{g_{H_c}}) = E\left(\int_0^{H_c} d\ell_u\, z_u\right).$$

The final formula follows by making the obvious time change in the integral with respect to $d\ell_u$. $\qquad\square$

We now explain how to exploit the above results to proceed with the solution of $(JL)^1$.

(a) Our aim is to compute the law of the triple (H_c, B_{H_c}, A_{H_c}). This problem is solved via the joint Laplace transform of this law, i.e.,

$$u(\alpha, k, \theta; c) \overset{def}{=} E\left[\exp\left(\alpha B_{H_c} - \frac{k^2}{2}H_c - \frac{\theta^2}{2}A_{H_c}\right)\right]$$

which, from Proposition 1 (i), is equal to:

$$u(\alpha, k, \theta; c) = u_b(k, \theta; c)\exp\left(-\frac{k^2 c}{2}\right)u_m(\alpha, \theta; c),$$

where

$$u_b(k, \theta; c) \overset{def}{=} E\left[\exp\left(-\frac{1}{2}(k^2 g_{H_c} + \theta^2 A_{g_{H_c}})\right)\right]$$

and

$$u_m(\alpha, \theta; c) \overset{def}{=} E\left[\exp\left(\alpha B_{H_c} - \frac{\theta^2}{2}\int_{g_{H_c}}^{H_c} du\, \exp(2B_u)\right)\right].$$

We now explain the mnemonic for u_b and u_m.

(i) We remark that the definition of u_b involves the trajectories of the Brownian motion between 0 and g_{H_c}, and that the law of the process $\left(\dfrac{1}{g_{H_c}}B_{ug_{H_c}}; u \le 1\right)$ is equivalent (i.e., mutually absolutely continuous with respect) to that of the Brownian bridge which may be represented as $\left(\dfrac{1}{g_t}B_{ug_t}; u \le 1\right)$; however, we shall not use, nor prove, this absolute continuity result here.

(ii) The definition of u_m involves the trajectories of the Brownian meander between g_{H_c} and H_c and the process $(|B_{g_{H_c}+u}|, u \le c)$ is a Brownian meander with length c.

Thus, again, problem $(JL)^1$ may be split into two subproblems, which we shall denote by $(JL)_m^1$ and $(JL)_b^1$.

(b) In order to solve problem $(JL)_b^1$, we use the multiplicative 'master formula' of excursion theory (See Revuz and Yor, chap. XII) which implies that, for any Borel function $f : \mathbb{R} \to \mathbb{R}_+$,

$$E[\mathbb{I}_{\Delta(\tau_s) \le c}\exp(-A_{\tau_s}^f)] = \exp(-sJ(c, f)),$$

where $A_t^f = \int_0^t du\, f(B_u)$ and[4]

$$J(c,f) \stackrel{\text{def}}{=} \int \mathbf{n}(d\epsilon)\left(1 - \mathbb{I}_{V(\epsilon)\leq c}\exp\left(-A_V^f(\epsilon)\right)\right). \tag{2}$$

As a consequence of Proposition 1(ii), one obtains the important result:

Corollary 1 *Using the previous notation, for any positive measurable function* f

$$\sqrt{\frac{\pi c}{2}}\, E[\exp(-A_{g_{H_c}}^f)] = 1/J(c,f). \tag{3}$$

4.1.1   On problem $(JL)_m^1$

This consists in the computation of $u_m(\alpha, \theta; c)$ which can be expressed in terms of the Brownian meander. Indeed,

$$\begin{aligned}
u_m(\alpha, \theta; c) &= \frac{1}{2}E\left[\exp\left(\alpha\sqrt{c}\,m_1 - \frac{\theta^2 c}{2}\int_0^1 du\,\exp(2\sqrt{c}\,m_u)\right)\right] \\
&\quad + \frac{1}{2}E\left[\exp\left(-\alpha\sqrt{c}\,m_1 - \frac{\theta^2 c}{2}\int_0^1 du\,\exp(-2\sqrt{c}\,m_u)\right)\right],
\end{aligned}$$

where now $(m_u, u \leq 1)$ denotes the standard Brownian meander (i.e. with length 1). We denote by M^c the law of the Brownian meander $(m_u^{(c)}, u \leq c)$ with length c, i.e.,

$$(m_u^{(c)} = \sqrt{c}\,m_{u/c}, u \leq c)$$

on the canonical space $(C(\mathbb{R}_+, \mathbb{R}_+), \mathcal{R}_\infty)$.

Notice that, in the particular case $\theta = 0$ ('Parisian' case), the computation of $u_m(\alpha, 0; c)$ follows from the expression of the law of m_1:

$$P(m_1 \in dx) = x\exp(-x^2/2)\mathbb{I}_{x>0}\,dx.$$

To compute $u_m(\alpha, \theta; c)$, we proceed by looking at the Laplace transform in the variable $\dfrac{k^2}{2}$ of $c \to \dfrac{u_m(\alpha, \theta; c)}{\sqrt{2\pi c}}$ ([5]); thus, we search for an expression of

$$\Phi_\pm(k, \alpha, \theta) \stackrel{\text{def}}{=} \int_0^\infty \frac{dc\,\exp(-k^2 c/2)}{\sqrt{2\pi c}}\, M^c\left(\exp\left[\pm\alpha R_c - \frac{\theta^2}{2}\int_0^c du\,\exp(\pm 2R_u)\right]\right),$$

where $(R_u, u \geq 0)$ is the canonical process. We know from excursion theory (see, e.g., Revuz and Yor, exercise 4.18, chap. XII) that

$$\begin{aligned}
\Phi_\pm&(k, \alpha, \theta) \\
&= \int_0^\infty da\, E_a\left[\exp -\left(\frac{k^2}{2}T_0 \mp \alpha a + \frac{\theta^2}{2}\int_0^{T_0} du\,\exp(\pm 2B_u)\right)\right] \\
&= \int_0^\infty da\,\exp(\pm\alpha a) E_{\pm a}\left[\exp\left(-\frac{1}{2}\left(k^2 T_0 + \theta^2\int_0^{T_0} du\,\exp(2B_u)\right)\right)\right],
\end{aligned}$$

---

[4]See the Appendix for the definitions of the Itô measure $\mathbf{n}$ and of the lifetime V.

[5]Dividing by $\sqrt{2\pi c}$ simplifies subsequent computations; this will be clear in the next lines.

where T_0 denotes the first time Brownian motion reaches 0 and E_a denotes the expectation under the law of Brownian motion starting at a. Note that the integral with $\exp(a\alpha)$ is finite for $\alpha < k$.

The crucial point in our solution of $(JL)_m^1$ is the following

Proposition 2 Let $f : \mathbb{R} \to \mathbb{R}_+$ be a locally bounded function. Then the function

$$u(a) \equiv u^f(k,a) \stackrel{\text{def}}{=} E_a\left[\exp -\left(\frac{k^2}{2}T_0 + \int_0^{T_0} du\, f(B_u)\right)\right] \qquad (4)$$

is the unique bounded solution of the Sturm-Liouville equation

$$\frac{1}{2}u'' = \left(\frac{k^2}{2} + f\right)u; \qquad u(0) = 1.$$

As particular examples, one has, for $a \geq 0$, $k \geq 0$, $\theta > 0$:

$$E_a\left[\exp -\frac{1}{2}(k^2 T_0 + \theta^2 A_{T_0})\right] = \frac{K_k(\theta e^a)}{K_k(\theta)} \qquad (5)_+$$

$$E_{-a}\left[\exp -\frac{1}{2}(k^2 T_0 + \theta^2 A_{T_0})\right] = \frac{I_k(\theta e^{-a})}{I_k(\theta)}, \qquad (5)_-$$

where I_k and K_k are modified Bessel functions.

Notation 1. In what follows, the two formulas $(5)_\pm$ will play an important role; related formulas (f) involving the positive (resp. negative) level a (resp. $-a$) and/or the function $\exp(2x)$ (resp. $\exp(-2x)$) will be presented as formulas $(f)_+$ (resp. $(f)_-$).

2. We denote by $P_\alpha^{(\nu)}$ (or, when more convenient, $^{(d)}P_\alpha$) the law of the Bessel process[6] $(R_u^{(\nu)}, u \geq 0)$ of index ν (of dimension d), starting at α.

Proof of Proposition 2: The general statement follows from the optional sampling theorem, Itô's formula, and/or the Feynman-Kac formula. Rather than deducing formula $(5)_\pm$ directly from the general case, we shall connect these formulas with computations done in Pitman and Yor (1981).

We denote by $P_\alpha^{(0)}$ the law of the 2-dimensional Bessel process $(R_u^{(0)}, u \geq 0)$, starting from $\alpha \geq 0$. We have exhibited in previous research (Geman and Yor (1993)) the power of Lamperti's representation of a geometric Brownian Motion as a time-changed Bessel process (see Revuz and Yor, chap. XI)

$$\exp(B_t + \nu t) = R_{A_t^{(\nu)}}^{(\nu)}. \qquad (6)$$

We also introduce $C_u^{(\nu)} = \int_0^u \frac{ds}{(R_s^{(\nu)})^2}$. Let us remark that $C^{(\nu)}$ and $A^{(\nu)}$ are inverses of each other, namely

$$C_u^{(\nu)} = \inf\{t \mid A_t^{(\nu)} > u\} \qquad (7)$$

---

[6]See the Appendix for some definitions.

and $R_u^{(\nu)} = \exp(B_{C_u^{(\nu)}} + \nu C_u^{(\nu)})$. Then, using the representation $\exp B_t = R_{A_t^{(0)}}$, and denoting by T_1 the hitting time $T_1 = \inf\{t : R_t = 1\}$, we deduce that the left-hand side of $(5)_+$ is equal to

$$E_{e^a}^{(0)}\left(\exp -\frac{1}{2}\left(k^2 \int_0^{T_1} \frac{du}{R_u^2} + \theta^2 T_1\right)\right) = \frac{K_k(\theta e^a)}{K_k(\theta)}, \tag{8}$$

the last equality being borrowed from Pitman and Yor (1981) (Proposition 2.3). The same argument leads to formula $(5)_-$. $\square$

Warning: In Yor (1993), points (c) and (d) of Lemma 1, the right-hand sides of $(5)_+$ and $(5)_-$ have erroneously been inverted. However, this does not affect the subsequent results in Yor (1993).

To proceed further, we shall use the Hartman distributions, the definition of which we now present. Let $0 < r < R < \infty$. P. Hartman (1976) showed from a purely analytical viewpoint, the existence of two positive integrable functions $h_{r,R}^{\uparrow}$ and $h_{r,R}^{\downarrow}$ such that

$$\frac{I_k(r)}{I_k(R)} = \int_0^\infty \exp\left(-\frac{k^2}{2}t\right) h_{r,R}^{\uparrow}(t)\, dt$$

and

$$\frac{K_k(R)}{K_k(r)} = \int_0^\infty \exp\left(-\frac{k^2}{2}t\right) h_{r,R}^{\downarrow}(t)\, dt.$$

Now, using Proposition 2 together with the definition of the Hartman densities $h_{r,R}^{\uparrow}$ and $h_{r,R}^{\downarrow}$, we obtain

$$\begin{aligned}
\Phi_+(k,\alpha,\theta) &= \int_0^\infty da \, \exp(\alpha a) \, \frac{K_k(\theta e^a)}{K_k(\theta)} \\
&= \int_0^\infty da \, \exp(\alpha a) \int_0^\infty dc \, \exp\left(-\frac{k^2 c}{2}\right) h_{\theta, \theta e^a}^{\downarrow}(c)
\end{aligned}$$

and

$$\begin{aligned}
\Phi_-(k,\alpha,\theta) &= \int_0^\infty da \, \exp(-\alpha a) \, \frac{I_k(\theta e^{-a})}{I_k(\theta)} \\
&= \int_0^\infty da \, \exp(-\alpha a) \int_0^\infty dc \, \exp\left(-\frac{k^2 c}{2}\right) h_{\theta e^{-a}, \theta}^{\uparrow}(c).
\end{aligned}$$

Next, recalling that $\Phi_\pm(k,\alpha,\theta)$ is a Laplace transform with respect to $k^2/2$, which involves $(M^c, c > 0)$, we obtain

$$\frac{1}{\sqrt{2\pi c}} M^c\left(\exp\left[-\alpha R_c - \frac{\theta^2}{2}\int_0^c du \, \exp(-2R_u)\right]\right) = \int_0^\infty da \, e^{-\alpha a} \, h_{\theta e^{-a}, \theta}^{\uparrow}(c)$$

and

$$\frac{1}{\sqrt{2\pi c}} M^c \left(\exp\left[\alpha R_c - \frac{\theta^2}{2} \int_0^c du \, \exp(2R_u) \right] \right) = \int_0^\infty da \, e^{\alpha a} \, h^{\downarrow}_{\theta, \theta e^a}(c) \,.$$

These formulas yield a complete solution of problem $(JL)^1_m$:

Proposition 3 *The function $u_m(\alpha, \theta; c)$ is given by*

$$u_m(\alpha, \theta; c) = \sqrt{\frac{\pi c}{2}} \int_0^\infty da \left(e^{-\alpha a} \, h^{\uparrow}_{\theta e^{-a}, \theta}(c) + e^{\alpha a} \, h^{\downarrow}_{\theta, \theta e^a}(c) \right) \,.$$

where the Hartman densities $h^{\uparrow}_{r,R}$ and $h^{\downarrow}_{r,R}$ are defined, via Laplace transform, by

$$\frac{I_k(r)}{I_k(R)} = \int_0^\infty \exp\left(-\frac{k^2}{2} t \right) h^{\uparrow}_{r,R}(t) \, dt$$

and

$$\frac{K_k(R)}{K_k(r)} = \int_0^\infty \exp\left(-\frac{k^2}{2} t \right) h^{\downarrow}_{r,R}(t) \, dt \,.$$

These formulas are still valid in the case $\theta = 0$, using equivalent expressions for the Bessel functions (See the Appendix).

4.1.2   More results about Hartman densities

In order to understand better the Hartman densities, we disintegrate formula (4) with respect to the law of T_0.

Proposition 4 *Using the notation introduced in Proposition 2, we have*

$$u^f(k, a) = \int_0^\infty dt \, \exp\left(-\frac{k^2 t}{2} \right) H_f(t, a) \,,$$

where

$$H_f(t, a) = \frac{a}{\sqrt{2\pi t^3}} \exp\left(-\frac{a^2}{2t} \right) \, {}^{(3)}E_0 \left[\exp - \int_0^t du \, f(R_u) \Big| R_t = a \right] \qquad (9)$$

and ${}^{(3)}P_0$ denotes the law of the 3-dimensional Bessel process starting from 0. As particular examples, one has, for any $a > 0$,

$$\frac{a}{\sqrt{2\pi c^3}} \exp\left(-\frac{a^2}{2c} \right) \, {}^{(3)}E_0 \left[\exp\left(-\frac{\theta^2}{2} \int_0^c du \, \exp(-2R_u) \right) \Big| R_c = a \right] = h^{\uparrow}_{\theta e^{-a}, \theta}(c)$$

$$(10)_-$$

$$\frac{a}{\sqrt{2\pi c^3}} \exp\left(-\frac{a^2}{2c} \right) \, {}^{(3)}E_0 \left[\exp\left(-\frac{\theta^2}{2} \int_0^c du \, \exp(2R_u) \right) \Big| R_c = a \right] = h^{\downarrow}_{\theta, \theta e^a}(c) \,.$$

$$(10)_+$$

Proof: The right-hand side of (4) may be written as

$$\int_0^\infty dt \frac{a}{\sqrt{2\pi t^3}} \exp\left(-\frac{a^2}{2t}\right) \exp\left(-\frac{k^2 t}{2}\right) E_a[\exp - \int_0^{T_0} du\, f(B_u)) | T_0 = t].$$

Furthermore, using Williams' time reversal and conditioning with respect to $L_a = \sup\{t : R_t = a\}$, we obtain

$$E_a\left[\exp - \int_0^{T_0} du\, f(B_u) \Big| T_0 = t\right] = {}^{(3)}E_0\left[\exp - \int_0^t du\, f(R_u) \Big| R_t = a\right]. \quad (\Box)$$

Now, we investigate limits in formulas (9) and (10)$_\pm$ as $a \to 0$. In particular, we find that

$$h_\theta^\uparrow(c) \stackrel{\text{def}}{=} \lim_{a \to 0+} \frac{1}{a} h_{\theta e^{-a}, \theta}^\uparrow(c) \quad \text{and} \quad h_\theta^\downarrow(c) \stackrel{\text{def}}{=} \lim_{a \to 0+} \frac{1}{a} h_{\theta, \theta e^a}^\downarrow(c)$$

exist and satisfy

$$h_\theta^\uparrow(c) = \frac{1}{\sqrt{2\pi c^3}} {}^{(3)}E_0\left[\exp\left(-\frac{\theta^2}{2}\int_0^c du\, \exp(-2R_u)\right) \Big| R_c = 0\right] \quad (11)_-$$

and

$$h_\theta^\downarrow(c) = \frac{1}{\sqrt{2\pi c^3}} {}^{(3)}E_0\left[\exp\left(-\frac{\theta^2}{2}\int_0^c du\, \exp(2R_u)\right) \Big| R_c = 0\right]. \quad (11)_+$$

More generally, we obtain the following

Proposition 5 *Using the notation of Propositions 2 and 4 , one obtains the existence of $K_f(t) \stackrel{\text{def}}{=} \lim_{a \to 0+} \dfrac{H_f(t, a)}{a}$ and the equality*

$$K_f(t) = \frac{1}{\sqrt{2\pi t^3}} {}^{(3)}E_0\left[\exp - \int_0^t du\, f(R_u) \Big| R_t = 0\right].$$

Furthermore, this function K_f is characterized via the following Laplace transform:

$$-\frac{\partial}{\partial a}\Big|_{a=0+} u^f(k, a) = k + \int_0^\infty \frac{dt}{\sqrt{2\pi t^3}} \exp\left(-\frac{k^2 t}{2}\right)\left(1 - \sqrt{2\pi t^3}\, K_f(t)\right)$$

$$= \int_0^\infty \frac{dt}{\sqrt{2\pi t^3}}\left(1 - \exp\left(-\frac{k^2 t}{2}\right) {}^{(3)}E_0\left[\exp - \int_0^t ds\, f(R_s) \Big| R_t = 0\right]\right).$$

In particular, one obtains

$$\int_0^\infty \frac{dt}{\sqrt{2\pi t^3}}\left(1 - \exp\left(-\frac{k^2 t}{2}\right) {}^{(3)}E_0\left[\exp\left(-\frac{\theta^2}{2}\int_0^t ds\, \exp(2R_s)\right) \Big| R_t = 0\right]\right)$$

is equal to

$$\frac{\theta K_{k+1}(\theta)}{K_k(\theta)} - k = \frac{\theta K_{k-1}(\theta)}{K_k(\theta)} + k, \tag{12}_+$$

and

$$\int_0^\infty \frac{dt}{\sqrt{2\pi t^3}} \left(1 - \exp\left(-\frac{k^2 t}{2}\right) \,^{(3)}E_0\left[\exp\left(-\frac{\theta^2}{2}\int_0^t ds \,\exp(-2R_s)\right) \Big| R_t = 0\right]\right)$$

is equal to

$$\frac{\theta I_{k-1}(\theta)}{I_k(\theta)} - k = \frac{\theta I_{k+1}(\theta)}{I_k(\theta)} + k. \tag{12}_+$$

Proof: We divide by a the two sides of the equality

$$u^f(k, a) - 1 = (u^f(k, a) - e^{-ka}) + (e^{-ka} - 1).$$

Then, using

$$e^{-ka} = \int_0^\infty \frac{dt}{\sqrt{2\pi t^3}} \, a \exp\left[-\frac{1}{2}\left(k^2 t + \frac{a^2}{t}\right)\right],$$

we obtain

$$-\frac{\partial}{\partial a}\Big|_{a=0^+} u^f(k, a) =$$

$$k + \int_0^\infty \frac{dt}{\sqrt{2\pi t^3}} \exp\left(-\frac{k^2 t}{2}\right) \left(1 - \,^{(3)}E_0\left[\exp\left(-\int_0^t ds \, f(R_s)\right) \Big| R_t = 0\right]\right).$$

We also note that $k = \int_0^\infty \frac{dt}{\sqrt{2\pi t^3}}(1 - \exp(-k^2 t/2))$, which leads to the second form of the derivative of $u^f(k, a)$. To obtain $(12)_\pm$, we apply the previous result together with the recurrence relations between Bessel functions and their derivatives (see Lebedev (1972), p. 110)

$$\begin{cases} \dfrac{\theta K_{k+1}(\theta)}{K_k(\theta)} - k &= \dfrac{\theta K_{k-1}(\theta)}{K_k(\theta)} + k \\[2mm] \dfrac{\theta I_{k-1}(\theta)}{I_k(\theta)} - k &= \dfrac{\theta I_{k+1}(\theta)}{I_k(\theta)} + k \end{cases}$$

and

$$\begin{cases} K_{k+1}(\theta) &= -K_k'(\theta) + \dfrac{k}{\theta} K_k(\theta) \\[2mm] I_{k-1}(\theta) &= I_k'(\theta) + \dfrac{k}{\theta} I_k(\theta) \end{cases}.$$

$\square$

In particular, we deduce from formulas $(11)_\pm$ and $(12)_\pm$ some analytic representation of $h_\theta^\uparrow$ and $h_\theta^\downarrow$:

Corollary 2 *One has for* $\theta, k > 0$

$$1 + \frac{\partial}{\partial k}\left(\theta \frac{K_{k-1}(\theta)}{K_k(\theta)}\right) = k \int_0^\infty dt \exp\left(-\frac{k^2 t}{2}\right) t\, h_\theta^\downarrow(t)$$

$$1 + \frac{\partial}{\partial k}\left(\theta \frac{I_{k+1}(\theta)}{I_k(\theta)}\right) = k \int_0^\infty dt \exp\left(-\frac{k^2 t}{2}\right) t\, h_\theta^\uparrow(t).$$

4.1.3   On problem $(JL)_b^{\downarrow}$

We are now interested in the computation of $u_b(k, \theta; c)$, which, from formula (3), amounts to the computation of

$$J_{\pm}(c, m) \stackrel{\text{def}}{=} \int \mathbf{n}_{\pm}(d\epsilon)\left(1 - \mathbb{I}_{V(\epsilon) \leq c} \exp\left(-\frac{m^2}{2}\int_0^{V(\epsilon)} du \exp(2\epsilon_u)\right)\right),$$

where $\mathbf{n}_{\pm}$ is the Itô measure of positive (negative) excursions (details of this reduction will be given below).

Proposition 6 *The following formulas hold, for* $\mu > 0$ *and* $m > 0$

$$\int_0^\infty dc \exp\left(-\frac{\mu^2 c}{2}\right) J_+(c, m) = \left(\frac{mK_{\mu+1}(m)}{K_\mu(m)} - \mu\right)\frac{1}{\mu^2} \qquad (13)_+$$

$$\int_0^\infty dc \exp\left(-\frac{\mu^2 c}{2}\right) J_-(c, m) = \left(\frac{mI_{\mu-1}(m)}{I_\mu(m)} - \mu\right)\frac{1}{\mu^2}. \qquad (13)_-$$

Proof: We first note that using Fubini's theorem, we obtain

$$\frac{\mu^2}{2}\int_0^\infty dc \exp\left(-\frac{\mu^2 c}{2}\right) J_{\pm}(c, m) =$$
$$\int \mathbf{n}_{\pm}(d\epsilon)\left(1 - \exp\left(-\frac{m^2}{2}\int_0^{V(\epsilon)} du \exp(2\epsilon_u) - \frac{\mu^2}{2}V(\epsilon)\right)\right). \qquad (14)_{\pm}$$

On the other hand, we use formulas (a) and (b) from Lemma 1 of Yor (1993)

$$E\left[\exp\left(-\frac{1}{2}(\mu^2 \tau_s^+ + m^2 A_{\tau_s}^+)\right)\right] = \exp\left(-\frac{s}{2}\left(\frac{mK_{\mu+1}(m)}{K_\mu(m)} - \mu\right)\right)$$

$$E\left[\exp\left(-\frac{1}{2}(\mu^2 \tau_s^- + m^2 A_{\tau_s}^-)\right)\right] = \exp\left(-\frac{s}{2}\left(\frac{mI_{\mu-1}(m)}{I_\mu(m)} - \mu\right)\right),$$

where

$$A_t^{\pm} = \int_0^t ds \exp(2B_s)\, \mathbb{I}_{B_s \in \mathbb{R}^{\pm}} \quad , \quad \tau_s^{\pm} = \int_0^{\tau_s} du\, \mathbb{I}_{B_u \in \mathbb{R}^{\pm}}.$$

Then, the master multiplicative formula implies that the right-hand sides of $(14)_{\pm}$ are respectively equal to

$$\frac{1}{2}\left(\frac{mK_{\mu+1}(m)}{K_\mu(m)} - \mu\right) \, , \, \frac{1}{2}\left(\frac{mI_{\mu-1}(m)}{I_\mu(m)} - \mu\right).$$

This proves, in particular, formulas $(13)_{\pm}$. $\square$

We now explain how the computation of $u_b(k, \theta; c)$ may be reduced to that of $J_\pm(c, m)$. Indeed, from formula (3), one has $u_b(k, \theta; c) = 1/J(c; k, \theta)$ where

$$J(c; k, \theta) = \int \mathbf{n}(d\epsilon) \left(1 - \mathbb{I}_{V(\epsilon) \leq c} \exp\left[-\frac{1}{2}\left(\theta^2 \int_0^{V(\epsilon)} du \exp(2\epsilon_u) + k^2 V(\epsilon)\right)\right]\right),$$

then, we write

$$\exp\left(-\frac{k^2 V}{2}\right) = \frac{k^2}{2} \int_V^\infty dx \exp\left(-\frac{xk^2}{2}\right)$$

which, plugged into the previous formula, yields

$$J(c; k, \theta) = \frac{k^2}{2} \int_0^\infty dx \exp\left(-\frac{xk^2}{2}\right) \times$$
$$\int \mathbf{n}(d\epsilon) \left(1 - \mathbb{I}_{V(\epsilon) \leq c} \mathbb{I}_{V(\epsilon) \leq x} \exp\left[-\frac{\theta^2}{2} \int_0^{V(\epsilon)} du \exp(2\epsilon_u)\right]\right).$$

Therefore, $J(c; k, \theta)$ may be written in terms of $J_\pm(c, \theta)$

$$J(c; k, \theta) = \frac{k^2}{2} \int_0^\infty dx \exp\left(-\frac{xk^2}{2}\right) [J_+(c \wedge x, \theta) + J_-(c \wedge x, \theta)].$$

Therefore, the $(JL)_b^1$ problem is solved:

Proposition 7 *The function $u_b(k, \theta; c)$ is defined by*

$$u_b(k, \theta; c) = \left[\frac{k^2}{2} \int_0^\infty dx \exp\left(-\frac{xk^2}{2}\right) [J_+(c \wedge x, \theta) + J_-(c \wedge x, \theta)]\right]^{-1},$$

where $J_\pm(c, \theta)$ are given, via Laplace transform by

$$\int_0^\infty dc \exp\left(-\frac{\mu^2 c}{2}\right) J_+(c, m) = \left(\frac{mK_{\mu+1}(m)}{K_\mu(m)} - \mu\right)\frac{1}{\mu^2}$$
$$\int_0^\infty dc \exp\left(-\frac{\mu^2 c}{2}\right) J_-(c, m) = \left(\frac{mI_{\mu-1}(m)}{I_\mu(m)} - \mu\right)\frac{1}{\mu^2}.$$

Let us note that the particular case $\theta = 0$ follows from the equalities

$$J_-(x, 0) = J_+(x, 0) = \frac{1}{\sqrt{2\pi x}}.$$

4.1.4 A general remark

Let us consider again formulas $(13_\pm)$, and define

$$J(c, m) = J_+(c, m) + J_-(c, m)$$
$$= \int \mathbf{n}(d\epsilon) \left(1 - \mathbb{I}_{V(\epsilon) \leq c} \exp\left[-\frac{m^2}{2} \int_0^{V(\epsilon)} du \exp(2\epsilon_u)\right]\right).$$

With the help of the formula

$$\left(\frac{mK_{\mu+1}(m)}{K_\mu(m)} + \frac{mI_{\mu-1}(m)}{I_\mu(m)} - 2\mu\right)^{-1} = I_\mu(m)\,K_\mu(m),$$

(see, e.g., bottom of p. 29 in Yor (1993)) we obtain

$$\int_0^\infty dc \exp\left(-\frac{\mu^2 c}{2}\right) J(c, m) = \frac{1}{\mu^2 I_\mu(m) K_\mu(m)}. \tag{15}$$

On the other hand (cf. Yor (1993), formula (17)), we have

$$\frac{\mu^2}{2} \int_0^\infty dt \, \exp\left(-\frac{\mu^2 t}{2}\right) E\left[\exp\left(-\frac{m^2}{2} A_{g_t}\right)\right] = 2\mu I_\mu(m) K_\mu(m). \tag{16}$$

Therefore,

$$\left(\frac{\mu^2}{2} \int_0^\infty dt \exp\left(-\frac{\mu^2 t}{2}\right) E\left[\exp\left(-\frac{m^2}{2} A_{g_t}\right)\right]\right)$$
$$\times \left(\frac{\mu^2}{2} \int_0^\infty dc \exp\left(-\frac{\mu^2 c}{2}\right) J(c, m)\right) = \mu.$$

The purpose of the next few lines is to show (quite simply, indeed) that this identity holds in a general setting.

Lemma 1 *For any Borel function $f : \mathbb{R} \to \mathbb{R}_+$,*

$$\frac{\mu^2}{2} \int_0^\infty dt \, \exp\left(-\frac{\mu^2 t}{2}\right) E[\exp(-A_{g_t}^f)] = \mu \int_0^\infty ds \, E\left[\exp\left(-\left(A_{\tau_s}^f + \frac{\mu^2}{2}\tau_s\right)\right)\right].$$

This equality follows directly from excursion theory. Nevertheless, instead of giving an excursion theoretical proof, we present a proof based on 'balayage' formula with the help of the following lemma.

Lemma 2 *Let $(z_t, t \geq 0)$ be a positive $(\mathcal{F}_t)$-predictable process and S_k an exponential variable with parameter $k^2/2$, independent of $(\mathcal{F}_t)$. Then,*

$$E(z_{g_{S_k}}) = kE\left(\int_0^\infty ds \exp\left(-\frac{k^2}{2}\tau_s\right) z_{\tau_s}\right).$$

Proof of Lemma 2: We assume that z is bounded. Then, from the balayage formula

$$E\left(z_{g_{S_k}}|B_{S_k}|\right) = E\left(\int_0^{S_k} d\ell_u \, z_u\right) = E\left(\int_0^{\ell_{S_k}} ds \, z_{\tau_s}\right)$$
$$= E\left(\int_0^\infty ds \, \mathbb{I}_{S_k \geq \tau_s} \, z_{\tau_s}\right),$$

and it follows that

$$E\left(z_{g_{S_k}}|B_{S_k}|\right) = E\left(\int_0^\infty ds\, z_{\tau_s} \exp\left(-\frac{k^2}{2}\tau_s\right)\right).$$

Since $|B_{S_k}|$ is independent of $\mathcal{F}_{g_{S_k}}$ and $|B_{S_k}| \overset{\text{law}}{=} \ell_{S_k}$ with common exponential distribution with parameter k, the left-hand side of the previous equality equals

$$E\left(z_{g_{S_k}}\right) E\left(|B_{S_k}|\right) = E\left(z_{g_{S_k}}\right) E(\ell_{S_k}) = \frac{1}{k} E\left(z_{g_{S_k}}\right).$$

The result follows. □

Proof of Lemma 1: It follows from Lemma 2 by taking $z_t = \exp - A_t^f$. □

Recall that from (2)

$$J(c, f) = \int \mathbf{n}(d\epsilon) \left(1 - \mathbb{I}_{V(\epsilon) \le c} \exp(-A_V^f(\epsilon))\right).$$

The Laplace transform of J is given by

$$\frac{\mu^2}{2} \int_0^\infty dc \exp\left(-\frac{\mu^2 c}{2}\right) J(c, f) = \int \mathbf{n}(d\epsilon) \left(1 - \exp\left(-\left(A_V^f(\epsilon) + \frac{\mu^2}{2} V(\epsilon)\right)\right)\right).$$
(17)

On the other hand, from the master multiplicative formula

$$\int_0^\infty ds\, E[\exp - (A_{\tau_s}^f + \frac{\mu^2}{2}\tau_s)] = \left[\int \mathbf{n}(d\epsilon) \left(1 - \exp - (A_V^f(\epsilon) + \frac{\mu^2}{2} V(\epsilon))\right)\right]^{-1},$$
(18)

so that the comparison between (18) and (17) yields

$$\left(\frac{\mu^2}{2} \int_0^\infty dt \exp\left(-\frac{\mu^2 t}{2}\right) E\left[\exp\left(-A_{g_t}^f\right)\right]\right)$$
$$\times \left(\frac{\mu^2}{2} \int_0^\infty dc \exp\left(-\frac{\mu^2 c}{2}\right) J(c, f)\right) = \mu.$$

4.1.5 Separating positive and negative excursions

We now go back to the initial case of Parisian options, where we consider only excursions above level 0.

For the reader's convenience, we recall that

$$H_{c,d} \overset{\text{def}}{=} \inf\{s : \mathbb{I}_{B_s^{(\nu)}>0}(s - g_s) \ge c \ \text{ or } \ \mathbb{I}_{B_s^{(\nu)}<0}(s - g_s) \ge d\},$$

and

$$H_c^+ = H_{c,\infty} = \inf\{s : \mathbb{I}_{B_s^{(\nu)}>0}(s - g_s) \ge c\}.$$

Proposition 1(i) extends to the stopping time H_c^+:

Proposition 8 *The process $\{B_{g_{H_c^+}+u}, u \le c\}$ is a Brownian meander with length c, independent of the σ-field $\mathcal{F}_{g_{H_c}}^+$.*

Our aim is to compute

$$u^+(\alpha, k, \theta; c) \overset{def}{=} E\left[\exp\left(\alpha B_{H_c^+} - \frac{k^2}{2}H_c^+ - \frac{\theta^2}{2}A_{H_c^+}\right)\right]$$

which, from Proposition 8, is equal to:

$$u^+(\alpha, k, \theta; c) = u_b^+(k, \theta; c)\exp\left(-\frac{k^2 c}{2}\right)u_m^+(\alpha, \theta; c),$$

where

$$u_b^+(k, \theta; c) \overset{def}{=} E\left[\exp\left(-\frac{1}{2}(k^2 g_{H_c^+} + \theta^2 A_{g_{H_c^+}})\right)\right]$$

and

$$u_m^+(\alpha, \theta; c) \overset{def}{=} E\left[\exp\left(\alpha B_{H_c^+} - \frac{\theta^2}{2}\int_{g_{H_c^+}}^{H_c^+} du\,\exp(2B_u)\right)\right].$$

The corresponding $(JL)_m^{1+}$ problem reduces to the computation of

$$E[\exp(\alpha\sqrt{c}\,m_1 - \frac{\theta^2 c}{2}\int_0^1 du\,\exp(2\sqrt{c}\,m_u))],$$

which is done in Subsection 4.1.1. Thus, we are looking at the $(JL)_b^{1+}$ problem.

Taking care to the positive (resp. negative) excursions, the proof of Proposition 1 leads us to the $(\mathcal{F}_{g_t}^+)$-martingales

$$\mu_t^\pm z_{g_t} - \frac{1}{2}\int_0^t d\ell_u z_u,$$

where (z_t) is any bounded $(\mathcal{F}_t)$-previsible process, and (μ_t^+) (resp. (μ_t^-)) is the positive (resp. negative) part of Azéma's martingale. We now use the $(\mathcal{F}_{g_t}^+)$-stopping time $H_{c,d}$, and, since $\mu_{H_{c,d}}^+ = 0$ if $H_{c,d} = H_d^-$, we obtain

$$\sqrt{\frac{\pi c}{2}}E\left(z_{g_{H_{c,d}}}\mathbb{I}_{H_c^+=H_{c,d}}\right) = \frac{1}{2}E\left(\int_0^{H_{c,d}}d\ell_u z_u\right)$$

$$\sqrt{\frac{\pi d}{2}}E\left(z_{g_{H_{c,d}}}\mathbb{I}_{H_d^-=H_{c,d}}\right) = \frac{1}{2}E\left(\int_0^{H_{c,d}}d\ell_u z_u\right).$$

The usual change of time yields

$$E\left(\int_0^{H_{c,d}}d\ell_u z_u\right) = \int_0^\infty ds E\left(z_{\tau_s}\mathbb{I}_{\Delta^+(\tau_s)\le c}\,\mathbb{I}_{\Delta^-(\tau_s)\le d}\right),$$

where $\Delta^{\pm}(\tau_s)$ is the maximum length of positive (resp. negative) excursions up to time τ_s.

We now apply the above computation to $z_u = \exp(-A_u^f)$.

The multiplicative master formula implies that

$$E\left(z_{\tau_s} \mathbb{I}_{\Delta^+(\tau_s) \le c}\, \mathbb{I}_{\Delta^-(\tau_s) \le d}\right) = \exp(-sK(c,d;f)),$$

where

$$K(c,d;f) \stackrel{\text{def}}{=} \int \mathbf{n}(d\epsilon)\left(1 - \mathbb{I}_{V^+(\epsilon) \le c,\, V^-(\epsilon) \le d}\exp\left(-A_V^f(\epsilon)\right)\right)$$

and V^+ (resp. V^-) is the lifetime for positive (resp. negative) excursions. We split $\mathbf{n}$ into $\mathbf{n}_+$ and $\mathbf{n}_-$; thus,

$$K(c,d;f) =$$
$$\int \mathbf{n}_+(d\epsilon)\left(1 - \mathbb{I}_{V(\epsilon) \le c}\exp(-A_V^f(\epsilon))\right) + \int \mathbf{n}_-(d\epsilon)\left(1 - \mathbb{I}_{V(\epsilon) \le d}\exp(-A_V^f(\epsilon))\right).$$

It follows that

$$E(\mathbb{I}_{H_c^+ = H_{c,d}}\,\exp(-A_{g_{H_{c,d}}}^f)) = \frac{1}{\sqrt{2\pi c}\,K(c,d;f)}$$

$$E(\mathbb{I}_{H_d^- = H_{c,d}}\,\exp(-A_{g_{H_{c,d}}}^f)) = \frac{1}{\sqrt{2\pi d}\,K(c,d;f)},$$

so, adding the members of these equalities, we get

$$E(\exp(-A_{g_{H_{c,d}}}^f)) = \frac{1}{K(c,d;f)}\left(\frac{1}{\sqrt{2\pi c}} + \frac{1}{\sqrt{2\pi d}}\right).$$

It remains to observe that

$$K(c,d;f) = J_+(c,f) + J_-(d,f).$$

In particular, the value of u_b^+ is obtained by choosing $f(x) = \frac{1}{2}(k^2 + \theta^2 \exp(2x))$ and letting $d \to \infty$.

4.2   On problem $(JL)^2$

We return to the general case of a Brownian motion with drift equal to ν; we recall that here $\Sigma = T_{-a,b}$. Let us denote $m = \exp(-a)$ and $M = \exp b$, and assume that $m < M$. We introduce the stopping times for the Bessel process $T_\rho = \inf\{t \ge 0 \mid R_t^{(\nu)} = \rho\}$ and we denote $T^* = T_m \wedge T_M$.

From the representation (6), and from (7), we deduce that, in this case,

$$(\Sigma, \exp B_\Sigma^{(\nu)}, A_\Sigma^{(\nu)}) \stackrel{(law)}{=} (C_{T^*}^{(\nu)}, R_{T^*}^{(\nu)}, T^*).$$

The strong Markov property and the following proposition (see Revuz and Yor, chap. XI, 1.22 p. 433 for a proof) will lead us to the solution.

Proposition 9 *Let $\alpha > 0$ and T be an $(\mathcal{R}_s)$-stopping time such that $E_\alpha^{(0)}(T^\nu) < \infty$ for any $\nu > 0$. Then, for each $\mu, \nu \geq 0$, and for each $(\mathcal{R}_{T+})$-measurable random variable $Y \geq 0$,*

$$E_\alpha^{(\nu)}\left[Y\left(\frac{\alpha}{R_T}\right)^\nu \exp\left(-\frac{k^2}{2}C_T\right)\right] = E_\alpha^{(\lambda)}\left[Y\left(\frac{\alpha}{R_T}\right)^\lambda\right], \tag{19}$$

where $\lambda = \sqrt{\nu^2 + k^2}$.

The strong Markov property enables us to compute

$$E_\alpha^{(\nu)}\left[\exp\left(-\frac{\theta^2}{2}T^*\right), R_{T^*} = m\right]$$

using the Laplace transforms of T_m and T_M which are obtained from (8), namely,

$$E_\alpha^{(\nu)}\left[\exp\left(-\frac{\theta^2}{2}T^*\right), R_{T^*} = m\right] = \left(\frac{m}{\alpha}\right)^\nu \frac{I_\nu(M\theta)K_\nu(\alpha\theta) - I_\nu(\alpha\theta)K_\nu(M\theta)}{I_\nu(M\theta)K_\nu(m\theta) - I_\nu(m\theta)K_\nu(M\theta)}.$$

In the same way, we obtain $E_\alpha^{(\nu)}[\exp(-(\theta^2/2)T^*), R_{T^*} = M]$ by interverting the parameters m and M in the right-hand side. Then, from (19) (see also Pitman and Yor (1980))

$$\left(\frac{\alpha}{m}\right)^\nu E_\alpha^{(\nu)}\left[\exp\left(-\frac{k^2}{2}C_{T^*} - \frac{\theta^2}{2}T^*\right), R_{T^*} = m\right]$$

$$= \left(\frac{\alpha}{m}\right)^\lambda E_\alpha^{(\lambda)}\left[\exp\left(-\frac{\theta^2}{2}T^*\right), R_{T^*} = m\right],$$

which characterizes the Laplace transform of the joint law of $(C_{T^*}^{(\nu)}, R_{T^*}^{(\nu)}, T^*)$ and, since

$$E_0\left(\exp\left(-\frac{k^2}{2}\Sigma - \frac{\theta^2}{2}A_\Sigma^{(\nu)}\right); B_\Sigma^{(\nu)} = -a\right) =$$

$$E_1^{(\nu)}\left[\exp\left(-\frac{k^2}{2}C_{T^*} - \frac{\theta^2}{2}T^*\right), R_{T^*} = e^{-a}\right],$$

we obtain

Proposition 10 *In the case $\Sigma = T_{-a,b}$, when $\exp(-a) < \exp b$, the joint law of $(\Sigma, \exp B_\Sigma^{(\nu)}, A_\Sigma^{(\nu)})$ is given by*

$$E_0\left(\exp\left(-\frac{k^2}{2}\Sigma - \frac{\theta^2}{2}A_\Sigma^{(\nu)}\right); B_\Sigma^{(\nu)} = -a\right) =$$

$$\exp(-\nu a)\frac{I_\lambda(e^b\theta)K_\lambda(\theta) - I_\lambda(\theta)K_\lambda(e^b\theta)}{I_\lambda(e^b\theta)K_\lambda(e^{-a}\theta) - I_\lambda(e^{-a}\theta)K_\lambda(e^b\theta)},$$

where $\lambda = \sqrt{\nu^2 + k^2}$, and analogous formula for $B_\Sigma^{(\nu)} = b$.

These formulas are still valid in the case $\theta = 0$, using equivalent expressions for the Bessel functions (see the Appendix).

5 A Solution of Problem (V)

5.1 The joint law of $(B_t^{(\nu)}, A_t^{(\nu)})$

Thanks to formula (6), we obtain

$$v^{(\nu)}(\lambda; a, b, k) = E\left[\int_0^\infty \frac{du}{(R_u^{(\nu)})^2} \exp(-\lambda C_u^{(\nu)}) \left(a R_u^{(\nu)} + bu - k\right)^+\right].$$

Next, we use the absolute continuity relationship between the laws of different Bessel processes, i.e.,

$$P_\alpha^{(\nu)}|_{\mathcal{R}_t} = \left(\frac{R_t}{\alpha}\right)^\nu \exp\left(-\frac{\nu^2}{2}\int_0^t \frac{ds}{R_s^2}\right) \cdot P_\alpha^{(0)}|_{\mathcal{R}_t},$$

which was already recalled in Proposition 9, and where $P_\alpha^{(\nu)}$ denotes the law (on the canonical space) of the Bessel process of index ν, starting at α. Thus if we define $\theta = \sqrt{2\lambda + \nu^2}$, we obtain

$$v^{(\nu)}(\lambda; a, b, k) = \int_0^\infty du\, E_1^{(\theta)}\left[\frac{(aR_u + bu - k)^+}{R_u^{2+\theta-\nu}}\right],$$

In order to compute this quantity, we need to know

$$E_1^{(\theta)}\left(\frac{\mathbb{I}_{R_u > q}}{R_u^{2m}}\right) = \int_q^\infty \frac{d\rho}{u\rho^{2m}}\,\rho^{\theta+1}\exp\left(-\frac{1+\rho^2}{2u}\right) I_\theta\left(\frac{\rho}{u}\right), \qquad (20)$$

where $q = (k - bu)^+/a$ for $a > 0$. We have no closed form formulas for these integrals except in the Asian case, i.e., for $a = 0$, which simplifies the computation since $q = 0$. In that case, we have the alternative expression to (20)

$$
\begin{aligned}
E_1^{(\theta)}\left(\frac{1}{R_u^{2m}}\right) &= \frac{1}{\Gamma(m)}\int_0^{1/2u} dv\, e^{-v} v^{m-1}(1 - 2uv)^{\theta-m} \\
&= \frac{1}{\Gamma(m)}\int_0^1 \frac{dw}{(2u)^m}\exp\left(-\frac{w}{2u}\right) w^{m-1}(1-w)^{\theta-m} \qquad (21) \\
&= \frac{\Gamma(1+\theta-m)}{\Gamma(1+\theta)(2u)^m}\Phi\left(m, 1+\theta; -\frac{1}{2u}\right)
\end{aligned}
$$

where $\Phi(\alpha, \gamma; z)$ denotes the hypergeometric function with parameters α and γ (see Lebedev (1972), p. 266); formula (21) is found in Yor (1993), p. 28, and the identity between (20) and (21) for $q = 0$, follows from exercise 12, p. 278 in Lebedev (1972) together with the relation $\Phi(\alpha, \gamma; z) = e^z\Phi(\gamma - \alpha, \gamma; -z)$.

In the case $a > 0, b > 0$, using obvious changes of variables, we obtain

$$v^{(\nu)}(\lambda; a, b, k) = a\int_0^\infty du\, \Theta(\nu, u) + b\int_0^\infty du\, u\, \Theta(\nu-1, u) - k\int_0^\infty du\, \Theta(\nu-1, u),$$

where

$$\Theta(\nu, u) = \begin{cases} \exp\left(-\dfrac{1}{2u}\right) \displaystyle\int_0^\infty d\rho\, \rho^\nu \exp\left(-\dfrac{\rho^2}{2u}\right) I_\theta\left(\dfrac{\rho}{u}\right) & \text{for } u > \dfrac{k}{b} \\[3ex] \exp\left(-\dfrac{1}{2u}\right) \displaystyle\int_{\frac{k-bu}{au}}^\infty d\rho\, \rho^\nu \exp\left(-\dfrac{\rho^2}{2u}\right) I_\theta\left(\dfrac{\rho}{u}\right) & \text{for } u < \dfrac{k}{b}. \end{cases}$$

In the case $u > k/b$, $\Theta(\nu, u)$ can be evaluated from (21), in the other case, we are lead to use the incomplete Gamma function $\Gamma(x, \lambda) = \displaystyle\int_x^\infty dt\, e^{-t} t^{\lambda-1}$ and the series expansion of the Bessel function I_θ.

In the general case the integral may be evaluated, for some numerical values of the parameters, by computer system (e.g., Maple or Mathematica).

5.2 Further research

To illustrate the flexibility of this method, we remark that it also allows us to obtain the price of spread Asian options, or 'Asian chooser' options, whose pay-off is $(aA_t^{(\nu)} + \tilde{a}\tilde{A}_t^{(\tilde{\nu})})$ where the two Asian payments are related to different underlying assets, with independent Brownian motions. In fact, our method enables us to obtain the joint Laplace transform

$$v^{(\nu,\tilde{\nu})}(\lambda, \tilde{\lambda}, a, \tilde{a}, k) = E\left(\int_0^\infty dt \int_0^\infty ds\, e^{-\lambda t} e^{-\tilde{\lambda} s}\left(aA_t^{(\nu)} + \tilde{a}\tilde{A}_s^{(\tilde{\nu})} - k\right)^+\right)$$
$$= \int_0^\infty du \int_0^\infty dv\, E_1^{(\theta)}\left(\frac{1}{R_u^{2+\theta-\nu}}\right) E_1^{(\tilde{\theta})}\left(\frac{1}{R_v^{2+\tilde{\theta}-\tilde{\nu}}}\right)(au + \tilde{a}v - k)^+,$$

hence, this double integral can be evaluated (at least, in principle ...) thanks to formula (21).

6   Conclusion

The probabilistic tools used in this chapter (Bessel processes and excursion theory) have been intensively studied by mathematicians for at least thirty years. Although fairly new in finance, their power appears in various applications such as exotic options pricing and interest rate derivative securities in particular in the Cox–Ingersoll–Ross framework.

The final numerical results associated with the formulas derived throughout the article depend on the inversion of Laplace transforms. This problem has been solved in Geman and Eydeland (1995) for Asian options, in Geman and Yor (1996) for double barrier options. An approximation has been proposed for Parisian options in Cornwall et al. (1997).

As illustrated briefly at the end of Section 5, the versatility of diffusion and excursion theories makes it possible to consider a much larger class of options; to illustrate further, our method could be extended to more general stopping times than $H_{c,d}$, e.g.,

$$H_\phi = \inf\{t : \mu_t^+ \geq \phi_+(\ell_t) \quad \text{or} \quad \mu_t^- \geq \phi_-(\ell_t)\},$$

where μ_t denotes the Azéma martingale (related with $t - g_t$), (ℓ_t) is the local time of the Brownian motion at 0, and $\phi_\pm$ are Borel functions, but this is perhaps getting a little too far ahead of currently traded options.

Appendix: Some Definitions

We recall here some definitions and results about Bessel processes, Brownian meander and excursion theory. For a precise study of Bessel processes and Brownian excursions, the reader can refer to chapter XI and chapter XII of Revuz and Yor (1994), some results about Brownian meander are given in exercise 3.8 chapter XII of that book. The seminal paper of Chung (1976) contains a number of results about Brownian meander. Most of the results about slow filtrations and 'balayage' formulas are to be found in Dellacherie, Maisonneuve and Meyer (1992).

A.1 Bessel processes

A Bessel process with index[7] $\nu \geq 0$ is a diffusion process which takes values in $\mathbb{R}_+$ and has infinitesimal generator

$$\frac{1}{2}\frac{d^2}{dx^2} + \frac{2\nu + 1}{2x}\frac{d}{dx}.$$

The number $d = 2(\nu + 1)$ is called the dimension of the Bessel process.

The Bessel process with dimension d starting at α satisfies the equation

$$R_t = \alpha + B_t + \frac{d-1}{2}\int_0^t \frac{1}{R_s}\,ds\,,$$

where B_t is a Brownian motion. On the canonical space $\Omega = C(\mathbb{R}_+, \mathbb{R}_+)$, we denote by R the canonical map $R_t(\omega) = \omega(t)$, by $\mathcal{R}_t = \sigma(R_s, s \leq t)$ the canonical filtration and by $P_\alpha^{(\nu)}$ (or $^{(d)}P_\alpha$) the law of the Bessel process of index ν (of dimension d), starting at α, i.e., such that $P_\alpha^{(\nu)}(R_0 = \alpha) = 1$. The transition density of the Bessel process of index ν is given by

$$p_t^{(\nu)}(\alpha, \rho) = \frac{\rho}{t}\left(\frac{\rho}{\alpha}\right)^\nu \exp\left(-\frac{\alpha^2 + \rho^2}{2t}\right) I_\nu\left(\frac{\alpha\rho}{t}\right),$$

where I_ν is the usual modified Bessel function with index ν.

---

[7]In this paper, when we consider a negative index ν, the corresponding Bessel process is only taken care of up to its first hitting time of zero; for simplicity, we do not discuss this case here.

Both functions I_ν and K_ν satisfy the Bessel differential equation

$$x^2 u''(x) + x u'(x) - (x^2 + \nu^2) u(x) = 0$$

and are given by:

$$I_\nu(z) = \left(\frac{z}{2}\right)^\nu \sum_{n=0}^{\infty} \frac{z^{2n}}{2^{2n}\, n!\, \Gamma(\nu + n + 1)} \quad , \quad K_\nu(z) = \frac{\pi(I_{-\nu}(z) - I_\nu(z))}{2 \sin \pi\nu} .$$

In particular, $I_\nu(z) \overset{z\to 0}{\sim} \dfrac{1}{\Gamma(\nu+1)} \left(\dfrac{z}{2}\right)^\nu$ and $K_\nu(z) \overset{z\to 0}{\sim} \dfrac{\Gamma(\nu)}{2} \left(\dfrac{z}{2}\right)^{-\nu}$.

(More results on I_ν and K_ν can be found in Lebedev (1972).) For each pair $(\nu, \theta) \in \mathbb{R}_+ \times \mathbb{R}$, and for each $t > 0$, one has

$$E_\alpha^{(\nu)}\left[\exp -\frac{\theta^2}{2} \int_0^t \frac{ds}{R_s^2}\, \Big| R_t = \rho\right] = \frac{I_\lambda\left(\dfrac{\rho\alpha}{t}\right)}{I_\nu\left(\dfrac{\rho\alpha}{t}\right)} ,$$

where $\lambda = \sqrt{\nu^2 + \theta^2}$.

A.2 Some σ-fields associated with random times

Let us denote by $(\mathcal{F}_t, t \geq 0)$ the natural filtration of the Brownian motion $(B_t, t \geq 0)$.

If L is an almost surely strictly positive random variable, we define the σ-field $\mathcal{F}_L^-$ of the past up to L as the σ-algebra generated by the variables ζ_L, where ζ is a predictable process.

Let t be given. The excursion which straddles t evolves between its left extremity $g_t = \sup\{s \leq t : B_s = 0\}$ and its right extremity $d_t = \inf\{s \geq t : B_s = 0\}$. Recall that g_t is not a $\mathcal{F}_t$-stopping time. We denote by $(\mathcal{F}_{g_t}^+, t \geq 0)$ the extended 'slow Brownian filtration' $(\mathcal{F}_{g_t}^+ = \mathcal{F}_{g_t}^- \vee \sigma(\mathrm{sgn}(B_t)), t \geq 0))$, a subfiltration of $\mathcal{F}_t$.

A.3 Brownian meander

We denote by $g = g_1 = \sup\{s \leq 1 : B_s = 0\}$ the left extremity of the excursion which straddles time 1. A Brownian meander is defined as a process

$$m_u = \frac{1}{\sqrt{1-g}} \left| B_{g + u(1-g)} \right| ; \, u \leq 1.$$

The process m is a Brownian scaled part of the (normalized) Brownian excursion which straddles time 1. The process m is independent of $\mathcal{F}_g^-$.

Using Brownian scaling, we remark that for fixed time t, the process

$$m_u^{(t)} = \frac{1}{\sqrt{t - g_t}} \left| B_{g_t} + u(t - g_t) \right| \; ; u \le 1$$

is a Brownian meander independent of the σ-field $\mathcal{F}_{g_t}^+$; in particular, the law of $m^{(t)}$ does not depend on t. More generally, for each $(\mathcal{F}_{g_t}^+)$-stopping time τ, in particular for $\tau = H_c$, $m^{(\tau)}$ is a meander. Remark that the definition of H_c implies that $H_c - g_{H_c} = c$.

The Azéma martingale $\mu_t \overset{\text{def}}{=} \text{sgn}(B_t)\sqrt{\frac{\pi}{2}(t - g_t)}$ is the projection of B_t on the filtration $\mathcal{F}_{g_t}^+$. Let ℓ_t be the local time at 0 of the Brownian motion. The projection of $|B_t| - \ell_t$ on the filtration $(\mathcal{F}_{g_t}^+)$, and, in fact, also on the filtration $(\mathcal{F}_{g_t}^-)$ is equal to $|\mu_t| - \ell_t$.

A.4 The Itô measure of Brownian excursions

Let $(B_t, t \ge 0)$ be a Brownian motion and (τ_s) be the inverse of the local time ℓ_t at level 0. The set $\{\cup_{s \ge 0}]\tau_{s-}(\omega), \tau_s(\omega)[\}$ is (almost surely) equal to the complement of the zero set $\{u : B_u(\omega) = 0\}$. The excursion process $(e_s, s \ge 0)$ is defined as

$$e_s(\omega)(t) = \mathbb{I}_{\{t \le \tau_s - \tau_{s-}\}} B_{\tau_{s-} + t}, t \ge 0.$$

This is a path-valued process $e : \mathbb{R}_+ \to \Omega_*$, where

$$\Omega_* = \{\epsilon : \mathbb{R}_+ \to \mathbb{R} : \exists V(\epsilon) < \infty, \text{ with } \epsilon(V(\epsilon) + t) = 0, \forall t \ge 0$$
$$\epsilon(u) \ne 0, \forall 0 < u < V(\epsilon), \epsilon(0) = 0, \epsilon \text{ is continuous}\}.$$

hence $V(\epsilon)$ is the lifetime of (ϵ).

The excursion process is a Poisson Point Process, and $\mathbf{n}(\Gamma)$ is defined as the intensity of the Poisson process

$$N_t^\Gamma \overset{\text{def}}{=} \sum_{s \le t} \mathbb{I}_{e_s \in \Gamma},$$

i.e., the positive real γ such that $N_t^\Gamma - t\gamma$ is an $(\mathcal{F}_{\tau_t})$-martingale.

The Itô-Williams' description of the measure $\mathbf{n}$ is

$$\mathbf{n}(d\epsilon) = \int_0^\infty \mathbf{n}_V(dv) \frac{1}{2}(\Pi_+^v + \Pi_-^v)(d\epsilon)$$

where $\mathbf{n}_V(dv) = dv/\sqrt{2\pi v^3}$ is the law of the lifetime V under $\mathbf{n}$ and Π_+^v (resp. Π_-^v) is the law of the Bessel Bridge (resp. the law of the opposite of the Bessel Bridge) with dimension 3 and length v.

References

Chesney, M., M. Jeanblanc-Picqué and M. Yor (1997) 'Brownian excursions and Parisian barrier options', to appear in *Adv. Appl. Prob.* March 1997.

Cornwall, M.J., G.W. Kentwell, M. Chesney, M. Jeanblanc-Picqué and M. Yor (1997) 'Parisian pricing', *Risk* 10(1), 77–79.

Chung, K.L. (1976) 'Excursions in Brownian motion', *Ark. für Math.* 14, 155–177.

Dellacherie, C., B. Maisonneuve and P.A. Meyer (1992) *Probabilités et Potentiel, Processus de Markov (fin), Compléments de Calcul Stochastique.* Hermann.

Geman, H. and A. Eydeland (1995) 'Domino effect: inverting the Laplace transform', *Risk*, March.

Geman, H. and M. Yor (1993) 'Bessel Processes, Asian options and perpetuities', *Mathematical Finance* 3, 349–375.

Geman, H. and M. Yor (1996) 'Pricing and hedging double-barrier options: a probabilistic approach', *Preprint.*

Goldman, M., H. Sosin and M. Gatto (1979) 'Path dependent options: buy at the low, sell at the high', *Journal of Finance* 34, 111–127.

Hartman, P. (1976) 'Completely monotone families of solutions of n-th order linear differential equations and infinitely divisible distributions', *Ann. Scuola Norm. Sup. Pisa* IV, 267–287.

He, H., W. Keirstead and J. Rebholtz (1995) 'Double lookback options', *Preprint.*

Kemna, A.G.Z. and A.C.F. Vorst (1992) 'A pricing method for options based on average asset values', *Journal of Banking and Finance* 14, 373–387.

Kunimoto, N. and M. Ikeda (1992) 'Pricing options with curved boundaries', *Mathematical Finance* 4, 275–298.

Lebedev, N. (1972) *Special Functions and their Applications.* Dover Publications.

Merton, R. (1973) 'Theory of rational option pricing', *Bell Journal of Economics and Management Science* 4, 141–183.

Pitman, J.W. and M. Yor (1980) 'Inversion du temps et processus de Bessel', Unpublished manuscript.

Pitman, J.W. and M. Yor (1981) 'Bessel processes and infinitely divisible laws', in *Stochastic Integrals. Durham Proceedings*, D. Williams (ed.), Lecture Notes in Maths. Vol. 851., Springer, 285–369.

Reiner, E. and M. Rubinstein (1991) 'Breaking down the barriers', *Risk* September, 28–35.

Revuz, D. and M. Yor (1994) *Continuous Martingales and Brownian Motion.* Second edition. Springer Verlag.

Rogers, L.C.G. and Z. Shi (1995) 'The value of an Asian option', *Journal of Applied Prob.* 32, 1077–1088.

Yor, M. (1980) 'Loi de l'indice du lacet Brownien, et distribution de Hartman–Watson', *Zeitschrift für Wahr.* **53**, 71–95.

Yor, M. (1993) 'From planar Brownian windings to Asian options', *Insurance Mathematics and Economics* **13**, 23–34.

Co-movement Term Structure and the Valuation of Energy Spread Options

A. Mbanefo

Abstract

This chapter presents a framework which may be applied to valuing an option on the difference between two underlying asset prices. Although much of the the analysis is applied to the energy markets, the same approach may be adopted for the valuation of spread options in other markets. The basis of the framework put forward is the need to model three types of term structure in order to price the option:

- the term structure of forward prices of the underlying assets;

- the term structure of volatility of the underlying asset prices; and

- the term structure of co-movement of the underlying asset prices.

It is suggested that no one of these factors remains constant and that their dynamics and inter-relationships are reflected in the behaviour of the spread over time.

1 Introduction

Second generation options and solutions for their pricing and hedging have appeared to be at the forefront of research in both academia and within the research departments of a number of financial institutions over the last few years. Some of the solutions were implemented and found to work suitably, whilst a number lacked practical appeal. In terms of options on spreads (or spread options), the 'jury is still out'!

After currencies, the energy market is the largest market in the world. For crude oil alone, there are over four hundred different blends or grades (not all of which are actively traded); a large number of refined products and liquefied natural gases also exist. Oil reserves are often traded and physical options built into contracts are synthetically reproduced in order to manage real exposures. The markets continue to expand with the growth of the US markets – and now the European markets – for natural gas. New moves are currently underway for an electricity market in the US, with the New York Mercantile Exchange (NYMEX) launching two new electricity contracts in 1996 with delivery at the California-Oregon border and Arizona respectively.

As real supply and demand play a major role in the price of energy the markets are susceptible to a large array of factors, including the political and fundamental factors affecting the financial markets and other factors such as weather and the effects of cashflow given the investment preferences of market speculators. In the early 1970s, the price of crude oil was a little over $1 per barrel, by the mid-1980s the price was around $40 per barrel and now, it is in the $20.00 to $25.00 per barrel range. With all the optionality of financial markets with regards to structures and second and third order derivative risk, and with some markets displaying annualised volatility levels of over 85%, energy is an exciting area for research.

2 Energy Spreads

In the energy markets, a number of producers and consumers are exposed to spreads. A *spread* is defined in this context as the difference in price between two or more related commodities arising from differences due to:

- *timing* - for instance purchasing crude now, but delivery not being due for 15 days or more;

- *location* - natural gas supplied on the Coastal pipeline and natural gas supplied on the Trunkline pipeline in the US; or

- *product* - purchasing crude, refining it into gasoline and selling the gasoline.

Crude oil is not available as a spot commodity; it must be drilled for and then shipped or piped to a destination. Costs of storage and delivery can often be high ($0.20 to $0.25 per barrel storage per month is often the norm), resulting in forward sales of crude oil not being uncommon. In entering into a forward sale, many consumers use the futures markets to hedge their exposures. However, they often still remain exposed to *basis risk*, as the grade of crude which they are purchasing is not necessarily reflected in the price of the futures contract. Inter-month exposure can also be managed through a *spread trade* - buying (selling) futures contracts for delivery in month A and selling (buying) other futures contracts for delivery in month B. In this case, the consumer and the producer are both exposed to the shape of the forward price curve for crude oil. As is the case in a number of commodities, the forward curve may not always be at full carry and can display a phenomenon known as 'backwardation'. *Backwardation* refers to a case in which the spot price of a commodity trades at a premium to the forward price. *Contango* refers to the inverse case. Hence, the producer selling crude now for delivery in 15 days is exposed to the forward curve changing from backwardation to contango during the period (and it can frequently do so within a trading day), whilst the consumer of the crude oil is exposed to the opposite scenario.

Quite often, exactly the same commodity will be quoted at different prices at different ends of the same pipeline, due to supply/demand imbalances. This is an exposure which again is managed by traders through spread trading.

Another example is the so called 'crack spread'. The term 'crack' refers to the technological catalytic cracking process undertaken in petroleum refineries to break down larger hydrocarbon molecules into lighter ones in order to gain more use and economic value from the product. The *crack spread* can thus be recognised as the refiner's profit margin.

Options on spreads

Given the plethora of spreads and therefore possibilities for spread trading in the energy market, it should be no surprise that *options on spreads* become necessary tools for risk management. This paper seeks to highlight the issues involved in valuing these options.

Various pieces of work have been carried out on pricing options on two underlying processes (Garman 1992, Jarrow and Rudd 1982, Nash and Shimko 1995, Ravindran 1993, Rubin 1991, Shimko 1994). However, from the viewpoint of experience, none of the solutions put forward so far can be regarded as complete, given the complexity of the problem and the fact that it presents itself in different forms in different markets.

The payoff of a spread option

A *call option* on a spread gives the holder the right to buy the spread at the strike price, whilst a *put option* on the spread gives the holder the right to sell the spread at the strike price. Buying and selling the spread can often be confused and should be based on specific market conventions, rather than on any particular formal method. The *payoff* for a call option can be expressed as:

$$C = \max(F_1 - F_2 - X, 0), \tag{1}$$

where

F_1 - price of long futures (i.e. a futures contract that is long on exercise),

F_2 - price of short futures,

X - exercise price of the spread option.

In order to develop a framework for pricing spread options, three types of term structure must be considered:

- the term structures of *forward prices* of the underlying assets,

- the term structures of *volatility* of the underlying asset prices, and

- the term structure of *co-movement* of the underlying asset prices.

I suggest that none of these factors remain constant, and that their dynamics and inter-relationships are reflected in the behaviour of the spread between the underlying assets over time; hence they must enter any realistic pricing model.

The forward curves

In order to price a number of the traded options we must be able to create the forward curve. Energy forward curves are made up of individual monthly futures contracts and so – like interest rate term structures – are not naturally continuous. The expiration dates of the futures contracts for the components of a spread are rarely the same and this has to be considered in the pricing of an option on it.

Additionally, when some markets are in backwardation, others might be in contango, providing the spread with a forward curve of its own. Furthermore, the spread can become negative as a result of fundamental changes.

There can be seasonality in the spreads. For example, the International Petroleum Exchange's (IPE) implied gasoil crack spread is wider in the fourth and first quarters of a calendar year, whilst it narrows in the spring and summer months. Gasoil is the main heating fuel used in Europe and is sold at a premium in the colder months of the year. A number of energy spreads seem to display mean reverting behaviour and hence this needs to be factored into the pricing of the option.

The volatilities

The volatility of a spread is less than the volatilities of the individual components and this needs to be considered in pricing options on it. In the energy markets the volatility smile (in terms of Black-Scholes implied volatilities for the spread) is very prominent and can seriously affect pricing an option on the spread, particularly for out-of-the-money strikes. The difficulties in assessing the volatilities also stem from the fact that historical analysis of the price components of the spreads is often done on closing prices, and may not fully represent the true volatilities involved. Furthermore, the volatility of the spread is often assessed inappropriately using a bivariate lognormal distribution with a constant correlation factor. Typically, this volatility is calculated as

$$\sigma_s := \sqrt{\sigma_{f_1}^2 + \sigma_{f_2}^2 + 2\rho\sigma_{f_1}\sigma_{f_2}} \ , \tag{2}$$

where σ_s represents the volatility of the spread, σ_{f_1} and σ_{f_2} represent the volatilities of the individual components of the spread and ρ represents the constant correlation between the components of the spread, but this assumes that *the difference of two correlated lognormal variates is lognormal* which is *false!*

Another better way to look at the volatility of the spread, allowing an element of subjectivity, is to calculate the price volatility of the spread directly. Applying assumptions of normality to the spread price itself,

$$v := F\sqrt{e^{\sigma_\tau^2} - 1} \ , \tag{3}$$

where v represents the price volatility, F represents the current value of the spread and σ_τ^2 represents the variance of the spread per time step. Looking at the volatility of the spread in this manner can often suggest to the experienced trader whether the correlation input to a model is adequately conservative, given the level of the spread, the time period or the season in question, etc.

Co-movement of the components of the spread

What should have become obvious from the above discussion is the problem of assessing the correct measure of co-movement for the spread. Correlation is input as a constant term to most models. Just as practitioners had a problem with model assumptions regarding equity derivatives on stock prices following geometric Brownian motion, in which returns are assumed to follow a random path with constant volatility, so too do practitioners have a problem with the all-too-often overlooked assumption of *constant correlation*. Should a dynamic process be governed by such an input? I suggest below that in order to apply a correlation term to most two factor models being developed for pricing these options, there is a need to develop a full term structure of correlation, similar to the term structures developed for forward prices and volatilities. An all too common problem faced by traders of spread options is the sensitivity of most models to the correlation coefficient, presenting the energy trader with an additional option Greek which is difficult to hedge, since (s)he does not have the ability to trade cross commodities (a bonus for the currency option trader) in order to hedge the implied correlations observed in the market.

3   Spread Option Models

In this section I will briefly review existing spread option models emphasising, in light of the above discussion, features which energy traders find inadequate.

This will be followed by some suggestions for alleviating these inadequacies. Some of these I am following up in current research.

Simple models

Garman (1992) shows that a number of complexities arise in valuing spread options. Simple modelling of spread options often views the spread as an asset price which is lognormally distributed, i.e. in terms of the Black-Scholes model. This fails immediately, as the spread can often become negative. Furthermore, spread fluctuation sizes do not increase for large spreads and decrease for smaller spreads. The trader's problems only begin here, as valuing the option is only a first step in trading it. Calculation of the convenience yield of the spread in such a one factor model is difficult, and managing a position based on the difference of two processes using single option Greeks can be dangerous. Take, for example, the delta of an option on a spread composed of a high and a low volatility asset: the amount to re-hedge cannot be described by a one factor model.

Shimko's model

Shimko (1995) produced what is often regarded as the most comprehensive paper on the subject of spread options relating to commodities. He based his work on that of Jarrow and Rudd (1982), assuming that spot prices follow geometric Brownian motion with *constant* correlation. Shimko's model is consistent with the stochastic convenience yield models and with the multivariate Gaussian models published by other authors.

The notation is as follows:

F_1- price of long futures (i.e. a futures contract that is long on exercise),

F_2- price of short futures,

f_1, f_2- futures prices at the maturity date of the spread option,

$X-$ exercise price of the spread option,

$S-$ value of the spread,

$\tau-$ maturity of spread option,

σ_1- volatility implied by an option on a futures 1 contract with the maturity of the spread option,

σ_2- volatility implied by an option on a futures 2 contract with the maturity of the spread option,

$\rho-$ correlation between the returns of the futures contracts over the life of the spread option,

$r-$ risk-free rate of interest,

$B = \exp(-r\tau)$ – the price per \$1 face value of a discount bond maturing at the same time as the spread option.

Let $P(f_1, f_2)$ represent the bivariate normal joint probability density of the two futures prices at spread option expiration. Making use of risk neutral valuation, the price of a call option on the spread can be expressed as a double integral, *viz.*

$$C := B \int_0^\infty \int_0^\infty \max(f_1 - f_2 - X, 0) P(f_1, f_2) df_1 df_2, \qquad (4)$$

whose exact solution may be approximated with an analytic function along the lines of Jarrow and Rudd (1982). The maximum error of this approximation is approximately equal to $0.0084\sqrt{\mu_2}$, where μ_2 is defined below.

To this end define, in terms of the non-centred moments $m_{kj} := \mathbb{E}[f_1^k f_2^j]$, $k, j = 0, 1, \ldots, 4$,

$$
\begin{aligned}
m &:= m_{10} - m_{01} \\
\mu_2 &:= (m_{20} - 2m_{11} + m_{02}) - m^2 \\
\mu_3 &:= (m_{30} - 3m_{21} + 3m_{12} - m_{03}) - 3m(m_{20} - 2m_{11} + m_{02}) + 2m^3 \\
\mu_4 &:= (m_{40} - 4m_{31} + 6m_{22} - 4m_{13} + m_{04}) \\
&\quad\; -4m(m_{30} - 3m_{21} + 3m_{12} - m_{03}) \\
&\quad\; + 6m^2(m_{20} - 2m_{11} + m_{02}) - 3m^4 \\
\kappa_3 &:= \mu_3 \\
\kappa_4 &:= \mu_4 - 3\mu_2^2 .
\end{aligned}
$$

The call option price (4) may be then approximated by the following equation:

$$C := B \left[\sqrt{\mu_2} dN(d) + n(d) \left\{ \sqrt{\mu_2} - \frac{\kappa_3 d}{6\mu_2} + \frac{\kappa_4 \left((d^2/\mu_2) - 1 \right)}{24\mu_2^{3/2}} \right\} \right] , \qquad (5)$$

where μ_2 is in practice set equal to the implied volatility of the spread σ_τ^2, $d := ((F_1 - F_2 - X))/\sqrt{\mu_2}$, $N(d)$ is the standard cumulative normal distribution function evaluated at d and $n(d)$ is its density function.

The approximation (5) accounts for both the skewness and kurtosis of the spread distribution and is based on a normal approximation to the distribution of the difference between two lognormal distributions. The terms involving κ_3 and κ_4 are in practice respectively the skewness and kurtosis *estimates*. When these terms are large the approximation may be unreliable.

This model is philosophically appealing in that it minimises reliance on historical data and uses implied data. Furthermore, as the implied volatilities $\hat{\sigma}_1, \hat{\sigma}_2$ of the components of the spread can easily be observed, Shimko suggests that it is a simple matter for the trader to determine (approximately) the implied correlation of the spread as

$$\hat{\rho} := \frac{(\sigma_\tau^2 - \hat{\sigma}_1^2 - \hat{\sigma}_2^2)}{2\hat{\sigma}_1\hat{\sigma}_2} . \tag{6}$$

He asserts that this allows the trader to focus on the decision to buy or sell crack options relating it specifically to whether the implied correlation given by (6) looks cheap or expensive. It is in this last apparent advantage of the model that its largest disadvantage lies. Shimko recognises that correlation is not a constant number, but rather a dynamic component of the spread which changes in response to price levels and the passage of time – yet his model assumes otherwise!

Correlation is not generally observable and, unfortunately, spread option prices are fairly elastic with respect to the correlation specification. A difficulty also arises from the fact that we are measuring correlation between two futures contracts. The correlation is possibly time-dependent, maturity-dependent, state dependent, seasonal, and stochastic in its own right. Hence, applying a static concept – either historical or the implied correlation suggested by Shimko – to the dynamic movement of the spread process is a dangerous practice for the options market maker. How is (s)he supposed to extract value from a position relying entirely on an *assessment* of correlation?

4 Practical considerations

We consider further here spread price mean reversion, the effect of implied volatility smiles on spread spot component options and the dynamic nature of spot correlations, concluding with a suggested spread spot component stochastic process model which remains to be analysed.

Spread price mean reversion

I have used the following approach to test for mean reversion in a time series S_t for a spread. Run a linear regression on S_{t+1} versus S_t. If the regression is meaningful (high R^2, significant parameters) then

$$S_{t+1} = \alpha + \beta S_t + u_t , \tag{7}$$

where u_t is an error term assumed proportional to a standard normal variate, i.e. $u_t := \sigma\epsilon_t$. By repeated substitution

$$S_{t+n} = \beta^n S_t + \alpha \sum_{i=0}^{n-1} \beta^i + \sigma \sum_{i=0}^{n-1} \beta^i \epsilon_{t+n-i}. \tag{8}$$

If the errors ϵ_t are assumed to be serially uncorrelated, the variance of S_t over n periods is

$$\text{var}(S_{t+n}) = \sigma^2 \sum_{i=0}^{n-1} \beta^{2i} \tag{9}$$

$$= \sigma^2 \frac{1 - \beta^{2n}}{1 - \beta^2} \,,$$

which converges to $\sigma^2/(1-\beta^2)$ as n tends to infinity, providing that $0 < \beta < 1$. Thus we can use a one-sided t-test of the statistic $1 - \hat{\beta}$ to test for mean reversion. The NYMEX West Texas Intermediate crude oil to IPE Brent crude oil spread gives a t statistic of 2.81 when tested for mean reversion on 1991 to 1994 daily sample data.

An alternative approach to testing for mean reversion of a spread is in terms of differences as follows:

$$S_{t+1} - S_t = \delta(S_t - \gamma) + \mu_{t+1} \,, \tag{10}$$

where γ is the long run mean reversion *level*. There is mean reversion if $\delta < 0$ and, for discrete observations of a process with a finer time-scale, $\delta \geq -2$ should hold. Therefore,

$$S_{t+1} = (1 + \delta)S_t - \delta \cdot \gamma + \mu_{t+1}. \tag{11}$$

Hence, the estimate of the mean reversion coefficient $(1 + \delta)$ should be significantly smaller than 1.

Of course other more sophisticated tests for the stability of models such as (7) and (10) are available (Rao 1994), but the simple tests discussed here are easy to apply to spreads in a trading environment.

Most spread option models do not take mean reversion of the spread into consideration and result in overpricing options of greater than 90 days duration. A number of other considerations arise as a result of mean reversion. These include changes in the amounts of volatility applicable to the model structure and changes in the timing of the volatility of the spread (volatility is delayed). Inputs to the models are not usually volatilities by the month, but rather terminal volatilities taken from the volatility term structure.

Volatility smiles

The implied volatility profile (across energy option strikes and/or maturities) is constant in most models, but is curved in reality (see Figure 1). One way to address this problem is to estimate the higher order moments of the distribution for underlying spot prices. For example, negative (positive) skewness gives rise to a negatively (positively) sloped volatility profile for call options.

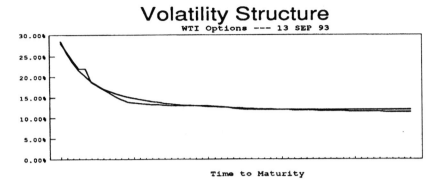

Figure 1. Black-Scholes implied volatility smile for West Texas Intermediate crude call options with respect to maturity.

Skewness relative to the lognormal distribution causes the volatility curve to 'smile', as traders adjust option prices to match the distribution which they observe in the market rather than that provided by the pricing models. Combinations of skewness and kurtosis can give rise to many different shapes of the implied volatility curve.

Skewness and kurtosis of spot prices can be estimated separately for each maturity. Estimating the smile relative to strike price for each maturity separately yields a flatter smile, but does not tell us how option prices of different maturities move together (see Figure 2). It is however possible to create a term structure of volatility, incorporating smiles across differing strikes and maturities for single energy options (see Derman and Kani (1994)).

Smiles also exist for spread options, but are not as pronounced as those for options on the components of the spread.

Seeking a term structure of co-movement

In analysing crack spreads I first performed a regression of Heating Oil against WTI Crude Oil (using NYMEX first nearby contracts). The results give $\beta = .8066$. Then the forward price curves and volatility curves were used to find the parameters for correlation ρ and volatility σ as $\rho = 0.9863$, and $\sigma^2 = 0.6355$.

Given these parameters, the variance of the spread and the correlation implied by this variance can be calculated by the method described in the previous section as shown in Figure 3. It is noticeable in Figure 3 that the implied correlation decreases with time. This may appear counter-intuitive since the slow growth of the spread variance would imply a correlation that increased with time. The explanation is that the volatility curves for both crude and product fall off so rapidly as to more than account for the slow growth of the variance.

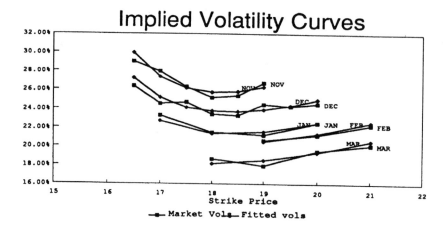

Figure 2. Black-Scholes implied volatility smiles with respect to strike price for the West Texas Intermediate crude call options of different maturities.

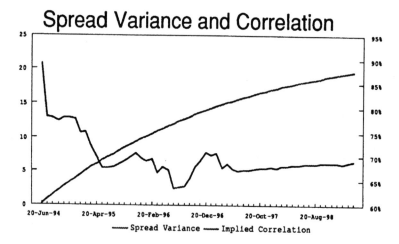

Figure 3. Spread implied variance and correlation for crack spread options.

One thing of interest in analysing this particular crack spread (also mentioned by Nash and Shimko (1995)) is the relationship between heating oil and crude oil. In crude oil high price environments, the correlation between crude and heating oil price changes is quite high. However, in low crude oil price environments, heating oil prices tend to stabilise at around $16 per barrel, and the correlation between heating oil and crude falls away significantly.

This is in part due to the 'price floor' placed upon the market by producers of heating oil in the knowledge that consumers need to keep warm.

Garman (1992) noted that an interesting phenomenon of spread options is that they can occasionally possess negative vegas, i.e. lower volatilities may result in higher option prices. This occurs, for example, when a constant correlation persists and one component of the spread has its volatility reduced, perhaps due to being price range bound for an unforeseeable period with market expectations low. If the more volatile component of the spread increases in price at such a speed that the less volatile component of the spread does not re-adjust quickly enough due to its falling volatility levels, the value of the spread option increases although the volatility of the less volatile component continues to be reduced. However, the question remains open as to whether the falling volatility in one component is consistent with a constant correlation in prices and whether, by adjusting for changing correlation, the problem of negative vegas would be avoided. By implementing a suitable co-movement structure this difficulty would thus be overcome.

The discussion above and option Greeks produced by spread option models – including cross gammas and cross vegas – suggest the need for a co-movement term structure. There are a number of considerations in developing such a term structure including the presence of mixed distributions; the presence and effect of volatility smiles and an assessment as to the appropriate measure as an input.

A *random* correlation coefficient between two assets could be interpreted as the result of asymmetric sensitivities or different degrees of reaction of the two factors involved towards the same piece of information. In many problems involving two correlated random variables, a bi-variate normal distribution is assumed. But as noted previously the parameters of the joint distribution of the two variables under consideration cannot be constant over time as it then will not capture the stochastic characteristics of the real world co-movement phenomenon.

In developing an appropriate term structure for correlation therefore, there is a need to assess what the correlation distribution can look like. Rao (1994) gives the density function of a *sample* correlation coefficient in terms of an infinite number of polynomial terms multiplied by the squares of gamma functions. Simulating a static system with a fixed correlation coefficient ρ and finding the distribution of the sample estimate r is thus a complex task; verifying a dynamic stochastic model for correlation is even more difficult, see for example Zhang (1994).

5 Towards a Satisfactory Pricing Model for Energy Crack-Spread Options

In this penultimate section of this chapter, I shall attempt to briefly sketch a sensible stochastic process model for a pair of forward prices (F_1, F_2) which hopefully can be used to price energy crack-spread options. In some cases, for example, if F_1 represents the future price of December corn and F_2 that of March gold, a bivariate geometric Brownian motion model with constant coefficients such as Shimko's model Shimko (1994) will be completely adequate. However, in other cases, for example if F_1 is the price of December West Texas Intermediate crude and F_2 is its March price, such a model will be inadequate, as I have argued above, perhaps mainly because it overlooks fundamental inequalities such as that F_2 cannot be significantly less than F_1 plus the cost of three months carry without creating arbitrage opportunities.

To be specific, I seek a realistic model for the process (F_1, F_2) of future prices which looks roughly like univariate geometric Brownian motion – possibly with variable volatility – for each individual price and moreover the inequality

$$aF_1 + bF_2 \le c \tag{12}$$

is always satisfied for fixed arbitrary constants a, b and c. Thus under the real-world probability distribution, one would have a model of the form

$$dF_1(t) = F_1(t) \left[\mu_i(t)dt + \sigma_{i1}(t)dW_1(t) + \sigma_{i2}(t)dW_2(t) \right] \tag{13}$$

for $i = 1, 2$, where W_1 and W_2 are independent Wiener processes.

It is useful to define the process

$$y(t) := c - aF_1 - bF_2 , \tag{14}$$

so that the constraint (12) becomes simply $y \ge 0$ (with probability one).

To achieve this constraint one might be tempted to adjust the drift coefficients μ_1 and μ_2 appropriately. For example, as y gets closer to zero, $|\mu_1|$ and $|\mu_2|$ could be made to become larger and larger with signs chosen in a natural way so as to cause y to drift away from the origin. This would be totally unsatisfactory for option pricing, however since under the risk-neutral probabilities these drifts do not enter the transformed versions of (13), in which both are replaced by the risk-free rate r, and hence the mechanism to enforce $y \ge 0$ has been removed. Thus, consistent with my previous arguments, one must work with the covariance matrix C of the bivariate process of (F_1, F_2).

To this end, consider the volatility matrix

$$\Sigma(t) := \begin{bmatrix} \sigma_{11}(t) & \sigma_{12}(t) \\ \sigma_{21}(t) & \sigma_{22}(t) \end{bmatrix} \tag{15}$$

implied by (13). The covariance matrix of (F_1, F_2) is given by

$$
\begin{aligned}
C(t) &:= \Sigma(t)\Sigma'(t) \\
&= \begin{bmatrix} \sigma_{11}^2(t) + \sigma_{12}^2(t) & \sigma_{11}(t)\sigma_{21}(t) + \sigma_{12}(t)\sigma_{22}(t) \\ \sigma_{11}(t)\sigma_{21}(t) + \sigma_{12}(t)\sigma_{22}(t) & \sigma_{21}^2(t) + \sigma_{22}^2(t) \end{bmatrix} .
\end{aligned}
\tag{16}
$$

Note that,

$$
\rho(t) := \frac{\sigma_{11}(t)\sigma_{21}(t) + \sigma_{12}(t)\sigma_{22}(t)}{\sqrt{\sigma_{11}^2(t) + \sigma_{12}^2(t)}\sqrt{\sigma_{21}^2(t) + \sigma_{22}^2(t)}}
\tag{17}
$$

and that ρ tends to 1 as σ_{12} and σ_{21} tend to σ_{11} and σ_{22} respectively. Thus, one possibility is to make the process ρ a function of y in such a way that, roughly speaking, ρ converges to one as y approaches 0. This can be achieved in a manner which gives some hope for the tractability of the resulting model as follows. Define

$$
\sigma_{12} := \frac{\sigma_{11}}{\sqrt{1+y}} \qquad \sigma_{21} := \frac{\sigma_{22}}{\sqrt{1+y}} .
\tag{18}
$$

Then the volatility matrix (15) becomes

$$
\Sigma(t) := \begin{bmatrix} \sigma_{11}(t) & \frac{\sigma_{11}(t)}{\sqrt{1+y(t)}} \\ \frac{\sigma_{22}(t)}{\sqrt{1+y(t)}} & \sigma_{22}(t) \end{bmatrix}
\tag{19}
$$

with corresponding covariance matrix (16)

$$
C(t) := \begin{bmatrix} \frac{2+y(t)}{1+y(t)}\sigma_{11}^2(t) & 2\frac{\sigma_{11}(t)\sigma_{22}(t)}{\sqrt{1+y(t)}} \\ 2\frac{\sigma_{11}(t)\sigma_{22}(t)}{\sqrt{1+y(t)}} & \frac{2+y(t)}{1+y(t)}\sigma_{22}^2(t) \end{bmatrix} .
\tag{20}
$$

The resulting bivariate stochastic differential equation in vector form is

$$
\begin{bmatrix} dF_1 \\ dF_2 \end{bmatrix} = \begin{bmatrix} F_1\mu_1 \\ F_2\mu_2 \end{bmatrix} dt + \begin{bmatrix} \sigma_{11}F_1 & \frac{\sigma_{11}F_2}{\sqrt{1+y(t)}} \\ \frac{\sigma_{22}F_1}{\sqrt{1+y(t)}} & \sigma_{22}F_2 \end{bmatrix} \begin{bmatrix} dW_1 \\ dW_2 \end{bmatrix} ,
\tag{21}
$$

which has a bivariate form apparently related to the squared Bessel processes as studied by Revusz and Yor (1991). If this observation proves valid it should be possible to obtain accurate European call and put spread option prices in closed form – a big step forward in pricing energy derivatives!

Conclusion

I have attempted to outline briefly some of the nuances of the energy markets and have suggested that an acceptable term structure of co-movement – be it a dynamic form of correlation or an alternative measure of co-movement – should be developed in order to better implement spread option models. Results of an analysis incorporating such techniques will be published elsewhere.

Acknowledgements

The author would like to thank the following individuals for their comments and support: Nigel Meade of Imperial College, London; Russell Newton and David Shimko of JP Morgan and Co. Inc.; Richard Flavell of Lombard Risk Systems Limited; his parents and Jennifer Warren for support, an anonymous referee and the editors. Of course all errors are those of the author alone and the views expressed are not those of Credit Suisse Holdings or of any of its subsidiaries or affiliates.

References

Derman, E. and I. Kani. (1994) 'Riding on a smile', *Risk* **7**, 32–37.

Garman, M. (1995) 'Spread the load', *Over the Rainbow*, Risk Publications **5**(11), 43–45.

Jarrow, R and A. Rudd. (1982) 'Approximate option valuation for arbitrary stochastic processes', *J. Financial Economics* **10**, 347–369.

Nash, D and D. Shimko. (1995) 'Coding the crack', *Energy Risk* **2**(5), 27–30.

Rao, B. B. (1994) *Co-integration for the Applied Economist*, Macmillan.

Ravindran, K. (1995) 'Low fat spreads', *Over the Rainbow*, Risk Publications **6**(12), 141–142.

Revusz, R. and M. Yor. (1991) *Continuous Martingales and Brownian Motion*, Springer Verlag.

Rubinstein M. (1991) 'Somewhere over the rainbow', *Risk* **4**(11), 63–66.

Shimko, D. (1994) 'Options on future spreads *J. Futures Markets* **14**(2), 183–213.

Zhang, P. (1994) 'Pricing correlation options with stochastic correlation coefficient'. MMS Working Paper.

Pricing and Hedging with Smiles

Bruno Dupire

Abstract

Black–Scholes volatilities implied from market prices exhibit a strike pattern, commonly termed a 'smile', as well as a term structure. This non-constancy of volatility contradicts the assumptions of the model and leads to the unpleasant situation in which a single spot process has many supposedly constant yet distinct volatilities. We show how to reconcile these seemingly incompatible assumptions with a single hypothesis on the spot process (instantaneous volatility which is a deterministic function of spot and time) which has the merit of preserving one-dimensionality and completeness. This process is used to price exotic options and hedge them robustly with standard European options.

1   Introduction

Option pricing consists mainly, after having specified a model and estimated its parameters, of deriving option prices (unique if the market model is complete) as a function of these parameters. A prototypical example is given by the Black–Scholes (Black and Scholes 1973) model which we will use as a guideline. It gives us options prices as a function of a parameter called *volatility*. We often have to invert this relationship, for what we know is the *price* of the option given by the market. We thus get the *implied* value of the parameter.

If the model were good, this implied value would be the same for all option market prices, a fact that reality crudely denies us. Implied Black–Scholes volatilities strongly depend on the maturity and the strike of the European option under scrutiny. If the implied volatilities of at-the-money options on the Nikkei are 20% for a maturity of 6 months and 18% for a maturity of 1 year, we are in the uncomfortable situation of assuming at the same time that the Nikkei vibrates with a constant volatility of 20% for six months and that the same Nikkei vibrates with a constant volatility of 18% for one year.

It is easy to solve this paradox by allowing volatility to be time-dependent, as Merton (1973) did long ago. The Nikkei would firstly exhibit an instantaneous volatility of 20% and subsequently a lower one, computed by a forward relationship to accommodate the one year volatility. We now have one unique process, compatible with the two option prices. From the term structure of implied volatilities we can infer a time-dependent instantaneous volatility, for

103

the former is the quadratic mean of the latter. The spot process S is then governed by the following stochastic differential equation:

$$\frac{dS}{S} = r(t)dt + \sigma(t)dW,$$

where r is the instantaneous forward rate implied from the yield curve. Some Wall Street houses incorporate this temporal information in their discretization schemes in order to price American or path-dependent options.

However, the dependence of implied volatility on the strike, for a given maturity (known as the *smile* effect) is trickier. Many researchers have attempted to enrich the Black–Scholes model to compute a theoretical 'smile'. Unfortunately they have to introduce a non-traded source of risk (jumps in the case of Merton (1976) and stochastic volatility in the case of Hull and White (1987)) thus losing the completeness of the model. Completeness is of the highest value; it allows for arbitrage pricing and hedging.

We address the following natural question: is it possible to build a spot process which:

(a) is compatible with the observed smiles at all maturities?

(b) keeps the model complete?

More precisely, given the prices of European Calls of all strikes K and maturities T: $C(K, T)$, is it possible to find a risk neutral process for the spot in the form of a diffusion,

$$\frac{dS}{S} = r(t)dt + \sigma(S, t)dW$$

where the instantaneous volatility σ is a deterministic function of the spot and time?

This would nicely extend the Black-Scholes model, to take full power of its diffusion setting, without increasing the dimension of the uncertainty. We would have the features of a one factor model (hence easily discretizable) to explain all European option prices. We could then price and hedge any American or path-dependent option[1]. We could thus answer questions like 'how to hedge a forward start option?' or 'what is the smile of Asian options?' or 'which strike to use to hedge the volatility risk on the intermediate date of a compound option?'

In the second section we review a few basic facts. We address the problem in a continuous time setting in the third section and in discrete time and price space in the fourth section. Hedging issues are tackled in the fifth section and concluding remarks take place in the final section.

---

[1] Even for European options, the knowledge of the whole process is compulsory in order to hedge.

2   The Problem

If the spot price follows a one dimensional diffusion process, then the model is complete and option prices can be computed by discounting an expectation with respect to a so-called 'risk neutral' probability under which the discounted spot has no drift (but retains the same diffusion coefficient).

More precisely, path-dependent options are priced as the discounted expected value of their terminal payoff over all possible paths. In the case of European options, it boils down to an expectation over the terminal values of the spot (which can be seen as bundling the paths which end at a same point).

It follows that the knowledge of the prices of all path-dependent options is equivalent to the knowledge of the full (risk neutral) diffusion process of the spot, while knowing all European option prices merely amounts to knowing the laws (state distributions) of the spot at different times, conditional on its current value.

The full diffusion contains much more information than the conditional laws, as distinct diffusions may generate identical conditional laws. However, if we restrict ourselves to risk neutral diffusions, the ambiguity is removed and we can retrieve from the conditional laws the unique risk neutral diffusion they come from. This result is interesting on its own but we will exploit its consequences in terms of hedging as well.

3   A Diffusion from Prices

In this section, we address the problem of existence, uniqueness and construction of a diffusion process compatible with observed option prices, in a continuous time setting. To gain considerably in clarity without losing much in generality, we assume that the interest rate is 0.

3.1   From prices to distributions

For a given maturity T, the collection of option prices $C(K,T)_K$ for different strikes yields the risk neutral density function ϕ_T of the spot at time T through the relationship:

$$C(K,T) = \int_0^\infty (x - K)^+ \phi_T(x) dx$$

which we differentiate twice with respect to K to obtain:

$$\phi_T(K) = \frac{\partial^2 C(K,T)}{\partial K^2}.$$

If we start from (S_0, T_0), we have $\phi_{T_0}(K) = \delta_{S_0}(K)$ for $C(K,T_0) := (S_0 - K)^+$.

We are then left with an interesting stochastic problem: Knowing all the state densities conditional on an initial fixed asset value x_0 and time t_0 is there a unique diffusion process which generates these densities?

The converse problem is well known: from the coefficients a and b (satisfying a slow growth assumption) of a general diffusion:

$$dx = a(x,t)dt + b(x,t)dW$$

we can deduce the conditional distributions ϕ_t thanks to the Fokker-Planck (or forward Kolmogorov) equation:

$$\frac{1}{2}\frac{\partial^2(b^2 f)}{\partial x^2} - \frac{\partial(af)}{\partial x} = \frac{\partial f}{\partial t}$$

where

$$f(x,t) = \left.\frac{\partial^2 C(K,T)}{\partial x^2}\right|_{K=x,T=t} = \frac{\partial^2 C(x,t)}{\partial x^2} = \phi_t(x).$$

However, a diffusion is more informative than the distributions it generates. It is easy to exhibit two distinct diffusions which generate the same distributions. For instance, with $x_0 = 0, t_0 = 0$:

$$dx = -\lambda x dt + \mu dW$$

and

$$dx = \mu e^{-\lambda t} dW$$

lead to the same Gaussian distribution for each t, with a mean equal to 0, and a variance equal to $(\mu^2/2\lambda)(1 - e^{-2\lambda t})$.

However, if we restrict ourselves to risk-neutral diffusions, we can recover, up to technical regularity assumptions, a unique diffusion process from the $f(x,t)$. The interest rate being 0, we only pay attention to martingale diffusions (i.e. $a = 0$), which in the case of our counterexample rules out the first candidate.

3.2   From distributions to the diffusion

The Fokker-Planck equation then takes the simple form (now f is known and b is the unknown!):

$$\frac{1}{2}\frac{\partial^2(b^2 f)}{\partial x^2} = \frac{\partial f}{\partial t}.$$

As f can be written as $\partial^2 C/\partial x^2$, we obtain, after changing the order of derivatives:

$$\frac{1}{2}\frac{\partial^2(b^2 f)}{\partial x^2} = \frac{\partial^2}{\partial x^2}\left(\frac{\partial C}{\partial t}\right).$$

Integrating twice in x for a constant t gives:

$$\frac{1}{2}b_{\alpha,\beta}^2 f = \frac{\partial C}{\partial t} + \alpha(t)x + \beta(t).$$

We *assume* that[2] $\lim_{x\to+\infty}(\partial C/\partial t) = 0$. Then the two integration constants, α and β, are actually zero because the lower limit of the LHS as x goes to infinity is 0[3]. Thus, $\frac{1}{2}b^2 f = \partial C/\partial t$ is the only possible candidate. Remembering that $f = \partial^2 C/\partial x^2$, we get:

$$\frac{1}{2}b^2 \frac{\partial^2 C}{\partial x^2} = \frac{\partial C}{\partial t}. \tag{3.1}$$

Both derivatives appearing in (3.1) are positive by arbitrage (butterfly for the convexity and conversion for the maturity). The definite candidate is then (we may impose that it be positive)

$$b(x,t) = \sqrt{\frac{2\dfrac{\partial C}{\partial t}(x,t)}{\dfrac{\partial^2 C}{\partial x^2}(x,t)}}. \tag{3.2}$$

To ensure that it is admissible (i.e. satisfies the slow growth condition) we impose: $\partial C/\partial t \le x^2(\partial^2 C/\partial x^2)$ for large x.

This condition makes sense: diffusions cannot generate everything. To see a counterexample, let us consider a diffusion process with a binary (martingale) jump at a fixed time t^*. The Call prices it generates will increase sharply at time t^* and cannot be reproduced by a diffusion; such kinks must be ruled out.

As x is a general diffusion, we can replace x by the spot process S in the form of generalised geometric Brownian motion, as defined in the introduction. Then we obtain the *instantaneous* volatility to be

$$\sigma(S,t) = \frac{b(S,t)}{S}.$$

Recalling that x actually denotes the underlying spot equal to the strike, we can rewrite (3.1) as:

$$\frac{1}{2}b^2 \frac{\partial^2 C}{\partial K^2} = \frac{\partial C}{\partial T}.$$

This equation has the same flavour as, but is distinct from, the classical Black–Scholes partial differential equation which involves, for a fixed option, derivatives with respect to current time and the value of the spot.

---

[2]This is somewhat reasonable since $\lim_{x\to+\infty} C = 0$.

[3]Otherwise, there would be a strictly positive real γ bounding from below $b^2 f$ which is in turn less than $\nu^2 x^2 f$ for a non-negative ν due to the slow growth assumption $xf \ge \gamma/\nu^2 x$, which contradicts the fact that f has a finite expectation (equal to x_0).

4 Discretization

It is indeed possible to compute b numerically from the relation (3.2) obtained from the continuous time and price analysis, and to discretize the associated spot process with explicit recombining binomial (Nelson and Ramaswamy 1990) or trinomial (Hull and White 1990) schemes. We prefer however to present a construction which makes use of a new technique widely used for interest rate model fitting: forward induction (Jamshidian 1991 and Hull and White 1992).

It is worthwhile stressing the following point: it is actually quite easy to find a set of coefficients which correctly price options since degrees of freedom are in superabundance compared to the constraints. The situation is analogous to the one encountered in the continuous case where various diffusions could generate the same densities. However, imposing the martingale condition (risk-neutrality) leads to uniqueness. In the discrete time setting, the martingale condition expressed at each node, gives additional constraints. This extra structure is a key point in our pricing/hedging approach, but existence and uniqueness are in general not achieved by a simplistic discretization. The trinomial scheme nicely meets these requirements.

We build a trinomial tree with equally spaced time steps and a price step consistent with the highest volatility[4]. Weights will be assigned to the connections, which will allow us to compute the discounted probability of each path, and hence to value any path-dependent option. It is actually possible to reduce the complexity of the computation in many cases.

At each discrete date, all profiles consisting of continuous piecewise linear functions with break points located at inner nodes of the tree must be correctly priced by the tree. At the nth step, the aforementioned space of profiles is of dimension $2n + 1$, for any such profile is uniquely characterized by the value it takes on the $2n + 1$ nodes. This space contains the zero-coupon bond, the asset itself and all Calls (and Puts) whose strikes are the inner nodes. To each node we associate an *Arrow–Debreu* profile whose value is 1 on this node and 0 on the others.

A node is labelled (n, i) with n denoting the time step and i the price step. Its associated Arrow-Debreu price is denoted by $A(n, i)$ and the probability weight of the connection between nodes (n, i) and $(n + 1, j)$, $j = i - 1$, i, or $i + 1$ is denoted by $w(n, i, j)$. Arrow–Debreu prices are computed from market prices as prices of portfolios of European Calls, spot and cash positions. The probability weights are computed through the tree in a forward fashion.

We can exploit two types of relations:

(1) Forward relations, which relate the Arrow–Debreu price of a node to the Arrow–Debreu prices of its immediate predecessors.

---

[4]This condition is equivalent to the stability condition in explicit finite difference schemes.

(2) Standard backward relations, which link the value of a contingent claim at a node to its value at the immediate successors. We apply this relation to two simple claims: a unit of the numeraire and one unit of the spot, both to be received one time step later.

The generic step of the algorithm is as follows:

Compute $w(n, i, i-1)$ from $A(n+1, i-1)$, $A(n, i)$, $A(n, i-1)$, $A(n, i-2)$, $w(n, i-1, i-1)$ and $w(n, i-2, i-1)$.

Compute $w(n, i, i)$ and $w(n, i, i+1)$ from the forward discount factors of cash and the spot.

5   Hedging

Knowledge of the whole process allows for the pricing of path-dependent options (by Monte Carlo methods) and American options (by Dynamic Programming). It also allows for hedging through an equivalent spot position because the sensitivity of the options with respect to the spot can be computed: knowing the full process, it is possible to shift the initial value and to infer the process which starts from this new value and the new price it incurs. Delta hedging can then be achieved which will be effective throughout the life of the option – if the spot behaves according to the inferred process.

It probably will not, which leads us to a more sophisticated method of hedging. We can build a robust hedge which will be efficient even if the spot does not behave according to the instantaneous inferred volatilities of the diffusion process.

The idea is to associate with every contingent claim X a portfolio of European options which will be tangent to X in the sense that it will change in value identically to changes in the value of X up to the first order for changes in the volatility manifold $\sigma(K, t)_{K,T}$.

We proceed as follows: A local move of the volatility manifold around (K_0, T_0) will lead to a new diffusion process, hence to a new value of X. We can then compute the sensitivity of X to a change of the volatility $\sigma(K_0, T_0)$ and the equivalent $C(K_0, T_0)$ position. Repeating for all (K, T), we obtain a spectrum of sensitivities $\text{Vega}(K, T)_{K,T}$ and the associated (continuous) portfolio of Calls with values $C(K, T)$, which can be seen as a projection of X onto the Calls. This portfolio will behave as X up to the first order, even if the market evolves so as to transgress the induced forward volatilities computed above.

6   Conclusion

The contribution of this article is twofold:

On the theoretical side, it shows that under certain conditions it is possible to recover from the conditional laws a full diffusion process whose drift is imposed. This means that from option prices observed in the market we can induce a unique diffusion process.

On the practical side, it tells how to elaborate sound pricing for path-dependent and American options. Moreover, it finely assesses the risk of such options by performing a risk analysis along both strikes and maturities. This enables the full and correct integration of these options into a book of standard European options, which is clearly a key point for many financial institutions.

Acknowledgements

I am happy to mention fruitful conversations with Nicole El Karoui, Marc Yor, Emmanuel Bocquet and my colleagues from the SORT (Swaps and Options Research Team) at Paribas. All errors are indeed mine. Since this article was written, two related approaches to some of the questions treated here have appeared (Derman and Kani 1996, Rubinstein 1994).

References

Black, F. and M. Scholes (1973). 'The pricing of options and corporate liabilities', *Journal of Political Economy* **81**, 637–654.

Breeden, D. and R. Litzenberger (1978). 'Prices of state-contingent claims implicit in option prices', *Journal of Business* **51**, 621–651.

Derman, E. and I. Kani, (1996) Implied trinomial trees of the volatility smile. *The Journal of Derivatives*, 7–22.

Duffie, D. (1988). *Security Markets, Stochastic Models* Academic Press.

Dupire, B. (1992). 'Arbitrage pricing with stochastic volatility', in *Proceedings of AFFI Conference in Paris*, June 1992.

El Karoui, N., R. Myneni, and R. Viswanathan (1992). 'Arbitrage pricing and hedging of interest rates claims with state variables', Working Paper.

Harrison, J.M. and D. Kreps (1979). 'Martingales and arbitrage in multiperiod securities markets, *Journal of Economic Theory* **20**, 381–408.

Harrison, J.M. and S. Pliska (1981). 'Martingales and stochastic integrals in the theory of continuous trading, *Stochastic Processes and their Applications* **11**, 215–260.

Hull, J. and A. White (1987). 'The pricing of options on assets with stochastic volatilities, *The Journal of Finance* **3**, 281–300.

Hull, J. and A. White (1990). 'Valuing derivative securities using the explicit finite difference method, *Journal of Financial and Quantitative Analysi* **25**, 87–100.

Hull, J. and A. White (1992) 'One factor interest-rate models and the valuation of interest-rate contingent claims', Working Paper, University of Toronto.

Jamshidian, F. (1991). 'Forward induction and construction of yield curve diffusion models', *Journal of Fixed Income* 1.

Karatzas, I. and S.E. Shreve (1988). *Brownian Motion and Stochastic Calculus*, Springer-Verlag.

Merton, R. (1973). 'The theory of rational option pricing', *Bell Journal of Economics and Management Science* 4, 141–183.

Merton, R. (1976). 'Option pricing when underlying stock returns are discontinuous', *Journal of Financial Economics* 3, 124–144.

Nelson, D. and K. Ramaswamy (1990). 'Simple binomial processes as diffusion approximations in financial models', *The Review of Financial Studies* 3, 393–430.

Rubinstein, M.E. (1994). 'Implied binomial trees', *Journal of Finance* 69, 771–818.

Option Pricing in the Presence of Extreme Fluctuations

Jean-Philippe Bouchaud, Didier Sornette and Marc Potters

Abstract

We discuss recent evidence that B. Mandelbrot's proposal to model market fluctuations as a Lévy stable process is adequate for short enough time scales, crossing over to a Brownian walk for larger time scales. We show how the reasoning of Black and Scholes should be extended to price and hedge options in the presence of these 'extreme' fluctuations. A comparison between theoretical and experimental option prices is also given.

1 Introduction

The efficiency of the statistical tools devised to address the problems of security pricing and portfolio selection strongly depends on the adequacy of the stochastic model chosen to describe the market fluctuations. Historically, the idea that price changes could be modelled as a Brownian motion dates back to Bachelier [1]. This hypothesis, or some of its variants (such as the Geometrical Brownian motion, where the log of the price is a Brownian motion) is at the root of most of the modern results of mathematical finance, with Markowitz portfolio analysis, the Capital Asset Pricing Model (CAPM) and the Black-Scholes formula [2] standing out as paradigms. The reason for success is mainly due to the impressive mathematical and probabilistic apparatus available to deal with Brownian motion problems, in particular Ito's stochastic calculus.

An important justification of the Brownian motion description lies in the Central Limit Theorem (CLT), stating that under rather mild hypothesis, the sum of N elementary random changes is, for large N, a Gaussian variable. In physics or in finance, where these changes occur as time is evolving, the number of elementary changes observed during a time interval t is given by $N = \frac{t}{\tau}$ where τ is an elementary correlation time, below which changes of velocity (for the case of a Brownian particle) or changes of 'trend' (in the case of the stock prices) cannot occur. The use of the CLT to substantiate the use of Gaussian statistics then requires that $t \gg \tau$. In financial markets, τ cannot be smaller than a few seconds which is not that small compared to the relevant time scales (days), in particular when one has to worry about the *tails* of the distribution, corresponding to large shocks (crashes). The Black-Scholes model and many subsequent developments suppose that $\tau \equiv 0$, which enables one to use Ito's stochastic calculus.

Finite values of N thus lead to corrections to Gaussian statistics which one would like to estimate and control. There are however other cases where the Brownian motion model fails, even when $\tau \to 0$. These cases occur either when intertemporal correlation cannot be neglected (the 'fractional Brownian motion' [3] is an example) or when fluctuations are so strong that the second moment of the distribution is infinite – leading to Lévy statistics and stable laws. There is quite a large amount of work making a case for the use of Lévy distributions in finance, starting by B. Mandelbrot's famous 1963 study of cotton prices [3, 4]. As we shall argue, we believe that the situation is more complicated (as in fact foreseen by Mandelbrot himself). In short (but see below), price changes seem to be Lévy-like for short enough time lags and become more and more Brownian as time grows. The *crossover* time T^* between the two regimes depends on the market (currencies, major stock indices, emerging markets..) and is typically a few days for currencies.

The aim of this contribution is first to summarize various important properties of *power-law* tailed distributions, which encompass Lévy stable laws. We then review recent empirical evidence for the 'mixed' behaviour alluded to above, which we shall refer to as the 'truncated' Lévy process. We shall then present a theory for option pricing and hedging in the case of a genuine Lévy process, and finally summarize the theory of option pricing for a general process with a finite variance (but not necessarily Gaussian). In both these cases, perfect hedging is in general impossible, but optimal strategies can be found (analytically or numerically) and the associated residual risk can be estimated – leading to option pricing formulae containing a risk premium.

2　A Few Results on Power-law/Lévy Distributions

We shall denote in the following the value of the stock at time t as $x(t)$, and δ_t the variation of the stock on a given time interval Δt: $\delta_t = x(t+\Delta t) - x(t)$. The probability density of δ is supposed to be of the form:

$$\rho(\delta) = \frac{\delta_0^\mu}{|\delta|^{1+\mu}} \qquad \text{for} \qquad \delta_0 \ll |\delta| \ll \delta^* \tag{1}$$

where μ is a certain exponent describing how fast the distribution decays to zero, and δ^* an upper cut-off value beyond which ρ decays much faster, say exponentially – see below. These power-law distributions are *scale invariant* (when $\delta^* = \infty$), in the sense that the relative frequency $\frac{\rho(\lambda\delta)}{\rho(\delta)}$ is independent of the chosen scale δ. For $\mu < 1$, the average of δ is of order $E(\delta) \simeq \delta_0^\mu \delta^{*1-\mu}$, and is thus *infinite* when $\delta^* = \infty$. Similarly, when $\mu < 2$, the second moment is of order $E(\delta^2) \simeq \delta_0^\mu \delta^{*\,2-\mu}$ and also diverges when $\delta^* = \infty$. More generally, only moments $E(\delta^\nu)$ such that $\nu < \mu$ give information on *typical* fluctuations which are of order δ_0.

Where do these power laws come from ? A large number of physical systems actually exhibit truncated power-law distributions of the form given in Eq. (1) [5]. These systems are called 'critical' because they are close to (δ^* large) or right on an instability point ($\delta^* = \infty$). A well known example is the *percolation* problem, where the size of the connected clusters become power-law distributed close to the percolation point [6]. Another trivial example is the probability of first return to the origin after a time t for a (one dimensional) random walk, which decays as $t^{-3/2}$ (or $\mu = 1/2$). More interesting for financial applications are the models exhibiting 'Self Organized Criticality', that is, *spontaneously* evolving towards a critical point [7]. Models of avalanches, earthquakes, crack propagation, etc... have been the subject of intense study in the recent physical literature, and might be relevant to describe bubbles and crashes in the financial markets [8].

Let us now describe a few remarkable properties of power-law distributed variables. Much more precise mathematical statements can of course be given [9]– we deliberately restrict here to a qualitative discussion of the salient features which are useful to our purpose.

- *Extreme values ('Range')*. If a set of N of these variables is considered, than the largest value encountered is of order:

$$\delta_{\max} = \max(\delta_1, \delta_2, \ldots, \delta_N) \propto \delta_0 N^{\frac{1}{\mu}}. \tag{2}$$

Note that $\delta_{\max}$ grows faster for smaller μ's, as expected intuitively.

- *Rank Ordering*. More generally, if one orders these N variables according to their rank, as:

$$y(1) = \delta_{\max}, \quad y(2) = \delta_{\max-1}, \ldots \; y(N) = \delta_{\min},$$

one obtains the following order of magnitude for $y(n)$:

$$y(n) \propto \delta_0 \left(\frac{N}{n}\right)^{\frac{1}{\mu}}. \tag{3}$$

This property is actually very useful for empirical characterization of the tail of a distribution (see [13]).

- *Sums* (Total return of a portfolio containing N shares, variation of price over N days, etc.). The order of magnitude of the sum of N independent power-law variables is given by[1]:

$$S = \sum_{i=1}^{N} \delta_i \propto \begin{cases} N\,E(\delta) & \text{if } \mu > 1 \\ N^{\frac{1}{\mu}} \simeq \delta_{\max} \; ! & \text{if } \mu < 1 \end{cases} \tag{4}$$

---

[1]The cases $\mu = 1$ or 2 are special: logarithmic corrections need to be included [9]

Note that when $\mu < 1$, the whole sum is of the same order of magnitude as the largest of its terms. This is the most striking aspect of these wildly fluctuating situations which one should keep in mind: few events (rare but important) completely dominate the phenomenon. If $\mu > 1$, on the other hand, the sum is 'democratic': all elementary moves contribute equally to the overall move.

More precisely, when $\mu < 1$, one should rescale S as $u = S/N^{1/\mu}$. The limiting distribution of u for large N is then a symmetric Lévy stable law $L_\mu(u)$ (generalizing the normal law) [9]. An important property of $L_\mu(u)$ is that it decays for large u precisely as the elementary distribution ρ (Eq. (1)): $L_\mu(u) \simeq \delta_0^\mu/|u|^{1+\mu}$. For $1 < \mu < 2$, the mean value $m = E(\delta)$ is finite, and one should consider the rescaled variable $u = (S - mN)/N^{1/\mu}$. Again, the limiting distribution of u is a Lévy stable law [9]. When $\mu > 2$, one recovers the usual CLT: the rescaled variable $u = (S - mN)/\sqrt{N}$ becomes Gaussian for large N.

However, it should be emphasized that, for any value of $\mu \in]0, \infty[$, the sum S of individual power-law variables δ_i, all distributed as in Eq. (1), but with possibly *different* 'tail amplitudes' $C_i \equiv \delta_{0i}^\mu$, is also a power-law variable with a tail amplitude C given by

$$C = \sum_{i=1}^N C_i. \tag{5}$$

This tail amplitude generalizes the property of the variance, which is additive for independent random variables. The distinction between the cases $\mu < 2$ or $\mu > 2$ lies in the fact that the total weight contained in these power-law tails remain finite in the former case, and decays to zero (as $N^{1-(\mu/2)}/(\log N)^{\mu/2}$) in the latter case, 'eaten up' by the Gaussian distribution.

- *Truncated power-laws.* If the power-law distribution only extends up to a 'cut-off' or crossover value δ^*, then all the above statements remain qualitatively valid when N is not too large. The simple criterion consists of comparing the order of magnitude of the largest term encountered $\delta_{\max}(N)$ and δ^*. If $\delta_{\max}(N) \ll \delta^*$, or equivalently if $N \ll N^* \equiv (\frac{\delta^*}{\delta_0})^\mu$, then the previous statements apply. Sums of these truncated power-laws will (for $\mu < 2$) first approach a Lévy stable law, and then realize that they are in the attraction basin of the Gaussian for $N > N^*$ [10] – this is graphically represented in Figure 1.

3 Lévy Distributions and Market Fluctuations

As stated in the introduction, reliable estimates of e.g. option prices require one first to adopt a faithful representation of reality. How faithful is a Lévy

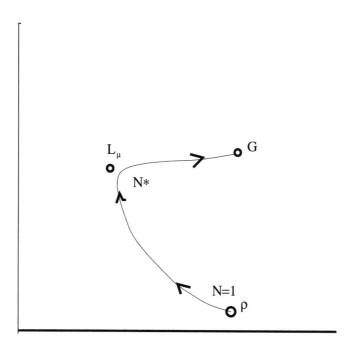

Figure 1. Graphical representation of the flow of the probability distribution under convolution. When $\delta^* = \infty$, ρ flows towards the fixed point (stable law) L_μ. For finite δ^*, the flow is first directed towards the 'phantom' fixed point L_μ, but after N^* iterations decides that it must flow towards the Gaussian fixed point G.

process description ? This is a much debated issue since Mandelbrot's seminal proposal [4, 3, 11], with pros and cons which we now summarize.

A point on which everyone agrees is the fact that the kurtosis $E(\delta^4)/E(\delta^2)^2$ is always larger – sometimes much larger – than the Gaussian value of 3. This strong 'leptokurticity' reveals the existence of fat tails, i.e. crashes which would be exceedingly improbable in a Gaussian world. Correspondingly, best fits to Lévy stable laws L_μ systematically favour values of $\mu \simeq 1.6 - 1.8$ rather than the Gaussian value $\mu = 2$. On the other hand, it has also been shown that the main property of stable laws, i.e. to be stable under aggregation, is not well obeyed by the data, and worsens as the time difference Δt increases. Correspondingly, the kurtosis decreases when Δt increases. Furthermore, the concept of an infinite variance seemed so daunting to many (see in particular [12]) that this possibility is often rejected on the basis that it is 'unreasonable' (the same 'common sense' argument was in fact used against these Lévy stable laws in physics for a long time). As we shall show in the next section, optimisation problems such as portfolio selection [13] or option hedging can be well defined even if the underlying process has an infinite variance.

However, we believe that a good representation of market fluctuations is the 'truncated' Lévy process. More precisely, the distribution of price variations at very small time scales τ (of the order of minutes) can be represented as:

$$\rho_\tau(\delta) \simeq \begin{cases} \frac{\delta_0^\mu}{\delta^{1+\mu}} & (\delta < \delta^*); \\ \exp -\frac{\delta}{\delta^*} & (\delta > \delta^*) \end{cases}. \tag{6}$$

We have obtained evidence for this power-law behaviour followed by an exponential cut-off using different techniques (rank ordered histograms, wavelet analysis) on different type of prices (shares, currencies, etc.) – a detailed account of this study goes beyond the scope of this paper and will be published elsewhere [14]. Interestingly, for many cases studied (although some exceptions exist), the value of μ is remarkably stable, $\mu \simeq 1.6-1.8$ (major currencies [14], CAC40 and MATIF [15]). A similar conclusion was recently reached by Mantegna and Stanley [16], who studied the SP Index and found a somewhat smaller value for $\mu = 1.4$, again followed by an exponential cut-off deep in the tails. We should also mention the recent work of Eberlein and Keller [17] which bears many similarities with the present work. In particular, the distribution describing market fluctuations is argued to be *hyperbolic*, which has the same exponential behaviour far in the tails, but a slightly different shape in the center compared to our choice.

As mentioned above, the existence of a cut-off δ^* removes the problem of an infinite variance, but implies the existence of a crossover time $T^* = N^*\tau$ separating a Lévy dominated regime followed by a slow 'creep' towards the Gaussian [10]. This resolves the problem of the 'instability' of the empirical distributions, which becomes manifest for large enough time differences Δt. However, for small Δt, the rescaling of $u = \delta/\Delta t^{1/\mu}$ allows one to "collapse" different histograms on a unique curve, which is indeed very well approximated by a Lévy distribution $L_\mu(u)$: see Figure 2. The finiteness of δ^* also allows one to rationalize the findings of Olsen et al. [18] who studied the growth of the moments of δ as Δt is increased. For different currencies, they find that

$$E(|\delta|) \propto \Delta t^{0.59} \qquad \text{but} \qquad \sqrt{E(\delta^2)} \propto \Delta t^{0.51}$$

This is precisely what one expects for a truncated Lévy process[2]: it is not difficult to show that in this case

$$E(|\delta|^\nu) \propto \begin{cases} \delta_0 \, \Delta t^{\frac{\nu}{\mu}} & \text{for } \nu < \mu \\ (\delta^*)^{\nu-\mu} \, \Delta t & \text{for } \nu > \mu \end{cases}. \tag{8}$$

---

[2] Such a behaviour is however less natural to interpret within the context of hyperbolic distributions [17]

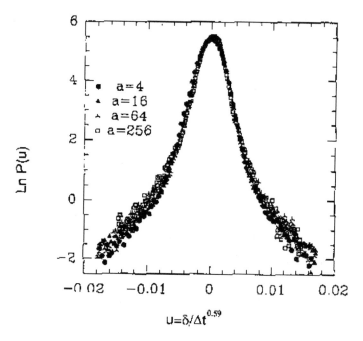

Figure 2. Rescaled distribution of the price differences, for different time lags, as $\frac{\delta}{\Delta t^{\frac{1}{\mu}}}$, with $\Delta t = a\tau$ and $\mu = 1.7$. The rescaled histogram is very well fitted by a symmetric Lévy distribution L_μ, with the same value of μ (see [14] for details).

Hence, the results of ref.[18] are in excellent agreement with a truncated Lévy process assumption, with $\mu = 1/0.59 \simeq 1.7$. It would be very interesting to understand the 'microscopic' origin of such a power-law, and to explain in particular why the value of μ seems to be so 'universal'. The exponential cut-off signals the break down of scale invariance, and is presumably related to external factors, such as allowed bands for currency fluctuations, quotation suspensions, etc.

4 Options in the Presence of Large Fluctuations

4.1 Infinite variance

We now turn to the problem of option pricing and hedging in a 'dangerous' world described by strongly non-Gaussian fluctuations, where crashes are allowed. As mentioned in section 2, the very characteristic of Lévy fluctuations is the dominance of the largest events. Since these events are by definition unpredictable, Lévy markets are necessarily *incomplete* and perfect hedging is impossible. In order to proceed, let us write down the global wealth balance $\Delta W|_0^T$ associated with the writing of a call option:

$$
\begin{aligned}
\Delta W|_0^T = {} & \mathcal{C}(x_0, x_c, T) \exp(rT) - \max(x(T) - x_c, 0) \\
& + \textstyle\sum_i \phi(x, t_i) \exp(r(T - t_i))[\delta_i - rx(t_i)\tau],
\end{aligned}
\tag{9}
$$

where $\mathcal{C}(x_0, x_c, T)$ is the price of the call, T is the maturity, x_c the striking price, $x_0 = x(t = 0)$ and $\phi(x, t)$ the trading strategy. Finally, r the (constant) interest rate and $t_i = i\tau$ is the discrete time. The first term is the gain from pocketing from the buyer the option price at $t = 0$, appreciated to time T. The second term gives the potential loss equal to $-(x(T) - x_c)$ if $x(T) > x_c$ (i.e. if the option is exercised) and zero otherwise. The third term quantifies the effect of the trading and interest between $t = 0$ and $t = T$: the extra variation of wealth W (due to trading) between t and $t + \tau$ is the result of the fluctuations of the stock ($\phi(x, t)\delta_i$), corrected by the fact that $x\phi(x, t)$ has not benefited from the risk-free interest rate.

We assume that δ_i's are identically distributed power-law variables[3] ($\delta^* = \infty$) with in general different tail amplitudes δ_0^+ and δ_0^- for positive (resp. negative) variations.

Since the sum of 'power-law' variables is a power law variable, then the distribution of 'large losses', given by Eq. (9), is a power-law:

$$
\rho(\Delta W|_0^T) \simeq_{\Delta W|_0^T \longrightarrow -\infty} \frac{(W_0[\phi(x, t)])^\mu}{|\Delta W|_0^T|^{1+\mu}}
\tag{10}
$$

with a tail amplitude $W_0[\phi(x, t)]$ which depends on the strategy ϕ. In other words, the probability that the total loss incurred due to trading the option is greater than a certain acceptable loss level $\mathcal{L}$ is given by:

$$
P(\Delta W|_0^T < -\mathcal{L}) \simeq_{\mathcal{L} \to \infty} \frac{(W_0[\phi(x, t)])^\mu}{\mu \mathcal{L}^\mu}
\tag{11}
$$

which serves as a meaningful measure of risk for $\mu < 2$, since in this case the variance of $\Delta W|_0^T$, which we shall use in next section, is infinite. The interesting point about Eq. (11) is that the minimization of risk implies that W_0 should be as small as possible, independently of the value of $\mathcal{L}$. This remark suggests a natural and objective procedure to determine the hedging strategy: the minimisation of large losses selects $\phi^*(x, t)$ such that:

$$
\left. \frac{\delta W_0^\mu}{\delta \phi(x, t)} \right|_{\phi = \phi^*} = 0
\tag{12}
$$

where a *functional minimisation* is implied.

---

[3]Note that we assume in the following that price differences and not their logs are power-law distributed.

Let us study a simple case first, where the trading strategy is trivial $\phi^*(x,t) \equiv \phi^*$ (no rehedging), as can be the case in the presence of very large trading costs. Then $\sum_i \phi^* \exp(r(T - t_i))[\delta_i - rx(t_i)\tau] \simeq \phi^* \left(x(T) - x_0 e^{rT} \right)$ (when $r\tau \ll 1^4$). Large losses occur in two cases:

• $x(t)$ drops dramatically: then $\max(x(T) - x_c, 0) = 0$ but there is a loss of $\left(x(T) - x_0 e^{rT} \right) \phi^*$ due to hold.

• $x(t)$ increases much above x_c: then the option is exercised, inducing a loss of $-(x(T) - x_c)$ with is partially compensated by the hedge. In this case, $\Delta W|_0^T = -(1 - \phi^*)x(T) + x_c - \phi^* x_0 e^{rT}$.

The resulting value of W_0 is then easy to compute, using Eq. (5):

$$W_0^\mu = \frac{T}{\tau} \left[\delta_0^{-\mu} \phi^{*\mu} + \delta_0^{+\mu}(1 - \phi^*)^\mu \right] \tag{13}$$

The optimal ϕ^* is thus given, for $\mu > 1$, by

$$\phi^* = \frac{\delta_0^{+\varsigma}}{\delta_0^{+\varsigma} + \delta_0^{-\varsigma}} \qquad \varsigma \equiv \frac{\mu}{\mu - 1} \tag{14a}$$

More generally, one can minimize $P(\Delta W|_0^T < -\mathcal{L})$ for values of $\mathcal{L}$ which are not infinitely large compared to $x_c - x(t)$. One finds in this case that

$$\phi^*(x,t) = \frac{\delta_0^{+\varsigma}}{\delta_0^{+\varsigma} + [\delta_0^- \mathcal{Q}(x,t)]^\varsigma} \qquad \mathcal{Q}(x,t) = \left(1 + \frac{x_c - x}{\mathcal{L} + \mathcal{C}[x, x_c, T - t]} \right), \tag{14b}$$

where we have set $r = 0$.

Once the optimal strategy is known, one can compute the option price by demanding that the *average* gain of the writer of the option must cover part of his potential losses, whose order of magnitude is precisely $W_0^* \equiv (E(W_0^\mu[\phi^*(x,t)]))^{\frac{1}{\mu}}$. In the simple case where $m = E(\delta) = 0$, $\mathcal{C}(x_0, x_c, T)$ is given by:

$$\mathcal{C}(x_0, x_c, T) = e^{-rT} \int_{x_c}^{\infty} \frac{dy}{\delta_0 T^{\frac{1}{\mu}}} (y - x_c) L_\mu \left(\frac{y - x_0}{\delta_0 T^{\frac{1}{\mu}}} \right) + \beta W_0^*. \tag{15}$$

β is a number of order one, depending on how risk adverse is the writer of the option, fixing a risk premium which could be thought of as a bid-ask spread. Note that when $m \neq 0$, the term $+mE(\phi^*(x,t))$ in Eq. (9), representing the average gain (or loss) due to trading, must be subtracted from the option price. In the Black-Scholes case, this term exactly compensates the difference between $E_{m\neq 0}(\max(x(T) - x_c, 0)$ and $E_{m=0}(\max(x(T) - x_c, 0)$ [20], and one recovers the well known result that $\mathcal{C}(x_0, x_c, T)$ is independent of m.

---

[4]Note that for $r = 10\%$ per year and $\tau = 1$ day, $r\tau = 2.6 \ 10^{-4}$

4.2   Non Gaussian fluctuations of finite variance

Suppose now that the maturity time scale T which is of interest becomes comparable or larger than the crossover value T^* – imposed by a finite δ^*. In this case, the variance of the wealth variation is a relevant measure of risk (although other ones are possible, such as the fourth moment, etc., depending on the weight that one wishes to give to the tails). The optimal strategy is then such that the variance of $\Delta W|_0^T$ is minimal [19, 22, 21]:

$$\frac{\delta E(\Delta W|_0^T[\phi]^2)}{\delta \phi(x,t)} = 0 \tag{16}$$

For a general uncorrelated process (i.e. $E(\delta_i \delta_j) = 0$ for $i \neq j$), the explicit solution of Eq. (16) is relatively easy to write if $m = 0$ and $r = 0$ (the generalisation to other cases is rather more cumbersome):

$$\phi^*(x,t) = \int_{x_c}^{\infty} dx' \langle \delta \rangle_{(x,t) \longrightarrow (x',T)} \frac{(x' - x_c)}{D(x,t)} P(x',T|x,t) \tag{17}$$

where $D(x,t) = E(\delta_i^2)|_{x,t}$ is the 'local volatility' – which may depend on x, t – and $\langle \delta \rangle_{(x,t) \longrightarrow (x',T)}$ is the mean instantaneous increment conditioned to the initial condition (x,t) and a final condition (x',T). The *minimal* residual risk, defined as $\mathcal{R}^* = E(\Delta W|_0^T[\phi^*]^2)$ is in general strictly positive, except for Gaussian fluctuations *in the continuous limit* $\tau = 0$, where one recovers the usual Black-Scholes results ($\mathcal{R}^* = 0$). For $0 < \tau \ll T$, however, the residual risk does not vanish and is given by [19] $\mathcal{R}^* = \frac{1}{2} D \tau \mathcal{P}(1 - \mathcal{P})$, where $\mathcal{P}$ is the probability that the option will be exercised at maturity.

Let us stress that our theory, based on Eq. (16), obviously reproduces the Black-Scholes results in the corresponding limit. Indeed, our starting point, the global balance equation Eq. (9), is nothing but the integrated version of the usual instantaneous balance equation used by Black and Scholes. Note also that approaches related to the minimisation of $E(\Delta W|_0^T[\phi]^2)$ were considered before in the mathematical literature [22], although the optimal strategy, Eq. (17), was not given in explicit form.

Now, in the spirit of the CAPM model, the option price should include a risk premium proportional to the residual risk, and thus be fixed by the equation

$$E(\Delta W|_0^T) = \beta \sqrt{E(\Delta W|_0^T[\phi^*]^2)} \tag{18}$$

where the expectation values are calculated using the empirical distribution $P(x,t|x',t')$[5].

The usefulness of Eqs. (17,18) comes from the fact that $P(x,t|x',t')$ can be rather easily reconstructed from empirical data, under the (reasonable)

---

[5]This is, again, similar in spirit to the work of Eberlein and Keller [17], except that the optimal strategy and the residual risk were not considered in their paper.

assumption of uncorrelated increments. This has enabled us to calculate numerically the price for real world options – we give an 'experimental' test of our method in Figure 3 [23] in the case of Bund options of short maturities. It is reasonable to assume that in such a liquid market, the risk premium is small (i.e. $\beta = 0$). Figure 3 shows that Eq. (18) with $\beta = 0$ reproduces the market prices very well: the regression gives a slope of 0.9993 ± 0.0009, whereas the Black-Scholes formula (not shown) gives a slope of 1.02 ± 0.002 (and a rather large intercept), which reflects the fact that the latter theory systematically misprices out-of-the-money options.

Other interesting cases, such as correlated Gaussian fluctuations (such as the 'Fractional Brownian motion') or option books can be handled with our formalism. We refer the reader to [19, 23] for more details.

5   Conclusion

Let us summarize the main messages of this chapter:

☐ We believe that power-law fluctuations $\mu \sim 1.6 - 1.8$ is a faithful representation of the financial market dynamics *but* only in a finite interval, below a certain cut-off δ^* which depend on the asset. A theory based on Lévy stable laws is thus expected to be most relevant for small enough time scales. For intermediate time scales (weeks), one is right in a *crossover* region, where no simple description is possible, and where formulae such as Eqs. (16,17) are most useful.

☐ In the case where the variance is infinite, the correct way to measure the fluctuations and thus the risk is through the 'tail parameter' W_0^μ, which is not very hard to handle analytically thanks to the additivity property Eq. (5). We have shown how option pricing and hedging could be established through minimisation of W_0, yielding formulae for the strategy extending in an interesting way the Black-Scholes recipe. Finally, the same idea of 'tail chiseling' as a way to control the extreme fluctuations was recently applied to portfolio selection in [13].

☐ More generally, the precise estimate of the residual risk associated with an option leads to a rational way of fixing a bid-ask spread around the fair price value, which turns out to be a very good estimate of real options, as exemplified in Figure 3.

Acknowledgements

We want to thank many collaborators for sharing with us their skills: J.P. Aguilar, A. Arneodo, E. Aurell, N. Mc Farlane, L. Mikheev, J. Miller, J.F. Muzy, C. Walter and in particular G. Iori. We want to thank Benoit Mandelbrot for many interesting comments and kind encouragements. JPB also

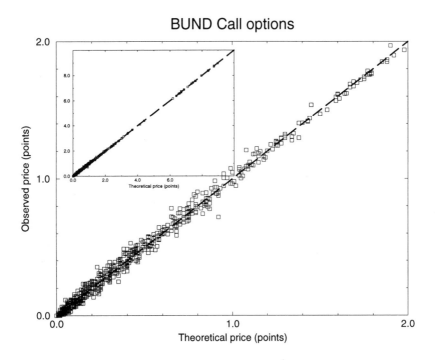

Figure 3. 'Experimental' prices for Bund call options of different maturities (all less than a month) and strikes between January and June 1995. The data has been extracted from LIFFE's CD-ROM. The coordinate of each point is the theoretical price given by Eq. (18) with $\beta = 0$ on the x axis, and the observed price. $P(x, t | x', t')$ was reconstructed using historical data in the same period. The overall agreement is gratifying, and shows that: (i) a truncated Lévy process description is suited for describing (to a first approximation) the whole 'implied volatility' surface; i.e., the way the 'smile' deforms with maturity; and (ii) the risk premium is small on very liquid markets. The inset shows the same results on a larger scale.

wants to thank the organisers and participants of the workshop at the Newton Institute for their warm welcome.

References

[1] L. Bachelier, 'Theory of Speculation' (translation of 1900 French edition), 17–78, in *The Random Character of Stock Market Prices*, P. H. Cootner (ed.), MIT Press, (1964).

[2] R. C. Merton, *Continuous-Time Finance*, Blackwell, (1990).

[3] B. B. Mandelbrot, *The Fractal Geometry of Nature*, Freeman, (1983).

[4] B. B. Mandelbrot, *Journal of Business* **36**, 394 (1963); **40**, 394 (1967).

[5] 'Lévy flights and related topics in physics', in *Lecture Notes in Physics* **450**, G. Zaslavsky, U. Frisch and M. Shlesinger (eds.), Springer (1995).

[6] D. Stauffer and A. Aharony, *Introduction to Percolation Theory*, 2nd edition, Taylor and Francis, (1992).

[7] P. Bak, C. Tang and K. Wiesenfeld, *Phys. Rev. Lett.* **59**, 381, (1987); D. Sornette, *Les phénomènes critiques auto-organisés*, Images de la Physique, édition du CNRS, January, 1994.

[8] P. Bak, K. Chen, J.A. Scheinkman and M. Woodford, *Ricerche Economiche* **47**, 3, (1993); H. Takayasu, H. Miura, T. Hirabayashi and K. Hamada, *Physica* **A 184**, 127, (1992); T. Hirabayashi, H. Takayasu, H. Miura and K. Hamada,*Fractals,* **1**, 29, (1993).

[9] B. V. Gnedenko and A. N. Kolmogorov, *Limit Distributions for Sums of Independent Random Variables*, Addison Wesley, (1954); P. Lévy, *Théorie de l'addition des Variables Aléatoires* Gauthier Villars, 1937–1954; E. J. Gumpel, *Statistics of Extremes*, Columbia University Press, (1960).

[10] R. Mantegna, and H. E. Stanley, *Phys. Rev. Lett.* **73**, 2946, (1994).

[11] for a review, see e.g. Ch. Walter, *Bulletin de l'Institut des Actuaires Français,* **349**, 3, (1990); **350**, 4, (1991); Thèse, 1994 (unpublished).

[12] P. Cootner, in *The Random Character of Stock Market Prices*, P. H. Cootner (ed.), MIT Press, 1964.

[13] J.P. Bouchaud, D. Sornette, Ch. Walter and J.P. Aguilar, preprint (October 1994), submitted to *Math. Finance.*

[14] A. Arneodo, J. P. Bouchaud, R. Cont, J. F. Muzy, M. Potters and D. Sornette, in cond-mat/9607120, available at http://xxx.lanl.gov, and in preparation.

[15] Ch. Walter, 'Lévy-stable distributions and fractal structure on the Paris market : an empirical examination', *Proc. of the first AFIR colloquium* (Paris, April 1990), **3**, 241.

[16] R. Mantegna and H. E. Stanley, *Nature* **376**, 46, (1995).

[17] E. Eberlein and U. Keller, *Bernoulli,* **1**, 281, (1995).

[18] U. A. Muller, M. M. Dacorogna, R. B Olsen, O. V. Pictet, M. Schwartz, and C. Morgenegg, *J. Banking and Finance* **14**, 1189, (1990).

[19] J. P. Bouchaud and D. Sornette, *J. Phys. I France* **4**, 863, (1994).

[20] L. Mikheev, *J. Phys. I France* **5**, 217, (1995); J. P. Bouchaud and D. Sornette, *J. Phys. I France* **5**, 219, (1995).

[21] Aurell E. Życzkowski K., "Option pricing and Partial Hedging: Theory of Polish Options" (1995), *Journal of Financial Abstracts: Series D (Working Paper Series)* **3**, January 26 (1996) [abstract], also available as ewp-fin/9601001 at http://econwpa.wustl.edu/wpawelcome.html.

[22] H. Follmer and D. Sondermann, in *Essays in honor of G. Debreu*, W. Hildenbrand, and A. Mas-Collel (eds.), North Holland (1986); C.E.R.M.A, *C. R. Acad. Sci. Paris* **307**, 625 (1988); H. Follmer and M. Schweitzer, in *Applied Stochastic Analysis*, M. H. A. Davis and R. J. Elliott (eds.), Gordon and Breach Science Publishers, (1990); M. Schweitzer, *Stochastic Processes and Their Applications* **37**, 339–363, (1991).

[23] J. P. Bouchaud, G. Iori and D. Sornette, 'Real world options', *RISK* **9** 61. (1996); J. P. Bouchaud, M. Potters and D. Sornette, book in preparation. Potters M., Cont R. and Bouchaud J.P., preprint cond-mat/9609172, available at http://xxx.lanl.gov/, submitted to *Science*.

Convergence of Snell Envelopes and Critical Prices in the American Put

N.J. Cutland, P.E. Kopp, W. Willinger and M.C. Wyman

Abstract

The numerical approximation of American option prices has a large and growing literature, whose theoretical basis lies in the approximation of functionals defined in continuous time pricing models by their discrete counterparts. For example, Kushner's method [14] of approximating diffusions by certain Markov chains has been exploited by Lamberton [15] to obtain convergence results for the critical price of American options.

In this paper, we present a new approach to convergence results in the theory of optimal stopping, such as those obtained in [15], by using the concept of D^2-*convergence* which was developed in [8, 9]. We show that the optimal stopping times for an American put option in a sequence of Cox–Ross–Rubinstein option pricing models D^2-converge to the unique optimal time in the Black–Scholes model.

Being stronger than weak convergence, D^2-convergence leads to a new convergence result for the associated Snell envelopes of the discounted return processes, and enables us to give a more direct proof of the main result of [15] by fully exploiting our understanding of the properties of discrete pricing models.

Keywords: American Put Option, Critical Price, D^2-convergence, Optimal Stopping, Nonstandard Analysis, Snell Envelope.

1   Introduction

Finite-horizon American options enable the holder to trade in the market at any time over a finite time period: the American call allows the holder to buy a stock at any time for a certain fixed price, while the American put similarly entitles the holder to sell stock at a fixed price at any time. It was pointed out in [20] that a zero-coupon American call will have the same price as the European call if the return on the stock is above the interest rate. The American put is not so easy to price.

In the well-known framework of the Black–Scholes model, Bensoussan [5], and Karatzas [12], treat the American put option as an optimal stopping problem. Their approach is based on the concept of the Snell envelope. This, and other methods, for example, the free boundary problem of [19] and the variational inequalities of [6, 11] are all discussed in detail in the survey by Myneni [21]. For a treatment of the American put option in a discrete setting, we refer to the seminal works by Lamberton and his co-workers, [11, 15, 16]. Of particular interest in this context is [15]; along a sequence of discrete Cox–Ross–Rubinstein models, it gives convergence results for quantities associated with the American put problem, such as the critical prices, optimal stopping times and Snell envelopes to their counterparts in the Black–Scholes model.

Since the limiting continuous stochastic pricing models and their discrete approximations are typically defined on different probability spaces, convergence in distribution, i.e. weak convergence, is all one can hope for when establishing convergence results in the theory of option pricing. However, as already observed by Aldous, [2], weak convergence is not preserved under under certain functionals, especially those arising in the American put option (for example, optimal stopping times and Snell envelopes). Thus, when deriving convergence results for the American put option, Lamberton in [15] relies on fairly technical results of [14] that obscure the apparently very close connection between the approximating discrete pricing models and their continuous counterpart (see also [17]). In this paper, we wish to derive such convergence results more explicitly using different methods.

We recall the concept of D^2-convergence introduced and developed in [8, 9]. The idea of D^2-convergence comes from the concept of an SL^2-lifting in Nonstandard Analysis. (See [1] for an introduction to the subject, or see the primer in [7].) In more standard terms, D^2-convergence is based on the idea of generating approximating random walks that are intimately related to a limiting Brownian motion by discretising the path of the underlying Brownian motion in an adapted and measure preserving fashion. This procedure allows one to link L^2-convergence via the discretising map with the "weak convergence along the graph" of square integrable functions. This results in an L^2-like convergence concept that is clearly stronger than weak convergence. For example, it was shown in [9] that this mode of convergence is preserved by the operations of the stochastic calculus (such as stochastic integrals and differentials). We show that the mode of D^2-convergence is also preserved under the functionals that are of particular importance for pricing the American option. Although our approach provides an alternative route to the convergence problem in the modern theory of option pricing and our results add to the attractiveness of the concept of D^2-convergence, the extent to which the approach here is "more transparent" than, for example, Lamberton's method based on Kushner's approximation technique, remains probably a matter of individual taste.

We begin by describing the theory of the American option in the standard Black–Scholes model. Section 3 describes the American put option in a family of discrete Cox–Ross–Rubinstein models constructed in [8]. Section 4 takes a hyperfinite Cox–Ross–Rubinstein model, as in [7], and details the lifting results that we shall need for Section 5, where the convergence results are stated without the terminology of nonstandard analysis.

Acknowledgements

This work was partially completed during a visit by two of us (Kopp and Willinger) to the Financial Mathematics Programme hosted by the Isaac Newton Institute for Mathematical Sciences, Cambridge, in January–June 1995. We wish to express our thanks to the Institute for its support.

Christophe Stricker suggested to us the use of the predictable representation property in Section 2.

2 The American put option for Black–Scholes Prices

Optimal stopping and the critical price

Let $b : \Omega \times [0, T] \longrightarrow \mathbb{R}$ for some finite $T > 0$ denote a standard Brownian motion on a filtered probability space $(\Omega, \mathcal{F}, P, \mathbb{B})$, where $\mathbb{B} = \{\mathcal{B}_u\}_{u \in [0,T]}$ is the filtration generated by b. The Black–Scholes model assumes a price process $s = (s^0, s^1) : \Omega \times [0, T] \longrightarrow \mathbb{R}^2$ for some finite time horizon $T > 0$, comprising a riskless bond s^0 and a risky stock s^1 such that

$$
\begin{aligned}
ds_u^0 &= s_u^0 r \, du, \quad s_0^0 = 1 \\
ds_u^1 &= s_u^1(\mu \, du + \sigma \, db_u), \quad s_0^1 > 0.
\end{aligned}
$$

Here $u \in [0, T]$, $r, \sigma > 0$ and $\mu \in \mathbb{R}$. Hence

$$
\begin{aligned}
s_u^0 &= e^{ru} \\
s_u^1 &= s_0^1 \exp\left(\left(\mu - \frac{1}{2}\sigma^2\right)u + \sigma b_u\right).
\end{aligned}
$$

The Girsanov transformation

$$
\frac{dQ}{dP}(\omega) = \exp\left(-\left(\frac{\mu - r}{\sigma}\right)b(\omega, T) - \frac{1}{2}\left(\frac{\mu - r}{\sigma}\right)^2 T\right)
$$

yields the risk-neutral probability measure $Q \sim P$ which is the unique measure making the discounted price process s^1/s^0 a Q-martingale. Writing $w_u = b_u + \frac{\mu - r}{\sigma}u$, w is a Q-Brownian motion and the price process s^1 can be written

$$
s_u^1 = s_0^1 \exp\left(\left(r - \frac{1}{2}\sigma^2\right)u + \sigma w_u\right).
$$

We shall work exclusively with Q and w in the sequel, hence let $\mathbb{E}$ denote expectation with respect to Q.

An *American put option* on one share with *expiry* T and *strike price* K yields a return process $y_u = (K - s_u^1)^+$ at time $u \in [0, T]$. Let $\tilde{y} = y/s^0$ be the *discounted return process*. The basic properties of the American option are summarised in [13, 15, 21]:

1. The *discounted value* $\tilde{v}$ of the option is given, at time $u \in [0, T]$ by the Snell envelope of the discounted return process, i.e.

$$\tilde{v}_u = \mathrm{ess}\sup_{\tau \in \mathcal{T}_{u,T}} \mathbb{E}[\tilde{y}_\tau \mid \mathcal{B}_u],$$

where $\mathcal{T}_{a,b} = \mathcal{T}_{a,b}^{\mathbb{B}}$ denotes the set of all $\mathbb{B}$-stopping times τ such that $a \le \tau \le b$. The process $\tilde{v}$ is the smallest $\mathbb{B}$-supermartingale dominating $\tilde{y}$.

2. A stopping time $\tau \in \mathcal{T}_{u,T}$ is *optimal* if the supremum is attained using τ, i.e. if $\tilde{v}_u = \mathbb{E}[\tilde{y}_\tau \mid \mathcal{B}_u]$. In the Black–Scholes model there is a *unique* optimal stopping time on $[u, T]$, defined by

$$\tau_u^* = \inf\{\eta \in [u, T] : \tilde{v}_\eta = \tilde{y}_\eta\}$$

and the stopped process $(\tilde{v}_{u \wedge \tau_u^*})_{u \in [0,T]}$ is a martingale.

3. As observed in [13], the predictable representation property of Brownian martingales implies that the process $\tilde{v}$ has continuous paths Q-almost surely.

4. Let $v = s^0 \tilde{v}$ be the *value process* for the option. As the model is Markov, we can write $v_u = p(u, s_u^1)$ where for $x > 0$

$$p(u, x) = \sup_{\tau \in \mathcal{T}_{0,T-u}} \mathbb{E}\left[e^{-r\tau}\left(K - x\exp\left(\left(r - \frac{1}{2}\sigma^2\right)\tau + \sigma w_\tau\right)\right)^+\right].$$

The *auxiliary function* p is decreasing in u and decreasing and convex in x. For each $u \in [0, T[$, the *critical price* (or *optimal stopping boundary*) is defined by

$$s^c(u) = \sup\{x > 0 : p(u, x) = K - x\}.$$

5. The function s^c is C^∞, nondecreasing and $\lim_{u \to T} s^c(u) = K$.

6. Write $\tau^* = \tau_0^*$. On the set $\{\tau^* < T\}$, $\tau^* = \inf\{u \in [0, T] : s_u^1 \le s^c(u)\}$, and if $s_0^1 > s^c(0)$, then the support of τ^* is the whole of $[0, T]$.

Invariance under change of filtration

In the proof of [15, Theorem 2.3] it is stated that the uniqueness of the optimal stopping time τ^* is independent of the filtration used. We clarify this statement, which is crucial for our main results. Suppose that $\mathbb{F} = \{\mathcal{F}_u\}_{u\in[0,T]}$ is a larger filtration under which w is still a Q-Brownian motion. We can define $\tilde{v}^{\mathbb{F}}$, $v^{\mathbb{F}}$, $\mathcal{T}_{a,b}^{\mathbb{F}}$ and $p^{\mathbb{F}}$ exactly as above, using the filtration $\mathbb{F}$ instead of $\mathbb{B}$. (We suppress the superscript when the natural filtration $\mathbb{B}$ is used.) Since $\mathbb{B} \subset \mathbb{F}$ we obtain immediately that

$$p(u, x) \leq p^{\mathbb{F}}(u, x)$$

as the latter is the supremum over a larger set of stopping times. Hence $v_u = p(u, s_u^1) \leq p^{\mathbb{F}}(u, s_u^1) = v_u^{\mathbb{F}}$, so $\tilde{v}_u \leq \tilde{v}_u^{\mathbb{F}}$. Conversely, any $L^2(\mathbb{B})$-supermartingale x can be represented in the form

$$x_u = x_0 + \int_0^u \theta_\eta dw_\eta - a_u$$

for a unique $\mathbb{B}$-predictable increasing process a and $\mathbb{B}$-predictable square-integrable process θ. The stochastic integral is an $\mathbb{F}$-martingale and a is also $\mathbb{F}$-predictable, so x is also an $\mathbb{F}$-supermartingale. In particular, $\tilde{v}$ is an $\mathbb{F}$-supermartingale dominating $\tilde{y}$, hence it follows that $\tilde{v}_u^{\mathbb{F}} \leq \tilde{v}_u$, so that $\tilde{v}^{\mathbb{F}}$is a version of $\tilde{v}$. Consequently the unique optimal stopping times obtained for the filtrations $\mathbb{B}$ and $\mathbb{F}$ coincide.

3   The American Put Option in the Cox–Ross–Rubinstein Model

The discrete counterpart of the Black–Scholes pricing model is provided by the Cox–Ross–Rubinstein model. The theory of the American put option has a natural counterpart in this setting, as described, for example, in [16]. We include a description of the main features for completeness and to introduce notation.

Fix positive constants T, r, σ, and $\mu \in \mathbb{R}$ as in Section 2. For each $n \in \mathbb{N}$, let $\Delta_n = \frac{T}{n}$ and define the nth finite time line $\mathbb{T}_n = \{0, \Delta_n, 2\Delta_n, \cdots, T\}$. Let $\Omega_n = (\Omega_n, \mathcal{A}^n, P_n, \mathbb{A}^n, S^n, Q_n)$ be the family of Cox–Ross–Rubinstein models as defined in [8]. In other words: $\Omega_n = \{-1, +1\}^{\mathbb{T}_n \backslash \{T\}}$, $\mathcal{A}^n = \mathcal{P}\Omega_n$, P_n is counting probability, and $\mathbb{A}^n = \{\mathcal{A}_t^n\}_{t\in\mathbb{T}_n}$ is the filtration generated by the random walk

$$B^n(\omega, 0) = 0$$
$$B^n(\omega, t + \Delta_n) = B^n(\omega, t) + \omega(t)\sqrt{\Delta_n}$$

for $t < T$ in $\mathbb{T}_n$. The market S^n is defined by:

$$
\begin{aligned}
S_t^{0,n} &= (1 + r\Delta_n)^{t/\Delta_n} \\
S_t^{1,n} &= S_0^1 \prod_{\eta < t} (1 + \sigma \Delta B_\eta^n + \mu \Delta_n),
\end{aligned}
$$

where $\Delta X_t^n = X_{t+\Delta_n}^n - X_t^n$ for $t \in \mathbb{T}_n$ and sums and products are taken over $\mathbb{T}_n$. The measure $Q_n \sim P_n$ is the unique measure making the process $S^{1,n}/S^{0,n}$ a martingale. Under Q_n, the process $W_t^n = B_t^n + (\mu - r/\sigma)t$ is a martingale and hence we can write

$$
S_t^{1,n} = S_0^1 \prod_{\eta < t} (1 + \sigma \Delta W_\eta^n + r\Delta_n).
$$

Let $\mathbb{E}_n$ denote expectation with respect to Q_n.

Denote by Y^n the payoff function for an American put option with strike K and expiry T in the model $\boldsymbol{\Omega}_n$. Then $Y_t^n = (K - S_t^{1,n})^+$, and the discounted payoff is given by $\tilde{Y}^n = Y^n/S^{0,n}$. The discounted value $\tilde{V}^n$ of the American put option is given by the discrete Snell envelope, which is defined by the backward recursion

$$
\begin{aligned}
\tilde{V}_T^n &= \tilde{Y}_T^n \\
\tilde{V}_t^n &= \max(\tilde{Y}_t^n, \mathbb{E}_n[\tilde{V}_{t+\Delta_n}^n \mid \mathcal{A}_t^n]),
\end{aligned}
$$

$t \in \mathbb{T}_n$, $t < T$. We can also write

$$
\begin{aligned}
\tilde{V}_t^n &= \sup_{\tau \in T_{t,T}^n} \mathbb{E}_n[\tilde{Y}_\tau \mid \mathcal{A}_t^n] \\
&= \mathbb{E}_n[\tilde{Y}_{\tau_{n,t}^*} \mid \mathcal{A}_t^n],
\end{aligned}
$$

where $T_{a,b}^n$ denotes the set of all $\mathbb{A}^n$-stopping times τ such that $a \le \tau \le b$ and where

$$
\tau_{n,t}^* = \inf\{\eta \in \mathbb{T}_n \cap [t, T] : \tilde{V}_\eta^n = \tilde{Y}_\eta^n\}
$$

is the first optimal stopping time. Let $\tau_n^* = \tau_{n,0}^*$.

Let $V^n = S^{0,n}\tilde{V}^n$ be the *value process*. As the model is Markov, we can write $V_t^n = P^n(t, S_t^{1,n})$ where the *auxiliary function* $P^n(t, x)$ is given by the backward recursion:

$$
\begin{aligned}
P^n(T, x) &= (K - x)^+ \\
P^n(t, x) &= \max\left((K - x)^+, \frac{1}{1 + r\Delta_n} \mathbb{E}_n[P^n(t + \Delta_n, x(1 + \sigma\Delta W_0^n + r\Delta_n))\right)
\end{aligned}
$$

for $t < T$ in $\mathbb{T}_n$, and also by the expression

$$
P^n(t, x) = \sup_{\tau \in T_{0,T-t}^n} \mathbb{E}_n\left[\frac{1}{S_\tau^{0,n}}(K - x\prod_{\eta < t}(1 + \sigma\Delta W_\eta^n + r\Delta_n))^+\right].
$$

The function $(t, z) \longmapsto P^n(t, z)$ is nonincreasing in t, nonincreasing and convex in z, and dominates $(K - z)^+$.

It is easy to see, following [15], that for sufficiently large n and $t \in \mathbb{T} \setminus \{T\}$ there exists $S_n^c(t) \in]0, K[$ such that if $x \leq S_n^c(t)$, then $P^n(t, x) = K - x$, and if $x > S_n^c(t)$, then $P^n(t, x) > K - x$. In fact, in this simple binomial model we have:

$$
\begin{aligned}
P^n(T - \Delta_n, K) &= \frac{K}{1 + r\Delta_n} \mathbb{E}_n[(1 - (1 + \sigma\Delta W_0^n + r\Delta_n))^+] \\
&= \frac{K}{1 + r\Delta_n}(\sigma\sqrt{\Delta_n} - \mu\Delta_n)Q_n\{\omega(0) = -1\}
\end{aligned}
$$

which is positive for large n, since $\sqrt{\Delta_n}$ dominates Δ_n. Since $t \longmapsto P^n(t, K)$ is nonincreasing, we see that $P^n(t, K) > 0$ for all $t < T$ in $\mathbb{T}_n$, hence the existence of S_n^c follows at once.

We call this non-decreasing function $t \longmapsto S_n^c(t)$ the nth *critical price* or *optimal stopping boundary*. The stopping time

$$
\tau_n' = \inf\{t \in \mathbb{T}_n : S_t^{1,n} \leq S_n^c(t)\} \wedge T
$$

belongs to $\mathcal{T}_{0,T}^n$ and clearly majorises τ_n^* on the set $\{\tau_n' < T\}$.

In [15], it is shown that τ_n' is also optimal for $\tilde{Y}^n$, as follows: if $\tau_n'(\omega) > \tau_n^*(\omega) = t$ then $S_t^{1,n}(\omega) > S_n^c(t)$, so that $P^n(t, S_t^{1,n}(\omega)) > K - S_t^{1,n}(\omega)$, while, by definition of τ_n^*, $P^n(t, S_t^{1,n}(\omega)) = (K - S_t^{1,n}(\omega))^+$. It follows that $(K - S_{\tau_n^*}^{1,n})^+ = 0$ a.s. on the set $\{\tau_n' > \tau_n^*\}$. Hence we have, Q_n-a.s.:

$$
\tilde{Y}_{\tau_n'}^n = \frac{1}{S_{\tau_n'}^{0,n}}(K - S_{\tau_n'}^{1,n})^+ \geq \frac{1}{S_{\tau_n^*}^{0,n}}(K - S_{\tau_n^*}^{1,n}) = \tilde{Y}_{\tau_n^*}^n
$$

This shows that τ_n' is optimal and the inequality becomes an identity.

The results of the next section will be used to show that in an appropriate sense *all* objects in the Cox–Ross–Rubinstein theory converge to the appropriate object in the Black–Scholes theory – with the mode of convergence depending on the type of object under consideration. For example, the critical prices S_n^c converge uniformly on $[0, T]$ to s^c (as shown in [15]), whereas the appropriate mode of convergence for random variables, (e.g. optimal stopping times) and processes (e.g. Snell envelopes) is that of D^2-convergence. These results are phrased in the language of liftings.

4 Lifting Theorems for the Hyperfinite Pricing Model

The notion of a *hyperfinite* Cox–Ross–Rubinstein model was first introduced in [7] to unify in a natural way the discrete and continuous theories of option pricing. We briefly review this notion, which uses basic ideas from Nonstandard Analysis and Loeb measure theory. We assume the reader is familiar with these: see [7] for a primer, or [1] for more detail.

Fix an infinite hyperinteger N. We will suppress dependence on N where convenient. So, for real $T > 0$, let $\Delta = \Delta_N t$ and let the hyperfinite time line $\mathbb{T} = \mathbb{T}_N$ be the Nth discrete time line. The Nth Cox–Ross–Rubinstein model $\overline{\Omega} = \Omega_N = (\overline{\Omega}, \mathcal{A}, \overline{P}, \mathbb{A}, S, \overline{Q})$ is defined exactly as in Section 3: The internal probability space $(\overline{\Omega}, \mathcal{A}, \overline{P})$, with $\overline{\Omega} = \Omega_N$, $\overline{P} = P_N$, carries the hyperfinite random walk $B = B^N$, whose standard part is Anderson's Brownian motion (see [1, 7]). The process B drives the internal price process $S = S^N = (S^0, S^1) : \Omega_N \times \mathbb{T} \longrightarrow {}^*\mathbb{R}^2$. Let $\mathcal{A}$ denote the internal algebra of all internal subsets of $\overline{\Omega}$ and let $\mathbb{A} = \mathbb{A}^N = \{\mathcal{A}_t\}_{t\in\mathbb{T}}$ be the filtration generated by B. In [7], the following facts were proved:

1. The unique internal probability measure $\overline{Q} \sim \overline{P}$ under which the discounted price process $\frac{S^1}{S^0}$ is a martingale is given by

$$\overline{Q}\{\omega\} = \prod_{t<T} \frac{1}{2}\left(1 - \frac{\mu - r}{\sigma}\Delta B(\omega, t)\right).$$

   Set $W = W_t^N = B_t + \frac{\mu-r}{\sigma}t$, so that $S_t^1 = S_0^1 \prod_{u<t}(1 + \sigma\Delta W_u^N + r\Delta_N t)$. The standard part w of W is an $L(Q_N)$-Brownian motion on the Loeb space $(\overline{\Omega}, \mathcal{F}, Q) \equiv (\overline{\Omega}, L(\mathcal{A}), L(\overline{Q}))$.

2. The hyperfinite Cox–Ross–Rubinstein market model S^N is an $\mathcal{S}$-continuous $\mathcal{S}L^2(\overline{Q})$-lifting of the Black–Scholes market s. We shall use this representation of w and $(\overline{\Omega}, \mathcal{F}, Q)$ when applying the results outlined in Section 2.

Let $Y_t = (K - S_t^1)^+$ be the return process for an internal American put option. Write $\tilde{Y} = \frac{Y}{S^0}$ for the discounted process, so that $\tilde{Y}$ is an $\mathcal{S}$-continuous $\mathcal{S}L^2(\overline{Q})$-lifting of the payoff function $\tilde{y}$ defined for the Black–Scholes model.

Let $\mathbb{F} = \{\mathcal{F}_u\}_{u\in[0,T]}$ be the associated stochastic filtration, i.e. for each $u \in [0, T]$

$$\mathcal{F}_u = \sigma(\bigcup_{t\approx u} L(\mathcal{A}_t)) \cup \mathcal{N}$$

where $\mathcal{N}$ is the collection of all $L(\overline{Q})$-null sets (see [1]). If $\mathbb{B}$ is the filtration generated by w it is clear that $\mathbb{B} \subset \mathbb{F}$ and w is a $(Q, \mathbb{F})$-Brownian motion, since $Q = L(\overline{Q})$. Hence the earlier remarks about the invariance of the optimal times under changes of filtration apply here.

We use this to show that the optimal stopping time τ_N' is an $\mathcal{S}L^2(\overline{Q})$-lifting of the optimal time τ^* defined for the Black–Scholes model in Section 2. This requires the following general lifting lemma for stopping times with respect to the stochastic filtration $\mathbb{F}$.

Lemma 4.1 *The map $\tau : \overline{\Omega} \longrightarrow [0, T]$ is a $\mathbb{F}$-stopping time if and only if $\tau = {}^{\circ}\rho$ for an $\mathbb{A}$-stopping time $\rho : \overline{\Omega} \longrightarrow \mathbb{T}$.*

Proof. This is a special case of [10, Theorem 4.7]. □

Remark: Since s^1 is a continuous $\mathbb{F}$-adapted process, with $s_0^1(\omega) = s_0^1$ for all ω, the $\mathbb{A}$-adapted $\mathcal{S}$-continuous $SL^2(\overline{Q})$-lifting S^1 satisfies for Q-almost all $\omega \in \Omega$:
$$\forall t \in \mathbb{T} \left(s^1(\omega, {}^{\circ}t) = {}^{\circ}S^1(\omega, t) \right)$$
Hence if an $\mathbb{A}$-stopping time ρ lifts a bounded $\mathbb{F}$-stopping time τ, it follows that S_ρ^1 is an $SL^2(\overline{Q})$-lifting of s_τ^1 and similarly for the discounted return process $\tilde{Y}$ and $\tilde{y}$.

Now since $\mathbb{B} \subset \mathbb{F}$, we can apply Lemma 4.1 to the optimal stopping time τ^* obtained in the Black–Scholes model. Since τ^* is bounded by T, we can obtain an $\mathbb{A}$-stopping time ρ as an $SL^2(\overline{Q})$-lifting of τ^*, and ρ takes values almost surely in $\mathbb{T} \subset {}^*[0, T]$, hence $\rho \in \mathcal{T}_{0,T}$.

Proposition 4.2 *Any internal optimal stopping time ρ in the hyperfinite Cox–Ross–Rubinstein model is an $SL^2(\overline{Q})$-lifting of the unique optimal stopping time τ^* for the Black–Scholes model.*

Proof. The internal optimal stopping time ρ is bounded and hence is an $SL^2(\overline{Q})$-lifting of an $\mathbb{F}$-stopping time $\tau \in \mathcal{T}_{0,T}$, so that $\tau = {}^{\circ}\rho$ Q-a.s.. Since the stopping time τ^* defined in Section 2 is also optimal in the set of $\mathbb{F}$-stopping times, i.e. optimal in the family $\mathcal{T}_{0,T}^{\mathbb{F}}$,
$$\mathbb{E}[\tilde{y}_{\tau^*}] \geq \mathbb{E}[\tilde{y}_\tau].$$

On the other hand, the optimal $\mathbb{B}$-stopping time τ^*, has a lifting $\hat{\rho} \in \mathcal{T}_{0,T}^N$ by the above Lemma. Since $\tilde{Y}$ is an $SL^2(Q)$-lifting of $\tilde{y}$, $\tilde{Y}_{\hat{\rho}}$ lifts $\tilde{y}_{\tau^*}$, and since ρ is optimal in $\mathcal{T}_{0,T}^N$ by hypothesis, it follows that
$$\mathbb{E}[\tilde{y}_\tau] = {}^{\circ}\overline{\mathbb{E}}[\tilde{Y}_\rho] \geq {}^{\circ}\overline{\mathbb{E}}[\tilde{Y}_{\hat{\rho}}] = \mathbb{E}[\tilde{y}_{\tau^*}],$$

using the remark following Lemma 4.1. Hence, τ is optimal for the Black–Scholes model. Since the optimal time in the Black–Scholes model is unique, it follows that $\tau = \tau^*$ a.s. and so ρ is an $SL^2(\overline{Q})$-lifting of τ^*. □

The price process s^1 is Markov, so the behaviour of s^1 over the time interval $[u, T]$ depends on the initial value s_u^1 and not on the previous history of the process. By homogeneity, the properties from time 0 will carry over to properties starting at time u. Thus there is a unique stopping time $\tau_u^* \in \mathcal{T}_{u,T}$ which optimises

$$\tilde{v}_u = \text{ess} \sup_{\tau \in \mathcal{T}_{u,T}} \mathbb{E}[\tilde{y}_\tau \mid \mathcal{F}_u] = \mathbb{E}[\tilde{y}_{\tau_u^*} \mid \mathcal{F}_u].$$

Similar remarks will apply to the internal price process S^1 : properties described above for the process starting at s_0^1 at time 0 will carry over to properties for the corresponding process starting at S_t^1 at time $t \in \mathbb{T}$.

Extending the functions P and S^c to the whole of $^*[0,T]$ by filling in linearly between the points of $\mathbb{T}$ we obtain:

Theorem 4.3 *The function* $P : {}^*[0,T] \times {}^*\mathbb{R}^{>0} \longrightarrow {}^*\mathbb{R}$ *is* S-*continuous and its standard part is the function* $p : [0,T] \times \mathbb{R}^{>0} \longrightarrow \mathbb{R}$ *defined for the Black–Scholes model, i.e.* $^\circ P(t,z) = p(^\circ t, ^\circ z)$ *for all* $t \in^* [0,T]$ *and all finite* z *in* $^*\mathbb{R}^{>0}$.

Proof: Suppose that $u = {}^\circ t$ and $x = {}^\circ z$ for t, z as stated. We need to show that $P(t,z) \approx p(u,x)$. The optimal stopping time τ_u^* has a lifting $\hat{\rho} \in \mathcal{T}_{0,T-t_0}^N$ for some $t_0 \in \mathbb{T}$ with $t_0 \approx u$. Truncating $\hat{\rho}$ at $T - t$ (if needed) produces a stopping time $\rho \in \mathcal{T}_{T-t}^N$ for which

$$\overline{\mathbb{E}}\left[\frac{1}{S_\rho^0}\left(K - z\prod_{\eta<\rho}(1 + \sigma\Delta W_\eta + r\Delta)\right)^+\right] \approx$$

$$\overline{\mathbb{E}}\left[\frac{1}{S_{\hat{\rho}}^0}\left(K - z\prod_{\eta<\hat{\rho}}(1 + \sigma\Delta W_\eta + r\Delta)\right)^+\right],$$

since S^1 is $\mathcal{S}L^2$ and by the remark following Lemma 4.1. It follows that

$$
\begin{aligned}
{}^\circ P(t,z) &\geq {}^\circ\overline{\mathbb{E}}\left[\frac{1}{S_\rho^0}\left(K - z\prod_{\eta<\rho}(1 + \sigma\Delta W_\eta + r\Delta)\right)^+\right] \\
&= \mathbb{E}\left[\left(K - x\exp\left(\left(r - \frac{1}{2}\sigma^2\right)\tau_u^* + \sigma w_{\tau_u^*}\right)\right)^+\right] \\
&= p(u,x).
\end{aligned}
$$

Hence $^\circ P(t,z) \geq p(u,x)$ whenever $t \approx u$ and $z \approx x$.

The internal optimal stopping time ρ_t^* has as its standard part some $\mathbb{F}$-stopping time $\tau_u \in \mathcal{T}_{0,T-u}$. We then have, with t, z as above

$$
\begin{aligned}
{}^\circ P(t,z) &= {}^\circ\overline{\mathbb{E}}_N\left[\frac{1}{S_{\rho_t^*}^0}\left(K - z\prod_{\eta<\rho_t^*}(1 + \sigma\Delta W_\eta + r\Delta)\right)^+\right] \\
&= \mathbb{E}\left[e^{-r\tau_u}\left(K - \exp\left(\left(r - \frac{1}{2}\sigma^2\right)\tau_u + \sigma w_{\tau_u}\right)\right)^+\right] \\
&\leq p(u,x).
\end{aligned}
$$

This completes the proof. $\qquad\qquad\qquad\qquad\qquad\qquad\qquad\qquad\qquad\square$

Lemma 4.4 *The Snell envelope of the hyperfinite discounted return process $\tilde{Y}$ is an $\mathbb{A}$-adapted S-continuous $SL^2(\overline{Q})$-lifting of the Snell envelope of the discounted return process $\tilde{y}$ for the Black–Scholes price.*

Proof: Note that $S_t^0 \tilde{V}_t = V_t = P(t, S_t^1) \approx p(u, s_u^1) = v_u = s_u^0 \tilde{v}_u$ when $u = {}^\circ t$ and $s_u^1 = {}^\circ S_t^1$. Since the function is uniformly bounded by K, we have $\max_{t \in \mathbb{T}} |\tilde{V}_t| \le K$, hence $\tilde{V}$ is $SL^2(\overline{Q})$. $\square$

We have a result similar to Theorem 4.3 for the critical prices in the Black–Scholes and the hyperfinite Cox–Ross–Rubinstein models:

Theorem 4.5 *The critical price function S^c defined in the hyperfinite Cox–Ross–Rubinstein model is S-continuous and its standard part is the critical price function s^c in the Black–Scholes model.*

Proof. Let $t \in \mathbb{T}$ and let $z = S^c(t)$. Then $P(t, x) = K - x$ for all $x \le z$, and so by Theorem 4.3, $p({}^\circ t, \xi) = K - \xi$ for all $\xi \le {}^\circ z$. Thus by definition of s^c, we have $s^c({}^\circ t) \ge {}^\circ z = {}^\circ S^c(t)$.

To prove the reverse inequality, fix $t \in \mathbb{T}$ and let $u = {}^\circ t$. Recall that by [15, Lemma 1.2], the support of τ^* is the whole of $[0, T]$, so that for all $m \in \mathbb{N}$ the set

$$A_m = \left\{ \tau^* \in \left] u - \frac{1}{m}, u - \frac{1}{2m} \right[\right\}$$

is non-null. The following subsets of Ω have full Q-measure:

$$
\begin{aligned}
B &= \{\tau^* = {}^\circ \tau_N^*\} \\
C &= \{\forall t \in \mathbb{T}({}^\circ S_t^1 = s_{{}^\circ t}^1)\} \\
D &= \{s_{\tau^*}^1 = s^c(\tau^*)\}.
\end{aligned}
$$

Choose $\omega_m \in A_m \cap B \cap C \cap D$ and let $t_m = \tau_N'(\omega_m)$, so that $t - \frac{1}{m} < t_m < t - \frac{1}{2m}$. Then

$$
\begin{aligned}
S^c(t_m) &\ge S_{t_m}^1 \text{ by definition of } \tau_N' \\
&\approx s_{{}^\circ t_m}^1 \\
&= s_{\tau^*(\omega_m)}^1 \text{ since } \omega_m \in B \\
&= s^c(\tau^*(\omega_m)) \text{ since } \omega_m \in D.
\end{aligned}
$$

Thus, s^c then ensures that $S^c(t_m) \ge {}^* s^c(t_m) - \frac{1}{m}$ for each finite m. This is an internal statement, so by overflow it holds for some infinite M with $t \approx t_M < t$. As S^c is nondecreasing, we have ${}^\circ S^c(t) \ge {}^\circ S^c(t_M) \ge {}^{\circ *} s^c(t_M) = s^c({}^\circ t)$ by the continuity of s^c. This completes the proof. $\square$

5 Convergence

We now validate the claim made earlier that the theory of the American put in the Cox–Ross–Rubinstein model converges to that in the Black–Scholes model. First we consider the deterministic functions P_n and S_n^c, again filling in linearly to define them for $t \in [0, T]$ and $x \in \mathbb{R}^{>0}$.

Theorem 5.1

1. $P_n \to p$ *uniformly on compact sets*

2. $S_n^c \to s^c$ *uniformly on $[0, T]$.*

Proof:

1. The nonstandard characterisation of uniform convergence on compact sets is that, for any infinite N, $P_N(t, z) \approx p({}^\circ t, {}^\circ z)$ for all $t \in^* [0, T]$ and finite z, which is precisely what we showed in Theorem 4.3.

2. This follows similarly from Theorem 4.5. □

For convergence of stochastic entities such as optimal stopping times and the Snell envelope, we use D^2-convergence as in [8, 9]. We recall the idea briefly. Let $\{\Psi_n\}_{n \in \mathbb{N}}$ be a sequence of random variables each on Ω_n, $n \in \mathbb{N}$ and ψ a random variable on Ω. Assume that ψ is $\mathbb{B}$-measurable. We may regard Ψ_n and ψ as functions of the paths B^n and b respectively, which we indicate by writing $\Psi_n(B^n)$ and $\psi(b)$. Then we have

Definition 5.2 Ψ_n *is D^2-convergent to ψ if any of the following equivalent conditions holds:*

(a) $(\Psi_n(B^n), B^n) \to (\psi(b), b)$ *weakly and $\mathbb{E}_n(\Psi_n^2) \to \mathbb{E}(\psi^2)$.*

(b) Ψ_N *is an $SL^2(Q_N)$-lifting of ψ for all infinite N.*

(c) $\Psi_n(d_n(b)) \to \psi(b)$ *in $L^2(Q)$-norm, where (d_n) is an adapted Q-discretisation scheme.*

To define the idea of a discretisation scheme we identify Ω_n with the path space $\mathcal{C}_n$ of paths $B^n(\omega, \cdot)$ $\omega \in \Omega_n$ with points joined polygonally and Ω with $\mathcal{C} = \{f \in C[0, T] : f(0) = 0\}$ under Wiener measure. Then an *adapted Q-discretisation scheme* is a sequence of measurable maps $d_n : \mathcal{C} \mapsto \mathcal{C}_n$ satisfying the conditions:

(i) d_n is adapted; i.e for each $t \in \mathbb{T}_n$ $d_n(\cdot)(t)$ is $\mathcal{A}_t$-measurable;

(ii) d_n is measure-preserving; i.e. for each $X \in \mathcal{C}_n$

$$Q(d_n^{-1}(X)) = Q_n(\{X\});$$

(iii) $d_n(b) \to b$ in Q-probability; i.e. $\forall \epsilon > 0, Q(|d_n(b) - b| < \epsilon) \to 1$ as $n \to \infty$. (Here $|\cdot|$ denotes the sup norm in $\mathcal{C}$).

It was shown in [8] how such schemes may be constructed and that (a)-(c) of Definition 5.2 are all equivalent. It was also shown in [9] that the notion of D^2-convergence extends naturally to random variables with values in a normed space M – here we will apply this to the case where $M = C[0, T]$ (with the uniform norm). From the previous section we have:

Theorem 5.3 *Let (τ_n) be a sequence of optimal stopping times in the Cox–Ross–Rubinstein models $\boldsymbol{\Omega}_n$. Then (τ_n) D^2-converges to the unique optimal stopping time τ^* for the Black–Scholes model.*

Proof: Take any infinite N; then ρ_N is optimal in the Nth Cox–Ross–Rubinstein model and is, by Proposition 4.2, an $\mathcal{SL}^2(Q_N)$-lifting of τ^*.   □

Turning to Snell envelopes, we can again fill in linearly in the discrete models (as we did for P_n) and we may therefore regard $\tilde{V}^n$ and $\tilde{v}$ as random variables with values in $C[0, T]$. Then we have:

Theorem 5.4 *$(\tilde{V}^n)$ D^2-converges to $\tilde{v}$ (as $C[0, T]$-valued random variables).*

Proof: From Lemma 4.4, we know that for any infinite N, $\tilde{V}^N$ is a uniform lifting of $\tilde{v}$, that is, for Q_N-a.a. $\omega \in \Omega_N$, $\tilde{V}^N(\cdot, B^N(\omega)) \approx \tilde{v}(\cdot, {}^\circ B^N(\omega))$, where $\approx$ denotes the uniform norm of the differences between the quantities is infinitesimal. It is again clear that $\tilde{V}^N(B^N)$ is $\mathcal{SL}^2(Q_N)$: in fact, since the function P_N is by definition uniformly bounded by K, the same holds for the internal Snell envelope, i.e. $\max_{t \in \mathbb{T}} |\tilde{V}^N(t, \cdot)| \leq K$. This completes the proof.
□

References

[1] Albeverio, S., Fenstad, J.E., Høegh-Krohn, R., Lindstrøm, T. *Nonstandard Methods in Stochastic Analysis and Mathematical Physics*, Academic Press (Orlando) (1986).

[2] Aldous, D. *Weak Convergence and the General Theory of Processes*, University of California, Berkeley monograph (1981).

[3] Anderson, R.M. 'Star-Finite Representations of Measure Spaces', *Trans. Amer. Math. Soc.* **271**, 667–687 (1982).

[4] Anderson, R.M., Rashid, S. 'A Nonstandard Characterisation of weak convergence', *Proc. Amer. Math. Soc.* **69**, 327–332 (1978).

[5] Bensoussan, A. 'On the Theory of Option Pricing', *Acta Applicandæ Mathematicæ* **2**, 139–158 (1984).

[6] Bensoussan, A., Lions, J.L. *Applications of Variational Inequalities in Stochastic Control*, North Holland (New York) (1982).

[7] Cutland, N.J., Kopp, P.E., Willinger, W. 'A Nonstandard Approach to Option Pricing', *Mathematical Finance* **1**, 1–38 (1991).

[8] Cutland, N.J., Kopp, P.E., Willinger, W. 'From Discrete to Continuous Financial Models: New Convergence Results for Option pricing', *Mathematical Finance* **3**, 101–123 (1993).

[9] Cutland, N.J., Kopp, P.E., Willinger, W. 'From Discrete to Continuous Stochastic Calculus', *Stochastics* **52**, 173–192 (1995).

[10] Hoover, D.N., Perkins, E. 'Nonstandard Construction of the Stochastic Integral and Applications to Stochastic Differential Equations I and II', *Trans. Amer. Math. Soc.* **275**, 1–58 (1983).

[11] Jaillet, P., Lamberton, D., Lapeyre, B. 'Variational Inequalities and the Pricing of American Options', *Acta Applicandæ Mathematicæ* **21**, 263–289 (1990).

[12] Karatzas, I. 'On the pricing of American Options', *Appl. Math. Optimization* **17**, 37–60 (1988).

[13] Karatzas, I. 'Optimization Problems in the Theory of Continuous Trading', *Siam J. Control and Optimization* **27**, 1221–1259 (1989).

[14] Kushner, H.J. *Probability Methods for Approximations in Stochastic Control and for Elliptic Equations*, Academic Press (New York) (1977).

[15] Lamberton, D. 'Convergence of the Critical Price in the Approximation of American Options', *Mathematical Finance* **3**, 179–190 (1993).

[16] Lamberton, D., Lapeyre, B. *Introduction au Calcul Stochastique Appliqué à la Finance*, SMAI (1991).

[17] Lamberton, D., Pagès, G. 'Sur l'Approximation des Réduites', *Ann. Inst. H. Poincaré* **26**, 331–355 (1990).

[18] Loeb, P.A. 'Weak Limits of Measures and the Standard Part Map', *Proc. Amer. Math. Soc.* **77**, 128–135 (1979).

[19] McKean, Jr, H.P. 'Appendix: A Free Boundary Problem for the Heat Equation Arising from a problem in Mathematical Economics', *Industrial Management Review,* **6**, 32–39 (1965).

[20] Merton, R.C. 'Theory of Rational Option Pricing', *Bell J. Econom. Management Sci.* **4**, 141–183 (1973).

[21] Myneni, R. 'The Pricing of the American Option', *Ann. of Appl. Probability* **2**, 1–23 (1992).

[22] van Moerbeke, P.L.J. 'On Optimal Stopping and Free Boundary Problems', *Arch. Rational Mech. Anal.* **60**, 101–148 (1976).

Filtering Derivative Security Valuations from Market Prices

Robert J. Elliott, Charles H. Lahaie, and Dilip B. Madan

Abstract

Option pricing models are viewed as evaluating the theoretical value of an asset given by its expected discounted cash flow under a martingale measure. Observed market prices are seen however as inclusive of model error, possible mispricing and errors of observation. Both stock and option prices are modelled for positivity as being equal to their theoretical values times a unit mean random term that accounts for the price deviations from their theoretical values. The construction of theoretical valuations, contingent on the observation of a particular price history of primary and derivative assets, is then viewed as a nonlinear filtering problem in the context of a Hidden Markov Model describing the evolution of market fundamentals. By way of an example, a filtering based modification of the Black–Scholes option pricing model is developed and implemented for the valuation of $S\&P500$ index options. It is observed that filtering the underlying has the potential to improve pricing quality and particularly to alleviate the well known smile effect with respect to strike in the unfiltered application of Black–Scholes. The filtering method is applicable to, and can potentially improve, the application of all pricing models.

Introduction

Derivative securities are often valued by arbitrage. Such valuation procedures are based on demonstrating that one may replicate the payoff to the derivative security by the payoff to a trading strategy that invests an initial amount in a set of replicating assets and subsequently strategically rebalances this portfolio. If arbitrage opportunities are absent, the market value of the derivative must equal the initial replication cost. Typically, under Markovian assumptions governing the evolution of market prices for the replicating assets, this initial replication cost is a nonlinear function of the current market prices of these replicating assets. This function is the derivative pricing formula. The most famous example of such a formula is the Black–Scholes (1973) formula for the price of a call or put option on a stock.

141

Standard applications of such formulas derive valuations for derivative securities by first estimating the parameters of the pricing function and then substituting the observed market prices for the prices of the primary replicating assets. Assuming the validity of the replicating strategy, the derivative asset must then in the absence of arbitrage opportunities trade at the formula price. The difference between the market price of the derivative asset and the formula price, if any, is then viewed as indicative of under or over pricing of the derivative asset.

This practice implicitly supposes that the prices of the primary replicating assets are correct and, in particular, observed without error, so that any discrepancies may be viewed as representing mispricing in the derivative markets. There is, however, evidence (French and Roll (1986)) that part of the volatility of stock returns is due to security mispricing. Further, there is now evidence supporting the view that derivative markets may lead in the price discovery process and that, perhaps due to the leverage advantages of derivative trading, information flows at first to these markets (Skinner (1990), Chan (1992)). The potential slowness in certain markets to adjust to changes in market fundamentals we term market mispricing.

At a theoretical level, Jarrow and Madan (1995) have recently shown that the absence of arbitrage does not necessarily imply equality of market prices with the expected discounted cash flow obtained under a martingale measure, for economies with either an infinite sample space or infinitely many trading dates. The prices given by option pricing formulas, and this includes the stock price as well, viewed as a zero strike call option, are however precisely such expected discounted cash flow computations. The difference between arbitrage free market prices and theoretical prices determined by pricing models is shown in Jarrow and Madan (1995) to be due to the market valuation of asymptotic cash flow positions in the time or state direction that they term rational bubbles.

A related view of the relationship between market prices and the valuations obtained from option pricing models is adopted in Jacquier and Jarrow (1995). Jacquier and Jarrow (1995) identify in addition to observation error or measurement error, the error induced by parameter uncertainty as explicitly modelled in Lo (1986), and model error that is also recognized by Bossaerts and Hillion (1993) in the form of GMM overidentifying restrictions. Jacquier and Jarrow (1995) allow for model extensions that are linear in current information set variables to cope with possible model misspecifications. We follow Bossaerts and Hillion (1994) and use these overidentifying restrictions as a diagnostic on the quality of the models pricing abilities.

There are then in all three sources for deviations between market prices and model prices. These are **errors of observation, model errors** be they due to the presence of rational bubbles, parameter estimation or the consequence of approximation strategies adopted in modelling, and **market**

mispricing due to possibly slow speeds of market adjustment in certain markets to changes in market fundamentals.

These theoretical and empirical observations motivate the working hypothesis of this chapter: all observed market prices of assets, be they primary replicating assets or derivatives thereof, are subject to deviations from theoretical valuations based on expected discounted cash flow computations. These deviations could be due to the effects of model error, mispricing or observational error. We view these components of observed prices as making them relatively noisy. Equivalently, we suppose that the part of value or price that corresponds to the model computation of expected discounted cash flows, that we henceforth refer to as the theoretical value, is never observed.

Consider from this perspective the classical results on the Black–Scholes (1973) option pricing formula. Theoretical stock prices are supposed to follow a geometric Brownian motion, and the dynamic arbitrage argument establishes that the theoretical call and put option prices obey the Black–Scholes formula. In the terminology of the theory of filtering, the geometric Brownian motion model for the theoretical stock price is a Hidden Markov process, while both the market stock price and call and put option price processes are relatively noisy processes. We know that the theoretical option prices are predetermined nonlinear functions of the theoretical stock price that constitutes a hidden or unobserved Markov model. To determine the theoretical value of these options we need to construct the expectation of these predetermined nonlinear functions of the theoretical stock price, conditional on all relatively noisy observations to date on stock, call and put option prices. If attention was restricted to just the stock price as opposed to option prices then one could avoid issues of nonlinearity and follow Bassett, France and Pliska (1991) in their use of Kalman filtering to infer the cash price of stocks that were not traded and the MMI index, from the prices of securities that were transacted during the crash of 1987.

The construction of the conditional expectation of a function of a Hidden Markov Model (HMM), by recursive updating coupled with the simultaneous updating of any parameters in these nonlinear functions, is precisely the filtering and estimation problem studied and solved in Elliott, Aggoun and Moore (1995). The objective of this article is to illustrate the application of these HMM filtering and estimation methods to the simultaneous theoretical valuation of primary and derivative assets. The illustration is conducted in the context of the classical Black–Scholes option pricing formula with empirical applications to stock and stock index options.

We expect that part of the empirical problems of Black–Scholes mispricing, for example the smile effect, are possibly due to the lack of recognition of the possibility of noise in the observed underlying asset price. The intuition underlying this conjecture can be explained by considering various strategies for inferring the unobservable theoretical price of the stock. One approach is to use the observed stock price as a proxy for the theoretical price. An alternative is the indirect inference of this theoretical value from the prices of at the money options. With the first approach the stock is priced without error while with the second the at the money option is priced without error. In both cases securities other than the one used as a proxy for the underlying theoretical price, are priced with errors that may tend to get larger as one moves away in payoff structure from the security used as the proxy for the theoretical price. Such error patterns can be compared to the smile effects in Black–Scholes option pricing. Now when we filter the theoretical price instead of proxying it, we essentially use all option prices as an indirect proxy for determining the theoretical price. This procedure makes it difficult to move away in payoff structure from the set of securities used collectively as a proxy for the theoretical price and we therefore expect a more balanced error structure, possibly free of smiles.

We could also have postulated a more complex stochastic volatility framework in which the stock price is observable but volatility follows a square root law as in for example Heston (1993). In this case we would end up constructing filtered prices as mixtures of Black–Scholes prices over a distribution on the volatility, a structure reminiscent of jump diffusion models, like Madan and Chang (1995), that are known to capture smiles. The essential improvement however comes from the mixing involved, and not from what is being mixed.

We illustrate the application of these methodologies on data for $SandP500$ futures options for the year 1992. Filtered option prices are constructed from market price observations on two extreme parameter settings for the volatility of the underlying asset. Graphs of the evolution over the year of the conditional density of the underlying asset price, conditional on option price observations to date, are presented. The quality of filtered option prices is then compared with that of the unfiltered prices by investigating the pricing errors of the two systems with respect to orthogonality to the current information set by employing a generalized method of moments (GMM) test statistic on overidentifying restrictions. It is observed in this GMM framework that unfiltered Black–Scholes prices exhibit the well known strike related smile effect. It is observed that this effect tends to be corrected by the use of filtered option prices.

The models estimated are mixtures of models obtained by filtering prices at unoptimized extreme parameter settings. In the presence of unobservable variables parameter estimation by standard maximum likelihood methods becomes computationally intractable and for this context Dempster, Laird and Rubin (1977) have proposed the EM algorithm that reverses the role of the expectation, E, and maximization, M, operators implicit in maximum likelihood estimation. Future research will incorporate the EM algorithm and exploit more fully the information contained in current market prices by extending the updating of the conditional density of the underlying asset price to simultaneously learning about all the parameter values as well. The updating of the conditional density for the parameters may be usefully compared with the approach taken in Lo (1986) and Jacquier and Jarrow (1995). Lo (1986) uses the sampling distribution of volatility from maximum likelihood estimates of volatility to approximate the standard deviation of the option price estimator and assumes that this is normally distributed. Jacquier and Jarrow (1995) infer the implied distribution of the option price estimate by using Monte Carlo methods to determine the Bayesian posterior distribution of the volatility parameter. The EM algorithm also constructs Bayesian posterior distributions, but this is done by an analytical periodic recursive updating instead.

Section 1 summarizes the general results of the HMM theory as they pertain to the derivative asset pricing problem. The specific filtering based pricing model for the Black–Scholes context is developed in Section 2. Filtering results on pricing stock and stock index options are presented and analysed in Section 3. Section 4 concludes.

1   Filtering Functionals of Hidden Markov Models

The theoretical price of a derivative asset, like a European call or put option price, is in many specific instances a given non-linear function of the theoretical price of the underlying asset. Furthermore, the underlying asset's theoretical price process is, in many applications, a Markov process in discrete or continuous time. The perspective of this paper regards these theoretical prices as unobservable and the asset valuation problem is then viewed as one of inferring the conditional expectation of these theoretical unobserved prices from observations on market prices that are viewed as relatively noisy. We envisage the construction of these conditional expectations or theoretical valuations to take place at regular discrete time points, as for example at the end of each trading day, using as inputs the market closing prices for a variety of securities actively traded during the day. Since actual asset prices are positive with values on the positive half line, we wish to consider the filtering problem in a discrete time continuous state context. In this regard we follow Chapter 5 on the General Recursive Filter in Elliott, Aggoun and Moore (1995). An alternative strategy is to discretize the state space of asset prices into tick sizes and then formulate the model performing integrations of the theoretical transition densities over various intervals to infer a Markov chain representation with some boundary approximations to ensure it is finite state. This approach leads us directly and immediately into issues of numerical approximation of continuous state processes that we avoid here by employing the continuous state model.

For purposes of summarizing the general results, let the time index be the set $\mathbb{N}$ of natural numbers and let $(\Omega, \mathcal{F}, P)$ be a probability space on which all processes considered are defined. The hidden Markov model is defined as a d-dimensional discrete time stochastic process $\{x_t\}, t \in \mathbb{N}$, taking values in $\mathbb{R}^d$. The dynamics of the hidden Markov process are given by a non-linear transition function and a sequence $\{v_t\}$, $t \in \mathbb{N}$, of $\mathbb{R}^d$-valued independent random variables with density $\psi_t(v)$ for the variable v_t. We suppose the existence of a measurable transition function

$$a : \mathbb{R}^d \times \mathbb{R}^d \to \mathbb{R}^d$$

such that

$$x_{t+1} = a(x_t, v_{t+1}) \tag{1}$$

The example of a hidden Markov model that we investigate in the section is the discretized form of one dimensional Geometric Brownian motion model for a stock index.

We suppose that the transition function in (1) has an inverse of the form

$$v_{t+1} = d(x_{t+1}, x_t) \tag{2}$$

where d is differentiable in x_{t+1}. This condition is satisfied for many of the applications envisaged.

In addition to the hidden Markov process, we have at each time point an m-dimensional vector of noisy observations on functions of the process x_t. We denote these observations by $y_t, t \in \mathbb{N}$, where y_t takes values in $\mathbb{R}^m$. The observation noise is supposed to be given by a sequence of m-dimensional independent random variables w_t with density $\phi_t(w)$ for w_t. The relationship between the hidden Markov process and the observations is given by the functions

$$\begin{aligned} c_t : \mathbb{R}^d \times \mathbb{R}^m &\to \mathbb{R}^m \\ y_t = c_t(x_t, w_t) \end{aligned} \tag{3}$$

where the dependence of c on t is allowed for to account for the possibility that we may have noisy observations on different securities on different days. Again we suppose that these relationships are invertible to yield for some function g_t

$$w_t = g_t(y_t, x_t) \tag{4}$$

where g_t is differentiable in y.

Finally we require the matrices of derivatives

$$D(x_{t+1}, x_t) = \left. \frac{\partial d(\xi, x_t)}{\partial \xi} \right|_{\xi = x_{t+1}} \tag{5}$$

and

$$G_t(y_t, x_t) = \left. \frac{\partial g_t(y, w_t)}{\partial y} \right|_{y = y_t} \tag{6}$$

to be nonsingular.

The strategy for constructing conditional expectations is one of recursively updating an unnormalized conditional density, conditional on observations to date with respect to a reference measure, for the position of the hidden Markov process x_t. In fact this unnormalized conditional density, when stored as a discretized vector, serves as a sufficient vector statistic that synthesizes all information from past data needed to keep the updating of theoretical valuations going.

Let this unnormalized conditional density be given by a positive function

$$q_t(x_t; y_1, \cdots, y_t)$$

with respect to a reference measure $\bar{P}$ that is equivalent to P.

The reference measure is chosen to simplify the dynamics of the observation process. In particular, under $\bar{P}$ the process y_t behaves by construction like w_t and is a sequence of independent random variables with density $\phi_t(y)$ for y_t. Expectations under P may be constructed from those under $\bar{P}$ by using the reverse change of measure density process, from that employed to get to $\bar{P}$ from P. The complexities of expectation calculations under P are thereby relegated to or substituted into the change of measure density process. An analogy can be made with the use of reference measures in importance sampling in simulation where one simulates from the wrong or reference density but then accounts for the change by multiplying the object of interest by the density of the true measure with respect to the reference measure.

Explicitly, let expectations under $\bar{P}$ be denoted by $\bar{E}$ and let the $\bar{P}$ martingale $\bar{\Lambda}_t$ be defined by

$$\bar{\Lambda}_t = \bar{E}\left[\frac{dP}{d\bar{P}} \mid \mathcal{F}_t \right] \tag{7}$$

The explicit form of the process $\bar{\Lambda}_t$ is given in Elliott, Aggoun and Moore (1995, page 105).

The unnormalized conditional density q_t is defined so that

$$
\begin{aligned}
E\left[I(x_t \in dz) \mid \mathcal{Y}_t\right] &= \frac{1}{E[\bar{\Lambda}_t | \mathcal{Y}_t]} E\left[\bar{\Lambda}_t I(x_t \in dz) \mid \mathcal{Y}_t\right] \\
&= \frac{1}{E[\bar{\Lambda}_t | \mathcal{Y}_t]} q_t(z; y_1, \cdots, y_t) dz,
\end{aligned}
\tag{8}
$$

where $I(A)$ denotes the indicator of a set A and $\mathcal{Y}_t = \sigma\{y_1, \cdots, y_t\}$ is the σ-field generated by the observations to time t. It follows that the conditional expectation of a function of x_t, say $f(x_t)$, is given by an application of the conditional Bayes Theorem, that is,

$$E\left[f(x_t) \mid \mathcal{Y}_t \right] = \frac{\bar{E}[\bar{\Lambda}_t f(x_t) \mid \mathcal{Y}_t]}{\bar{E}[\bar{\Lambda}_t \mid \mathcal{Y}_t]} = \frac{\int f(z) q_t(z; y_1, \cdots, y_t) dz}{\int q_t(z; y_1, \cdots, y_t) dz}. \tag{9}$$

The particular functions of interest to us are given by pricing functions for the prices of derivative securities when the underlying true asset price process is consistent with the dynamics provided by equation (1). Our derivative valuations will be an application of equation (9) to appropriate choices for f once we have constructed the unnormalized conditional density q_t.

A straightforward generalization of a proof of theorem 4.2 of Elliott, Aggoun and Moore (1995, page 106) provides the procedure for updating recursively the functions q_t. According to this generalization we may write

$$q_{t+1}(\xi; y_1, \cdots, y_{t+1}) = \Delta_1(y_{t+1}, \xi)\Gamma_1(\xi; y_1, \cdots, y_t), \qquad (10)$$

where

$$\Delta_1(y_{t+1}, \xi) = \frac{\phi_{t+1}(g(y_{t+1}, \xi))}{\phi_{t+1}(y_{t+1})} \mid \det(G_{t+1}(y_{t+1}, \xi)) \mid \qquad (10a)$$

and

$$\Gamma_1(\xi; y_1, \cdots, y_t) = \int_{\mathbb{R}^d} \psi_{t+1}(d(\xi, z)) \mid \det(D(\xi, z)) \mid q_t(z; y_1, \cdots, y_t)dz. \quad (10b)$$

We will refer to equation (10a) as defining a calibration factor as it incorporates the impact of the most recent observations on the density of the hidden Markov process. Equation (10b) on the other hand is the theoretical factor and involves the use of the prior unnormalized conditional density updated in accordance with the theoretical model for the evolution of the process. For example, we would be using (10b) even if there were no observations.

A typical valuation exercise, using the methods of filtering Markov models, involves the repeated application of equations (9) and (10), where we use market price information to update q_t periodically in accordance with (10) and then employ the output of this procedure to value derivatives using (9). The next two sections develop and implement a detailed application to stock index options.

2   Black–Scholes Filtering of Stock Index Option Values

The hypothesis on the dynamics of the underlying stock index consistent with the Black–Scholes formula for stock index options is that the index follows a geometric Brownian motion. The volatility rate of this motion is given by a parameter σ^2 and there is also a constant drift, say at rate $\mu - q$, where μ is the mean rate of return on the asset and q is the dividend rate. The discrete time dynamics under the statistical probability law P is therefore given by a one dimensional process $(d = 1)$ and

$$S_{t+1} = S_t e^{(\mu-q)h + \sigma\sqrt{h}v_{t+1} - \frac{\sigma^2 h}{2}} \qquad (11)$$

where v_{t+1} is normally distributed with zero mean and unit variance, and hence has the density $\psi_{t+1}(v) = n(v) = (1/\sqrt{2\pi})e^{-v^2/2}$, and h is the length of the discrete time interval.

The values of derivative assets like options on the stock index are given by functions of the value of the underlying index, with the functional relationship given by the famed Black–Scholes formula. This formula represents the discounted expected value of the option payoff under an equivalent martingale measure Q under which the discrete time dynamics of the underlying index is given by

$$S_{t+1} = S_t e^{(r-q)h + \sigma\sqrt{h}v_{t+1} - \frac{\sigma^2 h}{2}} \tag{12}$$

where r is the supposed constant risk free rate of interest.

An important question to be answered before proceeding is whether we wish to filter conditional expectations under P or Q. The answer is obtained by considering the difference between the measures, which is in the underlying drift. If we filter under Q, the drift in the filtered derivative prices will be r. The actual drift in the market prices of say an option with a zero strike will be $\mu - q$. If the latter is above r then over time our filtered values of all the derivative assets will end being below the quoted market prices and this is clearly not what is intended. Hence we see that the filtering exercise must be conducted under the statistical measure P. The measure Q has been accounted for in the option pricing formula used to price the derivative, for estimating the market price of the call unbiasedly we now have to filter the expected value of the call price under P.

It is also helpful to formulate the hidden Markov model in terms of the logarithm of the stock index value as taking values in $\mathbb{R}^1$. Hence letting $x_t = \ln S_t$ we may write the hidden Markov model for this application as

$$x_{t+1} = a(x_t, v_{t+1}) = x_t + (\mu - q)h + \sigma\sqrt{h}v_{t+1} - \sigma^2 h/2 \tag{13}$$

with inverse function

$$v_{t+1} = d(x_{t+1}, x_t) = \frac{x_{t+1} - x_t - (\mu - q)h + \sigma^2 h/2}{\sigma\sqrt{h}} \tag{14}$$

and Jacobian

$$D(x_{t+1}, x_t) = 1/(\sigma\sqrt{h}). \tag{15}$$

For our observations we take daily closing values on the underlying index and on a collection of actively traded call and put options on the underlying futures index. The options are American, but we will employ European pricing formulas. We attempt to mitigate the problems associated with this difference by considering near money low maturity options. At the end of each trading day we collect market close price information on the index and on $m = 16$ options, 8 calls and 8 puts of maturities of approximately one to two months, with four options in each maturity category having strike prices corresponding to the more actively traded options during the day. Letting i index the observed options we have for each i at the end of each trading day t, a type indicator u_{ti} that is unity for a call and zero for a put, a maturity τ_{ti} and a strike k_{ti} for $i = 1, \ldots, m = 16$. Let U_t denote the observed value of the index at the close of day t and let y_{ti} denote the observed market price of option i at the end of trading day t. Note as commented earlier that we shall view the underlying index as a zero strike call option.

We suppose that U_t and y_{ti} are observed with error ϵ_t and w_{ti} respectively, that are mutually independent as well as being independent across t and i. Though such a hypothesis may be reasonable from the perspective of errors of observations, from the viewpoint of model errors or errors of mispricing one might be inclined to expect some correlation across options and possibly time. At the stage of this preliminary investigation we leave aside these issues of correlation. Specifically we take the observation process to be given by

$$U_t = c(x_t, \epsilon_t) = e^{x_t + \nu_1 \epsilon_t - \nu_1^2/2} \qquad (16a)$$

and

$$y_{ti} = c_t(x_t, w_{ti}) = V(x_t, k_{ti}, \tau_{ti}, u_{ti}, r_t, \sigma, q)e^{\nu_2 w_{ti} - \nu_2^2/2} \qquad (16b)$$

where $V(x, k, \tau, u, r, \sigma, q)$ is the Black–Scholes formula for pricing a stock index futures option, with the option type being embedded in the formula and ν_1, ν_2 are the supposed error volatilities in the index and option pricing models respectively. Note that from the perspective of viewing the stock as a zero strike call option, the variable U_t or the observed index level is just a particular option price and is in this sense just a distinguished observation with both equations (16a) and (16b) jointly corresponding to the system of equations (3) of section 1.

We suppose that the observation errors are lognormally distributed, that is, ϵ_t and w_{ti} are normally distributed with zero mean and unit variance and hence have density $\phi_t(z) = n(z) = (1/\sqrt{2\pi})e^{-z^2/2}$. The variance of the percentage error in the ith option price could be modelled as a function of the parametric features of the contract, specifically the degree of moneyness and maturity. In this preliminary investigation we control our observations for variations in moneyness and maturity and then take this volatility to be constant across contracts.

The value of a futures option is a function of the underlying futures prices. For a futures delivery at time $t + \tau$, the time t price of the underlying futures index $F_{t\tau}$ may be related to the spot price S_t, by the spot index arbitrage relationship

$$F_{t\tau} = e^{(r-q)\tau} S_t. \qquad (17)$$

Substitution of (17) into the Black–Scholes futures index option pricing formula (Stoll and Whaley (1993), page 353) yields the specific form of the function V as given by

$$V(x, k, \tau, u, r, q, \sigma) = (-1)^{(1+u)}[e^{x-q\tau}N(d_1) - ke^{-r\tau}N(d_2)], \qquad (18)$$

where

$$d_1 = (-1)^{(1+u)}\left[\frac{x-\ln k}{\sigma\sqrt{\tau}} + \left(\frac{r-q}{\sigma} + \frac{\sigma}{2} \right)\sqrt{\tau} \right] \qquad (18a)$$

while

$$d_2 = (-1)^{(1+u)}\left[\frac{x-\ln k}{\sigma\sqrt{\tau}} + \left(\frac{r-q}{\sigma} - \frac{\sigma}{2} \right)\sqrt{\tau} \right] \qquad (18b)$$

The observation process equations (16) have an inverse in the required form and these are

$$\epsilon_t = g(U_t, x_t) = \frac{\ln(U_t) - x_t + v_1^2/2}{v_1} \qquad (19a)$$

$$w_{ti} = g_{ti}(y_{ti}, x_t) = \frac{\ln(y_{ti}/V(x_t, k_{ti}, \tau_{ti}, u_{ti}, r_t, q, \sigma)) + v_2^2/2}{v_2}, \qquad (19b)$$

and the Jacobian element for this transformation, the determinant of (6), is

$$\det[G(U_t, y_{ti}, x_t)] = \frac{1}{v_1}\frac{1}{v_2^m}\prod_{i=1}^{m}\frac{1}{y_{ti}}. \qquad (20)$$

The calibration component function of the recursive update equation (10), Δ_1, is given in this case up to a normalizing constant with respect to integration over ξ by

$$\Delta_1(y_{t+1}, \xi) = \\ \exp\left[-\tfrac{1}{2}(g(U_{t+1}, \xi))^2\right] \times \prod_{i=1}^{m}\exp\left[-\tfrac{1}{2}(g_{t+1,i}(y_{t+1,i}, \xi))^2\right], \qquad (21)$$

where the terms $[v_1(v_2^m)y_{t+1,i}]^{-1}$ and the term $\phi_{t+1}(U_{t+1})\phi_{t+1}(y_{t+1,i})$ are absorbed in the normalizing constants and the functions $g, g_{t+1,i}$ are as defined in equations (19a) and (19b), respectively. The theoretical component Γ_1 of the recursive update is given up to a normalizing constant by

$$\Gamma_1(\xi) = \int_{\mathbb{R}}\exp\left[-\tfrac{1}{2}\left(\frac{\xi-z-(\mu-q)h+\sigma^2h/2}{\sigma\sqrt{h}} \right)^2 \right]q_t(z)dz. \qquad (22)$$

We may then define $q_{t+1}(\xi)$ for this application to stock index options by

$$q_{t+1}(\xi) = \Delta_1(y_{t+1}, \xi)\Gamma_1(\xi) \qquad (23)$$

with Δ_1 and Γ_1 now given by (21) and (22). For a specific implementation we need to formulate the form in which the functions q_t and its updates are maintained or stored. Moreover, we need to specify the exact procedure for the evaluation of the integral in (22).

Before taking up these implementation details it is instructive to consider the role of the model and its parameters in the evolution of the conditional densities described by equation (23). For this purpose consider first the case where we believe the data or observation processes are of low quality. This would be the case if the variance of the noise components was large and ν_1 and ν_2 were large. These parameters only enter the term Δ_1, and for large values of ν_1 and ν_2 the functions g and g_t reduce to constants independent of the data. Essential the term Δ_1 is absorbed into the normalizing constant itself and $q_{t+1}(\xi)$ is given $\Gamma_1(\xi)$. The term $\Gamma_1(\xi)$ involves no data and essentially states that ξ is made up of $z + \epsilon$ where, conditional on z, ϵ is normally distributed with mean $(\mu - q)h - \sigma^2 h/2$ and variance $\sigma^2 h$. Since the stock price is a geometric Brownian motion with drift $(\mu - q)$ and volatility rate σ^2, in the absence of any information, we just let the density of ξ, the log of the stock price, evolve on its theoretical path.

Consider now the opposite case where we believe the data are of very high quality and correspondingly the values of ν_1 and ν_2 are small. We may now think of Δ_1 as essentially made up of the product of the negative exponential of the squared log price relative (actual price to model price) raised to the powers of $(1/\nu_1^2)$ and $(1/\nu_2^2)$ respectively for the index and option prices respectively. This raises the weight given in the log of q_{t+1} to the calibration term Δ_1 relative to the theoretical term Γ_1. Furthermore the weight is higher for the prices with greater prior quality or lower noise variances.

For any particular observation price the log of q_{t+1} reaches its highest value at the lowest value for g or g_{ti}, and this occurs at values for ξ that make the price relative (actual to model) unity or their logarithm zero. Hence the greatest weight in the update goes to what may be termed the 'implied index or underlying stock value,' or the index value that makes the option pricing formula exactly correct, given knowledge of σ^2, the underlying variance rate.

For the exact implementation procedure, the support for the function q_t is taken to be a symmetric interval around the current value of the logarithm of the spot index. Hence, the support for q_t is the interval $[L_t, U_t]$, where $L_t = \ln(S_t) - \delta$ and $U_t = \ln(S_t) + \delta$. The values of q_t evaluated at

$$z_i = L_t + 2(i - 1)\delta/K, \tag{24}$$

for $i = 1, \ldots, K + 1$, are stored as a vector of length $K + 1$.

In the normalization of q_t, and its use in valuation and updating, we need to perform integrations of the form,

$$A = \int_{L_t}^{U_t} f(x) q_t(x) dx \tag{25}.$$

These integrations will be performed by adopting Simpson's rule, and letting f_i be the value of f at x_i defined by (24), for $i = 1, \ldots, K + 1$ for K even we approximate A by

$$A \cong \frac{2\delta}{3K} \left[f_1 q_1 + \sum_{i=2}^{K} (3 + (-1)^i) f_i q_i + f_{K+1} q_{K+1} \right]. \tag{26}$$

In particular the integral (22) in the updating of q_t to q_{t+1} is replaced by the summation

$$\hat{\Gamma}_1(\xi) = \frac{2\delta}{3K} \left[\sum_{i=0}^{K+1} w_i q_{ti} \exp\left[-\frac{1}{2} \left(\frac{\xi - z_i - (\mu - q)h + \sigma^2 h/2}{\sigma\sqrt{h}} \right)^2 \right] \right], \qquad (27)$$

where $w_0 = w_{K+1} = 1$ and $w_i = (3 + (-1)^i)$ for $i = 1, \cdots, K$.

The normalized density is constructed from the unnormalized one by dividing by its integral. Letting qn_{ti}, for $i = 1, \ldots, K + 1$, be the value of the normalized density evaluated at z_i, we have that,

$$qn_{ti} = q_{ti}/I_t, \qquad (28)$$

where I_t is defined by

$$I_t = \frac{2\delta}{3K} \left[q_{t1} + \sum_{i=1}^{K} (3 + (-1)^i)q_{ti} + q_{t,K+1} \right]. \qquad (29)$$

The filtering based option pricing model values put and call options on day t using the equation

$$\begin{aligned} C(k, \tau, u, r, \sigma, q) &= \frac{2\delta}{3K} \Big[V(x_1, k, \tau, u, r, \sigma, q)qn_{t1} \\ &+ \sum_{k=2}^{K}(3 + (-1)^k)V(z_i, k, \tau, u, r, \sigma, q)qn_{ti} \qquad (30) \\ &+ V(z_{K+1}, k, \tau, u, r, \sigma, q)qn_{t,K+1} \Big], \end{aligned}$$

where the function V is as defined in equation (18) and equation (30) is to be employed to value an option with strike k, maturity τ, with $u = 1$ for calls and zero otherwise and the interest rate is r, the volatility is σ, and the dividend rate is q.

Equations (21), (23) and (27) define the updating of q_t, equations (28) and (29) define the normalization procedure, and the filtering based option valuation formula is given by equation (30). The results of implementing these equations on data for prices of options on the $S\&P500$ index are presented in the next section.

3   Results of Filtering Based Option Valuation for *S&P*500 Index Options

An investigation of the filtering based option valuation procedure described in section 2 was conducted for the pricing of $S\&P500$ index options. Daily data on closing prices of actively traded futures options on the $S\&P500$ index was obtained from the Financial Futures Institute in Washington D.C. for the year 1992. The maximum number of actively traded options, employed in this study, on any day was 16, while the minimum number was 9. The daily updating of the unnormalized density is based on 9 to 16 observations on puts and calls on the underlying futures index with varying strike prices and maturities around 1, 2 or 3 months. Daily data on the interest rate (3 month Treasury Bill rate) was obtained from the Federal Reserve Bank of Washington D.C. Data on the daily closing levels for the index was obtained from the CRSP tapes. The options were treated as if they were written on the underlying index by employing the spot forward arbitrage relation (17). Estimates of dividend rates were backed out using the spot forward arbitrage relation (17). There were in all 254 trading days in the year and 2824 option prices employed in the filtering exercise.

For an application of the filtering algorithm we need to set the parameters of the model. Ideally we would combine the updating of filtered prices and the updating of parameter values. In this preliminary investigation we filter prices on two extreme parameter settings and then consider modelling the theoretical price as a linear combination or affine mixture of the extreme value filtered prices. For our extreme values we fit the Black–Scholes formula to all the option pricing data for the year 1992. The estimate of σ minimizing the average squares percentage pricing error was .15. For our extreme parameter settings for filtering we chose the values $\sigma = .1$ and $\sigma = .3$. The parameters μ and $\nu_1 = \nu_2 = \nu$ were set at .1 and .2 respectively.

The initial q_t on January 2 1992 was taken to be uniform in the interval [5.95,6.05]. This places the underlying index between 383 and 424, when the observed value was 417.26. This assumption of uniformity in the log is inconsistent with an assumption of uniformity in the level with the latter distribution being asymmetric. However, the asymmetry is slight with the unnormalized density falling from 1/383 to 1/424. In any case, we have observed that the results are not sensitive to this initial distributional assumption.

At every update the interval for the support of the conditional density was set at $\ln(S_t)\pm.05$. Hence the value of δ used was .05. The support of length .1, was partitioned into 100 subintervals for the purposes of storing information on the unnormalized density and for performing integrations with respect to q_t using Simpson's rule. Hence the value of K was 100.

The results of the investigation are presented in two forms. We first present graphs on the evolution of the unnormalized density through time. Next we present comparisons of filtering based option valuations and the straight application of the Black–Scholes formula.

3.1   The Evolution of the Conditional Density of the Index over Time

Figure 1 presents four graphs of the mid month conditional densities for the three months in each of the four quarters of the year at the setting $\sigma = .3$.

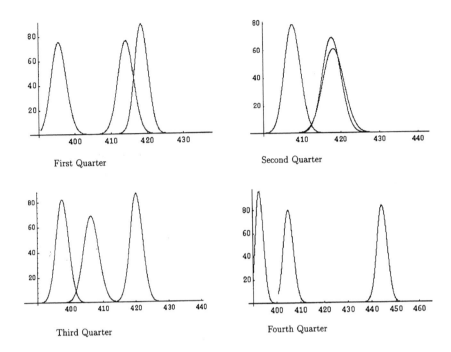

Figure 1. Filtered Mid-Month Densities for the $S\&P500$ Index Conditional on past Closing Index Level and Futures Option Price Observations for 1992

Figure 2 presents a three dimensional plot of the evolution of this density over the last quarter of the year.

The conditional density shifts slightly upward between January and February with a substantial downward movement in March. In May the density is back up at the February levels but down again to March levels in June and July. There is then a slow rise through August and September with a drop in October back to below February levels. November and December see the first substantial increases with the year closing at a peak. Figure 2 shows this revival in the final quarter of 1992. The time axis is labelled to go from .75 to 0.9 while the level of the index ranges from 355 to 445. The vertical axis shows the level of conditional density interpolated at each time point between the preceding and succeeding day. One may notice the upward shift in this density towards the latter half of the last quarter of 1992 that coincided with the possible resolution of the uncertainty surrounding the presidential election of 1992.

The spread of the conditional density also varies over the year. It is the general pattern of movement of the spread that we comment on here as the

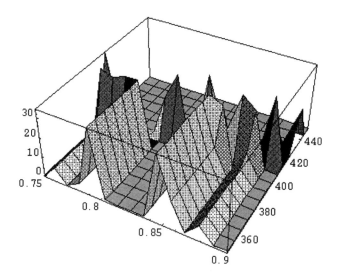

Figure 2. Three Dimensional Plot of the daily movement of the Conditional density for the $S\&P500$ Index Level in the Fourth Quarter of 1992

magnitude of the reflects the unoptimized parameter settings of .2 for ν_1 and ν_2. The spread is much wider in the first half of the year, lower in third quarter and lower still in the last quarter of the year. This observation is also broadly consistent with a pattern of uncertainty resolution connected with the presidential election. At the start of the year there are many candidates, by the third quarter there were in this election three candidates and by November and December we knew the result.

3.2   Comparison of Filtered and Unfiltered Black Scholes Option Pricing Models

For a comparison of the quality of Black–Scholes option prices with those implied by the filtered unconditional densities of the underlying level of the index we first obtained two sets of filtered prices in accordance with equation (30) for our two parameter settings of $\sigma = .1$ and $\sigma = .3$. This gives us two vectors of filtered option prices of length 2,824 for all the options for which we had market closing price information for the year 1992. For comparative purposes we also obtained two sets of Black–Scholes prices using the same σ settings. The Black–Scholes formula was applied in the usual way with the closing value of the index as input price of the underlying asset. We denote by f_i^1 and f_i^2 the filtered price of option i obtained with the parameter setting $\sigma = .1$ and $\sigma = .3$, respectively. The option index runs from 1 to 2824 and the prices for each day are stacked with $t(i)$ denoting the day on which the price of option i was observed. The corresponding Black–Scholes prices are b_i^1 and b_i^2. Let the closing market prices of these options on day $t(i)$ be p_i.

Consider the model formulation that theoretical prices are a linear combination or mixture of the filtered or Black–Scholes prices. For example, if the two parameter settings had a probability of q and $1 - q$ respectively then we would expect the Black–Scholes price to be $qb_i^1 + (1 - q)b_i^2$ and the filtered price to be $qf_i^1 + (1 - q)f_i^2$. One may then define the Black–Scholes and filtered pricing errors by

$$u_i = p_i - a_1 b_i^1 - a_2 b_i^2 \tag{31}$$

and

$$v_i = p_i - \alpha_1 f_i^1 - \alpha_2 f_i^2 \tag{32}$$

where we expect, due to the hypothesis that the theoretical prices are a probabilistic mixture of the parameter setting, that a_1, a_2, α_1 and α_2 are positive with $a_1 + a_2 = \alpha_1 + \alpha_2 = 1$.

For the true model, the pricing errors u_i, v_i by virtue of the definition of the true model, should be orthogonal to the information set just prior to $t(i)$ when the prices p_i are determined. If $\mathcal{F}_t$ denotes the information set at time t then for the true model we must, in particular, have that

$$E[u_i \mid \mathcal{F}_{t(i)-1}] = 0 \tag{33}$$

and

$$E[v_i \mid \mathcal{F}_{t(i)-1}] = 0. \tag{34}$$

If these conditions are not satisfied then one may have model prices that exhibit smiles or a strike bias, or there may be a maturity bias. More generally, there may be other information available at the time of pricing that may be used to reduce pricing errors and hence violate the hypothesis that the theoretical model provides us a with an unbiased conditional expectation of the asset's price.

The conditions (33) and (34) may be used to estimate the parameters a_1, a_2, α_1 and α_2 of equations (31) and (32) by the Generalized Method of Moments Estimator of Hansen (1982). Equations (33) and (34) are referred to as orthogonalities as they essentially assert that u_i and v_i are orthogonal to the information set $\mathcal{F}_{t(i)-1}$. To observe this, let y_{ij} for $j = 1, \ldots, q$ be a choice of instruments from the set $\mathcal{F}_{t(i)-1}$. Note on applying the law of iterated expectations, to (33) and (34) respectively conditioning first on $\mathcal{F}_{t(i)-1}$, that we must have

$$E[y_{ij} u_i] = 0, \tag{35}$$

and

$$E[y_{ij}v_i] = 0. \tag{36}$$

The instruments we employed were six in number and included, in addition to the constant term corresponding to just (33) and (34) directly, the value of the underlying index at the date the option price is calculated, the value of the option price under the alternate model (i.e. filtered option prices in the Black–Scholes case and vice versa for the filtered model) with the two parameter settings for σ, the strike price of the option, and the time to maturity of the option.

The Hansen GMM procedure minimizes, for the Black–Scholes parameters, over a_1, a_2a quadratic in the q averages $\frac{1}{N}\Sigma_i y_{ij}u_i, j = 1, \ldots, q$, where i runs over all the 2,824 option prices and $N = 2,824$. For the filtered price model parameter estimation the averages employed are $\frac{1}{N}\Sigma_i y_{ij}u_i, j = 1, \ldots, q$. The optimized objective is asymptotically distributed as a χ^2_{q-m}, where m is the number of parameters estimated. Hence, for q exceeding m, the optimized objective provides a test of the model using over-identifying restrictions.

The results for the GMM estimation of equations (35) and (36) for the selected instruments are presented in table 1. We observe from table 1 that for both the unfiltered and filtered Black–Scholes model, the parameter estimates are positive and essentially summing to unity. Hence the mixture of the two possible σ values is a reasonable interpretation for the estimated coefficients. All the coefficients are also significant. Both models are however rejected by the χ^2_4 test statistic based on the four overidentifying restrictions. The test statistic for the unfiltered model is more than double that of the filtered model and this is suggestive of a possibly better quality of filtered prices. There is therefore improvement reflected in the filtering procedure, even though at this stage we have only implemented a crude form of parameter estimation. We expect to be able to make further improvements once we implement the EM algorithm for recursive updates on the asset price volatility σ and the error volatilities ν_1 and ν_2.

With a view to discovering the nature of the bias in the pricing errors indicated by the level of the test statistics associated with our instrument choice, we expanded the model to be estimated and included in addition to the unfiltered and filtered prices at the σ values of .1 and .3, a constant term, the option strike and the square of the option strike. This is an ad hoc reformulation undertaken to allow for model error and in particular to allow for smile effects. A similar approach has been recently taken by Jacquier and Jarrow (1995). A negative coefficient on the strike and a positive coefficient for the squared strike is indicative of the smile effect requiring higher model prices for deep in and out of the money options and therefore higher implied volatilities. The instrument list was altered to remove variables on the right hand side and add variables permitting a test of overidentifying restrictions with four degrees of freedom. The specific instrument list includes in addition

Table 1. GMM results for Unfiltered and Filtered Black Scholes Option Pricing Models

Unfiltered Model: $E[y_{ij}(p_i - a_1 b_i^1 - a_2 b_i^2)] = 0$

$$y_{ij} = (1, S_{t(i)}, f_i^1, f_i^2, k_{t(i),i}, \tau_{t(i),i})$$

Filtered Model: $E[\tilde{y}_{ij}(p_i - \alpha_1 f_i^1 - \alpha_2 f_i^2)] = 0$

$$\tilde{y}_{ij} = (1, S_{t(i)}, b_i^1, b_i^2, k_{t(i),i}, \tau_{t(i),i})$$

Parameter	Unfiltered Black Scholes	Filtered Black Scholes
a_1/α_1 (t-val)	.9029 (63.64)	.8131 (35.19)
a_2/α_2 (t-val)	.1449 (37.40)	.1710 (32.14)
χ_4^2	800.6	318.38

to the constant term, the index, its square, the alternative model prices at the two parameter settings and their squares, the maturity and its square.

The results for the investigation of the smile effect in the filtered and unfiltered models are presented in Table 2. The smile effect is present in the unfiltered Black–Scholes model with the strike and its square both being significant of negative and positive signs as expected by the smile hypothesis. With the smile correction, the smile adjusted Black–Scholes is not rejected by the overidentifying restrictions.

For the filtered model however, the strike and its square though individually significant have the wrong sign and we have at best a frown. Furthermore the frown adjusted filtered Black–Scholes model is rejected by the overidentifying restrictions.

These observations lead us to conclude that filtering the underlying tends to improve the performance of the model. The smile effect in the unfiltered model is partly corrected for, though there may be, at this crude estimation level, an overcorrection. Ideally we would like to compare the models after implementing a daily updating of parameters based on the EM algorithm that permits a more sophisticated exploitation of market price observations by incorporating an on-line updating of all the parameters of the model.

Table 2. GMM results for Unfiltered and Filtered Black Scholes
Option pricing Models with Smile effects

Unfiltered Model: $E[y_{ij}(p_i - a_0 - a_1 b_i^1 - a_2 b_i^2 - a_3 k_{t(i),t} - a_4 k_{t(i),i}^2)] = 0$

$$y_{ij} = (1, S_{t(i)}, f_i^1, f_i^2, S_{t(i)}^2, (f_i^1)^2, (f_i^2)^2, \tau_{t(i),i}, \tau_{t(i),i}^2)$$

Filtered Model: $E[\tilde{y}_{ij}(p_i - \alpha_0 - \alpha_1 f_i^1 - \alpha_2 f_i^2 - \alpha_3 k_{t(i),t} - \alpha_4 k_{t(i),i}^2)] = 0$

$$\tilde{y}_{ij} = (1, S_{t(i)}, b_i^1, b_i^2, S_{t(i)}^2, (b_i^1)^2, (b_i^2)^2, \tau_{t(i),i}, \tau_{t(i),i}^2)$$

Parameter	Unfiltered Black Scholes	Filtered Black Scholes
a_0/α_0	162.25	-106.69
(t-val)	(6.67)	(-2.54)
a_1/α_1	.7819	.3489
(t-val)	(44.60)	(4.03)
a_2/α_2	.2616	.3702
(t-val)	(30.99)	(11.75)
a_3/α_3	-.7188	.5483
(t-val)	(-6.23)	(2.71)
a_4/α_4	.00079	-.00091
(t-val)	(5.73)	(-2.90)
χ_4^2	6.97	21.60

Conclusion

Market prices are viewed as noisy relative to theoretical valuations based expected discounted cash flow calculations. The theoretical values of derivative
securities are seen as nonlinear functions of market fundamentals given by
a Hidden Markov Model. Observed prices of assets are viewed as noisy signals on the underlying assets' theoretical value. Theoretical prices of assets
conditional on all noisy price observations to date are then obtained by the
application of nonlinear filtering methods in the context of Hidden Markov
models.

The filtering methodology is illustrated by an application to *S&P*500
futures options over the year 1992 for the underlying Hidden Markov Model

taken to be geometric Brownian motion for the index, applying for the option filtering exercise the nonlinear Black–Scholes formula. It is observed that filtered option prices appear to correct for the smile effect bias, with respect to the strike price, known to exist, and observed in our application in the unfiltered application of the Black–Scholes formula.

References

Bassett, G., V. France and S. Pliska (1991), 'Kalman Filter Estimation for Valuing Nontrading Securities, with applications to the MMI cash-futures spread on October 19 and 20, 1987,' *The Review of Quantitative Finance and Accounting* **1/2**, 135–151.

Black, F. and M. Scholes (1973), 'The Pricing of Options and Corporate Liabilities,' *Journal of Political Economy*, **81**, 637–654.

Bossaerts, P. and P. Hillion (1993), 'A Test of a General Equilibrium Stock Option Pricing Model,' *Mathematical Finance*, **3**, 311–347.

Chan, K. (1992), 'A Further Analysis of the Lead-Lag Relationship between the Cash Market and Stock Index Futures Markets,' *Review of Financial Studies*, **5**, 123–152.

Dempster, A.P., N.M. Laird and D.B. Rubin (1977), 'Maximum Likelihood from Incomplete Data via the EM algorithm,' *Journal of the Royal Statistical Society of London,* Series B, **39**, 1–38.

Elliott, R.J., Aggoun, L. and J.B. Moore (1995), *Hidden Markov Models*, Springer-Verlag, New York.

French, Kenneth R. and Richard Roll, (1986), 'Stock return variances: the arrival of information and the reaction of traders,' *Journal of Financial Economics*, **17**, 5–26.

Hansen, L.P. (1982), 'Large Sample Properties of Generalized Method of Moments Estimators,' *Econometrica*, **50**, 1029–1054.

Heston, Steve (1993), ' A Closed-Form Solution for Options with Stochastic Volatility with Applications to Bond and Currency Options,' *The Review of Financial Studies*, **6**, 327–343.

Jacquier, E. and R.A. Jarrow (1995), 'Dynamic Evaluation of Contingent Claim Models (An analysis of Model Error)', *Working Paper, Cornell University*, Ithaca, NY, 14853.

Jarrow, R.A. and D.B. Madan (1995), 'Arbitrage, Rational Bubbles and Martingale Measures,' *Working Paper, Cornell University*, Ithaca, NY, 14853.

Lo, A. (1986), 'Tests of Contingent Claims Asset Pricing Models,' *Journal of Financial Economics*, **17**, 143–173.

Madan, D.B. and E. Chang (1995), 'Volatility Smiles, Skewness Premia and Risk Metrics: Applications of a Four Parameter Closed Form Generalization of Geometric Brownian Motion to the Pricing of Options,' *Working Paper*, University of Maryland.

Skinner, Douglas J. (1990), 'Options Markets and the Information content of Accounting Earnings,' *Journal of Accounting and Economics*, **13**, 191–212.

Stoll, H.R. and R.E. Whaley (1993), *Futures and Options, Theory and Application*, South-Western Publishing Company, Cincinnati, Ohio.

Part III
Valuation and Hedging with Market Imperfections

Hedging Long Maturity Commodity Commitments with Short-dated Futures Contracts

M.J. Brennan and N.I. Crew

Abstract

The problem of hedging a long term commitment to deliver a fixed amount of a commodity by trading in short term futures contracts is analyzed. Conditions under which a simple or tailed stack and roll strategy will yield a perfect hedge are developed, and the relation between the 'rollover gains' from a stack and roll policy and the economic returns from the policy is clarified. The stack and roll policy presumes very restricted variation in the behavior of futures prices. Models of Brennan (1987, 1991) and Gibson and Schwartz (1990) allow for a richer range of behavior in futures prices by introducing a stochastic model of the dynamics of the commodity convenience yield. The futures hedging strategies implied by these models are developed and compared with the stack and roll hedge and a minimum variance hedge, using monthly price data on NYMEX light oil futures contracts.

Introduction

The Metallgesellschaft incident of December 1993 has led to interest in the related issues of the extent to which long dated commitments to deliver (or receive) a fixed amount of a commodity can be hedged by rolling over a series of short term futures contracts, and how this can best be accomplished.[1] In this article we analyze the hedging problem in terms that are familiar to economists, and consider the ability of different models of futures prices to yield trading strategies in short dated futures contracts that provide effective hedges for long-term commitments.

In a series of papers, Culp and Miller (1994, 1995) analyze this problem with special reference to Metallgesellschaft, and present numerical examples that suggest that a perfect hedge is possible. As we shall show, their analysis rests on the implicit assumption that the convenience yield on the underlying commodity is deterministic, an assumption that is in conflict with the data. They argue correctly that the funding required to meet the variation margin

---

[1]See Culp and Miller (1994, 1995), Edwards and Canter (1995), Mello and Parsons (1995), Neuberger (1995), and Ross (1995).

on the hedge is an inappropriate measure of the interim success of the strategy, but their analysis, and that of Edwards and Canter (1995), is presented in terms of 'rollover gains', a concept which is foreign to economists and which, we shall argue, does not correspond to the economic concept of gain or profit. In Section 1 we show how the final profit on a commitment matched by a 'stack-and roll' hedge is related to the sum of the 'rollover gains' over the life of the commitment, but show that the rollover gain in any single period bears no relation to the economic profit earned in that period.

The difficulty in measuring the interim success of a hedging program is that in general there exists no market price for the underlying commitment that is being hedged. It is therefore necessary to have a *model* of how the price is determined, both in order to determine the appropriate hedge strategy, and to measure the current position at a point in time prior to the maturity of the commitment[2]. Such models of futures or forward prices[3] have been developed by Brennan (1987)[4] and Gibson and Schwartz (1990)[5]. These models and the hedging strategies implied by them are presented in Section 2.

In Section 3 we describe the data that we use in our empirical evaluations of hedging effectiveness. In Section 4 we evaluate the hedge strategies derived from the models of futures prices, by examining their ability to provide trading strategies in short dated futures contracts to hedge forward commitments in oil, with maturities corresponding to those of extant long-dated traded futures contracts. We choose these particular commitments to hedge because the availability of the long dated futures price allows us to calculate the present value of the commitment each period, which permits us to measure the periodic hedging errors under the strategies. Of course, we would prefer to be able to assess directly the ability of the strategies to hedge much longer-term commitments[6]; while we are not able to do this, our results for commitments of maturities up to 24 months should provide useful indications of the efficacy of these strategies, since the evidence is that the futures price curve is essentially flat for longer maturities. For comparison, we also analyze the stack and roll hedge proposed by Culp and Miller (1994, 1995), and a minimum variance hedge suggested by Edwards and Canter (1995). We find that hedge strategies that take account of stochastic variation in the convenience yield lead to monthly hedging errors with a standard deviation which is only about 25% of that obtained from the simpler stack and roll strategy.

---

[2]Strictly speaking, prior to the period in which the commitment corresponds to the longest available futures contract.

[3]In this paper we assumethat the interest rates are non-stochastic so that the futures and forward prices are the same. See Cox, Ingersoll and Ross (1981).

[4]Later published as Brennan (1991).

[5]Garbade (1993) presents a model which is a special case of Brennan (1991).

[6]The commitments undertaken by Metallgesellschaft had maturities of 5-10 years.

Acknowledgements

We would like to thank seminar participants at the 1995 INQUIRE-Europe Meetings, the University of Tilburg, Seoul National University, and the Conference in Honor of Fischer Black at UCLA. We are happy to acknowledge that this paper received the 1995 First Prize at the INQUIRE-Europe Meetings.

1 Rollover Gains and the Profits on a Hedged Commitment

Consider a firm that makes a forward commitment to deliver one unit of a commodity, say a barrel of oil, at time T, in return for a payment of \$K at that time. The present value of the profit of entering into this commitment at t=0 is $Ke^{-rT} - PV_0(P_T)$, where r is the interest rate, P_T is the spot price at time T and $PV_t(P_T)$ denotes the present value at time T of the (uncertain) future amount P_T to be received at time T. If this commitment is held to maturity, uncertainty about P_T means that the total profit realized from entering into and meeting the commitment will be uncertain. This uncertainty could in principle be eliminated by entering into a futures contract of maturity T. However, in practice, there is no market for long term futures contracts. One suggested solution is to roll over a series of long positions in a short maturity futures contract which, without loss of generality, we shall take as the nearby futures contract: such a hedging strategy is often referred to as a 'stack-and roll' strategy.

To analyze the profit under a stack and roll strategy it will be convenient initially to assume that the interest rate is zero. Let $F_{t,\tau}$ denote the futures price at time t for delivery at time $t + \tau$. Then π_t, the profit that is realized at time t from holding the commitment from time $t - 1$, matched by an offsetting long position of n 1-period futures contracts, is equal to the change in the value of the futures position less the change in the present value of the delivery commitment:

$$\pi_t = -[PV_t(P_T) - PV_{t-1}(P_T) + n[F_{t,0} - F_{t-1,1}] \tag{1}$$

The cumulative profit from entering into the commitment and hedging it in this fashion till maturity is obtained by summing equation (1) over $t = 1, \ldots, T$, and adding the expression for the profit realized when the contract is entered into, $Ke^{-rT} - PV_0(P_T)$. Recognizing that $F_{T,0} = PV_T(P_T) = P_T$, the cumulative profit, Π, when the hedge ratio, n, is set equal to unity, may be written as:

$$\Pi = K - F_{0,1} + \sum_{t=1}^{T-1}(F_{t,0} - F_{t,1}) \tag{2}$$

The quantity $(F_{t,0} - F_{t,1})$ is referred to by Culp and Miller (1994, 1995) and Edwards and Canter (1995) as the 'rollover gain'; thus the cumulative profit is the difference between the contract price K and the one-period forward price, $F_{0,1}$, plus the cumulative rollover gains realized over the life of the strategy. The intuition behind the expression is that if it were possible to enter a futures contract at time 0 at a price $F_{0,1}$ and hold the contract until maturity at time T when its price would be equal to the spot price, the realized profit would simply be $K - F_{0,1}$, the gains on the future contract exactly offsetting changes in the price of the underlying commodity. The stack and roll strategy does this, except for the price 'gaps' that arise when one contract is closed out and another is entered into; it is these price gaps that give rise to the rollover gains or losses.

Equation (2) is no more than an accounting identity, and although it is tempting to identify the amount of the rollover gain realized in a period as the part of the final aggregate profit that is realized in that period, such an identification is wrong. Note first that the rollover gain does not appear in expression (1) for the economic profit realized in the period ending at time t; hence the rollover gain is not the same as the economic profit. In fact, economic gains and losses in futures markets can arise only from *changes* in the prices of the same futures contract, and not from *differences* between the prices of *different* futures contracts at a point in time which is what the 'rollover gain' is.

Thus far we have assumed that the interest rate is zero. To allow for a non-zero interest rate, r, it is necessary to 'tail' the hedge by adjusting the number of futures contracts, so that n, the number of one-period futures contracts entered into at time T is equal to $e^{-r(T-t-1)}$.[7] Thus with a tailed stack and roll hedge, the second term in equation (1) is multiplied by $e^{-r(T-t)}$, and the aggregate realized profit will still be given by equation (2).

It is apparent that the profit from a tailed stack and roll hedge will be riskless only if the rollover gains are certain when the contract is entered into. This requires that the basis between the price of the one period futures contract and the maturing contract be predictable. Figure 1 shows the end of month rollover gains for the 2 month NYMEX light oil futures contract[8] between 1983 and 1994. It is apparent that the rollover gains are highly volatile and, while the mean is positive, it is only $0.14 per barrel per month, and the standard deviation is $0.34. Moreover, since the serial correlation of

---

[7]More precisely, to account for daily settlement, the number of contracts should be adjusted daily to e^{-ry}, where y is the exact number of years to maturity of the commitment from the following day. The formula in the test implicitly assumes that settling up takes place only at the maturity of each futures contract; that is, it treats the futures as short-dated forwards.

[8]The rollover gain is defined as the difference between the price of the nearby futures contract and the next shortest contract, and corresponds to the gains from a strategy of going long in the second nearby contract and rolling over at the end of every month.

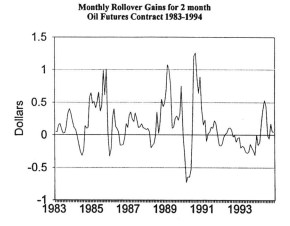

Figure 1. Monthly Rollover Gains for the 2 month NYMEX Light Oil Futures Contract, March 1983–December 1994. The figure shows the time series of the month-end differences between the prices of the nearby futures contract and the 2 month contract.

the rollover gains is 0.76, it would be rash to suppose that the rollover risk could be diversified away over time for a long-term commitment. Figure 2 shows the corresponding rollover gains for a 3 month contract[9]. The mean of the monthly gains is now \$0.27, and the standard deviation is \$0.65, while the serial correlation is 0.79. As these figures show, there is considerable uncertainty about future rollover gains, and therefore about the final realized profit from a stack and roll hedging strategy.

Equation (1), adjusted for the tailing of the hedge by setting n equal to $e^{-r(T-t-1)}$, implies that the periodic *economic profit* will be riskless only if the change in the future value of the amount to be delivered under the contract (i.e. the change in the implicit futures or forward price for delivery at time T) is equal to the change in the price of the short-dated futures contract. This requires that the implied futures price curve shift up and down in a parallel fashion[10]. Figure 3 plots the futures price curves for the ends of alternate years from 1983 to 1994. It is apparent that the assumption of parallel shifts is not a good one, even within the limited maturity range of traded futures contracts: in some periods the term structure of futures prices slopes up, while in others it slopes down. In the following section we will consider two models

---

[9]Defined as the one month futures price less the three month futures price.

[10]The reader will note the analogy with the assumption of parallel shifts in the yield curve that underlies the derivation of the duration model of bond price hedging. Indeed, the stack and roll policy has much in common with a duration matched bond price hedge in Treasury Bill futures.

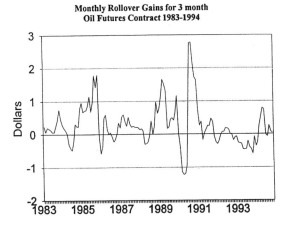

Figure 2. Monthly Rollover Gains for the 3 month NYMEX Light Oil Futures Contract, March 1983–December 1994. The figure shows the time series of the month-end differences between the prices of the nearby futures contract and the 3 month contract.

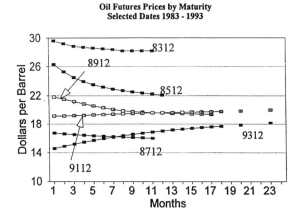

Figure 3. Term Structure of Oil Futures Prices. NYMEX Light Oil Contract, end of alternate years 1983-1994.

of the behavior of futures prices that allow for non-parallel shifts in the term structure of futures prices, and derive the hedging strategies that correspond to them.

2 Futures Prices and Hedging with Stochastic Convenience Yields

The convenience yield of a commodity is defined as the flow of services that accrues from possession a physical inventory but not to the owner of a contract for future delivery. The marginal convenience yield includes both the reduction in costs of acquiring inventory, and the value of being able to profit from temporary local (or grade specific) shortages of the commodity through ownership of an additional unit of inventory. The profit may arise either from local price variations, or from the ability to maintain a production process despite local shortages of a raw material. The convenience services yielded by an inventory depend upon the identity of the individual storing it; however, competition between potential storers will ensure that in equilibrium the marginal convenience yield net of storage costs will be equalized across all storers. Then, assuming that there exists a positive inventory of the commodity, the relation between spot and futures prices will reflect this marginal net convenience yield in a manner which we will now develop[11].

It will be helpful to make explicit the dependence of the futures price on the current spot price. Therefore, let $F(P, t, \tau)$ denote the futures price at time T for delivery at $t + \tau$, when the current spot price is P, and let $PVC(t, \tau)$ denote the present value of the marginal net convenience yield over the interval $t, \ldots, t + \tau$. Then the relation between the futures price, the current spot price and the convenience yield may be written as:

$$F(P, t, \tau)e^{-r\tau} = P - PVC(t, \tau) \tag{3}$$

The left hand side of (3) is the present value of a forward purchase commitment: this is equal to the current spot price less the value of the convenience services that would be available to a storer having physical possession of a marginal unit of the commodity up to the maturity of the forward commitment.

It follows from (3) that the rollover gain from a tailed stack-and-roll strategy is given by:

$$e^{-r(T-t)}[F(P, t, 0) - F(P, t, 1)] = e^{-r(T-t)}PVC(t, 1) \tag{4}$$

Therefore, unless the convenience yield is deterministic so that $PVC(t, 1)$ is known, the rollover gain will be uncertain. Thus, except in this special and unlikely case in which the futures price curve shifts only in a parallel fashion, a stack and roll strategy does not provide a perfect hedge.

In order to model the behavior of the structure of futures prices implied by (3) it is necessary to model the behavior of the marginal convenience yield.

---

[11]Previous authors who have discussed the convenience yield include Kaldor (1939), Working (1948, 1949), Brennan (1958), Telser (1958), and Fama and French (1987).

The (marginal net) convenience yield depends on the level of inventories; the higher the current level of inventories, the less will merchants and manufacturers be willing to pay to have an additional unit on hand. Since spot prices are also likely to be associated with the level of inventories, a natural simplifying assumption is that the instantaneous rate of convenience yield is a function of the spot price, $C(P)$, where P is the spot price. The simplest assumption is that the convenience yield is proportional to the spot price:

The Constant (proportional) Convenience Yield Model

If the convenience yield is proportional to the current spot price: $C(P) = cP$, then it is not difficult to show[12] that the term structure of futures prices is given by:

$$F(P, t, \tau) = Pe^{(r-c)\tau} \tag{5}$$

and the present value at time T of the uncertain future amount $P_{t+\tau}$ to be received at $t + \tau$ is

$$PV_t(P_{t+\tau}) = e^{-r\tau} F(P, t, \tau) = Pe^{-c\tau} \tag{6}$$

The only source of uncertainty in this setting is the spot price P. The derivative of the present value of a commitment to deliver one barrel of oil in T periods with respect to the spot price is e^{-cT}, and the derivative of a τ period futures price with respect to P is $e^{(r-c)\tau}$. Therefore to hedge the T period delivery commitment it is necessary to take a long position in $n = e^{-[r\tau+c(T-\tau)]}$ futures contracts.

However, as the results of Brennan (1991) show, the constant proportional convenience yield model is too simple to be descriptive of the behavior of futures prices for most commodities. Therefore, we turn to a more realistic model.

The Brennan Autonomous Convenience Yield Model

Brennan (1987, 1991) assumes that C, the instantaneous net marginal rate of convenience yield measured in dollars per unit of inventory per period, follows the simple mean-reverting process:

$$dC = \alpha(m - C)dt + \eta dz_C \tag{7}$$

where $\alpha > 0$, is the speed of adjustment, m is the long run mean rate of convenience yield, and dz_C is the increment to a standard Gauss-Wiener process. This assumption about the behavior of the convenience yield is motivated by the consideration that if the convenience yield is high because inventories are

---

[12]See Brennan and Schwartz (1985).

low, storage firms will tend[13] to have an incentive to increase their investment in inventories which, in turn, will tend to reduce the convenience yield. The commodity spot price is assumed to follow the exogenously given stochastic process:

$$\frac{dP}{P} = \mu dt + \sigma(P, C) dz_P \tag{8}$$

where dz_P is the increment to a Gauss-Wiener process and $dz_P dz_C = \rho dt$, and μ, the expected rate of price change, may be stochastic.

Under these assumptions, the futures price will depend upon the current instantaneous rate of convenience yield, as well as the current spot price, so we write it as $F(P, C, \tau)$. It is shown in the Appendix that the futures price under the Brennan model is given by:

$$F(P, C, \tau) = (P - PVC(C, \tau))e^{r\tau} \tag{9}$$

where

$$PVC(C, \tau) = \frac{m^*}{r}(1 - e^{-r\tau}) - \frac{m^* - C}{\alpha + r}(1 - e^{-(\alpha + r)\tau}) \tag{10}$$

and $m^* \equiv m - \lambda^*/\alpha$, where λ^* is a risk adjustment parameter, so that m^* is the risk-adjusted mean rate of convenience yield.

Consider now the problem of hedging a commitment to deliver one barrel of oil at a date T periods in the future. The present value of the commitment may be written as:

$$PV_0(P_T) = F(P, C, T)e^{-rT} = P - PVC(C, T) \tag{11}$$

Using equations (10) and (11), the derivatives of the present value of the commitment with respect to P and C are:

$$\frac{dPV}{dP} = 1$$
$$\frac{dPV}{dC} = \frac{-1}{\alpha + r}\left(1 - e^{-(\alpha + r)T}\right) \tag{12}$$

while, using equations (9) and (10), the derivatives of the price of a futures contract with maturity, τ, are:

$$\frac{dF}{dP} = e^{r\tau}$$
$$\frac{dF}{dC} = \frac{-1}{\alpha + r}\left(e^{r\tau} - e^{-\alpha\tau}\right) \tag{13}$$

In order to hedge a commitment of maturity T it is necessary to hold a portfolio of futures contracts with the same sensitivities to P and C. Using

---

[13]Only tend to because the investment decisions of storage firms will depend upon the expected rate of change in the commodity price as well as upon the convenience yield.

equations (12) and (13), the hedge portfolio consists of n_1 and n_2 contracts of maturities τ_1 and τ_2 respectively where n_1 and n_2 are the solutions to:

$$n_1 e^{r\tau_1} + n_2 e^{r\tau_2} = 1$$
$$n_1 e^{-\alpha\tau_1} + n_2 e^{-\alpha\tau_2} = e^{-(\alpha+r)T}. \tag{14}$$

The Gibson-Schwartz Model

Gibson and Schwartz (1990) follow Brennan in assuming a mean-reverting process for the instantaneous convenience yield; however, they define the instantaneous convenience yield in terms of dollars per dollar of inventory per period. This implies that $C(P)$ is written as δP, where δ follows the mean-reverting process:

$$d\delta = k(m_\delta - \delta)dt + \xi dz_\delta \tag{15}$$

with $dz_\delta dz_P = \rho dt$. Arbitrage arguments similar to those developed above imply that the forward or futures price, $F(P,t,\tau)$ can be written as[14]

$$F(P,\delta,\tau) = P \exp\left\{ -\frac{\delta\left(1 - e^{-k\tau}\right)}{k} + A\tau + \frac{1}{2}v^2\tau \right\} \tag{16}$$

where:

$$v^2 = \left(\sigma_p^2 + \frac{\xi^2}{k^2} + 2\frac{\rho\sigma_P\xi}{k}\right)\tau + \frac{\xi^2\left(1 - e^{-2k\tau}\right)}{2k^3} + 2\frac{\xi\left(\sigma_P\rho - \frac{\xi}{k}\right)\left(1 - e^{-k\tau}\right)}{k^2} \tag{17}$$

and

$$A(\tau) = \left(r - \frac{1}{2}\sigma_P^2 - \tilde{m}\right)\tau + \frac{\tilde{m}\left(1 - e^{-k\tau}\right)}{k} \tag{18}$$

$$\tilde{m} = m_\delta - \frac{\lambda_\delta\xi}{k} \tag{19}$$

The hedge portfolio for a commitment of maturity T is found by equating the derivatives of the present value of the commitment with respect to the spot price P and the convenience yield rate, δ, to the corresponding derivatives of the value of the hedge portfolio. Using equations (11) and (16), this yields the following equations for the numbers of futures contracts of maturities τ_1 and τ_2:

$$n_1 F(P,\delta,\tau_1) + n_2 F(P,\delta,\tau_2) = e^{-rT} F(P,\delta,T)$$
$$n_1 \left(1 - e^{-k\tau_1}\right) F(P,\delta,\tau_1) + n_2 \left(1 - e^{-k\tau_2}\right) F(P,\delta,\tau_2) \tag{20}$$
$$= \left(1 - e^{-kT}\right) F(P,\delta,T)e^{-rT}.$$

---

[14]Jamshidian and Fein (1990) appear to have been the first to develop a closed form expression for the futures price in this model.

The solutions to equations (14) and (20) determine the composition of the portfolio of futures contracts of maturities τ_1 and τ_2 which will hedge a fixed commitment of maturity T, when futures prices are described by the Brennan, and the Gibson-Schwartz, models of the convenience yield, respectively. Note that to determine the hedge portfolio weights in the Brennan model it is necessary to estimate only the single, speed of adjustment parameter, α. For the Gibson-Schwartz model it is necessary to estimate the speed of adjustment parameter, δ, and forward price of the commitment being hedged, $F(P, \delta, T)$.[15] This is estimated from equation (16) using the current estimated value of δ and an estimate of the risk-adjusted mean of the convenience yield process, m. In Section IV below we shall describe the estimation of the speed of adjustment parameters for the two models and the mean parameter, m, for the Gibson-Schwartz model, and investigate the effectiveness of their implied hedges empirically.

3 Data

The data that were used to evaluate the hedging strategies were end of month settlement prices on futures contracts for light crude oil quoted on NYMEX for the period March 1983 to December 1994[16]. We label the contracts as 'nearby', '2 month' , '3 month', etc., these designations corresponding roughly to the remaining time to maturity of the contracts. Table 1 reports the average of the number of days to maturity for each of the contracts as of the end of each month of the sample period. Note that the longer maturity contracts only became available towards the end of our sample period, and even then were not available each month; as a result, the number of observations for the 21 and 24 month contracts is only 17 and 15. Continuously compounded interest rates for monthly maturities of 1 to 12 months were taken from the Fama T-Bill files for the end of each month of the sample period; the rates for intermediate maturities were linearly interpolated according to the number of days. Interest rates for maturities beyond 12 months were estimated by assuming that the yield curve was flat beyond 12 months.

For each month from April 1985 to December 1994, α, the speed of adjustment parameter of the Brennan convenience yield model was estimated as follows. The value of the instantaneous convenience yield, c, for each of the prior months was approximated from the prices of the two nearby futures contracts by assuming that the rate of convenience yield was constant over this interval[17]. From this time series of instantaneous convenience yields, α, the

---

[15] Recall that we wish to evaluate a technique for hedging that does not rely on observation of the commitment value (though our evaluation of the technique does).

[16] On November 30, 1990 there was a limit move in futures prices which affected all but the nearby contract. Therefore we substituted the prices for November 29th.

[17] This is similar to the procedure followed by Gibson and Schwartz (1990). Note that

Nominal Contract Maturity (months)	Average number of days to maturity	Month of first price observation	Number of monthly observation
Nearby	20.5	1983.3	134
2	51.0	1983.3	134
3	81.4	1983.3	134
6	172.7	1983.3	134
9	262.0	1983.11	134
12	355.2	1983.11	94
15	446.6	1984.10	73
18	537.9	1989.9	55
21	629.5	1990.11	17
24	720.7	1990.11	15

Table 1. Summary Statistics on NYMEX Light Oil Futures Contracts from March 1983 to December 1994.

speed of adjustment parameter of the stochastic process (7), was estimated using the exact discrete model corresponding to (7). Note that the number of time series observations used to estimate α increases each month as one more historical observation becomes available; the smallest number of observations is 25 in April 1985.

Similarly, for each month from March 1985 to December 1994, k, the speed of adjustment parameter of the Gibson-Schwartz model, was estimated over the previous months, using estimates of the instantaneous convenience yield constructed under the assumption that δ was constant till the maturity of the second nearest futures contract[18]. Again, the number of time series observations used to estimate δ increases from a minimum of 24 in March 1985. The time series of estimates of α and δ are plotted in Figure 4. The estimates show considerable volatility in the early part of the sample period, reflecting in part the small number of observations used in estimation. By 1989 the estimates of the parameters for both models settle-down in the neighbourhood of 0.3 which corresponds to a half-life of 2.3 months for deviations of the convenience yield from its long run mean.

It is also necessary to estimate m to implement the hedge for the Gibson-Schwartz model. This was done for each month by finding the value that minimized the sum of the squared price prediction errors for the 3–6 month contracts using equations (16)–(18).

---

with a constant convenience yield, C, the futures price is given by $F(P, C, \tau) = Pe^{r\tau} + \frac{C}{r}(1 - e^{r\tau})$. Subtracting two adjacent futures prices yields an estimate of C.

[18]Under the Gand S model the expression for the futures price when the convenience yield is constant is $F(P, t, \tau) = Pe^{(r-\delta)\tau}$. The ratio of the two nearby futures prices yields a simple estimate of δ.

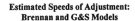

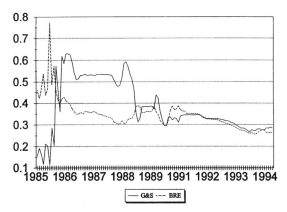

Figure 4. Time Series of Speed of Adjustment Estimates. The figure shows estimated values of α, the speed of adjustment parameter for the Brennan model, and δ, the speed of adjustment parameter for the Gibson-Schwartz model. The estimates are derived from monthly convenience yield estimates from March 1983 up to date, using the exact discrete form corresponding to the diffusion process.

4 Empirical Results

The hedge properties of four basic strategies were evaluated using the oil futures price data. These strategies are the Brennan, the Gibson-Schwartz, the stack and roll, and the Edwards-Canter minimum variance strategy. Two variants of the Brennan and Gibson-Schwartz strategies were implemented: the first using 2 and 3 month futures to construct the hedge portfolio, and the second using 2 and 6 month futures[19]. Two variants of the stack and roll and Edwards-Canter hedges were implemented: the simple version and the tailed version. These latter hedges were all implemented using the 2 month futures contract. The hedges were designed to hedge fixed commitments with maturities ranging from 6 to 24 months, and the present values of these commitments were computed each month using the observed futures prices and interest rates for these maturities. All the hedges were revised monthly and the monthly hedge errors were calculated as described below.

Specifically, for each month t, for each liability maturity τ, the hedging error under each of the policies was computed as follows. First, the present

---

[19]By a τ month contract we mean the contract which is τ^{th} closest to maturity: this will have a maturity which does not exceed τ months. See Table 1.

value of the liability under the commitment at the beginning of the month was computed by discounting the futures price for the corresponding maturity:

$$L_{t-1,T} = e^{-r_{t-1,T}T} F_{t-1,T} \tag{21}$$

where $r_{t,T}$ is the continuously compounded T period interest rate at the beginning of month t, $F_{t,T}$ is the T period futures price at the beginning of the month, and $L_{t,T}$ is the value of the liability. The change in the present value of the liability over the month is defined by $\Delta L_{t,T} \equiv L_{t,T-1} - L_{t-1,T}$. Suppose that the hedge consists of n_1 and n_2 futures contracts of maturities τ_1 and τ_2. Then the change in the hedger's wealth over the month, from hedging a T period liability, $\Delta W_{t,T}$ is given by

$$\Delta W_{t,T} = -\Delta L_{t,T} + n_1 \Delta F_1 + n_2 \Delta F_2 \tag{22}$$

where ΔF_s is the change in the τ_s period futures price during month t. Finally, the hedging error, $E_{t,T}$, is defined as the difference between the change in wealth and the riskless return on the present value of the liability:

$$E_{t,T} = \Delta W_{t,T} - \left(e^{r_{t-1,1}} - 1\right) L_{t-1,T} \tag{23}$$

For the Brennan and Gibson-Schwartz models, n_1 and n_2 were calculated from equations (14) and (20) respectively. For the simple stack and roll strategy, $n_1 = 1$, $n_2 = 0$, while for the tailed stack and roll strategy $n_1 = e^{-rT}$. For the Edwards-Canter strategy, $n_1 = \beta$, where β is the coefficient from the regression of the change in the six month futures price on the change in the nearby futures price estimated over all prior months; for the tailed version the coefficient is multiplied by e^{-rT}[20].

The means and standard deviations of the monthly hedging errors, measured in dollars per barrel of oil committed, for the various strategies are reported in Table 2 for commitment maturities of 6–24 months at 3 month intervals. In reviewing this table it is important to remember that these are the errors from hedging the same *fixed maturity* commitment each month.

First we observe that out to 18 months the standard deviation of the hedge errors increases monotonically with the maturity of the commitment; there is a discrete drop in the standard deviations beyond 18 months but this reflects the fact that the 21 and 24 month contracts have been available (and therefore the hedge errors could be calculated) only since December 1990, and even since then these contracts are not traded every month, so that the numbers of observations for these maturities are only 17 and 15, as compared with 54 for the 18 month maturity. Overall, for all maturities, the worst performing strategy is the simple stack and roll strategy, followed by the tailed version of the strategy. The next worst strategies are generally the

---

[20]The reason that we use the six month contract to calculate β is that we wish to assess the effectiveness of the hedge for a commitment whose market value cannot be observed.

	6 mnth	9 mnth	12 mnth	15 mnth	18 mnth	21 mnth	24 mnth
SandR	0.06	0.10	0.08	0.05	0.01	-026	-0.32
SandR$_{tailed}$	0.06	0.09	0.08	0.04	0.01	-0.23	-0.28
EandC	0.05	0.09	0.09	0.04	0.02	-0.18	-0.23
EandC$_{tailed}$	0.05	0.08	0.09	0.04	0.02	-0.16	-0.20
BRE$_{2and3}$	0.02	0.06	0.11	0.17	0.21	-0.02	-0.06
BRE$_{2and6}$	n.a.	0.02	0.04	0.08	0.06	-0.09	-0.12
GandS$_{2and3}$	0.04	0.07	0.11	0.13	0.16	-.003	-0.03
GandS$_{2and6}$	n.a.	0.02	0.05	0.08	0.07	-0.04	-0.05
No. of Obs.	116	116	85	67	54	17	15

Table 2a. Mean Hedging Errors. This table reports the means of the monthly errors in hedging a forward commitment of one barrel of oil deliverable at a fixed future maturity, under different strategies, over the period May 1985 to December 1994 or subperiods for which futures price data where available. SandR:Stack and Roll strategy; EandC: minimum variance hedge of Edwards-Canter (tailed strategies adjust the hedge for the time value of money. BRE$_{2and3}$ (BRE$_{2and6}$): strategy derived from Brennan model of futures prices implemented using 2 and 3 month (2 and 6 month) maturity futures contracts. GandS$_{2and3}$(GandS$_{2and6}$): strategy derived from Gibson-Schwartz model of futures prices implemented using 2 and 3 month (2 and 6 month) maturity futures contracts.

	6 mnth	9 mnth	12 mnth	15 mnth	18 mnth	21 mnth	24 mnth
SandR	0.70	0.99	1.30	1.44	1.66	0.92	0.99
SandR$_{tailed}$	0.64	0.90	1.16	1.27	1.43	0.81	0.86
EandC	0.44	0.65	0.89	0.97	1.14	0.66	0.71
EandC$_{tailed}$	0.43	0.60	0.79	0.85	0.97	0.58	0.62
BRE$_{2and3}$	0.31	0.54	0.77	0.90	1.08	0.38	0.41
BRE$_{2and6}$	n.a.	0.19	0.38	0.51	0.67	0.36	0.41
GandS$_{2and3}$	0.29	0.46	0.61	0.64	0.69	0.33	0.35
GandS$_{2and6}$	n.a.	0.17	0.29	0.34	0.40	0.23	0.26
No. of Obs.	116	116	85	67	54	17	15

Table 2b. Standard Deviation of Monthly Hedging Errors. This table reports the standard deviations of the monthly errors in hedging a forward commitment of one barrel of oil deliverable at a fixed future maturity, under different strategies, over the period May 1985 to December 1994 or subperiods for which futures price data where available. See note to Table 2a.

Edward-Canter 'minimum variance' strategies. The best performing strategy throughout is the Gibson-Schwartz strategy executed in the 2 and 6 month futures contracts, and the second best is the Brennan strategy in the same contracts. The Gibson-Schwartz strategy has a standard deviation of $0.40 per month for the 18-month maturity, compared with $1.66 for the simple stack and roll. Overall, it appears that substantial gains in hedging efficiency can be achieved by taking account of the variability in the convenience yield.

Ironically, the ranking in terms of the mean errors for the 18 month maturity is almost exactly reverse the ranking in terms of standard deviations! However, we are not inclined to accept these results at face value, but rather attribute to chance the fact that, for example, the mean error of the Stack and Roll hedge during this period was $0.01. Figures 5a–f show the performance of the different hedge strategies by aggregating over time the monthly hedging errors for selected strategies. The first three figures relate to the 12 month commitment hedge for which there are 85 monthly observations. The flat parts of the figures correspond to periods when there was no 12 month future outstanding to allow calculation of a hedge error. Figure 5a shows that the Stack and Roll and Edwards-Canter hedges performed similarly whether tailed or not, but that the hedge errors quickly cumulated to $5–10 per barrel. Figure 5b shows the Brennan and Gibson-Schwartz hedges. When implemented in the 2 and 3 month futures they perform badly. The reason for this is that the sensitivity of these futures prices to changes in the convenience yield is not very different; as a result it is necessary to take large offsetting positions in the two maturities to hedge out the convenience yield sensitivity, and this increases the importance of model specification and estimation errors.[21] By comparison, the 2 and 6 month hedge performs very well except in the second half of 1990 when the cumulative error of GandS(2, 6 months) increases by about $2.50. However, this was the time of the Iraqi invasion of Kuwait, an extremely turbulent period in oil markets as shown in Figure 7. The spot price more than doubled from $17.05 in June to $39.50 in September. Figure 5c compares the tailed Stack and Roll hedge with the Gibson-Schwartz (2, 6 months) on the same scale, and shows the substantial gains in effectiveness made possible by the GandS strategy.

Figures 5d-f repeat a similar analysis for hedging an 18 month commitment. Again, the Stack and Roll and Edwards-Canter hedges perform similarly, and the Brennan and Gibson-Schwartz (2,6 month) hedges are also similar, and perform well except during the period of extreme price turbulence in 1990. Figures 5g-h relate to a hedge for a 24 month commitment. Here there are only 15 monthly price changes to be hedged, so that the results cannot be regarded as in any way definitive. However, we note that the cumulative error

---

[21]For example, to hedge a 12 month liability the average position for the Gand S 2 and 3 month strategy would be a short position of 2.3 contracts in the 2 month maturity and long 3.2 contracts in the 3 month maturity. The corresponding hedging using 2 and 6 month contracts would be only short 0.3 2 month contracts and long 1.2 6 month contracts.

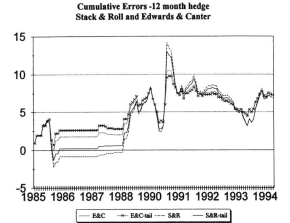

Figure 5a. Cumulated Hedge Errors: 12 month Hedge The figure shows the cumulated monthly errors in dollars from hedging a fixed maturity 12 month commitment using Stack and Roll (SandR) and Edwards-Canter (EandC) hedges.

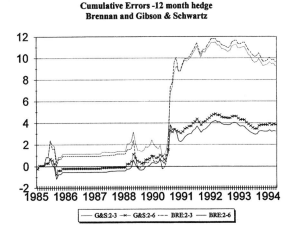

Figure 5b. Cumulative Hedge Errors: 12 month Hedge. The figure shows the cumulated monthly errors in dollars from hedging a fixed maturity 12 month commitment using Brennan (BRE) and Gibson-Schwartz (GandS) hedges.

Cumulative Errors -12 month hedge
Stack & Roll and Gibson & Schwartz

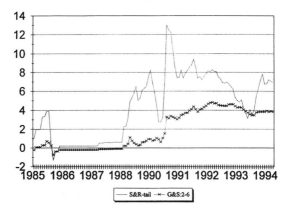

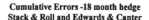

Figure 5c. Cumulative Hedge Errors: 12 month Hedge. The figure
shows the cumulated monthly errors in dollars from hedging a fixed
maturity 12 month commitment using Stack and Roll (SandR) and
Gibson-Schwartz (GandS) hedges.

Cumulative Errors -18 month hedge
Stack & Roll and Edwards & Canter

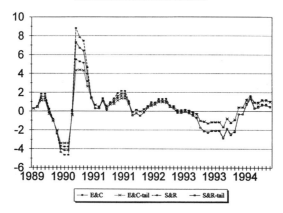

Figure 5d. Cumulated Hedge Errors: 18 month Hedge The figure shows the cumulated monthly errors in dollars from hedging a fixed maturity 18 month commitment using Stack and Roll
(SandR) and Edwards-Canter (EandC) hedges.

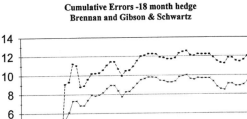

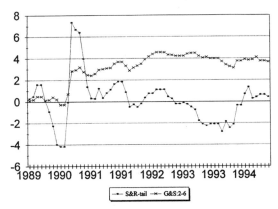

Figure 5e. Cumulative Hedge Errors: 18 month Hedge. The figure shows the cumulated monthly errors in dollars from hedging a fixed maturity 18 month commitment using Brennan (BRE) and Gibson-Schwartz (GandS) hedges.

Figure 5f. Cumulative Hedge Errors: 18 month Hedge. The figure shows the cumulated monthly errors in dollars from hedging a fixed maturity 18 month commitment using Stack and Roll (SandR) and Gibson-Schwartz (GandS) hedges.

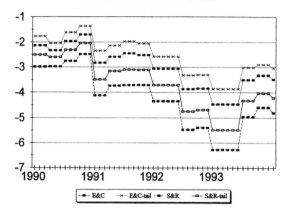

Figure 5g. Cumulated Hedge Errors: 24 month Hedge The figure shows the cumulated monthly errors in dollars from hedging a fixed maturity 24 month commitment using Stack and Roll (SandR) and Edwards-Canter (EandC) hedges.

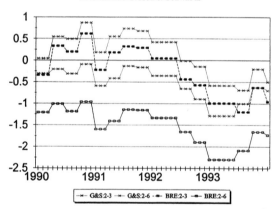

Figure 5h. Cumulative Hedge Errors: 24 month Hedge. The figure shows the cumulated monthly errors in dollars from hedging a fixed maturity 24 month commitment using Brennan (BRE) and Gibson-Schwartz (GandS) hedges.

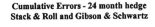

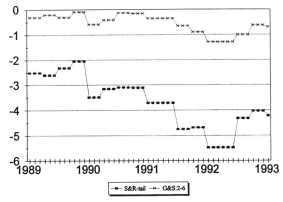

Figure 5i. Cumulative Hedge Errors: 24 month Hedge. The figure shows the cumulated monthly errors in dollars from hedging a fixed maturity 24 month commitment using Stack and Roll (SandR) and Gibson-Schwartz (GandS) hedges.

of the GandS (2, 6 month) hedge is less than $1.00, which is about one quarter that of the tailed Stack and Roll hedge.

Figures 6a and 6b show the hedge portfolios for the Edwards-Canter, Brennan and Gibson-Schwartz (2, 6 month) strategies for a 12 month commitment. The Edwards-Canter strategy takes a long position of about 0.7 barrels in the 2 month maturity; the volatility of the hedge in the first few months reflects the small number of observations used to estimate the hedge ratio. Note that the stack and roll hedge (before tailing) takes a long position in one barrel per barrel of commitment. The Brennan, and Gibson-Schwartz, strategies take very similar positions – roughly 1.3 barrels long in the 6 month maturity and 0.2 barrels short in the 2 month maturity. There is relatively little variation in the hedge position over time (once the initial volatility due to estimation error is passed), despite the wide variation we have noted in the level and slope of the futures pricing curves.

5   Conclusion

In this article we have developed two new models for constructing hedges for long term commodity commitments using short term futures contracts, and have compared their performance with that of the simpler Stack and Roll and minimum variance hedges, using monthly data on the prices of light crude oil. We find that the models of Brennan and of Gibson-Schwartz perform

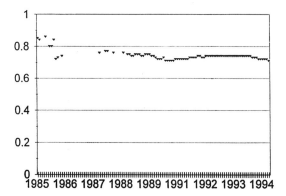

Figure 6a. Hedge Ratios for the Edwards-Canter 12 month hedge. The figure shows the estimated number of 2 month futures contracts to be held long to hedge a 12 month fixed maturity commitment using the Edwards-Canter tailed hedge.

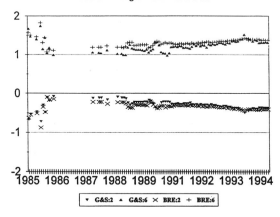

Figure 6b. Hedge Ratios for the Brennan and Gibson-Schwartz 12 month hedge. The figure shows the estimated number of 2 and 6 month futures contracts to be held in order to hedge a 12 month fixed maturity commitment using the Brennan (BRE) and Gibson-Schwartz (GandS) models.

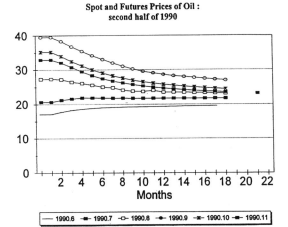

Spot and Futures Prices of Oil :
second half of 1990

Figure 7. Spot and Futures Prices of Oil, second half of 1990. The figure shows the spot and futures prices of the NYMEX Light Oil contract, month ends June–December, 1990.

significantly better than the two simpler models, the Gibson-Schwartz model reducing the standard deviation of the hedge error by about 75% relative to the Stack and Roll strategy.

Our analysis suffers from two limitations, one empirical and one theoretical. On the empirical side, we have had to limit the maturity of the commitments whose hedges were analyzed to the maturity of the longest available futures contract. Clearly it would be desirable to test the effectiveness of the strategies in hedging longer term commitments if data on the prices of the these could be obtained on a periodic basis. A significant limitation of our theoretical models is the assumption that interest rates are deterministic. For longer time horizons, uncertainty about future interest rates becomes important for the hedge strategies, and therefore it would be desirable to extend the Brennan and Gibson-Schwartz models to allow for stochastic interest rates. We leave this for subsequent work.

Appendix

Applying Ito's Lemma to the futures price, $F(P, C, \tau)$,

$$dF = \left[-F_\tau + \frac{1}{2} F_{PP} \sigma^2 P^2 + F_{PC} \rho \sigma \eta + \frac{1}{2} F_{CC} \eta^2 \right] dt + F_P dP + F_C dC \quad \text{(A-1)}$$

Consider a storage firm that invests one dollar in an inventory of the commodity, hedging its investment by shorting $(PF_P)^{-1}$ futures contracts. The return on this hedged investment, including the convenience yield is:

$$P^{-1}\left[C - F_P^{-1}\left[-F_\tau + \tfrac{1}{2}F_{PP}\sigma^2 P^2 + F_{PC}\rho\sigma\eta + \tfrac{1}{2}F_{CC}\eta^2 + F_C\alpha(m - C)\right]\right]dt$$
$$-(PF_P)^{-1}F_C\eta dz_C$$

(A-2)

The investment is not riskless because of the influence of the stochastic convenience yield on the futures price. We assume that the risk premium associated with any asset which is perfectly (positively) correlated with the stochastic change in the convenience yield is proportional to the standard deviation of the return on the asset. Then the equilibrium expected return on the above portfolio may be written as $r - (PF_P)^{-1}F_C\lambda_\eta$, where λ is a constant of proportionality. Equating this to the drift term in (A-2) and rearranging, we obtain the following partial differential equation for the futures price:

$$\frac{1}{2}F_{PP}\sigma^2 P^2 + F_{PC}\rho\sigma\eta + \frac{1}{2}F_{CC}\eta^2 + F_P(rP - C) + F_C(\alpha(m - C) - \lambda\eta) - F_\tau - 0$$

(A-3)

The solution to this equation may be written as:

$$F(P, C.\tau) = (P - PVC(C, \tau))e^{r\tau}$$

(A-4)

where

$$PVC(C, \tau) = \frac{m^*}{r}\left(1 - e^{-r\tau}\right) - \frac{m^* - C}{\alpha + r}\left(1 - e^{(\alpha+r)\tau}\right)$$

(A-5)

and $m^* \equiv \frac{\lambda^*}{\alpha}$.

References

Brennan, M.J. (1987) 'The cost of convenience and the pricing of commodity contingent claims', Working Paper, Columbia Futures Center.

Brennan, M.J., (1991) 'The price of convenience and the valuation of commodity contingent claims', in *Stochastic Models and Option Values*, D. Lund and B. Oksendal (eds.), North Holland.

Brennan, M.J., and E.S. Schwartz (1985) 'Evaluating natural resource investments', *Journal of Business* **58**, 133–155.

Cox, J.C., J.E. Ingersoll, and S.A. Ross (1981) 'The relation between forward prices and futures prices', *Journal of Financial Economics* **9**, 321–346.

Culp, C.L., and M.H. Miller (1994) 'Hedging a flow of commodity deliveries with futures: lessons from Metallgesellschaft', *Derivatives Quarterly* **1**, 7–15.

Culp, C.L., and M.H. Miller (1995) 'Metallgesellschaft and the economics of synthetic storage', *Journal of Applied Corporate Finance* **7**, 62–76.

Edwards, F., and M. Canter (1995) 'The collapse of Metallgesellschaft: unhedgeable risks, poor hedging strategy, or just bad luck?', *Journal of Futures Markets*, forthcoming.

Fama E.F., and K.R. French (1987) 'Commodity futures prices: some evidence on forecast power, premiums and the theory of storage', *Journal of Business* **60**, 55–74.

Garbade, K.D. (1993) 'A two-factor, arbitrage-free, model of fluctuations in crude oil futures prices', *Journal of Derivatives* **1**, 86–97.

Gibson, R., and E.S. Schwartz (1990) 'Stochastic convenience yield and the pricing of oil contingent claims', *Journal of Finance* **45**, 959–976.

Jamshidian, F., and M. Fein (1990) 'Closed-form solutions for oil futures and European options in the Gibson-Schwartz model: a note', Working Paper, Merrill Lynch Capital Markets.

Kaldor, N. (1939) 'Speculation and economic stability', *Review of Economic Studies* **7**, 1–27.

Mello, A.A., and J.E. Parsons (199lessons from the Metallgesellschaft debacle', *Journal of Applied Corporate Finance* **8**, 106–120.

Neuberger, A. (1995) 'How well can you hedge long term exposures with multiple short-term futures contracts?', unpublished manuscript, London Business School.

Ross, S.A. (1995) 'Hedging long run commitments: exercises in incomplete market pricing', preliminary draft.

Telser, L.G. (1958) 'Futures trading and the storage of cotton and wheat', *Journal of Political Economy* **66**, 233–244.

Working, H. (1948) 'Theory of the inverse carrying charge in futures markets', *Journal of Farm Economics* **30**, 1–28.

Working, H. (1949) 'The theory of the price of storage', *American Economic Review* **39**, 1254–1262.

Options Pricing and Hedging in Discrete Time with Transaction Costs

Fabio Mercurio and Ton C.F. Vorst

Abstract

If the stock price underlying a European call option evolves according to a binomial model, Boyle and Vorst (1992) derived a unique self-financing strategy replicating exactly the final payoff to a long (short) position under transaction costs. Bensaid *et al.* (1992) and Edirisinghe, Naik and Uppal (1993) lowered the cost of perfect replication (of a long position) by allowing super-replication. Relaxing the assumption of an infinite penalty if the final liability is not met, we derive tighter bounds for reasonable option prices. This is accomplished by two different procedures. The first is a super-replication approach. The latter is a local risk minimization criterion. Both of them rely on hedging as infrequently as possible. The main consequence is that even very risk-averse institutions will find it more attractive to hedge options according to the techniques we propose. The local risk minimization criterion, moreover, also works for American options and any claim whose final payoff is a function only of the underlying stock price at maturity.

Key words: Transaction costs, risky hedging, super-replication, local risk minimization.

1   Introduction

In the theory of pricing and hedging derivative securities, transaction costs are quite commonly neglected. This is not a realistic assumption, but the mathematical tractability of costless transaction problems improves enormously. If the stock price underlying the claim evolves according to a geometric Brownian motion, the introduction of transaction costs prevents the market from being complete, because not all contingent claims can be replicated in a self-financing manner. In the case of a European call option, Leland (1985) derived a strategy whose error of replication is a random variable with mean and variance approaching zero as the trading frequency is increased to infinity. His costs of replicating a long and a short position correspond to Black-Scholes prices with a volatility correction. A similar result has been achieved by Boyle and Vorst (1992), when the evolution of the stock price is described by a binomial model. Contrary to Leland, they derive self-financing strategies perfectly replicating the final payoffs to long and short positions. As a consequence of exact replication, their bounds for the option price are broader. Bensaid, Lesne, Pages and Scheinkman (1992) and Edirisinghe, Naik and Uppal (1993) have shown how super-replication, which requires a strategy whose payoff in all the final states of the world exceeds the liabilities due to long or short option positions, can reduce the cost of perfect replication of a long option position. However, the improvement is relevant only when transaction costs are high and the estimated volatility is small, which seems to rarely occur in practice.

The main purpose of this chapter is to derive tighter bounds for option prices with transaction costs. This is accomplished through two different approaches under the assumption of a finite penalty if the final liability is not met[1]. Both approaches depend on a binomial tree in which we do not hedge at every instant in time but only every fixed number of periods. The first approach is based on super-replication. The initial costs of the super-replicating strategy are higher if one does not change the hedging portfolio at each instant in time. However, one can calculate the expected value and standard deviation of the excess payoff. If in case of a long call position[2] one subtracts this expected value from the initial costs, a lower cost than in Boyle and Vorst's approach results. Moreover, the standard deviation of the error of replication can be small compared with the reduction in costs. This makes the alternative strategy particularly attractive.

The second approach is based on the theory of local risk minimization, a mean-variance criterion, to evaluate and hedge contingent claims. A mean-variance criterion has been introduced for the no transaction costs incomplete market case by Föllmer and Sondermann (1986) and subsequently further developed by Schweizer (1988, 1991, 1995) and Föllmer and Schweizer (1991). In this paper, their approach is extended to the case of proportional transaction costs. Once again, replication is not perfect, but the initial costs are lower while the standard deviations of the replication error are small. The results we obtain are even more satisfactory than those of the first procedure. This indicates that this second approach will have even more appeal to practitioners. In particular, it is possible to find an optimal number of hedging dates. This method also works for American options and any claim whose final payoff is a function only of the underlying stock price at maturity.

The article is organized as follows. In the next section, super-replicating strategies are studied for two cases: a one-period multinomial model and a general trinomial model. Moreover, a tight lower bound for the cost of super-replication of a long call position is provided for more general cases.

---

[1]This assumption contrasts with that of Boyle and Vorst (1992), Bensaid *et al.* (1992) and Edirisinghe *et al.* (1993).

[2]The case of a short call position is perfectly symmetric.

In Section 3, the local risk minimization method is described. Section 4 concludes the chapter.

2 Super-replication of Options

2.1 Basic assumptions

We consider a European call option with strike price X and maturity T.

Given the real constants $u > 1$, $0 < p < 1$, $S > 0$ and the natural number N, we assume that the price of the stock underlying the option follows the Cox-Ross-Rubinstein (1979) binomial model

$$\frac{S_k}{S_{k-1}} = \begin{cases} u & \text{with probability } p \\ d & \text{with probability } 1 - p \end{cases} \quad k = 1, \ldots, N, \qquad (2.1)$$
$$S_0 = S,$$

where $d := 1/u$ and p and $1 - p$ are the probabilities of an up-move and a down-move respectively. We set the risk free interest rate to be equal to zero in every period. In this way, our general problem improves in tractability, but without changing its main features and implications. Finally, we introduce proportional transaction costs in trading the stock, as in Merton (1990), Shen (1990), Boyle and Vorst (1992), Bensaid et al. (1992), Edirisinghe, Naik and Uppal (1993). That is, if S_k is the stock price at time k, buying one share implies the payment of $S_k(1 + \gamma)$, while selling one share involves receiving $S_k(1 - \gamma)$, where γ $(0 < \gamma < 1)$ is the proportional transaction cost coefficient. Hence, the transaction costs are $S_k \gamma$, which we assume include all costs such as direct costs payable to the exchange and indirect costs due to bid-ask spreads, liquidity and market impact. We explicitly assume there are no fixed costs per transaction, which, especially for large parties, is not too unrealistic.

If hedging is allowed at any period, Boyle and Vorst (1992) have proved the existence of a self-financing strategy which perfectly replicates the final payoff to a long position in a European call option with asset delivery. We call the initial value of this strategy the *cost of perfect replication*. Bensaid et al. (1992) and Edirisinghe, Naik and Uppal (1993) have shown how the super-replication[3] of the final payoff to a long position can lead to a lower initial cost of replication. The reverse holds for a short position. We refer to their costs as *costs of super-replication*. The cases of perfect replication and super-replication implicitly assume an infinite cost if we do not meet the final liability. In other words, hedging must be completely riskless. At the end of

---

[3]Replication requires the equality of payoffs and liabilities in all the possible states at the maturity of the option. Super-replication requires that the final payoff of a feasible strategy is equal to or exceeds the liabilities from the option position in all possible final states of the world.

this section we shall show how it is possible to decrease the expected cost of super-replication of a long call position by allowing the hedging strategy to bear a certain risk.

Assuming that hedging is only allowed at dates $\mathcal{D} = \{0 =: i_0 < i_1 < \cdots < i_{M-1}\}$ with $M < N$, the observed stock price follows the multinomial model (with $j = 1, \ldots, \alpha_k - 1; k = 1, \ldots, M$)

$$
\begin{aligned}
\frac{S_k}{S_{k-1}} &= \begin{cases} u^{\alpha_k} & \text{with probability } p^{\alpha_k} \\ \vdots & \vdots \\ u^{2j-\alpha_k} & \text{with probability } \binom{\alpha_k}{j} p^j (1-p)^{\alpha_k - j} \\ \vdots & \vdots \\ u^{-\alpha_k} & \text{with probability } (1-p)^{\alpha_k} \end{cases} \\
S_0 &= S,
\end{aligned}
\tag{2.2}
$$

where $\alpha_k := i_k - i_{k-1}$ and $i_M := N$. We denote by $(\Omega, \mathcal{F}, P)$ the probability space underlying the model (2.2), where $\Omega := \{\omega = (\omega_1, \ldots, \omega_M) : \omega_k \in \{0, \ldots, \alpha_k\} \; \forall k = 1, \ldots, M\}$. We put $\bar{\omega}_k := (\omega_1, \ldots, \omega_k)$, $\mathcal{F}_k := \sigma(S_j : j \leq k)$ and we write $\bar{\omega}_{k+1} := (\bar{\omega}_k, \omega_{k+1})$ with $(\bar{\omega}_0, \omega_1) := \omega_1$.

2.2 The optimization problem and its solution

We consider a similar problem to Bensaid *et al.* (1992) and Edirisinghe, Naik and Uppal (1993). We want to find the lowest cost of super-replicating the payoff to a long position in the given option under proportional transaction costs, where the stock price evolves according to (2.2)[4]. If Δ_k and B_k indicate respectively the *number of stock shares* and the *cash* held at time k, the strategy (Δ_k, B_k) is required to be *self-financing*. This means that the value of (Δ_k, B_k) at the following period $k + 1$ must be larger or equal than the value of (Δ_{k+1}, B_{k+1}) at time $k + 1$ plus the transaction costs due to the portfolio rebalancing. We call $\mathcal{P}_\mathcal{D}$ our optimization problem corresponding to the choice $\mathcal{D}$ of trading dates, and we write $\chi \in \mathcal{F}$ to indicate a stochastic process χ adapted to the filtration $\mathcal{F}_k$ whose value at time k is χ_k. $\mathcal{P}_\mathcal{D}$ is formally described by

$$
\min_{\Delta, B \in \mathcal{F}} \Delta_0 S_0 + B_0,
$$

subject to the self-financing constraints

$$
\begin{aligned}
&[\Delta_k(\bar{\omega}_{k-1}, \omega_k) - \Delta_{k-1}(\bar{\omega}_{k-1})] S_{k-1}(\bar{\omega}_{k-1}) u^{2\omega_k - \alpha_k} + B_k(\bar{\omega}_{k-1}, \omega_k) \\
&- B_{k-1}(\bar{\omega}_{k-1}) + \gamma |\Delta_k(\bar{\omega}_{k-1}, \omega_k) - \Delta_{k-1}(\bar{\omega}_{k-1})| S_{k-1}(\bar{\omega}_{k-1}) u^{2\omega_k - \alpha_k} \leq 0 \\
&\forall \bar{\omega}_{k-1} \in \mathcal{F}_{k-1}; \; \forall \omega_k \in \{0, \ldots, \alpha_k\}; \; \forall k = 1, \ldots, M-1,
\end{aligned}
\tag{2.3}
$$

---

[4]Notice that, since $M < N$, the perfect replication of options is not achievable also in case the transaction costs are neglected (see also Mercurio and Vorst (1995)).

where $\Delta_0(\bar{\omega}_0) := \Delta_0$, $B_0(\bar{\omega}_0) := B_0$, $S_0(\bar{\omega}_0) := S$, and subject to the terminal condition

$$
\begin{aligned}
&\Delta_{M-1}(\bar{\omega}_{M-1})S_{M-1}(\bar{\omega}_{M-1})u^{2\omega_M - \alpha_M} + B_{M-1}(\bar{\omega}_{M-1}) \\
&\quad -\gamma|\Delta_M(\bar{\omega}_{M-1},\omega_M) - \Delta_{M-1}(\bar{\omega}_{M-1})|S_{M-1}(\bar{\omega}_{M-1})u^{2\omega_M - \alpha_M} \\
&\geq (S_{M-1}(\omega^{M-1})u^{2\omega_M - \alpha_M} - X)^+ \\
&\qquad \forall\bar{\omega}_{M-1} \in \mathcal{F}_{M-1}; \ \forall\omega_M \in \{0,\dots,\alpha_M\}.
\end{aligned}
\tag{2.4}
$$

When the stock price is described by (2.1) and hedging can take place at any period, Edirisinghe, Naik and Uppal (1993) have shown that their super-replicating strategy is in general path-dependent. This leads to large scale computations, since the number of constraints grows exponentially with the number N of periods. However, Bensaid *et al.* (1992) proved the path-independence of this super-replicating strategy when transaction costs are not too high and the volatility is not too small, i.e. in the cases which usually occur in practice[5]. Hence, although we might sometimes expect the solution of problem $\mathcal{P}_\mathcal{D}$ to be path-dependent, we shall restrict our analysis to path-independent strategies. In this way, not only does the optimization problem remain substantially unchanged, but its tractability improves enormously, allowing explicit solutions and easier computations. We therefore modify $\mathcal{P}_\mathcal{D}$ by adding the new constraint

$$
S_k(\bar{\omega}_k') = S_k(\bar{\omega}_k'') \Rightarrow \Phi(\bar{\omega}_k') = \Phi(\bar{\omega}_k''),
\tag{2.5}
$$

for each $k = 1,\dots,M$ and each $\bar{\omega}_k'$, $\bar{\omega}_k'' \in \mathcal{F}_k$, where $\Phi \in \{\Delta, B\}$. The new problem $\mathcal{P}_\mathcal{D}$ becomes equivalent to the following recursive problem $\mathcal{P}_\mathcal{D}^*$:

$$
\min_{\Delta_k, B_k} \Delta_k S_k + B_k \quad \forall k = 0,\dots,M-1,
$$

given S_k and subject to

$$
\begin{aligned}
&\Delta_k S_k u^{2\omega_{k+1} - \alpha_{k+1}} + B_k - \Delta_{k+1}(S_k u^{2\omega_{k+1} - \alpha_{k+1}})S_k u^{2\omega_{k+1} - \alpha_{k+1}} \\
&\quad -B_{k+1}(S_k u^{2\omega_{k+1} - \alpha_{k+1}}) - \gamma|\Delta_{k+1}(S_k u^{2\omega_{k+1} - \alpha_{k+1}}) - \Delta_k|S_k u^{2\omega_{k+1} - \alpha_{k+1}} \geq 0 \\
&\qquad \forall\omega_{k+1} \in \{0,\dots,\alpha_{k+1}\}; \ k = 0,\dots,M-2
\end{aligned}
\tag{2.6}
$$

and

$$
\begin{aligned}
&\Delta_{M-1}S_{M-1}u^{2\omega_M - \alpha_M} + B_{M-1} \geq \gamma|\Delta_M(S_{M-1}u^{2\omega_M - \alpha_M}) \\
&\quad -\Delta_{M-1}|S_{M-1}u^{2\omega_M - \alpha_M} + (S_{M-1}u^{2\omega_M - \alpha_M} - X)^+; \ \forall\omega_M \in \{0,\dots,\alpha_k\}.
\end{aligned}
\tag{2.7}
$$

A strategy (Δ, B) solving $\mathcal{P}_\mathcal{D}^*$ is called *perfect super-replicating path-independent* (PSRPI). Its initial value is referred to as the *cost of path-independent super-replication*.

---

[5]Hence, in these cases, the perfect super-replicating strategy is equivalent to the perfect replicating strategy.

From now on, we shall assume that our option is exercised with the delivery of the underlying asset[6], i.e., we add to problem $\mathcal{P}_\mathcal{D}^*$ the terminal constraint

$$\Delta_M(S_M) = \begin{cases} 1 & \text{if } S_M \geq X \\ 0 & \text{otherwise} \end{cases} \qquad B_M(S_M) = \begin{cases} -X & \text{if } S_M \geq X \\ 0 & \text{otherwise.} \end{cases} \qquad (2.8)$$

As already shown by Edirisinghe *et al.* (1993) in the case of a binomial model, problem $\mathcal{P}_\mathcal{D}^*$ can be reduced to a linear programming problem also for more general choices of $\mathcal{D}$. However, in this section, our approach is different. In fact, we develop a method which yields explicit solutions and guarantees easier and faster numerical methods.

The possibility of deriving an easy explicit solution to problem $\mathcal{P}_\mathcal{D}^*$ strictly depends on the set $\mathcal{D}$ of trading dates. For a one-period multinomial model and a general trinomial model, this goal is achievable. In more general cases, a tight lower bound for the cost of path-independent super-replication is readily obtained.

2.3 The one-period multinomial model

The easiest characterization of $\mathcal{D}$ we can consider is $\mathcal{D} = \{0\}$, which corresponds to assuming that hedging takes place only at the initial date. Although this is not a case of practical relevance, it provides a useful insight which can be exploited in more general cases. Since $M = 1$, we observe the stock price at time 0 ($S_0 = S$) and at maturity, when its value is modelled by the following random variable

$$S_1 = \begin{cases} Su^N & \text{with probability } p^N \\ \vdots & \vdots \\ Su^{2k-N} & \text{with probability } \binom{N}{k} p^k (1-p)^{N-k} \quad k = 1, \ldots, N-1, \\ \vdots & \vdots \\ Sd^N & \text{with probability } (1-p)^N \end{cases}$$

$$(2.9)$$

The optimization problem $\mathcal{P}_\mathcal{D}^*$ in this case reduces to a one-stage problem which can be readily solved, i.e.,

$$\min_{\Delta, B \in \mathbb{R}} \Delta S + B$$

subject to

$$\Delta S u^{2k-N} + B - \gamma |1_{\{Su^{2k-N} \geq X\}} - \Delta| S u^{2k-N}$$
$$\geq (Su^{2k-N} - X)^+; \quad k = 0, \ldots, N, \qquad (2.10)$$

---

[6]A similar assumption is made in Boyle and Vorst (1992).

where 1_A is the indicator function of the set A. Since hedging is only allowed at the initial date, we expect the cost of super-replication to be prohibitively high and to converge to S when N tends to infinity and u is defined as

$$u := e^{\sigma \sqrt{\frac{T}{N}}}, \tag{2.11}$$

with σ a positive constant. This is confirmed by the following proposition, which provides a unique and explicit solution to problem $\mathcal{P}^*_{\{0\}}$.

Proposition 2.1 *Assuming that* $Sd^N < X < Su^N$ *to avoid trivialities, problem* $\mathcal{P}^*_{\{0\}}$ *has a unique solution* $(\hat{\Delta}, \hat{B})$ *given by*

$$\begin{aligned}
\hat{\Delta} &= \frac{Su^N(1+\gamma)-X}{Su^N(1+\gamma)-Sd^N(1-\gamma)} \\
\hat{B} &= -d^N(1-\gamma)\frac{Su^N(1+\gamma)-X}{u^N(1+\gamma)-d^N(1-\gamma)}
\end{aligned} \tag{2.12}$$

Proof It is evident that the optimal Δ must belong to $[0, 1]$. Therefore, we must simply notice that, when $0 \leq \Delta \leq 1$,

$$\begin{cases} (\Delta-1)Su^N(1+\gamma) \geq -B-X \\ \Delta Sd^N(1-\gamma)+B \geq 0 \end{cases} \Rightarrow \begin{cases} (\Delta-1)Su^{2k-N}(1+\gamma) \geq -B-X \\ \Delta Sd^{2k-N}(1-\gamma)+B \geq 0 \end{cases}$$

for every $k = 0, \ldots, N$. Since the reverse implication is obvious, these two systems of inequalities are equivalent. This implies that minimizing $\Delta S + B$ subject to (2.10) is equivalent to minimizing $\Delta S + B$ subject to the constraint represented by the first system. Applying Theorem 1 of Boyle and Vorst (1992) we obtain (2.12). $\square$

2.4   The trinomial model

Let us now suppose that N is even and that, starting from the initial model (2.1), hedging can take place only at even trading dates. That is, we set $\mathcal{D} = \{0, 2, \ldots, N-2\}$. We denote by $\mathcal{P}^*_{\text{trin}}$ the corresponding optimization problem. Also, in this specific case it is possible to derive an explicit solution to the optimization problem. The idea leading to the solution is very simple and is suggested by the previous case of a one-period model. At each node of the lattice describing the evolution of the stock price, we have three possible states in the subsequent period: an up-state, an equal-state and a down-state. If we forget the equal-state for a moment, the optimization problem reduces to that of Boyle and Vorst (1992), with the only difference that u is replaced by u^2. To prove the optimality of the strategy thus derived, it is be sufficient to verify that this strategy satisfies the self-financing constraint also when the stock price moves to an equal-state. This is done in the appendix.

Proposition 2.2 *Problem* $\mathcal{P}^*_{\text{trin}}$ *has a unique solution* (Δ^*, B^*) *given recursively by (2.8) and, for each* $k = 0, \ldots, M - 1$, *by*

$$
\begin{aligned}
\Delta^*_k = \Delta^*_k(S_k) &:= \frac{\Delta^*_{k+1}(S_k u^2) S_k \bar{u} - \Delta^*_{k+1}(S_k d^2) S_k \bar{d} + B_{k+1}(S_k u^2) - B_{k+1}(S_k d^2)}{S_k \bar{u} - S_k \bar{d}}, \\
B^*_k = B^*_k(S_k) &:= \frac{\bar{u}[\Delta^*_{k+1}(S_k d^2) S_k \bar{d} + B_{k+1}(S_k d^2)] - \bar{d}[\Delta^*_{k+1}(S_k u^2) S_k \bar{u} + B_{k+1}(S_k u^2)]}{\bar{u} - \bar{d}},
\end{aligned} \tag{2.13}
$$

where $\bar{u} := u^2(1 + \gamma)$ *and* $\bar{d} := d^2(1 - \gamma)$.

Proof See the appendix. □

The previous proposition implies that the cost of path-independent super-replication of a long position in a European call option, when the underlying stock price evolves according to a trinomial model, corresponds to the cost of perfect replication when the number of periods is $N/2$ and the size of an up-move is equal to u^2 .[7]. Furthermore, since Boyle and Vorst (1992) showed that their cost of perfect replication can be approximated by a Black-Scholes price with a corrected volatility, the same must also hold for the cost of path-independent super-replication as given by

$$
V^*_0 := \Delta^*_0(S)S + B^*_0(S). \tag{2.14}
$$

This is explained in the following

Proposition 2.3 *Define* u *as in (2.11), then for large values of* N *and small values of* γ, V^*_0 *can be approximated by the Black-Scholes price* $V_{\text{BS}}(S, X, \hat{\sigma}, T)$, *where* $\hat{\sigma} := \sigma\sqrt{2 + (2\gamma\sqrt{N})/(\sigma\sqrt{T})}$.

Proof Due to definition (2.11) the implied volatility in a binomial model like (2.1) with size of an up-move equal to u^2 and number of periods equal to $M = N/2$ is $\sigma\sqrt{2}$. Theorem 3 in Boyle and Vorst (1992) states that the cost of perfect replication can be approximated by a Black-Scholes price with volatility equal to $\sigma\sqrt{1 + (2\gamma\sqrt{N})/(\sigma\sqrt{T})}$, where σ is the volatility implied by the model and N is the number of periods. Combining these two results leads to the approximation. □

In the case of a binomial model, any super-replicating path independent strategy coincides with the unique perfect replicating strategy. Yet this property is no longer valid whenever the stock price evolves according to (2.2). Perfect replication, in fact, is only achievable in the binomial model case. The effect of path-independent super-replication in all the other cases can be summarized as follows. A PSRPI strategy not only assures a riskless hedge, but also produces an additional positive wealth with positive probability. In the previous case of a trinomial model, this leads to the following

---

[7]It is straightforward realizing that for $\gamma = 0$ we confirm the result by El Karoui *et al.* (1991).

Definition 2.1 *The gain* G *derived from a strategy* (Δ, B) *is the* $\mathcal{F}_M$-*measurable random variable defined by*

$$
\begin{aligned}
G(\omega) := \sum_{k=0}^{M-1} \Big[&\Delta_k(Su^{2W_k(\omega)-i_k})Su^{2W_{k+1}(\omega)-i_{k+1}} + B_k(Su^{2W_k(\omega)-i_k}) \\
&-\Delta_{k+1}(Su^{2W_{k+1}(\omega)-i_{k+1}})Su^{2W_{k+1}(\omega)-i_{k+1}} - B_{k+1}(Su^{2W_{k+1}(\omega)-i_{k+1}}) \\
&-\gamma|\Delta_{k+1}(Su^{2W_{k+1}(\omega)-i_{k+1}}) - \Delta_k(Su^{2W_k(\omega)-i_k})|Su^{2W_{k+1}(\omega)-i_{k+1}} \Big],
\end{aligned}
\tag{2.15}
$$

where $W_k(\omega) := \sum_{j=1}^{k} \omega_j$ ($W_0(\omega) := 0$).

If we want to calculate the mean and the variance of the gain derived from the optimal strategy (2.13), the previous definition (2.15) leads to large computational problems. For this purpose the following proposition is much more convenient, since it provides two formulae that can be readily calculated by backward recursion.

Proposition 2.4 *The mean and the second moment of the gain* G^* *derived from the optimal strategy* (2.13) *with respect to the probability measure* P *underlying our trinomial model are given by*

$$
E(G^*) = 2 \sum_{k=0}^{M-1} \sum_{j=0}^{2k} G_k^*(Su^{2(j-k)}) \binom{2k}{j} p^{j+1}(1-p)^{2k-j+1}
\tag{2.16}
$$

$$
\begin{aligned}
E((G^*)^2) &= 2 \sum_{k=0}^{M-1} \sum_{j=0}^{2k} \left[G_k^*(Su^{2(j-k)})\right]^2 \binom{2k}{j} p^{j+1}(1-p)^{2k-j+1} \\
&+ 8 \sum_{k=0}^{M-2} \sum_{j=k+1}^{M-1} \sum_{i=0}^{2k} \sum_{l=0}^{2(j-k-1)} G_k^*(Su^{2(i-k)})G_j^*(Su^{2(i-j+l+1)}) \times \\
&\qquad \binom{2k}{i} \binom{2(j-k-1)}{l} p^{i+l+2}(1-p)^{2j-i-l},
\end{aligned}
\tag{2.17}
$$

where $G_k^*(S_k) := V_k^*(S_k) - V_{k+1}^*(S_k) - \gamma|\Delta_{k+1}^*(S_k) - \Delta_k^*(S_k)|S_k$.

Some numerical examples are provided in Table 1, where p is defined by

$$
p := \frac{e^{\mu(T/N)} - d}{u - d},
\tag{2.18}
$$

with μ a positive constant such that $p < 1$.

We set the volatility σ to be 20% per annum, the rate μ of expected return to be 7% per annum, the spot price to be $50 and the maturity to be 0.5 years. We then consider three different values for γ (0.5%, 1%, 2%), three different values for the strike price X ($45, $50, $55) and four different values for the number N of periods (50, 100, 200, 500). Finally, we list the resulting costs (2.14) of path-independent super-replication (V_0^*) and the means and the

Table 1. Cost of path-independent super-replication of a long position in a European call option with delivery of the underlying asset. The trinomial model case.

$\sigma = 0.2$, $\mu = 0.07$, $S = 50$, $T = 0.5$								
X = 45								
	$\gamma=0.005$			$\gamma =0.01$			$\gamma=0.02$		
N	$V_0^{*\,(a)}$	$E(G^*)^{(b)}$	$\sigma(G^*)^{(c)}$	V_0^*	$E(G^*)$	$\sigma(G^*)$	V_0^*	$E(G^*)$	$\sigma(G^*)$
50	7.199	0.984	0.481	7.570	1.110	0.509	8.245	1.330	0.552
100	7.344	1.047	0.458	7.841	1.229	0.491	8.721	1.541	0.536
200	7.559	1.146	0.453	8.219	1.401	0.494	9.351	1.829	0.542
500	7.943	1.312	0.467	8.888	1.697	0.516	10.443	2.326	0.566
X = 50								
	$\gamma=0.005$			$\gamma =0.01$			$\gamma=0.02$		
N	V_0^*	$E(G^*)$	$\sigma(G^*)$	V_0^*	$E(G^*)$	$\sigma(G^*)$	V_0^*	$E(G^*)$	$\sigma(G^*)$
50	4.497	1.373	0.551	4.926	1.509	0.578	5.693	1.740	0.617
100	4.621	1.389	0.488	5.197	1.594	0.521	6.191	1.932	0.565
200	4.871	1.499	0.460	5.627	1.784	0.498	6.893	2.249	0.545
500	5.323	1.697	0.458	6.386	2.124	0.505	8.097	2.801	0.556
X = 55								
	$\gamma=0.005$			$\gamma =0.01$			$\gamma=0.02$		
N	V_0^*	$E(G^*)$	$\sigma(G^*)$	V_0^*	$E(G^*)$	$\sigma(G^*)$	V_0^*	$E(G^*)$	$\sigma(G^*)$
50	2.621	1.293	0.594	3.037	1.435	0.628	3.791	1.677	0.679
100	2.780	1.364	0.561	3.337	1.572	0.602	4.320	1.918	0.660
200	3.008	1.470	0.554	3.749	1.763	0.606	5.014	2.240	0.671
500	3.441	1.668	0.571	4.498	2.111	0.637	6.230	2.811	0.717

(a): cost of path-independent super-replication as defined by (2.14); (b),(c): mean and standard deviation of the gain G^*, computed according to (2.16) and (2.17).

standard deviations of the corresponding gains G^* (derived from the PSRPI strategy (2.13)). As expected, V_0^* increases with the number of periods and the coefficient of transaction costs. Moreover, V_0^* is always larger than the corresponding cost of perfect replication[8]. If we skip a certain hedging date, we must face the risk derived by the impossibility of portfolio rebalancing at that date. Hence, a larger initial endowment is required to fully cover the new exposure. However, we also observe that the difference between V_0^* and the expected gain $E(G^*)$ is always lower than the cost of perfect replication (of a long position), and it is always higher than the cost of perfect replication of the corresponding short position, which makes the value $V_0^* - E(G^*)$ free of any arbitrage opportunity. Therefore, a hedger might find following our strategy (2.13) more attractive than riskless hedging as given by the perfect replicating strategy. The comparison of these different strategies can be performed only after the definition of a suitable utility function. This will be done at the end of Section 3.

---

[8]We refer to Table 2 for a list of such values.

2.5   The general multinomial model

An explicit solution to problem $\mathcal{P}_{\mathcal{D}}^*$ is not always readily available for every choice $\mathcal{D}$ of the set of trading dates. Calling $(\Delta^{\mathcal{D}}, B^{\mathcal{D}})$ the obvious extension of (2.13) to the case described by $\mathcal{D}$, $(\Delta^{\mathcal{D}}, B^{\mathcal{D}})$ may fail to be optimal, since it may not satisfy the corresponding self-financing constraint. Therefore, $\Delta_0^{\mathcal{D}} S + B_0^{\mathcal{D}}$ provides only a lower bound for the cost of path-independent super-replication associated with $\mathcal{P}_{\mathcal{D}}^*$. In fact, if we exclude at each trading date all the possible moves of the stock price in the next period except for the two extreme ones, we face a minimization problem with a broader admissible region.

A straightforward generalization of Proposition 2.3 is given by the following proposition which yields a valuable characterization of this lower bound for the cost of path-independent super-replication.

Proposition 2.5 *Let us assume that price process corresponding to $\mathcal{D}$ describes an $(\alpha + 1)$-nomial model, with $\alpha \in \mathbb{N}$, $\alpha > 1$. If u and p are defined by (2.11) and (2.18), for large values of N and small values of γ, $\Delta_0^{\mathcal{D}} S + B_0^{\mathcal{D}}$ can be approximated by the Black-Scholes price $V_{\mathrm{BS}}(S, X, \hat{\sigma}, T)$, where $\hat{\sigma} := \sigma\sqrt{\alpha + (2\gamma\sqrt{N})/(\sigma\sqrt{T})}$.*

A natural question at this point is whether $(\Delta^{\mathcal{D}}, B^{\mathcal{D}})$ represents a close approximation to the PSRPI strategy. A possible way to evaluate the performance of $(\Delta^{\mathcal{D}}, B^{\mathcal{D}})$ is the definition of two gain random variables. The first, called G^+, must describe the (positive) gains accumulated by hedging till maturity. The second, called G^-, must count the (negative) losses which occur if $(\Delta^{\mathcal{D}}, B^{\mathcal{D}})$ is not self-financing. The closer the ratio $E(G^-)/E(G^+)$ is to zero, the closer $\Delta_0^{\mathcal{D}} S + B_0^{\mathcal{D}}$ is to the cost of path-independent super-replication[9]. However, also in cases when $(\Delta^{\mathcal{D}}, B^{\mathcal{D}})$ is (or is very close to be) self-financing, the variance of the random variable $G := G^+ + G^-$ might be too large. In other words, hedging might be quite risky. This is one of the reasons why we introduce a criterion of risk minimization in Section 3.[10]

---

[9]From an empirical point of view, the difference between $\Delta_0^{\mathcal{D}} S + B_0^{\mathcal{D}}$ and the cost of path-independent super-replication is often negligible, but not always equal to zero.

[10]In this section, we have only analyzed the problem of deriving the cost of path-independent super-replication of a long position in a European call option with asset delivery. The corresponding problem for the short position case does not admit, in fact, a general satisfactory solution. Assigning realistic values to the coefficients in our models, meaningless answers like 0 or $(S - X)^+$ are obtained in the case of a one-period multinomial model and in the case of a trinomial model. In the last case, our result is not really astonishing since a similar phenomenon has already been encountered by El-Karoui et al. (1991) where transaction costs are neglected. This impossibility to derive reasonable costs of path-independent super-replication of a short position is another reason why we shall introduce a different criterion to evaluate European call options in discrete time, incomplete markets and under transaction costs. This criterion will lead to more satisfactory results and it will be also applicable to more general claims.

3   A Local Risk Minimization Approach

The nature of the binomial model (2.1) determines the existence of a unique self-financing strategy which perfectly replicates the final payoff to a given claim. The same property, however, does not hold if we assume that the underlying stock price evolves according to models like (2.2). In these cases, in fact, the corresponding financial markets are incomplete. One possible way to tackle the problem of hedging claims under incompleteness and transaction costs has been illustrated in the previous section. The super-replication of the final payoff of the claim is a very promising criterion since it assures completely riskless hedging. Nevertheless, the initial costs of these super-replications are prohibitive, although they are partially offset in expectation by excess payoffs. In this section, we shall analyze a different approach, a mean-variance criterion for hedging general claims in incomplete markets and under proportional transaction costs. When transaction costs are neglected, a mean-variance criterion has been introduced by Föllmer and Sondermann (1986) and subsequently developed by Schweizer (1988, 1991, 1995), and Föllmer and Schweizer (1991)[11]. The difference between the payoffs of a given claim and of any self-financing strategy is a random variable, which cannot in general be set identically equal to zero, due to market incompleteness. A mean-variance criterion provides the strategy, called *total risk minimizing*, which minimizes the variance of this random variable. A slightly different problem consists in the minimization of what is called *local risk*, which is defined as the expected squared deviation from local perfect replication. In this section, we shall extend this latter approach to cases with proportional transaction costs[12].

We consider the payoff of a contingent claim H, a discrete finite random variable. In the examples and applications developed in this section, H will be mostly viewed as a European call option. However, the method we propose is applicable to all the claims which are functions of the value of the underlying stock at maturity[13]. It also works for American options.

We suppose that the stock price underlying the claim follows the multinomial model (2.2) for a given choice of the subdivision $\mathcal{D} = \{0 =: i_0 < i_1 < \cdots < i_{M-1}\}$. We shall simply write k to denote the trading time i_k.

A pair (Δ, B) is called a *strategy* if Δ and B are finite-valued and adapted with respect to the filtration $\mathcal{F}_k$ generated by (2.2).

---

[11]We just quote some references.

[12]A similar criterion has been derived by Hipp (1993).

[13]For example to claims like bull, bear, butterfly spreads, straddles, strips, straps, strangles.

Definition 3.1 *The* local risk *associated with the strategy* (Δ, B) *is defined, for each* $k = 0, \ldots, M - 1$, *by*

$$
\begin{aligned}
R_k(\Delta, B) &= \varphi(\Delta_k, V_k) \\
&:= E\left\{ [V_{k+1} - V_k - \Delta_k(S_{k+1} - S_k) + \gamma|\Delta_{k+1} \right. \\
&\qquad \left. -\Delta_k|S_{k+1}]^2 \Big| \mathcal{F}_k \right\},
\end{aligned}
\tag{3.1}
$$

where $V_k := \Delta_k S_k + B_k$ *for each* $k = 0, \ldots, M - 1$.

The random variable

$$
D_k(\Delta, V) := V_k - V_{k-1} - \Delta_{k-1}(S_k - S_{k-1}) + \gamma|\Delta_k - \Delta_{k-1}|S_k
$$

calculates the deviation of the strategy (Δ, B) from local perfect replication at time k[14]. Therefore, due to Definition 3.1, minimizing the local risk $R_k(\Delta, B)$ at time k is equivalent to minimizing the second moment of $D_k(\Delta, V)$.

Proposition 3.2 below states that $E[D_k(\Delta, V)|\mathcal{F}_{k-1}] = 0$ if (Δ, B) minimizes the local risk $R_k(\Delta, B)$. As a consequence, minimizing the local risk $R_k(\Delta, B)$ at time i_k is also equivalent to minimizing the variance of the deviation $D_k(\Delta, V)$ from local perfect replication, under the condition that this deviation must be zero in expectation. Formally, we face the recursive problem

$$
\min_{(\Delta_k, V_k) \in \mathbb{R}^2} R_k(\Delta, B)
$$

for each $k = 0, \ldots, M - 1$, subject to the following conditions at the final state

$$
\begin{aligned}
\Delta_M &= \bar{\Delta}_M, \\
V_M &= H,
\end{aligned}
\tag{3.2}
$$

where $\bar{\Delta}_M$ represents a given amount of shares the seller must hold at maturity M. For example, in the case of a European call option with strike price X and delivery of the underlying asset, $\bar{\Delta}_M = 1_{\{S_M \geq X\}}$.

A strategy solving this optimization problem is called *local risk minimizing*.

Proposition 3.1 *A* local risk minimizing strategy *exists*.

Proof See the appendix. □

Before providing some numerical simulations, we would like to stress the main justifications for our choice of the local risk minimization technique. When the given claim is not achievable by a self-financing strategy, a natural objective is to try to minimize the variance of the error of replication. This

---

[14]In case of market completeness, the random variable $D_k(\Delta, V)$ is identically equal to zero for suitable values of Δ and V.

task is performed by a total risk minimizing strategy, i.e., a strategy, if any, solving the problem

$$\min E\left[\left(H - y - \sum_{k=1}^{M}\Delta_{k-1}(S_k - S_{k-1}) + \gamma\sum_{k=1}^{M}|\Delta_k - \Delta_{k-1}|S_k\right)^2\right]$$

over all the initial endowments $y \in \mathbb{R}$ and over all the strategies (Δ, B). If transaction costs are neglected, Schweizer (1988, 1995) and Schäl (1994) proved the equality of the initial values of total and local risk minimizing strategies. Moreover, according to simulations where the stock price evolution is described by multinomial models, the riskiness of the former strategy turns out to be only slightly smaller than that of the latter. All these considerations make us suppose that even when the problem of total risk minimization with transaction costs has a solution, there is no valuable improvement over the solution to the local risk minimization problem. Furthermore, a local risk minimizing strategy is path-independent, in contrast with the path-dependence of any total risk minimizing strategy. From a computational point of view, this difference is of course quite significant.

Definition 3.2 *The loss $L(\Delta, B, Y)$ derived from the strategy (Δ, B) with respect to the initial endowment Y is the $\mathcal{F}_M$-measurable random variable defined by*

$$L(\Delta, B, Y) := H - Y - \sum_{k=1}^{M}\Delta_{k-1}(S_k - S_{k-1}) + \gamma\sum_{k=1}^{M}|\Delta_k - \Delta_{k-1}|S_k. \quad (3.3)$$

$L(\Delta, B, Y)$ can be viewed as the random variable which measures the deviation of (Δ, B) from the total (i.e. from the initial time until maturity) perfect replication of H, when the starting wealth is Y. This random variable can take both positive and negative values.

Proposition 3.2 *The loss derived from any local risk minimizing strategy (Δ, B), with respect to the initial endowment Y, has expectation*

$$E(L(\Delta, B, Y)) = \Delta_0 S_0 + B_0 - Y \quad (3.4)$$

and variance

$$\sigma^2(L(\Delta, B, Y)) = \sum_{k=0}^{M-1} E(R_k(\Delta, B)). \quad (3.5)$$

Proof See the appendix. □

In models where transaction costs are neglected, the existence of a local risk minimizing strategy implies also its uniqueness. But if $\gamma > 0$, it is not clear whether the uniqueness still holds true. However, this is not a main issue. In fact, the set of all the local risk minimizing strategies is finite. Hence, we can always select a strategy through, for instance, the criterion outlined in the sequel. The initial value of the selected strategy is referred to as the *cost of local risk minimization*.

Table 2. Costs of local risk minimization for long positions in European call options with delivery of the underlying asset. The case of a trinomial model.

$\sigma = 0.2$, $\mu = 0.07$, $S = 50$, $T = 0.5$								
$X = 45$								
	$\gamma = 0.005$			$\gamma = 0.01$			$\gamma = 0.02$		
N	V_0[a]	$\sigma(L)$[b]	V_{BV}[c]	V_0	$\sigma(L)$	V_{BV}	V_0	$\sigma(L)$	V_{BV}
50	6.168	0.284	6.375	6.449	0.313	6.810	7.046	0.349	7.579
100	6.271	0.214	6.553	6.680	0.238	7.131	7.540	0.261	8.119
200	6.432	0.162	6.798	7.028	0.180	7.556	8.217	0.188	8.808
500	6.772	0.111	7.250	7.703	0.122	8.310	9.393	0.125	9.989
1000	7.167	0.082	7.713	8.472	0.083	9.055	10.642	0.076	11.120
$X = 50$								
	$\gamma = 0.005$			$\gamma = 0.01$			$\gamma = 0.02$		
N	V_0	$\sigma(L)$	V_{BV}	V_0	$\sigma(L)$	V_{BV}	V_0	$\sigma(L)$	V_{BV}
50	3.176	0.403	3.446	3.542	0.430	3.988	4.279	0.460	4.898
100	3.324	0.294	3.680	3.840	0.318	4.381	4.861	0.335	5.523
200	3.534	0.217	3.984	4.262	0.234	4.880	5.638	0.236	6.299
500	3.955	0.146	4.524	5.075	0.151	5.740	7.004	0.143	7.601
1000	4.427	0.106	5.062	5.924	0.105	6.573	8.314	0.093	8.833
$X = 55$								
	$\gamma = 0.005$			$\gamma = 0.01$			$\gamma = 0.02$		
N	V_0	$\sigma(L)$	V_{BV}	V_0	$\sigma(L)$	V_{BV}	V_0	$\sigma(L)$	V_{BV}
50	1.430	0.378	1.669	1.753	0.413	2.164	2.436	0.454	3.031
100	1.550	0.284	1.873	2.019	0.312	2.528	2.994	0.334	3.638
200	1.734	0.214	2.150	2.412	0.234	3.005	3.749	0.241	4.408
500	2.122	0.145	2.661	3.196	0.152	3.851	5.120	0.145	5.724
1000	2.568	0.107	3.182	4.003	0.113	4.683	6.451	0.096	6.982

(a): cost of local risk minimization; (b): standard deviation of the loss derived from the selected local risk minimizing strategy with respect to V_0, computed according to (3.5); (c) cost of perfect replication.

3.1   Numerical examples

Some numerical examples of the solution of the problem of local risk minimization are listed in the Tables 2, 3, 4 and 5, where the cases of long and short positions in a European call option with delivery of the underlying asset are analyzed. This enables us to compare the results of our local risk minimization with those of super-replication (as developed in the previous section) and with those of perfect replication.

The tables are provided under the assumptions that

Table 3. Costs of local risk minimization for long positions in European call options with delivery of the underlying asset. The case of a 6-nomial model.

$\sigma = 0.2$, $\mu = 0.07$, $S = 50$, $T = 0.5$								
$X = 45$								
	$\gamma = 0.005$			$\gamma = 0.01$			$\gamma = 0.02$		
N	V_0 (a)	$\sigma(L)$ (b)	V_{BV} (c)	V_0	$\sigma(L)$	V_{BV}	V_0	$\sigma(L)$	V_{BV}
50	6.079	0.505	6.375	6.269	0.527	6.810	6.640	0.569	7.579
100	6.151	0.372	6.553	6.408	0.394	7.131	6.894	0.437	8.119
200	6.254	0.274	6.798	6.599	0.297	7.556	7.233	0.340	8.808
500	6.452	0.184	7.250	6.958	0.207	8.310	7.846	0.249	9.989
1000	6.664	0.138	7.713	7.331	0.160	9.055	8.460	0.201	11.120
$X = 50$								
	$\gamma = 0.005$			$\gamma = 0.01$			$\gamma = 0.02$		
N	V_0	$\sigma(L)$	V_{BV}	V_0	$\sigma(L)$	V_{BV}	V_0	$\sigma(L)$	V_{BV}
50	3.066	0.712	3.446	3.322	0.728	3.988	3.800	0.761	4.898
100	3.172	0.512	3.680	3.507	0.532	4.381	4.111	0.572	5.523
200	3.309	0.371	3.984	3.747	0.393	4.880	4.512	0.436	6.299
500	3.561	0.245	4.524	4.181	0.269	5.740	5.219	0.314	7.601
1000	3.824	0.181	5.062	4.621	0.206	6.573	5.913	0.250	8.833
$X = 55$								
	$\gamma = 0.005$			$\gamma = 0.01$			$\gamma = 0.02$		
N	V_0	$\sigma(L)$	V_{BV}	V_0	$\sigma(L)$	V_{BV}	V_0	$\sigma(L)$	V_{BV}
50	1.336	0.678	1.669	1.561	0.702	2.164	1.993	0.748	3.031
100	1.418	0.496	1.873	1.716	0.521	2.528	2.274	0.570	3.638
200	1.534	0.364	2.150	1.930	0.391	3.005	2.651	0.440	4.408
500	1.759	0.243	2.661	2.334	0.270	3.851	3.336	0.319	5.724
1000	1.999	0.181	3.182	2.754	0.208	4.683	4.022	0.256	6.982

(a): cost of local risk minimization; (b): standard deviation of the loss derived from the selected local risk minimizing strategy with respect to V_0, computed according to (3.5); (c) cost of perfect replication.

- the stock price evolves according to a $(r+1)$-nomial model. In Tables 2 and 5, $r = 2$, in Table 3, $r = 5$ and in Table 4, $r = 10$;

- the options are exercised with the asset delivery. The terminal conditions (3.2) then become

$$\Delta_M = 1_{\{S_M \geq X\}}$$
$$V_M = (S_M - X)^+,$$

for a long position and

$$\Delta_M = -1_{\{S_M \geq X\}}$$
$$V_M = -(S_M - X)^+,$$

for a short position;

- the optimal hedge ratio $\Delta(S_k)$ at time k when the stock price is S_k is taken inside the interval $[\Delta_{k+1}(S_k u^{-\alpha_{k+1}}), \Delta_{k+1}(S_k u^{\alpha_{k+1}})]$ in cases of long positions and inside the interval $[\Delta_{k+1}(S_k u^{\alpha_{k+1}}), \Delta_{k+1}(S_k u^{-\alpha_{k+1}})]$ in cases of short positions;

- if, in the backward derivation of a local risk minimizing strategy, $(\Delta_k'(S_k), V_k'(S_k))$ and $(\Delta_k''(S_k), V_k''(S_k))$ are two pairs minimizing the local risk at time k, we select the one with the smaller second component[15]. This procedure is used to tackle the theoretical problem of having more than one local risk minimizing strategy. We shall simply call local risk minimizing the unique strategy thus selected.

The costs of local risk minimization (V_0 in the tables) and the standard deviations of the losses derived from the local risk minimizing strategies with respect to these costs are compared with the corresponding costs of perfect replication. We first analyze the case of replication of a long position. As expected, for any r-nomial model, the cost of local risk minimization increases both with the number N of periods and with the transaction cost coefficient γ. Conversely, the standard deviation $\sigma(L)$ of the loss L decreases as N increases. Moreover, if we increase r keeping N fixed, the cost V_0 decreases, while the standard deviation of the loss increases. Comparing our cost of local risk minimization to that of perfect replication, we see that the former is always smaller than the latter.

Boyle and Vorst (1992) have defined the spread as the difference between the cost of perfect replication with transaction costs and the cost of perfect replication without transaction costs. They have noticed that, for increasing values of the strike price, the spread initially increases, reaches a maximum when the option is at-the-money, and then decreases. We observe a similar phenomenon if we define a new spread as the difference between the cost of perfect replication and our cost of local risk minimization. Only few exceptions are encountered, namely, for high transaction costs and large numbers of periods.

From an empirical point of view, our cost V_0 of local risk minimization gets closer and closer to Leland's cost of replication, if we increase the number N of periods while the number M of hedging dates is kept fixed. This is observed both for the long call value and for the short call value. However, we can not state any convergence result. This is also intuitive due to the quite different nature of our approach compared with that of Leland.

In the case of local risk minimization of a short position, some results are analogous to those of the long position cases. For instance, the standard deviation $\sigma(L)$ decreases with N and is larger for at-the-money options. However,

---

[15]If $V_k'(S_k) = V_k''(S_k)$, from our simulations it turns out that choosing either strategy does make no difference.

Table 4. Costs of local risk minimization for long positions in European call options with delivery of the underlying asset. The case of a 11-nomial model.

$\sigma = 0.2$, $\mu = 0.07$, $S = 50$, $T = 0.5$								
$X = 45$								
	$\gamma = 0.005$			$\gamma = 0.01$			$\gamma = 0.02$		
N	$V_0{}^{(a)}$	$\sigma(L)^{(b)}$	$V_{BV}{}^{(c)}$	V_0	$\sigma(L)$	V_{BV}	V_0	$\sigma(L)$	V_{BV}
50	6.016	0.721	6.375	6.157	0.741	6.810	6.439	0.782	7.579
100	6.070	0.534	6.553	6.257	0.556	7.131	6.621	0.599	8.119
200	6.141	0.394	6.798	6.391	0.418	7.556	6.865	0.463	8.808
500	6.279	0.263	7.250	6.649	0.288	8.310	7.335	0.332	9.989
1000	6.428	0.196	7.713	6.929	0.220	9.055	7.827	0.262	11.120
$X = 50$								
	$\gamma = 0.005$			$\gamma = 0.01$			$\gamma = 0.02$		
N	V_0	$\sigma(L)$	V_{BV}	V_0	$\sigma(L)$	V_{BV}	V_0	$\sigma(L)$	V_{BV}
50	2.979	1.024	3.446	3.170	1.038	3.988	3.537	1.066	4.898
100	3.061	0.741	3.680	3.308	0.760	4.381	3.772	0.798	5.523
200	3.160	0.538	3.984	3.483	0.560	4.880	4.075	0.602	6.299
500	3.340	0.354	4.524	3.806	0.379	5.740	4.630	0.424	7.601
1000	3.532	0.260	5.062	4.147	0.286	6.573	5.197	0.331	8.833
$X = 55$								
	$\gamma = 0.005$			$\gamma = 0.01$			$\gamma = 0.02$		
N	V_0	$\sigma(L)$	V_{BV}	V_0	$\sigma(L)$	V_{BV}	V_0	$\sigma(L)$	V_{BV}
50	1.264	0.985	1.669	1.434	1.006	2.164	1.767	1.047	3.031
100	1.324	0.719	1.873	1.543	0.744	2.528	1.966	0.791	3.638
200	1.404	0.527	2.150	1.693	0.554	3.005	2.238	0.604	4.408
500	1.560	0.350	2.661	1.982	0.378	3.851	2.761	0.429	5.724
1000	1.730	0.259	3.182	2.300	0.287	4.683	3.312	0.337	6.982

(a): cost of local risk minimization; (b): standard deviation of the loss derived from the selected local risk minimizing strategy with respect to V_0, computed according to (3.5); (c) cost of perfect replication.

the cost V_0 of local risk minimization is always higher than the corresponding cost of perfect replication. This result is opposite to the one obtained in the long position case essentially because the payoff to a short position has opposite sign to the payoff of the corresponding long position.

As already seen by Leland (1985) and Boyle and Vorst (1992), the problem of replicating a short position is not always solvable. We encounter a similar problem in the application of our local risk minimization criterion. However, we are able to derive short call values in a broader context, even in cases when the cost of perfect replication is not available.

Table 5. Costs of local risk minimization for short positions in European call options with delivery of the underlying asset. The case of a trinomial model.

$\sigma = 0.2$, $\mu = 0.07$, $S = 50$, $T = 0.5$								
$X = 45$								
	$\gamma = 0.002$			$\gamma = 0.005$			$\gamma = 0.01$		
N	V_0 [a]	$\sigma(L)$ [b]	V_{BV} [c]	V_0	$\sigma(L)$	V_{BV}	V_0	$\sigma(L)$	V_{BV}
50	5.778	0.232	5.686	5.606	0.208	5.362	5.320	0.163	N.A.
100	5.731	0.162	5.593	5.492	0.137	5.142	5.116	0.087	N.A.
200	5.669	0.110	5.465	5.336	0.082	N.A.	5.000	0.031	N.A.
500	5.546	0.062	5.215	5.059	0.029	N.A.	N.A.	N.A.	N.A.
$X = 50$								
	$\gamma = 0.002$			$\gamma = 0.005$			$\gamma = 0.01$		
N	V_0	$\sigma(L)$	V_{BV}	V_0	$\sigma(L)$	V_{BV}	V_0	$\sigma(L)$	V_{BV}
50	2.644	0.351	2.503	2.390	0.328	1.967	1.905	0.289	N.A.
100	2.590	0.239	2.377	2.222	0.212	1.507	1.441	0.166	N.A.
200	2.503	0.160	2.178	1.951	0.130	N.A.	0.250	0.093	N.A.
500	2.315	0.092	1.706	1.244	0.056	N.A.	N.A.	N.A.	N.A.
$X = 55$								
	$\gamma = 0.002$			$\gamma = 0.005$			$\gamma = 0.01$		
N	V_0	$\sigma(L)$	V_{BV}	V_0	$\sigma(L)$	V_{BV}	V_0	$\sigma(L)$	V_{BV}
50	0.981	0.314	0.871	0.780	0.283	0.484	0.439	0.224	N.A.
100	0.927	0.218	0.763	0.646	0.184	0.209	0.180	0.115	N.A.
200	0.853	0.150	0.609	0.453	0.115	N.A.	0.000	0.053	N.A.
500	0.707	0.085	0.302	0.096	0.040	N.A.	N.A.	N.A.	N.A.

(a): cost of local risk minimization; (b): standard deviation of the loss derived from the selected local risk minimizing strategy with respect to V_0, computed according to (3.5); (c) cost of perfect replication.

3.2 Choosing a hedging strategy

Let us now consider the case of a financial institution which has to hedge its short position in a European call option, with delivery of the underlying asset. We call $\hat{V}$ the traded price of the given option. Assuming that the underlying stock price evolves according to the binomial model (2.1), the unique perfect replicating strategy does not necessarily provide the most desirable hedging tool. In fact, if its initial value is larger than $\hat{V}$, perfect replication leads to a sure loss given by the difference between its costs and the actual price $\hat{V}$. As a consequence, a strategy replicating the option only in expectation could be more attractive to the financial institution. Furthermore, if fixed transaction costs are taken into account, it is clear that the hedging frequency should be decreased.

To compare the performances of different strategies we use a derived utility function which mirrors the investor's preferences. We define a derived utility function which possesses a clear probabilistic interpretation[16].

From now on, we consider only partitions $\mathcal{D}$ such that the corresponding price process (2.2) is a $(r+1)$-nomial model for a suitable choice of r. We indicate with (Δ^r, B^r) the corresponding local risk minimizing strategy, as derived according to the procedure previously outlined. We then define the derived utility function

$$U(\Delta^r, B^r) := E(-L(\Delta^r, B^r, \hat{V})) - A\sigma(L(\Delta^r, B^r, \hat{V})), \qquad (3.6)$$

where $A \in \mathbb{R}$ is a coefficient of risk-aversion[17]. We write $(\Delta^{r_1}, B^{r_1}) \succeq (\Delta^{r_2}, B^{r_2})$ if and only if $U(\Delta^{r_1}, B^{r_1}) \geq U(\Delta^{r_2}, B^{r_2})$.

For the long position case, we compare the performance of the perfect replicating strategy (Δ^1, B^1) with that of (Δ^r, B^r), for the values of r considered in Tables 2, 3, 4. If $N = 50$ and, say, $A = 2$, $(\Delta^1, B^1) \succeq (\Delta^r, B^r)$ for each $r \in \{2, 5, 10\}$, each $\gamma \in \{0.005, 0.01, 0.02\}$, and each $X \in \{\$45, \$50, \$55\}$. Therefore, if N is small, a sufficiently risk-averse institution will always choose to hedge at every period and the resulting ask-price will be the cost of perfect replication. However, as N increases, even a very risk-averse institution will have a convenience in hedging according to a local risk minimizing strategy, even when transaction costs are small. In fact, notice for example that $\Delta_0^2 S + B_0^2 + 4\sigma(L(\Delta^2, B^2, \Delta_0^2 S + B_0^2)) < \Delta_0^1 S + B_0^1$, for $N = 500$, $X = \$45$ and any considered γ. We also observe that the larger the coefficient of transaction costs, the more convenient a local risk minimizing approach is.

If the financial institution realizes that it is more profitable skipping some hedging dates, is there any optimal number of hedging dates? In the previously considered case with $\gamma = 0.005$, the best policy is to hedge every 5 periods, since $(\Delta^5, B^5) \succeq (\Delta^2, B^2)$ and $(\Delta^5, B^5) \succeq (\Delta^{10}, B^{10})$. However, for higher values of γ, it is preferable hedging every 10 periods. In other words, there always exist r^* and $M^* = 1/r^*$, optimal values for r and M respectively. As expected, r^* increases (M^* decreases) as γ increases.

---

[16]There are numerous papers on option pricing under transaction costs based on an expected utility maximization. See, for instance, Hodges and Neuberger (1989), Dumas and Luciano (1991), Davis, Panas and Zariphopoulou (1992).

[17]The utility function (3.6) has a precise probabilistic interpretation. From (3.4) and Chebyshev's inequality we get

$$Prob\{|L(\Delta^r, B^r, \Delta_0^r S + B_0^r)| \geq t\sigma(L(\Delta^r, B^r, \Delta_0^r S + B_0^r))\} \leq \frac{1}{t^2}.$$

Therefore, having a coefficient A of risk-aversion is equivalent to require that hedging must produce losses with a smaller probability than $\frac{1}{A^2}$. This probability is even lower in practice. In fact, not only is Chebyshev's inequality universally applicable, but also it concerns the absolute value of a random variable. In our case, large negative losses (i.e. large positive gains) are pleasingly welcomed. Furthermore, Chebyshev's inequality provides a constructive criterion for the definition of investors' coefficient of risk-aversion.

Whenever the financial institution has found its optimal r^*, the resulting ask-price must be

$$\hat{V}_a = \Delta_0^{r^*} S + B_0^{r^*} + A\sigma(L(\Delta^{r^*}, B^{r^*}, \Delta_0^{r^*} S + B_0^{r^*})) \qquad (3.7)$$

to be consistent with previous arguments. In this way, hedging according to (Δ^{r^*}, B^{r^*}) produces an expected positive gain of $A\sigma(L(\Delta^{r^*}, B^{r^*}, \Delta_0^{r^*} S + B_0^{r^*}))$ and will lead to a final loss only with a probability smaller than $1/A^2$.

A similar approach can be employed for the derivation of a bid-price. For example, if $N = 500$, $\gamma = 0.005$ and $X = \$50$, the bid-price for an institution with a coefficient of risk-aversion of 4 is 1.244-4*0.056=1.02. Notice that the cost of perfect replication is not available in this case.

In the first section we developed a super-replication approach under incompleteness, to derive an alternative to perfect replication. The question we need to address is whether the strategy given by Proposition 2.2 might be advisable under certain conditions. The answer seems to be negative in general. Our PSRPI strategy (2.13) can produce a higher utility than the corresponding (Δ^2, B^2), but is always improved by (Δ^{r^*}, B^{r^*}).

Conclusions

If the stock price underlying a given European call option evolves according to a binomial model, Boyle and Vorst (1992) derived a self-financing strategy perfectly replicating the final payoff of the option itself. In this article we provide tighter bounds for an option price inclusive of transaction costs. This is accomplished by allowing risky hedging. We develop two different criteria for pricing and hedging options in discrete time, incomplete markets and under proportional transaction costs. The first one is a super-replication approach for the case of a multinomial model. Its main feature is the possibility of deriving an initial cost which is independent of investors' preferences. Explicit formulae have been derived for one-period models and trinomial models. The second approach consists in the minimization of the squared local deviation from perfect replication. This method works also for American options and, more generally, for any claim whose payoff is a function of the underlying asset at maturity. Contrary to the first approach, it explicitly depends on investors' preferences and predictions. The conclusion from our numerical experiments is that even very risk-averse institutions will find it convenient to skip some trading dates for rebalancing and hedge according to our local risk minimizing strategies.

A   Appendix

To prove Proposition 2.2 we need the following

Lemma A.1 *If* (Δ^*, B^*) *is given by (2.8) and (2.13), then for each* $k = 0, \ldots, M - 1$,

$$\Delta_k^*(S_k)S_k + B_k^*(S_k) - \Delta_{k+1}^*(S_k)S_k - B_{k+1}^*(S_k) - \gamma|\Delta_{k+1}^*(S_k) - \Delta_k^*(S_k)|S_k \geq 0. \tag{A.1}$$

Proof The proof is by backwards induction on the number of steps in our trinomial model. In period $M - 1$, we have four different cases depending on the level S_{M-1} of the stock price. First, $\Delta_M^*(S_{M-1}u^2) = 0$. This implies $\Delta_{M-1}^*(S_{M-1}) = B_{M-1}^*(S_{M-1}) = 0$. Hence, the left-hand side in (A.1) is $0 \geq 0$. Second, $\Delta_M^*(S_{M-1}d^2) = 1$. This implies $\Delta_{M-1}^*(S_{M-1}) = 1$ and $B_{M-1}^*(S_{M-1}) = -X$. Hence, also in this case, the left-hand side in (A.1) is $0 \geq 0$. Third, $\Delta_M^*(S_{M-1}u^2) = 1$ and $\Delta_M^*(S_{M-1}) = 0$. This implies that $\Delta_{M-1}^*(S_{M-1}) = (S_{M-1}\bar{u} - X)/(S_{M-1}(\bar{u} - \bar{d}))$ and $B_{M-1}^*(S_{M-1}) = -\bar{d}(S_{M-1}\bar{u} - X)/(\bar{u} - \bar{d})$. Hence, the left-hand side in (A.1) is

$$\frac{1 - \bar{d}}{\bar{u} - \bar{d}}(S_{M-1}\bar{u} - X) - \gamma\frac{S_{M-1}\bar{u} - X}{\bar{u} - \bar{d}} = \frac{1 - \gamma - \bar{d}}{\bar{u} - \bar{d}}(S_{M-1}\bar{u} - X)$$

$$= \frac{(1 - \gamma)(1 - d^2)}{\bar{u} - \bar{d}}(S_{M-1}\bar{u} - X) \geq 0,$$

since $\Delta_M^*(S_{M-1}u^2) = 1$ is equivalent to $S_{M-1}u^2 \geq X$.

Finally, $\Delta_M^*(S_{M-1}) = 1$ and $\Delta_M^*(S_{M-1}d^2) = 0$. This implies again that

$$\Delta_{M-1}^*(S_{M-1}) = \frac{S_{M-1}\bar{u} - X}{S_{M-1}(\bar{u} - \bar{d})}$$

and

$$B_{M-1}^*(S_{M-1}) = -\bar{d}\frac{S_{M-1}\bar{u} - X}{\bar{u} - \bar{d}}.$$

Hence, the left-hand side in (A.1) is

$$\frac{1 - \bar{d}}{\bar{u} - \bar{d}}(S_{M-1}\bar{u} - X) - S_{M-1} + X - \gamma\left[1 - \frac{S_{M-1}\bar{u} - X}{S_{M-1}(\bar{u} - \bar{d})}\right]S_{M-1}$$

$$= \frac{-S_{M-1}\bar{u}\bar{d} - X(1 + \gamma) + S_{M-1}\bar{d}(1 + \gamma) + X\bar{u}}{\bar{u} - \bar{d}}$$

$$= \frac{1}{\bar{u} - \bar{d}}(1 + \gamma)(u^2 - 1)(X - S_{M-1}\bar{d}) \geq 0,$$

since $\Delta_M^*(S_{M-1}d^2) = 0$ is equivalent to $S_{M-1}d^2 \leq X$. The first step of the induction is thus concluded. We now suppose that (A.1) holds at period $k+1$ ($k = 0, \ldots, M - 2$) and we show the validity of (A.1) also at period k. To this aim, we notice that (A.1) is equivalent to the following system

$$\begin{cases} \Delta_k^*(S_k)S_k(1 + \gamma) + B_k^*(S_k) \geq \Delta_{k+1}^*(S_k)S_k(1 + \gamma) + B_{k+1}^*(S_k) \\ \Delta_k^*(S_k)S_k(1 - \gamma) + B_k^*(S_k) \geq \Delta_{k+1}^*(S_k)S_k(1 - \gamma) + B_{k+1}^*(S_k) \end{cases} \tag{A.2}$$

The first inequality in (A.2) is satisfied since

$$
\Delta_k^*(S_k)S_k(1+\gamma) + B_k^*(S_k)
$$
$$
= \frac{\Delta_{k+1}^*(S_k u^2)S_k \bar{u} + B_{k+1}^*(S_k u^2) - \Delta_{k+1}^*(S_k d^2)S_k \bar{d} - B_{k+1}^*(S_k d^2)}{\bar{u} - \bar{d}}(1+\gamma)
$$
$$
+ \frac{\bar{u}[\Delta_{k+1}^*(S_k d^2)S_k \bar{d} + B_{k+1}^*(S_k d^2)] - \bar{d}[\Delta_{k+1}^*(S_k u^2)S_k \bar{u} + B_{k+1}^*(S_k u^2)]}{\bar{u} - \bar{d}}
$$
$$
= \frac{1+\gamma-\bar{d}}{\bar{u}-\bar{d}}[\Delta_{k+1}^*(S_k u^2)S_k \bar{u} + B_{k+1}^*(S_k u^2)]
$$
$$
+ \frac{\bar{u}-1-\gamma}{\bar{u}-\bar{d}}[\Delta_{k+1}^*(S_k d^2)S_k \bar{d} + B_{k+1}^*(S_k d^2)]
$$
$$
\geq \frac{1+\gamma-\bar{d}}{\bar{u}-\bar{d}}[\Delta_{k+2}^*(S_k u^2)S_k \bar{u} + B_{k+2}^*(S_k u^2)]
$$
$$
+ \frac{\bar{u}-1-\gamma}{\bar{u}-\bar{d}}[\Delta_{k+2}^*(S_k d^2)S_k \bar{d} + B_{k+2}^*(S_k d^2)]
$$
$$
= \Delta_{k+1}^*(S_k)S_k(1+\gamma) + B_{k+1}^*(S_k),
$$

where, for the inequality, we have exploited the induction hypothesis and the positivity of the coefficients $(1+\gamma-\bar{d})/(\bar{u}-\bar{d})$ and $(\bar{u}-1-\gamma)/(\bar{u}-\bar{d})$. Similarly, also the second inequality in (A.2) is satisfied, since

$$
\Delta_k^*(S_k)S_k(1-\gamma) + B_k^*(S_k)
$$
$$
= \frac{1-\gamma-\bar{d}}{\bar{u}-\bar{d}}[\Delta_{k+1}^*(S_k u^2)S_k \bar{u} + B_{k+1}^*(S_k u^2)]
$$
$$
+ \frac{\bar{u}-1+\gamma}{\bar{u}-\bar{d}}[\Delta_{k+1}^*(S_k d^2)S_k \bar{d} + B_{k+1}^*(S_k d^2)]
$$
$$
\geq \frac{1-\gamma-\bar{d}}{\bar{u}-\bar{d}}[\Delta_{k+2}^*(S_k u^2)S_k \bar{u} + B_{k+2}^*(S_k u^2)]
$$
$$
+ \frac{\bar{u}-1+\gamma}{\bar{u}-\bar{d}}[\Delta_{k+2}^*(S_k d^2)S_k \bar{d} + B_{k+2}^*(S_k d^2)]
$$
$$
= \Delta_{k+1}^*(S_k)S_k(1-\gamma) + B_{k+1}^*(S_k),
$$

where, for the inequality, we have exploited the induction hypothesis and the positivity of the coefficients $(1-\gamma-\bar{d})/(\bar{u}-\bar{d})$ and $(\bar{u}-1+\gamma)/(\bar{u}-\bar{d})$, which concludes our proof.   □

Proof of Proposition 2.2   Let us call $\Gamma(S_k)$ the admissible region of problem $\mathcal{P}_{\text{trin}}^*$ at period k $(k = 0, \ldots, M-1)$ when the stock price is S_k given the optimal values $(\Delta_{k+1}^*, B_{k+1}^*)$ at time $k+1$, i.e.

$$
\Gamma(S_k) := \Big\{ (\Delta_k, B_k) : \Delta_k S_k u^{2v_k} + B_k
$$
$$
-\Delta_{k+1}^*(S_k u^{2v_k})S_k u^{2v_k} - B_{k+1}^*(S_k u^{2v_k}) - \gamma|\Delta_{k+1}^*(S_k u^{2v_k}) -
$$
$$
\Delta_k|S_k u^{2v_k} \geq 0; \ \forall v_k \in \{1, 0, -1\} \Big\} \tag{A.3}
$$

and let us define a new admissible region $\bar{\Gamma}(S_k)$ as follows

$$
\begin{aligned}
\bar{\Gamma}(S_k) \ := \ & \Big\{ (\Delta_k, B_k) : \Delta_k S_k u^{2v_k} + B_k - \Delta^*_{k+1}(S_k u^{2v_k}) S_k u^{2v_k} \\
& - B^*_{k+1}(S_k u^{2v_k}) - \gamma |\Delta^*_{k+1}(S_k u^{2v_k}) - \Delta_k| S_k u^{2v_k} \ge 0; \\
& \forall v_k \in \{1, -1\} \Big\},
\end{aligned} \tag{A.4}
$$

which differs from $\Gamma(S_k)$ for the possible values of v_k. Calling $\bar{\mathcal{P}}^*_{\text{trin}}$ the problem $\mathcal{P}^*_{\text{trin}}$ with admissible region $\bar{\Gamma}(S_k)$, we prove the equivalence of problems $\bar{\mathcal{P}}^*_{\text{trin}}$ and $\mathcal{P}^*_{\text{trin}}$. In fact, if $\bar{V}^*_k(S_k)$ and $V^*_k(S_k)$ are their respective optimal solution at time k when the stock price is S_k, $\bar{V}^*_k(S_k) \le V^*_k(S_k)$ since $\bar{\Gamma}(S_k) \supseteq \Gamma(S_k)$ and $\bar{V}^*_k(S_k) \ge V^*_k(S_k)$ since the optimal pair $(\bar{\Delta}^*_k, \bar{B}^*_k)$ for $\bar{\mathcal{P}}^*_{\text{trin}}$ belongs also to $\Gamma(S_k)$, due to the previous Lemma A.1 and Theorem 1 in Boyle and Vorst (1992).

Proof of Proposition 3.1 For each $k = 0, \ldots, M - 1$, we prove the existence of a global minimum for the function $\varphi(\Delta_k, V_k)$ in (3.1) given an optimal pair (Δ_{k+1}, V_{k+1}) for the period $k + 1$. To ease the notation, we omit the conditioning in the expectation, we drop the subscript k in Φ_k and we denote with Φ^* the optimal values in period $k + 1$, with $\Phi \in \{\Delta, B\}$. The same is accomplished relatively to the stock price S.

The function φ is differentiable with respect to V. In fact,

$$
\frac{\partial \varphi}{\partial V} = 2V - 2E(V^*) + 2\Delta[E(S^*) - S] - 2\gamma E(S^*|\Delta^* - \Delta|),
$$

which is zero if and only if $V = E(V^*) - \Delta[E(S^*) - S] + \gamma E(S^*|\Delta^* - \Delta|)$. Substituting this in $\varphi(\Delta, V)$ and calling $\varphi(\Delta)$ the resulting expression, we get

$$
\begin{aligned}
\varphi(\Delta) = \\
E\Big\{ &[V^* - E(V^*) - \Delta(S^* - E(S^*)) + \gamma S^*|\Delta^* - \Delta| - \gamma E(S^*|\Delta^* - \Delta|)]^2 \Big\}.
\end{aligned}
$$

The possible values of Δ^* divide the real line in a finite number of intervals, say $I_1, \ldots, I_n$. Internally in each I_k, $\varphi(\Delta)$ is quadratic, with coefficient of the second degree term

$$
\begin{aligned}
& \text{Var}\,(S^*) + \gamma^2 E(S^{*2}) - 2\gamma E(S^*)E(S^*\text{sign}\,(\Delta^* - \Delta)) \\
& \quad - \gamma^2 [E(S^*\text{sign}\,(\Delta^* - \Delta))]^2 + 2\gamma E(S^{*2}\text{sign}\,(\Delta^* - \Delta)) \\
& = \gamma^2 \text{Var}\,(S^*\text{sign}\,(\Delta^* - \Delta)) + 2\gamma \text{Cov}\,(S^*\text{sign}\,(\Delta^* - \Delta), S^*) + \text{Var}\,(S^*),
\end{aligned}
$$

where sign represents the function sign. The right hand-side of the last equality is always positive since it can be viewed as a second degree trinomial in γ whose discriminant (divided by four) is

$$
[\text{Cov}\,(S^*\text{sign}\,(\Delta^* - \Delta), S^*)]^2 - \text{Var}\,(S^*\text{sign}\,(\Delta^* - \Delta))\text{Var}\,(S^*) \le 0.
$$

This implies the existence of a global minimum for $\varphi(\Delta)$ on each interval I_k and hence on the whole real line.

Proof of Proposition 3.2 Given the local risk minimizing strategy (Δ, B), remembering that $V_M = H$ and $V_0 = \Delta_0 S_0 + B_0$, we have

$$E(L(\Delta, B, Y)) = E\left\{\sum_{k=1}^{M}[V_k - V_{k-1} - \Delta_{k-1}(S_k - S_{k-1}) + \gamma|\Delta_k - \Delta_{k-1}|S_k]\right\}$$
$$+ \Delta_0 S_0 + B_0 - Y$$
$$= \sum_{k=1}^{M} E\left\{E\left\{[V_k - V_{k-1} - \Delta_{k-1}(S_k - S_{k-1})\right.\right.$$
$$\left.\left. + \gamma|\Delta_k - \Delta_{k-1}|S_k]|\mathcal{F}_{k-1}\right\}\right\} + \Delta_0 S_0 + B_0 - Y$$
$$= \Delta_0 S_0 + B_0 - Y,$$

since all the conditional expectations in the last equality are zero, being (Δ, B) local risk minimizing. The same reasoning leads to

$$\sigma^2(L(\Delta, B, Y))$$
$$= E\left\{\left[\sum_{k=1}^{M}(V_k - V_{k-1} - \Delta_{k-1}(S_k - S_{k-1}) + \gamma|\Delta_k - \Delta_{k-1}|S_k)\right]^2\right\}$$
$$= \sum_{k=1}^{M} E\left\{E\left\{[V_k - V_{k-1} - \Delta_{k-1}(S_k - S_{k-1}) + \gamma|\Delta_k - \Delta_{k-1}|S_k]^2\Big|\mathcal{F}_{k-1}\right\}\right\}$$
$$= \sum_{k=1}^{M} E\left\{R_{k-1}(\Delta, B)\right\}.$$

$\square$

References

Bensaid, B., J. Lesne, H. Pages, and J. Scheinkman (1992),'Derivative Asset Pricing with Transaction Cost', *Mathematical Finance* **2**, 63–86.

Boyle, P., and T. Vorst (1992), 'Option pricing in Discrete Time with Transaction Costs', *Journal of Finance* **47**, 271–293.

Cox, J.C., S. Ross, and M. Rubinstein (1979),'Option Pricing: a Simplified Approach', *Journal of Financial Economics* **7**, 229–263.

Davis, M.H.A., V.G. Panas and T. Zariphopoulou (1992),'European Option Pricing with Transaction Costs', *SIAM Journal of Control and Optimization* **31**, 470–498.

Dumas, B., and E. Luciano (1991),'An Exact Solution to a Dynamic Portfolio Choice Problem under Transaction Costs', *Journal of Finance* **46**, 577–595.

Edirisinghe, C., V. Naik, and R. Uppal (1993),'Optimal Replication of Options with Transactions Costs and Trading Restrictions', *Journal of Financial and Quantitative Analysis* **28**, 117–138.

El Karoui, N., M. Jeanblanc-Picqué and R. Viswanathan (1991), 'On the Robustness of Black-Scholes equation', Laboratoire de Mathematiques et Modelisations, Ecole Nationale des Ponts and Chaussees-Ecole Normale Superieure de Cachan, Prepublication 91/1.

Föllmer, H., and M. Schweizer (1991),'Hedging of Contingent Claims Under Incomplete Information', in *Applied Stochastic Analysis*, M.H.A. Davis and R.J. Elliott (eds.), Stochastic Monographs **5**, Gordon and Breach, 389–414.

Föllmer, H., and D. Sondermann (1986),'Hedging of Non-Redundant Contingent Claims', in *Contribution to Mathematical Economics*, W. Hildenbrand and A. Mas-Colell (eds.), 205–223.

Hipp, C. (1993),'Portfolio Management and Transaction Costs', University of Karlsruhe, preprint No. 2/93.

Hodges, S., and A. Neuberger (1989),'Optimal Replication of Contingent Claims under Transaction Costs', *Review of Futures Markets* **8**, 222–239.

Leland, H.E. (1985),'Option Pricing and Replication with Transaction Costs', *Journal of Finance* **49**, 1283–1301.

Mercurio, F., and T. Vorst (1995),'Option Pricing with Hedging at Fixed Trading Dates'. To appear in *Applied Mathematical Finance*.

Merton, R.C. (1990), *Continuous Time Finance*, Chapter 14, Section 14.2, Basil Blackwell.

Schäl, M. (1994),'On Quadratic Cost Criteria for Option Hedging', *Mathematics of Operation Research* **19**.

Schweizer, M. (1988),'Hedging of Options in a General Semimartingale Model', Diss. ETHZ No 8615.

Schweizer, M. (1991),'Option Hedging for Semimartingales', *Stochastic Processes and their Applications* **37**, 339–363.

Schweizer, M. (1995),'Variance-Optimal Hedging in Discrete Time', *Mathematics of Operation Research* **20**, 1–32.

Shen, Q. (1990), 'Bid-Ask Prices for Call Options with Transaction Costs Part I: Discrete Time Case'. Working Paper. Finance Department, The Warton School, University of Pennsylvania.

Option Pricing in Incomplete Markets

Mark H.A. Davis

Abstract

In this chapter a general option pricing formula is proposed, using arguments based on marginal substitution value. By giving the investor an external objective in the form of a utility maximization problem we arrive at a unique price in situations where standard arbitrage arguments cannot be used. Further, we show using Markov process theory that the price can be expressed as a discounted expectation where both the measure and the discount rate are uniquely determined. Models with stochastic coefficients and transaction cost models are studied in detail.

1 A General Option Pricing Formula

The Black-Scholes option pricing formula depends on exact replication and is only applicable in complete markets. It expresses the option value as the expected discounted exercise value where the expectation is calculated using the uniquely defined "martingale measure". In incomplete markets, exact replication is impossible and holding an option is a genuinely risky business, meaning that no *preference independent* pricing formula is possible. In technical terms, the problem is that no *unique* martingale measure exists. A variety of approaches have been suggested to get round this problem, none of them perhaps entirely satisfactory. Here we show that if option pricing is imbedded in a utility maximization framework, i.e. the potential option purchaser's attitude to risk is specified, then a unique measure emerges in a very natural way.

An investor with concave utility function U and starting with initial cash endowment x forms a dynamic portfolio whose cash value at time t is $X_x^\pi(t)$ when he uses trading strategy $\pi \in \mathcal{T}$, where $\mathcal{T}$ denotes the set of admissible trading strategies. His objective is to maximize expected utility of wealth at a fixed final time T; we denote

$$V(x) = \sup_{\pi \in \mathcal{T}} E[U(X_x^\pi(T))]. \tag{1}$$

Throughout the paper it will be assumed that the utility function U is non-decreasing and C^2 on R_+ with $U' > 0, \lim_{x \to 0} U'(x) = \infty$ and

$\lim_{x \to \infty} U'(x) = 0$. We ask the question whether the maximum utility in (1) can be increased by the purchase (or short-selling) of a European option whose cash value at time T is some non-negative random variable B, the purchase price at time zero being p. We use a "marginal rate of substitution" argument: p is a fair price for the option if diverting a little of his funds into it at time zero has a neutral effect on the investor's achievable utility. This is an entirely traditional approach to pricing in economics — see [6] for references — but does not appear to have been used much in an option pricing context. To state the definition in precise terms, we need the function W given as

$$W(\delta, x, p) = \sup_{\pi \in T} EU \left(X^{\pi}_{x-\delta}(T) + \frac{\delta}{p} B \right).$$

Definition 1 *Suppose that for each (x, p) the function $\delta \mapsto W(\delta, x, p)$ is differentiable at $\delta = 0$ and there is a unique solution $\hat{p}(x)$ of the equation*

$$\frac{\partial W}{\partial \delta}(0, p, x) = 0.$$

Then $\hat{p}(x)$ is the fair option price at time 0.

This definition will clearly reproduce the Black-Scholes value if perfect hedging is possible. The argument is as follows: suppose p_0 is the perfect-replication value and the option is offered for p. The investor buys δ/p options with cash δ, investing the remaining $x - \delta$ in a portfolio. A moment's thought shows that his optimal procedure is to short the hedging portfolio, whose value is $\delta p_0/p$ and invest his cash fund of $x + \delta(p_0/p - 1)$ optimally, attaining an expected utility of $V(x + \delta(p_0/p - 1))$. (The option and short hedging fund have equal and opposite value at time T.) The marginal rate of substitution is therefore

$$\frac{d}{d\delta} V \left(x + \delta \left(\frac{p_0}{p} - 1 \right) \right) \bigg|_{\delta=0} = \left(\frac{p_0}{p} - 1 \right) V'(x).$$

Evidently, this is equal to zero exactly when $p = p_0$.

In general, if the investor diverts δ into options and uses trading strategy π then his expected utility is

$$E \left[U \left(X^{\pi}_{x-\delta}(T) + \frac{\delta}{p} B \right) \right] = E[U(X^{\pi}_{x-\delta}(T))] + \frac{\delta}{p} E[U'(X^{\pi}_{x-\delta}(T)B)] + o(\delta). \quad (2)$$

We now need the following lemma.

Lemma 2 *Let $f : A \times R \to R$ be a function, where A is some set, and for $\delta \in R$ define*

$$v(\delta) := \sup_{\pi \in A} f(\pi, \delta).$$

Suppose that, for some $\delta_0 \in R$, v is differentiable at δ_0, there exists $\pi^* \in A$ such that $v(\delta_0) = f(\pi^*, \delta_0)$ and the function $\delta \mapsto f(\pi^*, \delta)$ is differentiable at δ_0. Then

$$\frac{d}{d\delta}v(\delta_0) = \frac{\partial}{\partial\delta}f(\pi^*, \delta_0).$$

We can now give a general option pricing formula based on Definition 1.

Theorem 3 *Suppose that V is differentiable at each $x \in R_+$ and that $V'(x) > 0$. Then the fair price $\hat{p}(x)$ of Definition 1 is given by*

$$\hat{p} = \frac{E[U'(X_x^{\pi^*}(T))B]}{V'(x)}. \tag{3}$$

The proof is obtained by evaluating the derivative with respect to δ of the maximum utility at $\delta = 0$, using (2) and Lemma 2, giving a value of

$$-V'(x) + \frac{1}{p}E_x[U'(X_T^{\pi^*}B)],$$

from which (3) follows.

2 The Standard Model

This model is the one described in, for example, [7, 9]. Let $(\Omega, \mathcal{F}, P)$ be a complete probability space on which is defined a d-dimensional Brownian motion $(W_t)_{t\geq 0}$, with natural filtration $(\mathcal{F}_t)$. The available instruments in the market consist of a *bond* whose price $S_0(t)$ satisfies

$$dS_0(t) = r(t)S_0(t)dt, \quad S_0(0) = 1, \tag{4}$$

and m *stocks* $(m \leq d)$ whose prices satisfy the SDEs

$$dS_i(t) = S_i(t)[b_i(t)dt + \sum_{j=1}^{d} \sigma_{ij}(t)dW_j(t)], i = 1, \ldots, m. \tag{5}$$

The processes $r(t), b_i(t), \sigma_{ij}(t)$ are assumed to be bounded, measurable and $\mathcal{F}_t$- adapted, with the matrix $\sigma = \sigma_{ij}$ having full rank and being such that the entries of the matrix $(\sigma(t)\sigma^T(t))^{-1}$ are bounded. We now define the following processes (**1** is the m-vector each of whose entries is equal to 1, and superscript 'T' denotes transpose):

$$\theta(t) = \sigma^T(t)(\sigma(t)\sigma^T(t))^{-1}(b(t) - r(t)\mathbf{1})$$

$$\beta(t) = \exp\left(-\int_0^t r(s)ds\right)$$

$$Z_0(t) = \exp\left(-\int_0^t \theta^T(t)dW(t) - \frac{1}{2}\int_0^t ||\theta(s)||^2 ds\right)$$

$$W_0(t) = W(t) + \int_0^t \theta(s)ds.$$

The class $\mathcal{T}$ of admissible trading strategies consists of all adapted m-vector processes $\pi(t)$ satisfying

$$\int_0^t ||\pi(s)||^2 ds < \infty.$$

The interpretation here is that $\pi_i(t)$ represents the wealth invested in the ith stock at time t. The portfolio value process $X^\pi(t)$ corresponding to a strategy $\pi \in \mathcal{T}$ and with initial wealth x is the unique solution of the SDE

$$dX^\pi(t) = r(t)X^\pi(t)dt + \pi^T(t)\sigma(t)dW_0(t), X^\pi(0) = x. \tag{6}$$

Let us first consider the *complete market* case $m = d$. The solution of the utility maximization problem (1) is then described as follows [7]. Let I denote the inverse function of the gradient of the utility function, $I = (U')^{-1}$. Now define

$$\mathcal{H}(y) = E[\beta(T)Z_0(t)I(y\beta(T)Z_0(T))]$$

and $\mathcal{Y}(x) = \mathcal{H}^{-1}(x)$. The terminal wealth corresponding to the optimal strategy π^* is $X_x^{\pi^*}(T) = I(\mathcal{Y}(x)\beta(T)Z_0(T))$, and hence the value function is

$$V(x) = E\left[U\left(I(\mathcal{Y}(x)\beta(T)Z_0(T))\right)\right].$$

A simple computation using the fact that $U'(I(x)) = x$ shows that $V'(x) = \mathcal{Y}(x)$ and the pricing formula (3) becomes

$$\hat{p} = E[\beta(T)Z_0(T)B],$$

which, as claimed earlier, is just the Black-Scholes price.

Now consider the incomplete case $m < d$, which is analyzed by Karatzas *et al.* in [9]. They show that under certain conditions there is a *least favourable completion* of the market, i.e. a specification of $d - m$ fictitious stocks in such a way that the utility-maximizing investor would place zero investment in these stocks even if they were available. The maximizing strategy can then be expressed by formulas similar to the above in terms of the completed market, and the following result is readily shown; see Rabeau [10] for details.

Theorem 4 *For the standard model with $m < d$, the option price (3) coincides with the Black-Scholes price computed for the least favourably completed market.*

3   Multiplicative Functionals

3.1   The Markov Case

Returning to general formula market models, let us suppose that the optimally controlled portfolio process $X_x^{\pi^*}(t)$ is one component – say the first, denoted x_t^1 – of some Markov process x_t in R^n. The pricing formula (3) then has a simple interpretation as a multiplicative functional (MF) transformation of the Markov semigroup. A discussion of MFs adequate for this application can be found in [2]; the standard reference is Blumenthal and Getoor [1]. The treatment here is at an informal level, making various assumptions that would need checking in specific applications.

Let $\mathcal{A}_t$ denote the *extended generator* of (x_t) and $\hat{A}$ be the generator of the corresponding time-space process (t, x_t), i.e.

$$\hat{A} = \frac{\partial}{\partial t} + \mathcal{A}_t.$$

Let

$$V(t, x) = E_{t,x}[U(x_T^1)],$$

and suppose that V is differentiable in x and that $(\partial/\partial x^1)V(t, x) > 0$. Defining

$$h(t, x) = \log \frac{\partial}{\partial x^1} V(t, x),$$

we see that the pricing formula is expressed as

$$\hat{p}(x) = e^{-h(0,x)} E_{0,x}[e^{h(T, x_T)} B]. \tag{7}$$

If $P_{s,t}$ denotes the Markov semigroup of (x_t) then (7) is just integration of B with respect to the measure defined by a new semigroup

$$P_{s,t}^h f(x) = e^{-h(s,x)} P_{s,t}[e^h f](x)$$

obtained by transformation of $P_{s,t}$ by the MF $m_{st} = \exp(h(t, x_t) - h(s, x_s))$. We can obtain more insight into this transformation by factoring this MF. Recall that if f is a function in the domain $\mathcal{D}(\hat{A})$ of the extended generator $\hat{A}$ then the process $(f(t, x_t))$ is expressed as

$$f(t, x_t) = f(0, x) + \int_0^t \hat{A}(s, x_s) ds + C_t^f, \tag{8}$$

where C_t^f is a local martingale. For simplicity, let us assume that (C_t^f) *is a continuous local martingale for all* $f \in \mathcal{D}(\hat{A})$. We then have the following result.

Theorem 5 *Suppose that $h, h^2 \in \mathcal{D}(\hat{A})$. Then the price $\hat{p}$ of (3) is expressed as*

$$\hat{p} = \tilde{E}\left[\exp\left(-\int_0^t \gamma^h(s, x_s)ds\right)B\right]$$

where

$$\gamma^h = (h-1)\hat{A}h - \frac{1}{2}\hat{A}h^2 \tag{9}$$

and $\tilde{E}$ denotes expectation with respect to the measure $\tilde{P}$ defined by

$$\frac{d\tilde{P}}{dP} = \exp\left(C_T^h - \frac{1}{2}\langle C^h, C^h\rangle_T\right).$$

PROOF: Writing $h = h(t, x_t)$ etc., we have as in (8)

$$dh^2 = \hat{A}h^2 dt + dC_t^{h^2},$$

whereas applying the Ito formula to the product $h \cdot h$ we obtain

$$d(h \cdot h) = 2h(\hat{A}hdt + dC_t^h) + d\langle C^h, C^h\rangle_t.$$

Identifying the bounded variation and local martingale terms in these two decompositions shows that

$$\langle C^h, C^h\rangle_t = \int_0^t (\hat{A}h^2 - 2h\hat{A}h)ds. \tag{10}$$

This is in fact a standard result, the operator in the integrand being known as the *opérateur carré du champ*. Using (10) we see that

$$h(t, x_t) - h(0, x) = C_t^h - \frac{1}{2}\langle C^h, C^h\rangle_t - \int_0^t \gamma^h(s, x_s)ds$$

where γ^h is given by (9). The result follows.

Formula (5) expresses the price as a discounted expectation where both the measure and the discount rate are *uniquely specified* by the utility maximization problem.

3.2   Example: The Standard Model with Random Coefficients

As is well known, the utility maximization problem for the standard model of Section 2 is explicitly solvable for logarithmic utility $U(x) = \log x$. Let us consider a single risky asset, so that $m = 1$. The bond, stock and wealth equations are given by (4,5,6); define $b = b_1$, $\sigma = \sigma_{11}$ and $u = \pi/X^\pi$ (the proportion of wealth invested in the risky asset) to simplify the notation. Then (6) becomes

$$dX^u = (r + (b-r)u)X^u dt + uX^u \sigma \, dw,$$

so that by the Ito formula

$$d \log X^u = \left(r + (b - r)u - \frac{1}{2}\sigma^2 u^2 \right) dt + u\sigma \, dw. \tag{11}$$

Define

$$\lambda(t) = \frac{b(t) - r(t)}{\sigma(t)}.$$

The 'dt' integrand in (11) is maximized pointwise by $u = u^* := \lambda/\sigma$ and the value of the integrand is then $(r + \lambda^2/2)$. It follows easily that

$$V(t, x) = \log x + E \int_t^T \left(r(t) + \frac{1}{2}\lambda^2(t) \right) dt. \tag{12}$$

In Section 2 the coefficients $r(t)$ etc. were allowed to be random in an essentially arbitrary way. To apply the results of this section we need to assume that they are deterministic functions of some underlying Markovian 'state variable'. (From a modelling point of view this is hardly a restriction since computation would rarely be feasible in any other case.) Thus, assume there is a Markov process (y_t) with extended generator $\mathcal{G}$ such that $r(t) = r(y_t), \sigma(t) = \sigma(y_t)$ and $\lambda(t) = \lambda(y_t)$, where $r(\cdot), \sigma(\cdot), \lambda(\cdot)$ are bounded measurable functions. Further, assume that (y_t) and (w_t) are independent processes (this *is* a sigificant restriction – see below). Then the generator of the joint space-time process $(t, z_t) := (t, X_t^{u^*}, y_t)$ is

$$\hat{\mathcal{A}}f(t, x, y) = \frac{\partial f}{\partial t} + (r(y) + \lambda^2(y))x\frac{\partial f}{\partial x} + \frac{1}{2}\lambda^2(y)x^2\frac{\partial^2 f}{\partial x^2} + \mathcal{G}f(x, y).$$

To allow for dependence on the initial state of the y_t process, the utility maximimization value function V of (12) should be written as

$$V(t, x, y) = \log x + E_{t,y} \int_t^T \left(r(t) + \frac{1}{2}\lambda^2(t) \right) dt. \tag{13}$$

Thus

$$\begin{aligned} h(t, x, y) &= \log \frac{\partial}{\partial x} V(t, x, y) \\ &= -\log x, \end{aligned}$$

so that

$$\hat{\mathcal{A}}h = -(r + \frac{1}{2}\lambda^2) \tag{14}$$

and

$$\hat{\mathcal{A}}h^2 = (2r + \lambda^2) \log x + \lambda^2. \tag{15}$$

From (9), (14) and (15) we find that

$$\gamma^h(t) = r(y_t),$$

while from the Ito formula, or by applying (10),

$$\frac{d\tilde{P}}{dP} = \exp\left(C_T^h - \frac{1}{2}\langle C^h, C^h \rangle_T\right) = \exp\left(-\int_0^T \lambda\, dw_s - \frac{1}{2}\int_0^T \lambda^2\, ds\right).$$

Under measure $\tilde{P}$ the stock price process satisfies

$$dS_t = r(y_t)S_t dt + \sigma(y_t)S_t\, d\tilde{w}_t \tag{16}$$

where $\tilde{w}_t$ is a $\tilde{P}$-Brownian motion, and Theorem 5 gives the option price as

$$\hat{p}(x) = \tilde{E}_{0,x,y}\left[e^{-\int_0^T r(y_s)ds}B\right]. \tag{17}$$

This is a striking result: the $\tilde{P}$-stock price model (16) is just the original model with the drift b replaced by the riskless rate r, and (17) expresses the price as the $\tilde{P}$ expectation discounted at the riskless rate. This is the same expression as the Black-Scholes formula (valid when r etc are deterministic) even though the present model is not complete and there is no interpretation in terms of perfect hedging. Note, however, that the result depends on w_t and y_t being independent: if they are correlated then the second term in (13) also depends on x and the expressions given for $\hat{A}h$ etc will be sustantially different.

4 Transaction Cost Models

We can apply the pricing formula immediately to a model with transaction costs as considered in [3, 5, 4]. In this model the stock price S, number of shares held y and cash x satisfy the following equations:

$$dS = bSdt + \sigma Sdw$$

$$dy = dL - dM \tag{18}$$

$$dx = (rx - c)dt - (1 + \lambda)dL - (1 - \mu)dM.$$

Here w is a standard Brownian motion, c_t is the consumption rate at time t and L_t, M_t are the *cumulative* purchases and sales respectively of stock units on the interval $[0, t]$. For standard application of the formula (3) we take $c_t \equiv 0$. However, we can also obtain a modified formula based on the utility maximization problem considered in [3], namely calculating

$$v(x, y, S) = \sup_{(c,L,M)} \frac{1}{\gamma}E\int_0^\infty e^{-\eta t}c_t^\gamma dt.$$

Thus we are maximizing the infinite-horizon discounted utility of consumption as measured by the utility function c^γ/γ, where $\gamma < 1, \gamma \neq 0$. A pricing formula based on this utility maximization problem can be derived by the same approach as above, and it is

$$\hat{p}(x, y, S) = \frac{e^{-\eta T} v_x(x_T^*, y_T^*, S_T)}{v_x(x, y, S)}. \tag{19}$$

Here x_T^*, y_T^* is the composition of the optimal portfolio at time T, starting at x, y at time 0. By using the analysis of the utility maximization problem in [3], Rabeau obtains in [10] the following result.

Theorem 6 *For the transaction cost model* (18), *the option price $\hat{p}$ of* (19) *is given by*

$$\hat{p}(x, y, S) = \tilde{E}[e^{-rT} B]$$

where

$$\frac{d\tilde{P}}{dP} = \exp\left(\int_0^T \frac{1}{v_x}(\sigma y_t^* v_{xy}) dw_t - \frac{1}{2} \int_0^T \left(\frac{\sigma y_t^* v_{xy}}{v_x}\right)^2 dt\right)$$

As in the random coefficients model of Section 3.2, the discount rate specified by the formula is always the riskless rate r. It is not at all obvious from the general multiplicative functional decomposition of Theorem 5 why this should be so.

5   Concluding Remarks

5.1   Hedging

Having given a price, the trader will want to know how to hedge the portfolio. In incomplete markets this cannot be done exactly, but the utility function gives a measure of the riskiness of an imperfect hedge. If ϵ options are written at price $\hat{p}$ given by (3) then the best trading strategy from this point of view will be π^* maximizing

$$EU(X_{x+\epsilon\hat{p}}^\pi - \epsilon B).$$

As pointed out in Section 1, this strategy will amount to perfect replication of the option if this is possible.

For the transaction cost models discussed in Section 4 above, these utility maximization problems have been considered in [5, 4]. In particular, [4] gives a detailed account of discretization methods based on binomial approximation and sebsequent solution of the utility maximization problems by dynamic programming.

The algorithms described in [4] are computationally intensive. A promising topic for future research is fast computation of nearly optimal strategies, perhaps by parametrising a class of 'good' strategies in some simple way and searching for best parameters.

5.2 Arbitrage

An obvious question is whether an option priced at the value $\hat{p}$ of Definition 1 affords any arbitrage opportunities. This and many other matters are taken up in a recent paper by Karatzas and Kou [8], written since the first version of the present article. For the standard model with constraints on portfolio choice they show that for each option C there is an *interval* $I_C = [p_0, p_1]$ such that any $p \in I_C$ is a no-arbitrage price. I_C reduces to a single point in the case of a complete market. The price $\hat{p}$ lies in I_C if a certain superposition principle holds: if π_1, π_2 are admissible strategies with initial wealth x_1, x_2 respectively, then there is an admissible strategy π starting from $x_1 + x_2$ such that

$$X_{x_1}^{\pi_1}(T) + X_{x_2}^{\pi_2}(T) = X_{x_1+x_2}^{\pi}(T) \quad \text{a.s.}$$

This is a very mild condition, but it is possible to construct constraint sets for which $\hat{p}$ fails to be arbitrage-free.

References

[1] R.M. Blumenthal and R.K. Getoor, *Markov Processes and Potential Theory*, Academic Press, 1968.

[2] M.H.A. Davis, 'Pathwise nonlinear filtering', in *Stochastic Systems: the Mathematics of Filtering and Identification*, eds. M. Hazewinkel and J.C. Willems, D. Reidel, 1981.

[3] M.H.A. Davis and A.R. Norman, 'Portfolio selection with transaction costs', *Math. of Operations Research* **15**, (1990), 676–713.

[4] M.H.A. Davis and V.G. Panas, 'The writing price of a European contingent claim under proportional transaction costs', *Comp. Appl. Math.* **13**, (1994), 115–157.

[5] M.H.A. Davis, V.G. Panas and T. Zariphopoulou, 'European option pricing with transaction costs', *SIAM J. Control and Opt.* **31**, (1993), 470–493.

[6] D. Duffie, *Dynamic Asset Pricing Theory*, Princeton University Press, 1992.

[7] I. Karatzas, 'Optimization problems in the theory of continuous trading', *SIAM J. Control and Optimization* **27**, (1989), 1221–1259.

[8] I. Karatzas and S.-G. Kou, 'On the pricing of contingent claims under constraints', preprint, Columbia University, June 1995 [submitted to *Ann. Appl. Prob.*].

[9] I. Karatzas, J.P. Lehoczky, S.E. Shreve and G.L. Xu, 'Martingale and duality methods for utility maximization in an incomplete market', *SIAM J. Control and Optimization* **29**, (1991), 702–730.

[10] N.M. Rabeau, PhD Thesis, Imperial College, University of London, February 1996.

Nonlinear Financial Markets: Hedging and Portfolio Optimization

Jakša Cvitanić

Abstract

This is a survey article on the techniques and results of the theory of optimal trading for a single agent with a nonlinear wealth process, in a continuous-time model. Examples include the case of policy dependent prices, portfolio constraints and different interest rates for borrowing and lending. We study the hedging problem and the portfolio optimization problem for the investor in this market. Mathematical tools involved are those of continuous-time martingales, convex duality in the case of convex nonlinearities, and forward-backward SDEs and PDEs in the case of more general, but Markovian nonlinearities. We often present only the main ideas of the proofs, and refer the reader to the literature for details.

1 Introduction and Summary

We survey in this chapter the theory of portfolio optimization and hedging of contingent claims in financial markets for a single agent, whose wealth process is given by a nonlinear SDE. This includes the situations in which the agent's strategy can influence the asset prices, the case of different interest rates for borrowing and lending, and, at least formally, the case of portfolio constraints. In the complete markets case (linear SDE, no constraints), the martingale-duality methodology presented here has been developed in a number of papers, such as Harrison and Pliska (1981), Pliska (1986), Karatzas, Lehoczky and Shreve (1987), Cox and Huang (1989); in the incomplete markets (nonlinear and/or constrained markets) the technique has been further generalized by Karatzas, Lehoczky, Shreve and Xu (KLSX) (1991), He and Pearson (1991), Xu and Shreve (1992), El Karoui and Quenez (1995), Cvitanić and Karatzas (CK)(1992, 1993). We provide a framework that includes all of those problems, and offer simplified proofs. The idea for the unified approach comes from El Karoui, Peng and Quenez (1994), and the results on hedging in Section 3 of this work are mostly taken from that paper, although we use the stochastic control approach in order to establish directly the existence of hedging strategies, while they use the theory of Backward Stochastic Differential Equations (BSDEs), as developed in Pardoux and Peng (1990). Special cases of portfolio constraints and different interest rates have been considered by many other authors, using different, mostly Partial Differential Equations (PDE), methods in Markovian frameworks. We mention only a few: Fleming and Zariphopoulou (1991), Zariphopoulou (1989, 1994), Barron and Jensen (1990), Bergman (1991), Jouini and Kallal (1995), Korn (1992). Readers unfamiliar with the mathematical finance problems of portfolio optimization and hedging in the standard, generalized Black-Scholes-Merton linear model can consult Karatzas (1989).

In all of the above mentioned works, only the drift term of the wealth process becomes nonlinear. The case of a nonlinear diffusion term case is presented in the second part of this survey, based on the recent theory of Forward-Backward SDEs, as developed by Ma, Protter and Yong (1994) and applied in mathematical finance in Cvitanić and Ma (1996). We can only solve the hedging problem in this framework, and only under Markovian and smoothness assumptions.

The model with the drift term being nonlinear (concave) in the portfolio variable, is described in Section 2. Section 3 gives a formula for the minimal hedging process of a given contingent claim B, in such a model (Theorem 3.7). In Section 4 we define utility functions, and in Section 5 we solve the portfolio optimization problem (Theorem 5.15) by first solving an appropriate dual problem, and then using results on hedging from Section 3. This enables us to simplify greatly the original proof in Xu and Shreve (1992), CK (1992), albeit under somewhat more stringent conditions. We show in Section 6 that the markets with different interest rates and the markets with portfolio constraints can be regarded as special cases of the model of Section 2 (at least formally, in the case of constraints). We also consider an example of a market in which short-selling is "punished". In Section 7 we indicate analogous results for a market in which the nonlinearities also depend on the wealth (not just the portfolio) of the investor. Finally, in Section 8 we describe the PDE which provides the minimal hedging process for a European claim in a Markovian framework, with nonlinearities present in both the drift and the diffusion term of the price and wealth processes. This is the case of Forward-Backward SDEs, namely, of feedback interaction in both directions between the forward price process and the backward (hedging) wealth process.

2   The Model

We consider a financial market $\mathcal{M}$ which consists of one *bank account* and several (d) stocks. There is an investor in the market, whose portfolio process is denoted by $\pi = (\pi_1, \ldots, \pi_d)$ (precise definition is given below). His investment policy influences the behaviour of the prices $P_0, \{P_i\}_{1 \le i \le d}$ of the financial instruments. More precisely, these prices evolve according to the equations

$$dP_0(t) = P_0(t)[r(t) + f_0(\pi_t)]dt , \qquad P_0(0) = 1 \qquad (2.1)$$

$$dP_i(t) = P_i(t)[(b_i(t) + f_i(\pi_t))dt + \sum_{j=1}^{d} \sigma_{ij}(t)dW^{(j)}(t)],$$

$$P_i(0) = p_i \in (0, \infty) \qquad (2.2)$$

for $i = 1, \ldots, d$. Here $f_i : \mathbb{R}^d \mapsto \mathbb{R}, \ i = 0, \ldots, d$ are some given functions that describe the effect of the investor's strategy on the prices, $W = (W^{(1)}, \ldots, W^{(d)})^*$ is a standard Brownian motion in $\mathbb{R}^d$, defined on a complete probability space $(\Omega, \mathcal{F}, \mathbf{P})$, and we shall denote by $\{\mathcal{F}_t\}$ the $\mathbf{P}$-augmentation of the filtration $\mathcal{F}_t^W = \sigma(W(s); \ 0 \le s \le t)$ generated by W. The *coefficients* of $\mathcal{M}$ - i.e., the processes $r(t)$ (scalar interest rate), $b(t) = (b_1(t), \ldots, b_d(t))^*$ (vector of appreciation rates) and $\sigma(t) = \{\sigma_{ij}(t)\}_{1 \le i,j \le d}$ (volatility matrix) - are assumed to be progressively measurable with respect to $\{\mathcal{F}_t\}$ and *bounded* uniformly in $(t, \omega) \in [0, T] \times \Omega$. We shall also impose the following strong non-degeneracy condition on the matrix $a(t) := \sigma(t)\sigma^*(t)$:

$$\xi^* a(t)\xi \ge \varepsilon ||\xi||^2, \quad \forall \ (t, \xi) \ \in \ [0, T] \times \mathbb{R}^d \qquad (2.3)$$

almost surely, for a given real constant $\varepsilon > 0$. All processes encountered throughout the paper will be defined on the fixed, finite horizon $[0, T]$.

Let us remark that the interpretation of policy-dependent prices is not the only possible one; one could simply start with the economy in which the wealth process of the investor is given by equation (2.7) below, and forget about the prices.

For a given $\{\mathcal{F}_t\}$−progressively measurable process $\nu(\cdot)$ with values in $\mathbb{R}^d$ and $\mu(\cdot)$ with values in $\mathbb{R}$, we introduce the 'stochastic discount process' ('shadow state-price density')

$$\begin{aligned} Z_\nu(t) &:= Z_\nu(0, t), \\ Z_\nu(u, t) &:= \exp\left[\int_u^t (\sigma^{-1}(s)\nu(s))^* dW(s) - \tfrac{1}{2}\int_u^t ||\sigma^{-1}(s)\nu(s)||^2 ds\right] \end{aligned} \qquad (2.4)$$

and the discount process

$$\beta_\mu(t) := \beta_\mu(0, t), \quad \beta_\mu(u, t) := \exp\left\{-\int_u^t \mu(s)ds\right\}. \qquad (2.5)$$

2.1 Remark

It is a straightforward consequence of the strong non-degeneracy condition (2.3), that the matrices $\sigma(t), \sigma^*(t)$ are invertible, and that the norms of

$(\sigma(t))^{-1}, (\sigma^*(t))^{-1}$ are bounded above and below by δ and $1/\delta$, respectively, for some $\delta \in (1, \infty)$; cf. Karatzas and Shreve (1991). Notice also that the process Z_ν satisfies

$$dZ_\nu(t) = \sigma^{-1}(t)\nu(t)Z_\nu(t)dW(t) \tag{2.6}$$

and is a nonnegative local martingale, hence a supermartingale.

Let us describe now the investor's strategies in more details. He can decide, at any time $t \in [0, T]$,

1. what proportion $\pi_i(t)$ of his wealth $X(t)$ to invest in the ith stock ($1 \leq i \leq d$), and

2. what amount of money $c(t + h) - c(t) \geq 0$ to withdraw for consumption during the interval $(t, t + h]$, $h > 0$.

Of course these decisions can only be based, at any given time $t \in [0, T]$, on the current information $\mathcal{F}_t$, without anticipation of the future. With $\pi(t) = (\pi_1(t), \ldots, \pi_d(t))^*$ chosen, the amount $X(t)[1 - \sum_{i=1}^d \pi_i(t)]$ is invested in the bond. Thus, in accordance with the model set forth in (2.1), (2.2), the wealth process $X(t)$ satisfies the stochastic differential equation

$$dX(t) = X(t)g(t, \pi_t)dt + X(t)\pi^*(t)\sigma(t)dW(t) - dc(t) , \quad X(0) = x > 0, \tag{2.7}$$

where

$$g(t, \pi) := r(t) + f_0(\pi) + \sum_{i=1}^d \pi_i[b_i(t) + f_i(\pi) - r(t) - f_0(\pi)] , \tag{2.8}$$

the real number $x > 0$ represents the initial capital, and $c(t)$ is the cumulative consumption up to time t.

We formalize the above discussion as follows.

2.2   Definition

(i) An $\mathbb{R}^d$-valued, $\{\mathcal{F}_t\}$-progressively measurable process $\pi = \{\pi(t), 0 \leq t \leq T\}$ with $\int_0^T ||\pi(t)||^2 dt < \infty$, a.s., will be called a *portfolio process*.

(ii) A nonnegative, nondecreasing, $\{\mathcal{F}_t\}$ - progressively measurable process $c = \{c(t), 0 \leq t \leq T\}$ with RCLL paths, $c(0) = 0$ and $c(T) < \infty$, a.s., will be called a *consumption process*.

(iii) Given a pair (π, c) as above, the solution $X \equiv X^{x, \pi, c}$ of the equation (2.7) will be called the *wealth process* corresponding to the portfolio/consumption pair (π, c) and initial capital $x \in (0, \infty)$.

2.3 Definition

A portfolio/consumption process pair (π, c) is called *admissible* for the initial capital $x \in (0, \infty)$, if

$$X^{x,\pi,c}(t) \geq 0, \qquad \forall\, 0 \leq t \leq T \qquad (2.9)$$

holds almost surely. The set of admissible pairs (π, c) will be denoted by $\mathcal{A}_0(x)$. ◇

We will restrict ourselves in Sections 3–7 by imposing the following assumption:

2.4 Assumption

The function $g(t, \cdot)$ is concave for all $t \in [0, T]$, and uniformly (with respect to t) Lipschitz:

$$|g(t, x) - g(t, y)| \leq k \|x - y\|, \quad \forall\, t \in [0, T]; \; x, y \in \mathbb{R}^d,$$

for some $0 < k < \infty$.

3 Hedging

In this section we consider the problem of hedging random amounts of money for our investor.

3.1 Definition

A *Contingent Claim* is a nonnegative, $\mathcal{F}_T$-measurable random variable B, such that $E B^2 < \infty$. The *hedging price* of this contingent claim is defined by

$$h(0) := \inf\{x > 0; \; \exists (\pi, c) \in \mathcal{A}_0(x) \text{ s.t. } X^{x,\pi,c}(T) \geq B \text{ a.s.}\} . \qquad (3.1)$$

It is shown in Karatzas and Kou (1996) that $h(0)$ is the no-arbitrage upper bound for the price of the claim; in other words, the seller of the claim (our agent) cannot achieve an arbitrage opportunity ('free lunch') if and only if the claim is sold for at most $h(0)$ dollars. They also provide the (more or less symmetrical) analysis for the lower bound for the price.

In order to to find the minimal hedging process, i.e., a wealth process $X(\cdot)$ for which $X(T) \geq B$ and $h(0) = X(0)$, thus also a somewhat explicit expression for $h(0)$, we define, for each $t \in [0, T]$, the *convex conjugate* function $\tilde{g}(t, \cdot) : \mathbb{R}^d \mapsto \mathbb{R}$ of the concave function $g(t, \cdot)$ by

$$\tilde{g}(t, \nu) := \sup_{\pi \in \mathbb{R}^d} \{g(t, \pi) + \pi^* \nu\}, \qquad (3.2)$$

on its *effective domain* $\mathcal{D}_t := \{\nu \in \mathbb{R}^d; \ \tilde{g}(t, \nu) < \infty\}$. Introduce also the class $\mathcal{D}$ of $\{\mathcal{F}_t\}$−progressively measurable processes $\nu(\cdot)$ which satisfy $\nu(t) \in \mathcal{D}_t$ a.e. $t \times \omega$. We shall assume throughout the paper that

$$\mathcal{D} \text{ is not empty;} \tag{3.3}$$

$$\begin{array}{c} \text{The function } \tilde{g}(t, \cdot) \text{ of (3.2) is bounded} \\ \text{on its effective domain, uniformly in } t. \end{array} \tag{3.4}$$

Recall the notation of (2.4) and let

$$\gamma_\nu(t, u) := \exp\left\{-\int_t^u \tilde{g}(s, \nu_s)ds\right\}, \quad \gamma_\nu(t) := \gamma_\nu(0, t), \quad H_\nu(t) := Z_\nu(t)\gamma_\nu(t) . \tag{3.5}$$

For every $\nu \in \mathcal{D}$ we have (by Ito's rule)

$$H_\nu(t)X(t) + \int_0^t H_\nu(s)\left[X(s)(\tilde{g}(s, \nu_s) - g(s, \pi_s) - \pi^*(s)\nu(s))ds + dc(s)\right]$$

$$= x + \int_0^t H_\nu(s)X(s)\left[\pi^*(s)\sigma(s) + \sigma^{-1}(s)\nu(s)\right]dW(s). \tag{3.6}$$

In particular, the process on the right-hand side is a nonnegative local martingale, hence a supermartingale. Therefore we get the following *necessary condition for π to be admissible*:

$$\sup_{\nu \in \mathcal{D}} E\left[H_\nu(T)X(T) + \int_0^T H_\nu(s)X(s)\{\tilde{g}(s, \nu_s) - g(s, \pi_s) - \pi^*(s)\nu(s)\}ds\right]$$

$$\leq x .\tag{3.7}$$

3.2   Remark

The supermartingale property excludes arbitrage opportunities from this market: if $x = 0$, then from (3.7) necessarily $X(t) = 0$, $\forall \ 0 \leq t \leq T$, a.s.. If $f_i \equiv 0$, $i = 0, \ldots, d$, then $\mathcal{D}$ consists of only one process $\hat{\nu}(\cdot)$, given by $\hat{\nu}_i(t) = r(t) - b_i(t)$, $i = 1, \ldots, d$, i.e., we are in the standard Black–Scholes–Merton complete market model with the unique 'equivalent martingale risk neutral measure' $\mathbf{P}^{\hat{\nu}}$, defined as in (3.9) below. In general, there are several interpretations for the processes $\nu \in \mathcal{D}$ and the processes of (3.5): in economics jargon, they correspond to the shadow prices relevant to the incompleteness of the market introduced by nonlinearities; they can also be considered as the dual processes appearing in the stochastic maximum principle corresponding to the stochastic control problems we shall be considering.

Next, for a given $\nu \in \mathcal{D}$, introduce the process

$$W_\nu(t) := W(t) - \int_0^t \sigma^{-1}(s)\nu(s)ds , \tag{3.8}$$

as well as the measure

$$\mathbf{P}^\nu(A) := E[Z_\nu(T)1_A] = E^\nu[1_A], \quad A \in \mathcal{F}_T . \tag{3.9}$$

Notice that Assumption 2.4 implies that the sets $\mathcal{D}_t$ are uniformly bounded. Therefore, if $\nu \in \mathcal{D}$, then Z_ν *is a martingale.* Thus, for every $\nu \in \mathcal{D}$, the measure $\mathbf{P}^\nu$ of (3.9) is a probability measure and the process $W_\nu(\cdot)$ of (3.8) is a $\mathbf{P}^\nu$−Brownian motion, by the Girsanov theorem.

Finally, denote by $\mathcal{S}$ the set of all $\{\mathcal{F}_t\}$-stopping times τ with values in $[0, T]$, and by $\mathcal{S}_{\rho,\sigma}$ the subset of $\mathcal{S}$ consisting of stopping times τ s.t. $\rho \le \tau \le \sigma$, a.s., for any two $\rho \in \mathcal{S}, \sigma \in \mathcal{S}$ such that $\rho \le \sigma$, a.s. Given a contingent claim B, consider, for every $\tau \in \mathcal{S}$, the $\mathcal{F}_\tau$-measurable random variable

$$V(\tau) := \operatorname{ess\,sup}_{\nu \in \mathcal{D}} E^\nu[B\gamma_\nu(\tau, T)|\mathcal{F}_\tau]. \tag{3.10}$$

We show now, using the methods of CK (1993), that V solves the equation (2.7) regarded as a *Backward* SDE, with the given *terminal* condition $X(T) = B$, a.s., instead of the initial condition $X(0) = x$. That will imply $V(0) \ge h(0)$. We also show that, in fact, $V(0) = h(0)$. We first state two results which are proved as in CK (1993):

3.3 Proposition

For any contingent claim B, the family (3.10) of random variables $\{V(\tau)\}_{\tau \in \mathcal{S}}$ satisfies the equation of Dynamic Programming

$$V(\tau) = \operatorname{ess\,\,sup}_{\nu \in \mathcal{D}_{\tau,\theta}} E^\nu \left[V(\theta) \exp\left\{ -\int_\tau^\theta \tilde{g}(u, \nu_u)du \right\} |\mathcal{F}_\tau \right] ; \quad \forall \ \theta \in \mathcal{S}_{\tau,T} , \tag{3.11}$$

where $\mathcal{D}_{\tau,\theta}$ is the restriction of $\mathcal{D}$ to the stochastic interval $[\![\tau, \theta]\!]$.

3.4 Proposition

The process $V = \{V(t), \mathcal{F}_t; 0 \le t \le T\}$ of Proposition 3.5 can be considered in its RCLL modification and, for every $\nu \in \mathcal{D}$,

$$\left\{ Q_\nu(t) := V(t)e^{-\int_0^t \tilde{g}(u,\nu_u)du}, \mathcal{F}_t; \ 0 \le t \le T \right. \tag{3.12}$$
$$\left. \text{is a } \mathbf{P}^\nu\text{-supermartingale with RCLL paths} \right\}$$

Furthermore, V is the smallest adapted, RCLL process that satisfies (3.12) as well as

$$V(T) \ge B, \quad \text{a.s.} \tag{3.13}$$

3.5   Theorem

For an arbitrary contingent claim B, we have $h(0) = V(0)$. Furthermore, there exists a pair $(\hat{\pi}, \hat{c}) \in \mathcal{A}_0(V(0))$ such that $X^{V(0),\hat{\pi},\hat{c}}(\cdot) = V(\cdot)$.

Proof $h(0) \geq V(0)$ follows immediately from the supermartingale property (3.7). To show the rest, notice that from (3.12), the martingale representation theorem and the Doob-Meyer decomposition (e.g. Karatzas and Shreve (1991), §1.4), we have for every $\nu \in \mathcal{D}$:

$$Q_\nu(t) = V(0) + \int_0^t \psi_\nu^*(s)dW_\nu(s) - A_\nu(t), \quad 0 \leq t \leq T , \qquad (3.14)$$

where $\psi_\nu(\cdot)$ is an $\mathbb{R}^d$-valued, $\{\mathcal{F}_t\}$-progressively measurable and a.s. square-integrable process and $A_\nu(\cdot)$ is adapted with increasing, RCLL paths and $A_\nu(0) = 0, A_\nu(T) < \infty$ a.s. The idea then is to find a pair $(\hat{\pi}, \hat{c}) \in \mathcal{A}_0(V(0))$ such that $Q_\nu(t)\gamma_\nu^{-1}(\cdot) = V(\cdot) = X^{V(0),\hat{\pi},\hat{c}}(\cdot)$.

In order to do this, let us observe that for any $\mu \in \mathcal{D}, \nu \in \mathcal{D}$ we have from (3.12)

$$Q_\mu(t) = Q_\nu(t) \exp\left[\int_0^t \{\tilde{g}(s, \nu_s) - \tilde{g}(s, \mu_s)\}ds\right] ,$$

and from (3.14):

$$
\begin{aligned}
dQ_\mu(t) &= \exp[\textstyle\int_0^t \{\tilde{g}(s,\nu_s) - \tilde{g}(s,\mu_s)\}ds] \cdot [Q_\nu(t)\{\tilde{g}(t,\nu_t) - \tilde{g}(t,\mu_t)\}dt \\
&\qquad + \psi_\nu^*(t)dW_\nu(t) - dA_\nu(t)] \\
&= \exp[\textstyle\int_0^t \{\tilde{g}(s,\nu_s) - \tilde{g}(s,\mu_s)\}ds] \cdot [V(t)\gamma_\nu(t)\{\tilde{g}(t,\nu_t) - \tilde{g}(t,\mu_t)\}dt \\
&\qquad - dA_\nu(t) + \psi_\nu^*(t)\sigma^{-1}(t)(\mu(t) - \nu(t))dt + \psi_\nu^*(t)dW_\mu(t)] .
\end{aligned}
$$
$$(3.15)$$

Comparing this decomposition with

$$dQ_\mu(t) = \psi_\mu^*(t)dW_\mu(t) - dA_\mu(t) , \qquad (3.14)'$$

we conclude that

$$\psi_\nu^*(t) \, e^{\int_0^t \tilde{g}(s,\nu_s)ds} = \psi_\mu^*(t) \, e^{\int_0^t \tilde{g}(s,\mu_s)ds}$$

and hence that this expression is independent of $\nu \in \mathcal{D}$:

$$\psi_\nu^*(t) \, e^{\int_0^t \tilde{g}(s,\nu_s)ds} = V(t)\hat{\pi}^*(t)\sigma(t); \quad \forall \, 0 \leq t \leq T, \, \nu \in \mathcal{D} , \qquad (3.16)$$

for some adapted, $\mathbb{R}^d$-valued, a.s. square integrable process $\hat{\pi}$ (word of caution: if $V(t) = 0$ for some t, then $V(s) = 0$ for all $s \geq t$; in that case let $\hat{\pi}(s) = 0, \, s \geq t$).

Similarly, we conclude from (3.15), (3.16) and (3.14)′:

$$e^{\int_0^t \tilde{g}(s,\nu_s)ds}dA_\nu(t) - V(t)[\tilde{g}(t,\nu_t) - \hat{\pi}^*(t)\nu(t)]dt$$
$$= e^{\int_0^t \tilde{g}(s,\mu_s)ds}dA_\mu(t) - V(t)[\tilde{g}(t,\mu_t) - \hat{\pi}^*(t)\mu(t)]dt$$

and hence this expression is also independent of $\nu \in \mathcal{D}$:

$$\hat{c}(t) := \int_0^t \gamma_\nu^{-1}(s)dA_\nu(s) - \int_0^t V(s)[\tilde{g}(s,\nu_s) - g(s,\hat{\pi}_s) - \nu^*(s)\hat{\pi}(s)]ds , \quad (3.17)$$

for every $0 \le t \le T, \nu \in \mathcal{D}$. Now, Assumption 2.4 implies that there exists a process $\hat{\nu}(\cdot) \in \mathcal{D}$ such that

$$g(t,\hat{\pi}_t) + \hat{\nu}^*(t)\hat{\pi}(t) \equiv \tilde{g}(t,\hat{\nu}_t). \quad (3.18)$$

(see for example El Karoui, Peng and Quenez (1994)). From (3.17), with $\nu \equiv \hat{\nu}$, we obtain $\hat{c}(t) = \int_0^t \gamma_{\hat{\nu}}^{-1}(s)dA_{\hat{\nu}}(s), 0 \le t \le T$ and hence

$$\{\hat{c}(\cdot) \text{ is an increasing, adapted, RCLL process} \atop \text{with } \hat{c}(0) = 0 \quad \text{and} \quad \hat{c}(T) < \infty, \text{ a.s.}\}. \quad (3.19)$$

Now we can put together (3.14)–(3.17) to deduce

$$\begin{aligned} d(\gamma_\nu(t)V(t)) &= dQ_\nu(t) = \psi_\nu^*(t)dW_\nu(t) - dA_\nu(t) \\ &= \gamma_\nu(t)[-d\hat{c}(t) - V(t)\{\tilde{g}(t,\nu_t) - g(t,\hat{\pi}_t)\}dt \quad (3.20) \\ &\quad +V(t)\hat{\pi}^*(t)\sigma(t)dW(t)] , \end{aligned}$$

for any given $\nu \in \mathcal{D}$. This shows $V(\cdot) \equiv X^{V(0),\hat{\pi},\hat{c}}(\cdot)$, and hence $h(0) \le V(0)$.◇

So far we have found the minimal hedging price for a claim B; in fact, it is easy to see (using the same supermartingale argument) that the process $V(\cdot)$ is the minimal wealth process that hedges B. There remains the question of whether consumption is necessary. We show that, in fact, $\hat{c}(\cdot) \equiv 0$.

3.6 Definition

We say that a contingent claim B is *attainable*, if there exists a portfolio process π such that $(\pi,0) \in \mathcal{A}_0(V(0))$ and $X^{V(0),\pi,0}(T) = B$, a.s.

3.7 Theorem

Every contingent claim B is attainable, namely the process $\hat{c}(\cdot)$ from Theorem 3.7 is a zero-process. Moreover, if the process $Q_{\hat{\nu}}(\cdot)$ is of class $D[0,T]$ under $\mathbf{P}^{\hat{\nu}}$, where $\hat{\nu}(\cdot)$ is a process that satisfies (3.18), then $Q_{\hat{\nu}}$ is a $\mathbf{P}^{\hat{\nu}}-$martingale, which is equivalent to $\hat{\nu}(\cdot)$ being optimal for the optimization problem $V(0) = \sup_{\nu \in \mathcal{D}} E^\nu[B\gamma_\nu(T)]$.

Proof Let $\{\nu_n; n \in \mathcal{N}\}$ be a maximizing sequence for achieving $V(0)$, i.e., $\lim_{n\to\infty} E^{\nu_n} B\gamma_{\nu_n}(T) = V(0)$. The necessary condition (3.7) (with $X(\cdot) = V(\cdot)$) implies $\lim_{n\to\infty} E^{\nu_n} \int_0^T \gamma_{\nu_n}(t)d\hat{c}(t) = 0$ and, since the processes $\gamma_{\nu_n}(\cdot)$ are bounded away from zero (uniformly in n), $\lim_{n\to\infty} E[Z_{\nu_n}(T)\hat{c}(T)] = 0$. Using weak compactness arguments as in CK (1993, Theorem 9.1) we can show that there exists $\nu \in \mathcal{D}$ such that $\lim_{n\to\infty} E[Z_{\nu_n}\hat{c}(T)] = E[Z_\nu(T)\hat{c}(T)] = 0$ (along a subsequence). It follows that $\hat{c}(\cdot) \equiv 0$.

It follows then from (3.17) that $A_{\hat{\nu}} \equiv 0$, and hence that $Q_{\hat{\nu}}(\cdot)$ is a martingale, which is equivalent to $V(0) = E^{\hat{\nu}}[B\gamma_{\hat{\nu}}(T)]$.   ◇

3.8 Remark

In fact, optimality of $\hat{\nu} \in \mathcal{D}$ for the problem $\sup_{\nu \in \mathcal{D}} E^\nu B\gamma_\nu(T)$, being equivalent to the martingale property of $Q_{\hat{\nu}}$, implies, by (3.17), that $\hat{c} \equiv \tilde{g} - g - \hat{\nu}^*\hat{\pi} \equiv 0$ (and is even equivalent to the latter if $Q_{\hat{\nu}}$ is of class $D[0,T]$).

3.9 Remark

(i) In order to calculate explicitly $h(0)$ (which is, after all, the Black-Scholes price in the linear case), one has to impose Markovian assumptions on the price process, to obtain the appropriate PDE. This we will do in Section 8.

(ii) A careful inspection of Section 3 shows that the assumption of finite second moment in the definition of contingent claims is not necessary; however, the hedging price could be infinite without it.

4 Utility Functions

A function $U : (0, \infty) \to \mathbb{R}$ will be called a *utility function* if it is strictly increasing, strictly concave, of class C^1, and satisfies

$$U'(0+) := \lim_{x \downarrow 0} U'(x) = \infty , \quad U'(\infty) := \lim_{x \to \infty} U'(x) = 0 . \quad (4.1)$$

We shall denote by I the (continuous, strictly decreasing) inverse of the function U'; this function maps $(0, \infty)$ onto itself, and satisfies $I(0+) = \infty, I(\infty) = 0$. We also introduce the Legendre–Fenchel transform

$$\tilde{U}(y) := \max_{x>0}[U(x) - xy] = U(I(y)) - yI(y), \quad 0 < y < \infty \quad (4.2)$$

of $-U(-x)$; this function $\tilde{U}$ is strictly decreasing and strictly convex, and satisfies

$$\tilde{U}'(y) = -I(y), \quad 0 < y < \infty, \quad (4.3)$$

$$U(x) = \min_{y>0}[\tilde{U}(y) + xy] = \tilde{U}(U'(x)) + xU'(x) , \quad 0 < x < \infty . \tag{4.4}$$

The useful inequalities

$$U(I(y)) \geq U(x) + y[I(y) - x] \tag{4.5}$$

$$\tilde{U}(U'(x)) + x[U'(x) - y] \leq \tilde{U}(y) , \tag{4.6}$$

valid for all $x > 0, y > 0$, are direct consequences of (4.2), (4.4). It is also easy to check that

$$\tilde{U}(\infty) = U(0+), \quad \tilde{U}(0+) = U(\infty) \tag{4.7}$$

hold; cf. KLSX (1991), Lemma 4.2.

4.1 Remark

We shall have occasion, in the sequel, to impose the following conditions on our utility functions:

$$c \mapsto cU'(c) \text{ is nondecreasing on } (0, \infty) , \tag{4.8}$$

for some $\alpha \in (0, 1), \gamma \in (1, \infty)$ we have : $\alpha U'(x) \geq U'(\gamma x), \quad \forall x \in (0, \infty) .$
$$\tag{4.9}$$

4.2 Remark

Condition (4.8) is equivalent to

$$y \mapsto yI(y) \text{ is nonincreasing on } (0, \infty) , \tag{4.8}'$$

and implies that
$$x \mapsto \tilde{U}(e^x) \text{ is convex on } \mathbb{R} . \tag{4.8}''$$

(If U is of class C^2, then condition (4.8) amounts to the statement that $\frac{-cU''(c)}{U'(c)}$, the so-called 'Arrow–Pratt measure of relative risk-aversion', does not exceed 1.)

Similarly, condition (4.9) is equivalent to having

$$I(\alpha y) \leq \gamma I(y), \quad \forall y \in (0, \infty) \quad \text{for some } \alpha \in (0, 1), \quad \gamma > 1 . \tag{4.9}'$$

Iterating (4.9)', we obtain the apparently stronger statement

$$\forall \alpha \in (0, 1), \ \exists \ \gamma \in (1, \infty) \text{ such that } I(\alpha y) \leq \gamma I(y), \quad \forall y \in (0, \infty) . \tag{4.9}''$$

5 Portfolio Optimization Problem

In this section we consider the optimization problem of maximizing utility from terminal wealth for our investor, i.e., we want to maximize

$$J(x; \pi) := EU(X^{x,\pi}(T)) , \qquad (5.1)$$

over $\pi \in \mathcal{A}_0(x)$, provided that the expectation is well-defined. More precisely, we have

5.1 Definition

The *utility maximization problem* is to maximize the expression of (5.1) over the class $\mathcal{A}'_0(x)$ of processes $\pi \in \mathcal{A}_0(x)$ that satisfy

$$EU^-(X^{x,\pi}(T)) < \infty. \qquad (5.2)$$

(x^- denotes the negative part of x : $x^- = max\{-x, 0\}$.) The value function of this problem will be denoted by

$$V(x) := \sup_{\pi \in \mathcal{A}'_0(x)} J(x; \pi) , \quad x \in (0, \infty) . \qquad (5.3)$$

5.2 Assumption

$$V(x) < \infty, \quad \forall \, x \in (0, \infty) .$$

It is fairly straightforward that the function $V(\cdot)$ is increasing and concave on $(0, \infty)$.

5.3 Remark

It can be checked that the Assumption 5.2 is satisfied if the function U is nonnegative and satisfies the growth condition

$$0 \leq U(x) \leq \kappa(1 + x^\alpha) ; \quad \forall(x) \in (0, \infty) \qquad (5.4)$$

for some constants $\kappa \in (0, \infty)$ and $\alpha \in (0, 1) -$ cf. KLSX (1991) for details.

5.4 Definition

We introduce the function

$$\mathcal{X}_\nu(y) := E\Big[H_\nu(T)I(yH_\nu(T))\Big] , \quad 0 < y < \infty \qquad (5.5)$$

and consider the subclass of $\mathcal{D}$ given by

$$\mathcal{D}' := \{\nu \in \mathcal{D}; \ \mathcal{X}_\nu(y) < \infty, \ \forall y \in (0, \infty)\} . \qquad (5.6)$$

For every $\nu \in \mathcal{D}'$, the function $\mathcal{X}_\nu(\cdot)$ of (5.9) is continuous and strictly decreasing, with $\mathcal{X}_\nu(0+) = \infty$ and $\mathcal{X}_\nu(\infty) = 0$; we denote its inverse by $\mathcal{Y}_\nu(\cdot)$.

5.5   Remark

Suppose that $U(\cdot)$ satisfies condition (4.9). It is then easy to see, using (4.9)$''$, that $\mathcal{X}_\nu(y) < \infty$ for some $y \in (0, \infty)$ implies: $\nu \in \mathcal{D}'$.

Next, we prove a lemma, which provides sufficient conditions for optimality in the problem of (5.1).

5.6   Lemma

For any given $x > 0$, $y > 0$ and $\pi \in \mathcal{A}_0(x)$, we have

$$EU(X^{x,\pi}(T)) \leq E\tilde{U}(yH_\nu(T)) + yx, \ \forall \nu \in \mathcal{D}. \tag{5.7}$$

In particular, if $\hat{\pi} \in \mathcal{A}_0(x)$ is such that *equality* holds in (5.11), for some $\lambda \in \mathcal{D}$ and $\hat{y} > 0$, then $\hat{\pi}$ is optimal for our (primal) optimization problem, while λ is optimal for the *dual problem*

$$\tilde{V}(\hat{y}) = \inf_{\nu \in \mathcal{D}} E\tilde{U}(\hat{y}H_\nu(T)). \tag{5.8}$$

Furthermore, equality holds in (5.11) if and only if

$$
\begin{aligned}
X^{x,\pi}(T) &= I(yH_\nu(T)) \text{ a.s.,} &(5.9)\\
\tilde{g}(t, \nu_t) &= g(t, \pi_t) + \nu^*(t)\pi(t) \text{ a.e.,} &(5.10)\\
E[H_\nu(T)X^{x,\pi}(T)] &= x &(5.11)
\end{aligned}
$$

(the latter being equivalent to $y = \mathcal{Y}_\nu(x)$ if (5.10) holds and $\nu \in \mathcal{D}'$).

Proof   By definitions of $\tilde{U}, \tilde{g}$ we get

$$
\begin{aligned}
U(X(T)) \ \leq \ & \tilde{U}(yH_\nu(T)) + yH_\nu(T)X(T) \\
& + \int_0^T H_\nu(t)X(t)[\tilde{g}(t,\nu_t) - g(t,\pi_t) - \nu^*(t)\pi(t)]dt.
\end{aligned} \tag{5.12}
$$

The upper bound of (5.7) follows from the supermartingale property (3.7); condition (5.9) follows from (4.2), condition (5.10) is obvious, and condition (5.11) corresponds to equality holding in (3.7). $\diamond$

5.7   Remark

Lemma 5.6 suggests the following strategy for solving the optimization problem:

1. show that the dual problem (5.8) has an optimal solution $\lambda_y \in \mathcal{D}'$ for all $y > 0$;

2. using Theorem 3.5, find the minimal hedging price $h_y(0)$ and a corresponding portfolio $\hat{\pi}_y$ for hedging $B := I(yH_{\lambda_y}(T))$;

3. prove (5.10) for the pair $(\hat{\pi}_y, \lambda_y)$;

4. show that, for every $x > 0$, you can find $y = y_x > 0$ such that $x = h_y(0) = E[H_{\lambda_y}(T)I(yH_{\lambda_y}(T))]$.

Then (i)–(iv) would imply that $\hat{\pi}_{y_x}$ is the optimal portfolio process for the utility maximization problem of an investor starting with initial capital equal to x.

To verify that step (i) can be accomplished, we impose the following condition:

$$\forall y \in (0, \infty), \quad \exists \nu \in \mathcal{D} \quad \text{such that} \quad \tilde{J}(y; \nu) := E\tilde{U}(yH_\nu(T)) < \infty \tag{5.13}$$

We shall also impose the assumption

$$U(0+) > -\infty \ , \quad U(\infty) = \infty \ . \tag{5.14}$$

5.8   Remark

Under the conditions of Remark 5.3, the requirement (5.13) is satisfied. Indeed, the condition (5.4) leads to

$$0 \leq \tilde{U}(y) \leq \tilde{\kappa}(1 + y^{-\rho}) \ ; \quad \forall y \in (0, \infty) \tag{5.15}$$

for some $\tilde{\kappa} \in (0, \infty)$ and $\rho = \alpha/(1 - \alpha)$.

5.9   Theorem

Assume that (4.8), (4.9), (5.13) and (5.14) are satisfied. Then condition (i) of Remark 5.7 is true, i.e. the dual problem admits a solution in the set $\mathcal{D}'$, for every $y > 0$.

The fact that the dual problem admits a solution under the conditions of Theorem 5.9 follows almost immediately (by standard weak compactness arguments) from Proposition 5.10 below. The details, as well as a relatively straightforward proof of Proposition 5.10, can be found in CK(1992). Denote by $\mathcal{H}$ the Hilbert space of progressively measurable processes ν with norm $[\![\nu]\!] = E \int_0^T \nu^2(s)ds < \infty$.

5.10   Proposition

Under the assumptions of Theorem 5.9, the functional $\tilde{J}(y; \cdot) : \mathcal{H} \to \mathbb{R} \cup \{+\infty\}$ of (5.13) is (i) convex, (ii) coercive: $\lim_{[\![\nu]\!] \to \infty} \tilde{J}(y; \nu) = \infty$, and (iii) lower-semicontinuous: for every $\nu \in \mathcal{H}$ and $\{\nu_n\}_{n \in \mathbf{N}} \subseteq \mathcal{H}$ with $[\![\nu_n - \nu]\!] \to 0$ as $n \to \infty$, we have

$$\tilde{J}(y; \nu) \leq \lim_{n \to \infty} \tilde{J}(y; \nu_n) \ . \tag{5.16}$$

5.11 Remark

It can be shown that the optimal dual process λ_y satisfies $\lambda_y \in \mathcal{D}'$; see, for example, Karatzas, Lehoczky, Shreve and Xu (1991), proof of Theorem 12.3.

We move now to step (ii) of Remark 5.7: Given $y > 0$ and the optimal λ_y for the dual problem, let π_y be the portfolio of Theorem 3.5 for hedging the claim $B_{\lambda_y} = I(yH_{\lambda_y}(T))$ (recall Remark 3.9 (ii)). We have the following useful fact:

5.12 Lemma

For every $\nu \in \mathcal{D}$, $0 < y < \infty$, we have

$$E[H_\nu(T)B_{\lambda_y}] \leq E[H_{\lambda_y}(T)B_{\lambda_y}] . \tag{5.17}$$

5.13 Remark

In fact, (5.17) is equivalent to λ_y being optimal for the dual problem, but we shall not need that result; its proof is quite lengthy and technical (see CK (1992, Theorem 10.1). We are going to provide a simpler proof for Lemma 5.12, but under the additional assumption that

$$E[H_{\lambda_y}(T)I(yH_\nu(T))] < \infty, \ \forall \nu \in \mathcal{D}, y > 0. \tag{5.18}$$

Proof of Lemma 5.12 Fix $\varepsilon \in (0,1), \nu \in \mathcal{D}$ and define (suppressing dependence on t)

$$G_\varepsilon := (1-\varepsilon)H_{\lambda_y} + \varepsilon H_\nu, \quad \mu_\varepsilon := G_\varepsilon^{-1}((1-\varepsilon)H_{\lambda_y}\lambda_y + \varepsilon H_\nu \nu),$$
$$\tilde{\mu}_\varepsilon := G_\varepsilon^{-1}((1-\varepsilon)H_{\lambda_y}\tilde{g}(\lambda_y) + \varepsilon H_\nu \tilde{g}(\nu)). \tag{5.19}$$

Then $\mu_\varepsilon \in \mathcal{D}$, because of the convexity of sets $\mathcal{D}_t$. Moreover, we have

$$dG_\varepsilon = \sigma^{-1}\mu_\varepsilon G_\varepsilon dW - \tilde{\mu}_\varepsilon G_\varepsilon dt,$$

and convexity of $\tilde{g}$ implies $\tilde{g}(\mu_\varepsilon) \leq \tilde{\mu}_\varepsilon$, and therefore, comparing the solutions to the respective (linear) SDEs, we get

$$G_\varepsilon(\cdot) \leq H_{\mu_\varepsilon}(\cdot), \ \text{a.s.} \tag{5.20}$$

Since λ_y is optimal and $\tilde{U}$ is decreasing, (5.20) implies

$$\varepsilon^{-1}\left(E[\tilde{U}(yH_{\lambda_y}(T)) - \tilde{U}(yG_\varepsilon(T))]\right) \leq 0. \tag{5.21}$$

Next, recall that $I = -\tilde{U}'$ and denote by V_ε the random variable inside the expectation operator in (5.21). Fix $\omega \in \Omega$, and assume, suppressing the dependence on ω and T, that $H_\nu \geq H_{\lambda_y}$. Then $\varepsilon^{-1}V_\varepsilon = I(F)y(H_\nu - H_{\lambda_y})$, where $yH_{\lambda_y} \leq F \leq yH_{\lambda_y} + \varepsilon y(H_\nu - H_{\lambda_y})$. Since I is decreasing we get $\varepsilon^{-1}V_\varepsilon \geq yI(yH_\nu)(H_\nu - H_{\lambda_y})$. We get the same result when assuming $H_\nu \leq H_{\lambda_y}$. This and assumption (5.18) imply that we can use Fatou's lemma when taking the limit as $\varepsilon \downarrow 0$ in (5.21), which gives us (5.17).   ⋄

Lemma 5.12 implies that, in the notation of Section 3,

$$h_y(0) = V_y(0) = E[H_{\lambda_y}(T)I(yH_{\lambda_y}(T))] = \text{ initial capital for portfolio } \pi_y,$$

so (5.11) is satisfied for $x = h_y(0)$. It also implies, by Remark 3.8, that (5.10) holds for the pair (π_y, λ_y). Therefore we have completed both steps (ii) and (iii). Step (iv) is a corollary of the following result.

5.14 Proposition

Under the assumptions of Theorem 5.9, for any given $x > 0$, there exists $y_x > 0$ that achieves $\inf_{y>0}[\tilde{V}(y) + xy]$ and satisfies

$$x = \mathcal{X}_{\lambda_{y_x}}(y_x). \tag{5.22}$$

For the (straightforward) proof see CK (1992, Proposition 12.2). We now put together the results of this section:

5.15 Theorem

Under the assumptions of Theorem 5.9, for any given $x > 0$ there exists an optimal portfolio process $\hat{\pi}$ for the utility maximization problem of Definition 5.1. $\hat{\pi}$ is equal to the portfolio of Theorem 3.5 for minimally hedging the claim $I(y_x H_{\lambda_{y_x}}(T))$, where y_x is given by Proposition 5.14 and λ_{y_x} is the optimal process for the dual problem (5.8).

6 Examples

6.1 Example *(portfolio constraints)*

Let K be a nonempty, closed, convex set in $\mathbb{R}^d$ containing the zero vector, and let $\mathbf{1}_K$ be its indicator function in the sense of convex analysis, namely $\mathbf{1}_K(\pi) = 0$ if $\pi \in K$, $\mathbf{1}_K(\pi) = \infty$, otherwise. Let formally $f_i(\pi) = -\mathbf{1}_K(\pi), i = 0, \ldots, d$ in (2.1) and (2.2) (therefore, heuristically speaking, it does not pay to have $\pi \notin K$, since then both the bank account and the stocks are 'infinitely bad'; we should remark that even short positions in these 'bad' assets are then of no use, since *all* the assets are infinitely bad, and the wealth is instantaneously pushed to zero, where it stays). Assume for (notational) simplicity that $r \equiv b_i \equiv 0, i = 1, \ldots, d$. Then the function $\tilde{g}(\nu)$ of (3.2) is equal to

$$\tilde{g}(\nu) = \sup_{\pi \in K}(\pi^* \nu) \ : \ \mathbb{R}^d \to \mathbb{R} \cup \{+\infty\}, \tag{6.1}$$

the *support function* of the convex set K. This is a closed, positively homogeneous, proper convex function on $\mathbb{R}^d$ (Rockafellar (1970), p. 114), finite on its effective domain

$$\tilde{K} := \{\nu \in \mathbb{R}^d; \ \tilde{g}(\nu) < \infty\} = \{\nu \in \mathbb{R}^d \ ; \ \exists \beta \in \mathbb{R} \text{ s.t. } \pi^* \nu \leq \beta, \ \forall \pi \in K\} \ , \tag{6.2}$$

which is a convex cone (called the 'barrier cone' of K). Obviously $\tilde{K}$ is not compact, implying that we are in a situation different from the theory developed so far. In particular, the processes Z_ν are not necessarily martingales, but only local martingales. It can be seen that this does not make a difference for hedging, where we can restrict ourselves to those Z_ν that are martingales, but it makes a difference for optimization, because we do not know in general whether, for optimal dual process λ, Z_λ is a martingale. Nevertheless, almost all our results from Sections 3–5 still apply (assuming that $\tilde{g}$ is continuous on $\tilde{K}$). The only significant difference is that, in the case of hedging, the consumption process $\hat{c}$ of Theorem 3.5 will typically be different from the zero process, i.e., the corresponding part of Theorem 3.7 does not apply; furthermore, it can very well happen that the hedging price is infinite, $V(0) = \infty$, if the 'set of constraints' K is too restrictive for a particular claim B.

The problem of this example is in fact equivalent to the problem of optimal trading under the portfolio constraint $\pi \in K$, a.e. It is dealt with in great detail in Cvitanic and Karatzas (1992, 1993). Explicit solutions are given for the optimization problem for logarithmic and power utility functions under various kinds of constraints, such as prohibition of borrowing, short-selling and/or investing at all in certain stocks – the case of so called 'incomplete markets'. Notice that the condition (5.14), used to prove existence, is not satisfied by logarithmic function. Nevertheless, since one has explicit formulas in this case, one can prove directly that they correspond to the optimal solution. The hedging problem for European options is also analyzed beyond the general formula for the hedging price of Theorem 3.5.

6.2   Example *(different interest rates for borrowing and lending)*

We want to consider a model in which one is allowed to borrow money, but at an interest rate R greater than or equal to the bond rate r, $R(\cdot) \geq r(\cdot)$. We assume that the progressively measurable process $R(\cdot)$ is also bounded, and, for simplicity, as in the previous example, that $r \equiv b_i \equiv 0$, $i = 1, \ldots, d$. We get the model by setting $f_0(t, \pi_t) = R(t)\mathbf{1}_{\{\sum_{i=1}^d \pi(t) > 1\}}$ and $f_i(\pi) \equiv 0, i = 1, \ldots, d$

(assuming that the agent does not borrow and invest money in the bond at the same time). In the notation of Section 3 we get

$$\mathcal{D} := \{\nu \in \mathcal{H} \; ; \; 0 \leq \nu_1 = \ldots = \nu_d \leq R \; , \qquad \ell \otimes P - a.e.\} \qquad (6.3)$$

$$\tilde{g}(\nu(t)) := \nu_1(t) \qquad \text{for every } \nu \in \mathcal{D}, \; 0 \leq t \leq T \; . \qquad (6.4)$$

It is again possible to get explicit solutions for the optimization problem in the case of logarithmic utility function (solving the dual problem which is easy in this case)) and of power utility functions (in the case of deterministic coefficients, by solving the Hamilton–Jacobi–Bellman (HJB) equation for the dual problem). Also, it is possible to obtain a Black–Scholes formula for the price of hedging European options, by solving the HJB equation associated with the stochastic control problem of (3.10). We refer again to Cvitanic and Karatzas (1992, 1993); see also Section 8.

6.3   **Example** *(different stock rate when short-selling)*

The analogue to Example 6.2 would be a market in which short-selling is discouraged by increasing the stock return rate for the investor who is short-selling. To illustrate what happens, assume $d = 1, r \equiv 0$, and set $f_1(\pi_t) = R\mathbf{1}_{\{\pi_t < 0\}}$, for some $R > 0$, and $f_0 \equiv 0$. It is then easy to see that $\tilde{g}(t, \nu_t) = 0$, for $-R - b(t) \leq \nu(t) \leq -b(t)$, and equal to infinity otherwise. Consider the optimization problem of maximizing $E \log X(T)$. Then the dual problem boils down to minimizing ν^2, so that the optimal dual process is given by (suppressing dependence on t)

$$\lambda = \left\{ \begin{array}{lll} 0 & ; & \text{if } b < 0, -R - b < 0 \\ -b & ; & \text{if } b \geq 0 \\ -R - b & ; & \text{if } b < 0, -R - b \geq 0 \end{array} \right\}$$

and is independent of y. The optimal portfolio process $\hat{\pi}$ is then the one that minimally hedges $B = I(yH_\lambda(T)) = (yZ_\lambda(T))^{-1}$. In this case we have $V(t) = V(0)Z_\lambda^{-1}(t)$ for the process V of (3.10). Therefore, we have $\psi_\nu(t) = -\lambda(t)(\sigma Z_\lambda)^{-1}(t)$ in (3.16). Hence, as one would expect (from the classical Merton problem),

$$\hat{\pi}_t = \left\{ \begin{array}{lll} 0 & ; \text{if} & b < 0, -R - b < 0 \\ \sigma^{-2}b & ; & \text{if } b > 0 \\ \sigma^{-2}(b + R) & ; & \text{if } b < 0, -R - b > 0 \end{array} \right.$$

Assume now constant market parameters, and consider the problem of hedging a European option $B = (P(T) - q)^+$ in this market, where $P(T)$ is the value of the stock price at time T, evaluated from a perspective of the investor who does not invest in the stock, i.e., with $f_1 \equiv 0$. It is well-known that the minimal hedging portfolio, in the market in which short-selling is not discouraged (i.e., the Black-Scholes portfolio), never sells short (in fact, it always borrows), and therefore can be used in the market of this example as well. Thus, the minimal hedging price is still the Black-Scholes price.

6.4 Remark

If the interest rate r and the vector of stock return rates b are not equal to zero, all we have to do in the above examples is first to discount by r, and then change the underlying probability measure to the risk-neutral one.

6.5 Remark

The minimal hedging price $h(0)$ of a claim B is the no-arbitrage upper bound for the price, and is typically too high, in the sense that no one would be willing to pay that much for the claim; for example, with the no-borrowing constraint and a European call on stock $P(\cdot)$, the minimal hedging price $h(0)$ is equal to the initial stock price $P(0)$ (see Karatzas and Kou (1996)). There are various ways to define a reasonable price, below the upper bound. For a given utility function U and initial wealth x, Davis (1994) suggests to use the price $V(0)$ which makes the agent's utility neutral with respect to infinitesimal diversion of funds into the claim. His methods can be used to prove that this price is given by using the change of measure and discounting corresponding to the optimal dual process; in other words, if λ_x is the optimal dual process, take $V(0) = E[H_{\lambda_x} B]$. Karatzas and Kou (1996) provide rigorous proofs in a number of situations, and give sufficient conditions for $V(0)$ not to depend on x and/or U.

7 Dependence on the Wealth Process

We briefly consider in this section a financial market in which the prices can also depend on the wealth of the investor. Thus, we let f_i, $i = 0, \ldots, d$ in (2.1) and (2.2) depend also on the current wealth $X(t)$. In order to apply the approach of the previous sections, we change the notation and denote by π the vector of actual amounts invested in stocks, rather than proportions of wealth. We can then relax condition (2.9) to the boundedness from below by an integrable random variable (which may depend on $\pi(\cdot)$ but not on t), i.e., for some random variable ξ, $E\xi < \infty$,

$$X^{x,\pi,c}(t) \geq \xi, \quad \forall\, 0 \leq t \leq T. \tag{2.9}'$$

Equations (2.7) and (2.8) become

$$dX(t) = g(t, \pi_t, X_t)dt + \pi^*(t)\sigma(t)dW(t) - dc(t), \quad X(0) = x > 0, \tag{7.1}$$

$$g(t, \pi, X) := X[r(t)+f_0(\pi, X)]+\sum_{i=1}^{d} \pi_i[b_i(t)+f_i(\pi, X)-r(t)-f_0(\pi, X)]. \tag{7.2}$$

Assumption 2.4 extends now to the pair of variables (π, X). (3.2) becomes

$$\tilde{g}(t, \nu, \mu) := \sup_{(\pi, X) \in \mathbb{R}^{d+1}} \{g(t, \pi, X) + \pi^*\nu + X\mu\}. \tag{7.3}$$

Recall the notation (2.5) and replace everywhere H_ν of (3.5) by $H_{\nu,\mu} := Z_\nu \beta_{-\mu}$. The equality analogous to (3.6) is (suppressing obvious arguments)

$$H_{\nu,\mu}(t)X(t) - \int_0^t H_{\nu,\mu}\tilde{g}ds = x - \int_0^t H_{\nu,\mu}[(\tilde{g} - g - \pi^*\nu - X\mu)ds + dc(s)]$$
$$+ \int_0^t H_{\nu,\mu}\left[\pi^*\sigma + \sigma^{-1}\nu\right]dW(s). \quad (7.4)$$

Condition (3.7) becomes

$$\sup_{(\nu,\mu)\in\mathcal{D}} E[H_{\nu,\mu}(T)X(T) + \int_0^T H_{\nu,\mu}\{\tilde{g} - g - \pi^*\nu - X\mu\}ds]$$
$$\leq x + E\int_0^T H_{\nu,\mu}\tilde{g}ds \quad (7.5)$$

The process V of (3.10) becomes

$$V(\tau) := ess\sup_{(\nu,\mu)\in D} E^\nu\left[B\beta_\mu(\tau,T) - \int_\tau^T \beta_\mu(\tau,s)\tilde{g}(s,\nu_s,\mu_s)ds \mid \mathcal{F}_\tau\right]. \quad (7.6)$$

The dual problem is now of the form

$$\inf_{(\nu,\mu)\in\mathcal{D}} E\left[\tilde{U}(yH_{\nu,\mu}(T)) + \int_0^T H_{\nu,\mu}(t)\tilde{g}(t,\nu_t,\mu_t)dt\right], \quad (7.7)$$

and hence not generally convex in (ν,μ). However, the problem is convex in $H_{\nu,\mu}$ and one can use the fact that the processes ν and μ are uniformly bounded, to prove the existence in the dual, and hence in the primal problem, under appropriate assumptions; see Cuoco and Cvitanić (1995). That is not the case in the interesting example of a market with portfolio constraints that also depend on the wealth process, i.e., in the context of Example 6.2, $K = K(t, X_t)$. Here the dual domain is not bounded and, in fact, it can be seen from results of He and Pagès (1993) and El Karoui and Jeanblanc-Picqué (1994) in the special case of the nonnegativity constraint on wealth, that one has to enlarge the set of dual processes to get existence in the dual problem. See also Zariphopoulou (1994). The hedging part can still be done using (7.6), although the supremum will typically not be attained. In the case of the constraint $X(\cdot) \geq S(\cdot)$, this gives yet another representation for the price of an American option with payoff given by process $S(\cdot)$. In this case, with linear dynamics (and setting $d = 1$ for simplicity), there is only one $\nu = \sigma^{-1}(b-r)$ and $\tilde{g}(t,\nu,\mu) = \mu S(t)$, with μ ranging through nonpositive progressively measurable processes.

8 More General Nonlinearities and Forward–Backward SDEs

We denote by π the vector of amounts of money invested in stocks and change the model (2.1), (2.2) for the asset prices to

$$dP_0(t) = P_0(t)r(t, X_t, \pi_t)dt, \qquad P_0(0) = 1 \tag{8.1}$$

$$dP_i(t) = b_i(t, P_t, X_t, \pi_t)dt + \sum_{j=1}^{d} \sigma_{ij}(t, P_t, X_t, \pi_t)dW^{(j)}(t), \; P_i(0) = p_i \in (0, \infty) \tag{8.2}$$

for $i = 1, \ldots, d$. We require that the wealth replicates, at time T, the contingent claim with value $l(P(T))$, for a given function l, the assumptions on which are specified below. The wealth equation becomes

$$dX(t) = \hat{b}(t, P_t, X_t, \pi_t)dt + \hat{\sigma}(t, P_t, X_t, \pi_t)dW(t); \; X(T) = l(P(T)), \tag{8.3}$$

where

$$\begin{aligned}
\hat{b}(t, p, x, \pi) &= (x - \textstyle\sum_{i=1}^{d} \pi_i)r(t, x, \pi) + \textstyle\sum_{i=1}^{d} \frac{\pi_i}{p_i} b_i(t, p, x, \pi); \\
\hat{\sigma}_j(t, p, x, \pi) &= \textstyle\sum_{i=1}^{d} \frac{\pi_i}{p_i} \sigma_{ij}(t, p, x, \pi), \; j = 1, \cdots, d,
\end{aligned} \tag{8.4}$$

for $(t, p, x, \pi) \in [0, T] \times \mathbb{R}^d \times \mathbb{R} \times \mathbb{R}^d$. The system of SDEs (8.2) and (8.3) is called a *Forward–Backward* SDE; the forward component is the price process, having been assigned an initial value, whereas the backward component is the wealth process, having been assigned the terminal value $X(T) = l(P(T))$. An existence theory for such equations has been developed by Ma, Protter and Yong (1994).

The main differences compared to the model of Section 2 are: (i) more general, nonconvex nonlinearities, including the volatility term; (ii) the contingent claim value $l(P(T))$ is not given in advance, but depends on the portfolio strategy and wealth of the investor through P; (iii) Markovian structure of the model.

In this section we shall use the following notations throughout: we denote $\mathbb{R}_+^d = \{(x_1, \cdots, x_d) \in \mathbb{R}^d | x_i > 0, i = 1, \cdots, d\}$; the inner product in $\mathbb{R}^d$ by $\langle \cdot, \cdot \rangle$; the norm in $\mathbb{R}^d$ by $|\cdot|$ and that of $\mathbb{R}^{d \times d}$, the space of all $d \times d$ matrices, by $\|\cdot\|$ and the transpose of a matrix $A \in \mathbb{R}^{d \times d}$ (resp. a vector $x \in \mathbb{R}^d$) by A^T (resp. x^T). We also denote $\mathbf{1}$ to be the vector $\mathbf{1} := (1, \cdots, 1) \in \mathbb{R}^d$, and define a (diagonal) matrix-valued function $\Lambda : \mathbb{R}^d \mapsto \mathbb{R}^{d \times d}$ by

$$\Lambda(x) := \begin{bmatrix} x_1 & 0 & \cdots & 0 \\ 0 & x_2 & \cdots & 0 \\ \vdots & \vdots & \ddots & \vdots \\ 0 & 0 & \cdots & x_d \end{bmatrix}, \qquad x = (x_1, \cdots, x_d) \in \mathbb{R}^d. \tag{8.5}$$

It is obvious that $\|\Lambda(x)\| = |x|$ for any $x \in \mathbb{R}^d$, and whenever $x \notin \partial\mathbb{R}^d_+$, $\Lambda(x)$ is invertible and $[\Lambda(x)]^{-1}$ is of the same form as $\Lambda(x)$ with $x_1, \cdots, x_d$ being replaced by $x_1^{-1}, \cdots, x_d^{-1}$. We can then rewrite functions $\widehat{b}$ and $\widehat{\sigma}$ in (8.4) as

$$
\begin{aligned}
\widehat{b}(t, p, x, \pi) &= xr(t, x, \pi) + \langle \pi, b^1(t, p, x, \pi) - r(t, x, \pi)\mathbf{1} \rangle; \\
\widehat{\sigma}(t, p, x, \pi) &= \langle \pi, \sigma^1(t, p, x, \pi) \rangle,
\end{aligned}
\tag{8.6}
$$

where

$$
\begin{aligned}
b^1(t, p, x, \pi) &:= [\Lambda(p)]^{-1}b(t, p, x, \pi) = \left(\tfrac{b_1}{p_1}, \cdots, \tfrac{b_d}{p_d}\right)(t, p, x, \pi); \\
\sigma^1(t, p, x, \pi) &:= [\Lambda(p)]^{-1}\sigma(t, p, x, \pi) = \left\{\tfrac{\sigma_{ij}}{p_i}\right\}_{i,j=1}^d(t, p, x, \pi).
\end{aligned}
\tag{8.7}
$$

To be consistent with the standard model, we henceforth call b^1 the *appreciation rate* and σ^1 the *volatility matrix* of the stock market. We restrict ourselves to the portfolios for which $E \int_0^T |\pi(t)|^2 dt < \infty$ and $X(t) \geq 0$, $\forall t \in [0, T]$, a.s.. Let us impose the following

Standing Assumptions:

(A1) The functions b, $\sigma : [0, T] \times \mathbb{R}^d \times \mathbb{R} \times \mathbb{R}^d \mapsto \mathbb{R}$ and $l : \mathbb{R}^d \mapsto \mathbb{R}$ are twice continuously differentiable, such that $b(t, 0, x, \pi) = \sigma(t, 0, x, \pi) = 0$, for all $(t, x, \pi) \in [0, T] \times \mathbb{R} \times \mathbb{R}^d$. The functions b^1 and σ^1, together with their first order partial derivatives in p, x and π are bounded, uniformly in (t, p, x, π). Further, we assume that partial derivatives of b^1 and σ^1 in p satisfy

$$
\sup_{(t,p,x,\pi)} \left\{ \left|p_k \frac{\partial b^1}{\partial p_k}\right|, \left|p_k \frac{\partial \sigma^1_{ij}}{\partial p_k}\right| \right\} < \infty, \qquad i, j, k = 1, \cdots, d.
\tag{8.8}
$$

(A2) The function σ satisfies $\sigma\sigma^T(t, p, x, \pi) > 0$, for all (t, p, x, π) with $p \notin \partial\mathbb{R}^d_+$; and there exists a positive constant $\mu > 0$, such that

$$
a^1(t, p, x, \pi) \geq \mu I, \qquad \text{for all } (p, t, x, \pi),
\tag{8.9}
$$

where $a^1 = \sigma^1(\sigma^1)^T$.

(A3) The function r is twice continuously differentiable and such that the following conditions are satisfied:

(a) For $(t, x, \pi) \in [0, T] \times \mathbb{R} \times \mathbb{R}^d$, $0 < r(t, x, \pi) \leq K$, for some constant $K > 0$.

(b) The partial derivatives of r in x and π, denoted by a generic function ψ, satisfy

$$
\varlimsup_{|x|, |\pi| \to \infty,} (|x| + |\pi|)^2 |\psi(t, x, \pi)| < \infty.
\tag{8.10}
$$

Either

(A4.a) The function l is bounded, C^2 and nonnegative; Its partial derivatives up to second order are all bounded;

or

(A4.b) The function l is nonnegative and $\lim_{|p|\to\infty} l(p) = \infty$; moreover, l has bounded, continuous partial derivatives up to third order, and there exist constants $K, M > 0$ such that

$$\begin{cases} |\Lambda(p)l_p(p)| \leq K(1 + l(p)); \\ \sup_{p\in\mathbb{R}_+^d} \|\Lambda^2(p)l_{pp}\| = M < \infty. \end{cases}$$

(A5) The partial derivatives of σ^1 in x and π satisfy

$$\sup_{(t,p,x,\pi)} \left\{ \left|x\frac{\partial\sigma^1_{ij}}{\partial x}\right| + \left|x\frac{\partial\sigma^1_{ij}}{\partial\pi_k}\right| \right\} < \infty, \quad i,j,k = 1,\cdots,d. \tag{8.11}$$

8.1 Remark

The conditions are quite restrictive, which is largely due to the generality of our setting. The PDE method described below often works even if the assumptions are far from being satisfied. In particular, it works in the case of the model used prior to this section, if we restrict ourselves to the Markovian setting and to standard European options. We note that the assumptions (A1) and (A2) obviously contain those cases in which $b(t, p, x, \pi) = \Lambda(p)b_1(t, x, \pi)$ and $\sigma(t, p, x, \pi) = \Lambda(p)\sigma_1(t, x, \pi)$ where b_1 and σ_1 are bounded, continuously differentiable functions with bounded first order partial derivatives; and $\sigma_1\sigma_1^T$ is positive definite and bounded away from zero, as we often see in more classical models. The second conditions on l restricts it to have at most quadratic growth. An example of a function σ satisfying (A1), (A2) and (A4.b) could be $\sigma(t, p, x, \pi) = p(\sigma(t) + \arctan(x^2 + |\pi|^2))$ with $\sigma(\cdot)$ satisfying (A2).

All the results below are proved in Cvitanić and Ma (1996):

8.2 Lemma

Suppose that (A1), (A2) hold. Then for any portfolio π and initial wealth x, the price process P satisfies: $P_i(t) > 0$, $i = 1, \cdots, d$ for all $t \in [0,T]$, almost surely, provided the initial prices $p_1, \cdots, p_d$ are positive.

The Four Step Scheme of Ma, Protter and Yong (1994), in our setting, consists of the following (and consists of three steps only):

Step 1: Solve the Black–Scholes type (but nonlinear) PDE

$$\begin{cases} 0 = \theta_t + \frac{1}{2}\mathrm{tr}\{\sigma\sigma^T(t,p,\theta,\Lambda(p)\theta_p)\theta_{pp}\} + (\langle p, \theta_p \rangle - \theta)r(t,\theta,\Lambda(p)\theta_p), \\ \theta(T,p) = l(p), \qquad p \in \mathbb{R}_+^d. \end{cases}$$

$$(8.12)$$

Step 2: Setting

$$\begin{cases} \tilde{b}(t,p) &=& b(t,p,\theta(t,p),\Lambda(p)\theta_p(t,p)) \\ \tilde{\sigma}(t,p) &=& \sigma(t,p,\theta(t,p),\Lambda(p)\theta_p(t,p)), \end{cases}$$

$$(8.13)$$

solve the forward SDE

$$P(t) = p + \int_0^t \tilde{b}(s, P(s))ds + \int_0^t \tilde{\sigma}(s, P(s))dW(s). \qquad (8.14)$$

Step 3: Set

$$\begin{cases} X(t) &=& \theta(t, P(t)) \\ \pi(t) &=& \Lambda(P(t))\theta_p(t, P(t)), \end{cases}$$

$$(8.15)$$

8.3 Theorem

Suppose that the standing assumptions (A1)—(A3), (A4.b) and (A5) hold. Then the PDE (8.12) and the SDE (8.14) admit unique solutions. Moreover, for any given $p \in \mathbb{R}_+^d$, the FBSDE (8.2), (8.3) admits a unique adapted solution (P, X, π), given by (8.15) with θ being the solution of (8.12).

The theorem implies that the initial value $X(0)$ of the backward process provides an upper bound for the minimal hedging price $h(0)$ of the contingent claim $B = l(P(T))$, since the claim can indeed be hedged starting with $X(0)$. The next result shows that $X(0) = h(0)$, and that π given by (8.15) is the least expensive hedging portfolio.

8.4 Theorem *(Comparison Theorem)*

Suppose that (A1)-(A5) hold. Let initial prices $p \in \mathbb{R}_+^d$ be given, and let π be any admissible portfolio such that the corresponding price/wealth process (P, Y) satisfies $Y(T) \geq l(P(T))$. Then $Y(\cdot) \geq \theta(\cdot, P(\cdot))$, where θ is the solution to (8.12). In particular, $Y(0) \geq \theta(0, p) = X(0)$, where X is the solution to the FBSDE (8.2), (8.3), starting from $p \in \mathbb{R}_+^d$, constructed by the Four–Step Scheme.

We conclude by examples in which a model like the one of this section would be appropriate:

(i) *Large investor.* Suppose that our investor is really an important one, so that, if she invests too much in the bank, the government (or the market) decides to decrease the bank interest rate. For example, we can assume that $r(t, x, \pi)$ is a decreasing function of $x - \pi$, for $x - \pi$ large.

(ii) Borrowing rate could be decreasing in wealth.

(iii) *Several agents – equilibrium model.* In Platen and Schweizer (1994), an SDE for the stock price is obtained from equilibrium considerations; both its drift and volatility coefficients depend on the hedging strategy of the agents in the market in a rather complex fashion. As the authors mention, "it is not clear at all how one should compute option prices in an economy where agents' strategies affect the underlying stock price process". Our results provide the price that would enable the seller to hedge against all the risk, i.e., the upper bound for the price.

(iv) *Cheating does not always pay for the large investor.* Suppose that, up to time $t = 0$, we have the standard, Black-Scholes model, i.e., the interest rate r is constant and the volatility function is given by $\sigma(t, p, x) = \sigma p$, $\sigma > 0$ (the drift function does not matter). Then, at time $t = 0$, the large investor sells the option worth $l(P(T))$ at time $t = T$, for the price of $\gamma(0, P(0))$, where γ is a pricing function. Typically,

$$\gamma(\cdot, P(\cdot)) \geq \rho(\cdot, P(\cdot)), \tag{8.16}$$

where ρ is the Black–Scholes pricing function given by the solution to (8.12), with r and σ as above. Having strict inequality in (8.16) means that the large investor is trying to 'cheat', i.e., to sell the option for more than its worth, the Black–Scholes price. Suppose that the investor, (the seller), finds buyers for the option at this price. This would create instabilities in the market and arbitrage opportunities, if the volatility of the stock were to remain the same. Let us assume that the effect is felt as a change in the volatility, so that the function $\sigma(t, p, x)$ is not equal to σp any more. A natural example would be $\sigma(t, p, x) = p[\sigma + f(\gamma(t, p) - \rho(t, p))]$, with f increasing. Also assume that $\sigma(t, p, x)$ remains equal to σp, if $\gamma(\cdot, \cdot) = \rho(\cdot, \cdot)$; in other words, if no cheating is attempted at any time t, we remain in the Black-Scholes world, and therefore the hedging is possible if one starts with initial capital equal to the Black-Scholes price. If, on the contrary, the option sells for more, than the volatility increases and the minimal hedging price changes. For example, for European call option the initial value of the solution to PDE (8.12) is an increasing function of $\sigma \in (0, \infty)$. Therefore, selling the option for a price greater than the Black-Scholes price might increase the stock volatility, hence also the minimal hedging price, and hedging might not be possible, in which case the cheating does not pay. On the other hand, there are cases for which the hedging is possible, and the cheating would pay.

From the point of view of the market, this model can indicate *how the volatility has to change if the option is overpriced*, in order to exclude arbitrage opportunities for the seller. For example, in a simple model in which $\gamma(\cdot, \cdot) = \rho(\cdot, \cdot) + \varepsilon$, and $\sigma(t, p, x) = p(\sigma + \delta_\varepsilon)$, it is easy to calculate (for standard European call options) what δ_ε would have to be in order to exclude arbitrage profit, or, equivalently, in order to have $\gamma(0, P(0))$ equal to the Black–Scholes price of a stock with volatility $\sigma + \delta_\varepsilon$. If δ_ε is less than the critical value, the cheating pays, and if it is larger, the cheating does not pay. Here we have a phenomenon unknown in the classical models: *Selling the option for its fair, Black-Scholes price, ensures that hedging is possible, but selling for more than that price does not guarantee the hedging.*

In general, if the seller of the option has an idea how much the cheating will affect the volatility of the stock, then she would also have an idea of how much she can safely cheat, by solving the PDE (8.12) for different σ's (assuming, of course, that there are always buyers willing to buy the option).

Another example would be the case with $\sigma(t, p, x) = p[\sigma + f(x - \rho(t, p))]$, where, again, f is increasing, and $f(z) = \sigma p$, for $z \leq 0$. Assuming that the seller will always reinvest the profits rather than consume, $X(t) - \rho(t, P(t))$ could be thought of as a measure of her arbitrage profit at time t. It is clear that the Black–Scholes pricing function ρ is a solution to (8.12), even with this modified volatility function $\sigma(t, p, x)$. Therefore, if $\sigma(t, p, x)$ is such that Assumptions (A1)-(A5) are satisfied, then the Black–Scholes price $\rho(0)$ is the smallest one that still guarantees successful hedging. However, it is not clear that the hedging is guaranteed if the investor sells the option for more than the minimal hedging price $\rho(0)$; since there is no consumption, we will have $X(t) > \rho(t, P(t))$ for small t, so that the volatility possibly increases. If the increase is significant compared to the price of the option, there might be no hedging portfolio. For example, one could have $\sigma(t, p, x) = p[\sigma + \arctan\{(x - \rho(t, p))^2\}]$; with $l(p) = p$, we have then $\rho(t, p) = p$. Suppose, for example, that the seller sells for the price of $p + 1$ and invests this amount in the market, buying at least one whole share of the stock. If he/she does not consume, then the volatility will always be greater than σ.

Acknowledgements

I would like to thank my one-time adviser, frequent co-author, and permanent dear friend Yannis Karatzas for his valuable comments on the paper, as well as for introducing me to the beauties and the challenges of the field of mathematical finance, and for the years of encouragement and collaboration. Thanks are also due to the editors of this volume, Michael Dempster and Stan Pliska, especially to the latter editor, for his useful suggestions on the paper.

Research supported in part by the National Science Foundation under Grant NSF-DMS-95-03582. A preliminary version presented at the Bank of

England Conference at the Newton Institute of Mathematical Sciences, Cambridge, UK, in June 1995. I am grateful to the institute for their hospitality.

References

Barron, E. and Jensen, R. (1990) 'A stochastic control approach to the pricing of options', *Math. Oper. Res.* **15**, 49–79.

Bergman, Y.Z. (1991) 'Option pricing with different borrowing and lending rates'. Preprint.

Cox, J. and Huang, C.F. (1989) 'Optimal consumption and portfolio policies when asset prices follow a diffusion process', *J. Econ. Theory* **49**, 33–83.

Cuoco, D. and Cvitanić, J. (1995) 'Optimal consumption choices for a large investor'. Preprint.

Cvitanić, J. and Karatzas, I. (1992) 'Convex duality in constrained portfolio optimization', *Annals of Applied Probability* **2**, 767–818.

Cvitanić, J. and Karatzas, I. (1993) 'Hedging contingent claims with constrained portfolios', *Annals of Applied Probability* **3**, 652–681.

Cvitanić, J. and Ma, J. (1996) 'Hedging options for a large investor and forward-backward SDEs'. To appear in *Annals of Applied Probability*.

Davis, M.H.A. (1994) 'A general option pricing formula'. Preprint.

El Karoui, N. and Jeanblanc-Picqué, M. (1994) 'Optimization of consumption with labor income'. Preprint.

El Karoui, N., Peng, S. and Quenez, M.C. (1994) 'Backwards stochastic differential equations in finance and optimization'. To appear in *Mathematical Finance*.

El Karoui, N. and Quenez, M.C. (1995) 'Dynamic programming and pricing of contingent claims in an incomplete market', *SIAM J. Control Optim.* **33**, 29–66.

Fleming, W. and Zariphopoulou, T. (1991) 'An optimal investment/consumption model with borrowing', *Math. Oper. Res.* **16**, 802-822.

Harrison, J.M. and Pliska, S.R. (1981) 'Martingales and stochastic integrals in the theory of continuous trading', *Stoch. Processes and Appl.* **11**, 215–260.

He, H. and Pageés, H.F. (1993) 'Labor income, borrowing constraints and equilibrium asset prices: a duality approach', *Economic Theory* **3**, 663–696.

He, H. and Pearson, N. (1991) 'Consumption and portfolio policies with incomplete markets and short-sale constraints: the infinite-dimensional case', *J. Econ. Theory* **54**, 259–304.

Jouini E. and Kallal H. (1995) 'Arbitrage in securities markets with short-sales constraints', *Math. Finance* **5**, 197–232.

Korn, R. (1992) 'Option pricing in a model with a higher interest rate for borrowing than for lending'. Preprint.

Karatzas, I. (1989) 'Optimization problems in the theory of continuous trading', *SIAM J. Control and Optimization* **27**, 1221–1259.

Karatzas, I. and Kou, S. (1996) 'On the pricing of contingent claims under constraints'. To appear in *Annals of Applied Probability*.

Karatzas, I., Lehoczky, J.P. and Shreve, S.E. (1987) 'Optimal portfolio and consumption decisions for a 'small investor' on a finite horizon', *SIAM J. Control and Optimization* **25**, 1157–1586.

Karatzas, I. and Shreve, S.E. (1991) *Brownian Motion and Stochastic Calculus* (2nd edition), Springer-Verlag, New York.

Karatzas, I., Lehoczky, J.P. and Shreve, S.E. and XU, G.L. (1991) 'Martingale and duality methods for utility maximization in an incomplete market', *SIAM J. Control and Optimization* **29**, 702–730.

Ma, J., Protter, P. and Yong, J. (1994) 'Solving forward-backward stochastic differential equations explicitly—a four step scheme', *Probability Theory and Related Fields*, **98**, 339–359.

Pardoux, E. and Peng, S.G. (1990) 'Adapted solution of a backward stochastic differential equation', *Systems and Control Letters* **14**, 55–61.

Platen, E. and Schweizer, M. (1994) 'On smile and skewness'. Preprint.

Pliska, S.R. (1986) 'A stochastic calculus model of continuous trading: optimal portfolios', *Math. Oper. Res.* **11**, 371–382.

Rockafellar, R.T. (1970) *Convex Analysis.* Princeton University Press, Princeton.

Xu, G. and Shreve, S.E. (1992) 'A duality method for optimal consumption and investment under short-selling prohibition. I. General market coefficients. II. Constant market coefficients', *Ann. Appl. Probab.* **2**, 87-112; 314–328.

Zariphopoulou, T. (1989) *Optimal Investment/Consumption Models with Constraints.* Doctoral Dissertation, Brown University.

Zariphopoulou, T. (1994) 'Consumption-investment models with constraints', *SIAM J. Control Optim.* **32**, 59–85.

Semimartingales and Asset Pricing under Constraints

Marco Frittelli

Abstract

We extend the well known fundamental theorem of asset pricing to the case of security markets models with frictions. We provide necessary and sufficient conditions for the existence of martingale, supermartingale and quasimartingale laws equivalent to the reference one. These conditions have a natural interpretation as no free lunch conditions in financial markets models with frictions. In this context we also show that an adapted stochastic process is a semimartingale if and only if no-extreme-arbitrage opportunities are allowed.

1 Introduction

In recent literature, increasing attention has been devoted to the question of asset pricing in *frictional* markets. Many different models and approaches have been proposed in order to examine the properties of markets with peculiar frictions. Is there a general principle which provides a unifying framework for the different models of frictional markets?

The so called fundamental theorem of asset pricing (informally) states that in *frictionless* markets there exists an equivalent *martingale* measure (also frequently called risk neutral or risk adjusted measure) if and only if the market admits *no free lunch*; many authors ([5], [6], [7], [9], [12], [15], [18], [19], [22], [23]) have proposed different models of security markets (essentially assuming different hypotheses on the classes of admissible stochastic processes representing deflated price processes and of admissible trading strategies) in order to give the theorem a precise and rigorous formulation.

No one would expect the theorem to be true in *frictional* markets too (see for example Jouini and Kallal [16]), but in this work we analyze to what extent and under what restrictions the theorem may be generalized to cover market frictions.

We model a fairly general class of restrictions by requiring the trading strategies to belong to a convex set Θ. The selection of this set determines the class of *admissible* trading strategies. The generality of the specification of Θ provides sufficient flexibility in order to treat different forms of constraints. Note, however, that we are not modeling transaction costs and bid-ask spreads in asset prices.

Moreover, in our model *any* real adapted stochastic process may be selected for representing the (deflated) price process of an available asset; in particular we do not a priori require the processes to be semimartingales. This of course forces us to consider, as admissible trading strategies, only *elementary* (or even *simple*) bounded predictable processes, since otherwise the notion of a stochastic integral with a non semimartingale integrator is not well defined.

Usually one requires, in order to deal with a nice space of terminal payoffs, some sort of nice behavior of the price process X (assuming for example continuity, boundedness, L^2-integrability). In this paper we take a different approach: we consider the (adapted) process X as primitive (hence no further assumptions are made) and select consequently the appropriate functional space and topology.

The advantage of permitting great generality for the price processes that define the security market model isn't costless: indeed we need to use some abstract topology (introduced in Section 2.3.) to determine which random variable may be considered 'close' to an arbitrage opportunity and therefore to consistently define the notion of an *extreme* arbitrage opportunity.

However, this topology will become a more familiar one, whenever appropriate assumptions are made on X. Furthermore, as it will be shown later, this approach permits us to determine a new characterization of semimartingales in terms of no-extreme-arbitrage opportunities.

To be more specific, let us fix a time horizon $\mathcal{T} = [0, T]$ and consider a vector valued adapted stochastic process $X = (X^j, \ j = 1, \ldots, N)$, defined on a given filtered probability space $(\Omega, \mathcal{F}, (\mathcal{F}_t)_{t \in \mathcal{T}}, P)$, which represents the (deflated) price processes of N securities available for trading in the market.

Let $\Theta = (\Theta^j, \ j = 1, \ldots, N)$ be a family of *convex* subsets of $L^\infty(\Omega, \mathcal{F}, P)$ with $0 \in \Theta^j$ for all j. The set of cumulative gains resulting from trading in the available assets is $K[\Theta]$, where Θ determines the class of *admissible* strategies:

$$K[X^j, \Theta^j] \triangleq \left\{ \sum_{i=0}^{n} H_i^j (X_{t_{i+1}}^j - X_{t_i}^j) : \forall i = 0, \ldots, n \ H_i^j \in L^\infty(\Omega, \mathcal{F}_{t_i}, P) \cap \Theta^j, \right.$$

$$\left. \text{for some } 0 \le t_0 < \ldots < t_{n+1} \le T \text{ and } n \in \mathbb{N} \right\} \qquad (1)$$

$$K[\Theta] \triangleq \left\{ \sum_{j=1}^{N} k_j : \forall j = 1, \ldots, N \quad k_j \in K[X^j, \Theta^j] \right\} \qquad (2)$$

Notice that if Θ^j are convex then $K[X^j, \Theta^j]$ and $K[\Theta]$ are convex as well.

A trading strategy consists of the selection of a finite number of trading dates $0 \le t_0 < \ldots < t_{n+1} \le T$, and a finite number of a.s. bounded, $\mathcal{F}_{t_i}$-measurable r.v. $H_i^j \in \Theta^j$ which determine the amount invested in asset j between date t_i and t_{i+1}.

More realistic and flexible trading strategies may be defined by considering stopping times instead of deterministic time indices (i.e. considering *simple* predictable trading strategies instead of *elementary* ones; see Dellacherie and Meyer [8] Ch.VIII, 2); with minor modifications of the assumptions (right-continuity of the filtration $(\mathcal{F}_t)_{t \in T}$ and of the process X) the subsequent results essentially hold as well (this issue is discussed in a similar context in [12] and [11]).

If, for all j, Θ^j coincides with the whole space $L^\infty(\Omega, \mathcal{F}, P)$ then no restrictions are imposed on the class of trading strategies and the set $K[\Theta]$ of cumulative gains (or the set of contingent claims replicable with admissible trading strategies with zero initial cost) is a *linear* space and we are back to the standard frictionless case.

The limitation of no short selling is modeled by requiring, $\forall j$, $\Theta^j = L^\infty_+(\Omega, \mathcal{F}, P)$ and in this case $K[\Theta]$ is a convex cone. Further convex restrictions on trading strategies may be imposed simply by requiring that Θ is a family of convex sets. By selecting different sets Θ^j for different indices j one may model various constraints according to the characteristics of the j-asset. For example, the following constraints impose bounds on each component of the portfolio that one is allowed to hold:

$$\Theta^j = \Theta_{a^j b^j} \triangleq \{H \in L^\infty(\Omega, \mathcal{F}, P) : -a^j \le H \le b^j \quad (P - a.s.)$$
$$\text{where } 0 \le a^j \le +\infty, \ 0 \le b^j \le +\infty\}.$$

In the sequel we will denote with $K \subseteq L^0(\Omega, \mathcal{F}, P)$ an arbitrary (convex) set (of $P - a.s.$ finite random variables) that has to be interpreted as the set of cumulative gains; one may then replace the generic set K with any set of the form $K[\Theta]$ for any specific selection of the family Θ.

It has been already pointed out by many authors (see for example Jouini and Kallal [16]) that, when one admits frictions in security markets models, one may not expect to identify the appropriate pricing functional with an equivalent *martingale* measure; for example, when there are shortsale constraints it is reasonable to look, for the purpose of super-replication and no arbitrage pricing, for equivalent *supermartingale* measures. Indeed, one would like to bet the semimartingale goes down, but one is unable to do so because of the constraint.

In this general setting we show (Corollary 7) that a specific *No-Extreme-Arbitrage* condition (No EA) in security markets models allowing for the type of frictions just described is equivalent to the existence of an *equivalent quasimartingale measure*.

We then deduce (Theorem 8) that an adapted stochastic process is a semimartingale *if and only if* no-extreme-arbitrage opportunities are allowed; informally (see definition 5 for the exact statement) an EA opportunity consists of a *uniformly bounded* sequence of trading strategies that replicate a sequence of contingent claims which approximate (in some sense to be defined later) an *arbitrarily large* non negative random variable.

Note that the various versions of free lunches (see the above references) usually require that a non negative and non zero r.v. must be approximated by a sequence of contingent claims. However, it may happen that an arbitrarily large amount is needed in order to replicate these contingent claims, while in the notion of EA only uniformly bounded investments are allowed.

For the specific case of the fractional Brownian Motion – which is not a semimartingale for self similarity parameter $H \neq \frac{1}{2}$ – Rogers [21] and Cutland *et al.* [2] have explicitly constructed a similar sort of arbitrage opportunity; here we show that for *any non-semimartingale* process there will indeed be an EA opportunity.

In the special case of *shortsale constraints* the notion of No EA may be simplified, and we show that now it is equivalent to the existence of a *supermartingale* measure (Corollary 6), which is in accordance with the results of Jouini and Kallal (see [16]) that were obtained in an L^2 environment.

Analogously, in frictionless security market models we deduce Frittelli and Lakner's formulation (see [12]) of the fundamental theorem of asset pricing: there exists an equivalent martingale measure if and only if no global free lunches are allowed (Corollary 5).

The condition of No EA here proposed is invariant under substitution of the reference measure P with an equivalent one and, since X is assumed to be only an adapted stochastic process, the statements in Corollaries 7 , 6 and 5 may be interpreted as an almost sure characterization of quasi / super / sub / martingales.

In Section 2 we present the basic notations and definitions, we illustrate a peculiar property – called the taming property – of quasi / super / sub / martingales laws, and we introduce the topology used for the formulation of the No EA The taming property was studied in a previous paper (see Frittelli [11]) and is here used in the proof of the above mentioned results, which are formally presented in Section 3.

2 Definitions and Properties

2.1 Martingale, supermartingale and quasimartingale measures

Let $\mathcal{T} = [0, T]$ and consider a vector valued adapted stochastic process $X = (X^j, \ j = 1, \ldots, N)$ defined on a given filtered probability space $(\Omega, \mathcal{F}, (\mathcal{F}_t)_{t \in \mathcal{T}}, P)$ with $\mathcal{F}_0$ trivial and complete. Let $\mathbb{P}$ be the set of probabilities equivalent to P and $L^0(\Omega, \mathcal{F}, P)$ be the space of $P - a.s.$ finite random variables on $(\Omega, \mathcal{F})$. For any non empty set $K \subseteq L^0(\Omega, \mathcal{F}, P)$ define:

$$\mathcal{L}^K \triangleq \{Q << P : K \subseteq L^1(\Omega, \mathcal{F}, Q)\}; \quad \mathcal{L}_e^K \triangleq \mathcal{L}^K \cap \mathbb{P}$$
$$\mathcal{Q}^K \triangleq \{Q \in \mathcal{L}^K : \sup_K E_Q[k] < +\infty\}; \quad \mathcal{Q}_e^K \triangleq \mathcal{Q}^K \cap \mathbb{P}.$$

Note that if K is positively homogeneous ($k \in K, \lambda \geq 0 \Rightarrow \lambda k \in K$) then $\mathcal{Q}^K \equiv \{Q \in \mathcal{L}^K : \sup_K E_Q[k] = 0\}$, while if K is homogeneous ($k \in K, \lambda \in \mathbb{R} \Rightarrow \lambda k \in K$) then $\mathcal{Q}^K \equiv \{Q \in \mathcal{L}^K : E_Q[k] = 0 \ \forall k \in K\}$.

Definition 1 *We call a probability measure Q defined on $(\Omega, \mathcal{F})$ a semimartingale law (resp. quasimartingale, submartingale, supermartingale, martingale law) for X if all processes X^j $j = 1, \ldots, N$ are semimartingales (resp. quasimartingales, submartingales, supermartingales, martingales) under the probability measure Q.*

Recall the definition of the sets $K[\Theta]$ and $K[X^j, \Theta^j]$ given in (1) and (2). For any subset $V \subseteq L^0(\Omega, \mathcal{F}, P)$ we denote by $V_+ \triangleq \{f \in V : f \geq 0 \ P - a.s.\}$, $V_{++} \triangleq V_+ \setminus \{0\}$ and analogously for V_- and V_{--}.

Definition 2 $M \triangleq K[L^\infty(\Omega, \mathcal{F}, P), \ldots, L^\infty(\Omega, \mathcal{F}, P)]$;
$M_+ \triangleq K[L_+^\infty(\Omega, \mathcal{F}, P), \ldots, L_+^\infty(\Omega, \mathcal{F}, P)]$;
$M_- \triangleq K[L_-^\infty(\Omega, \mathcal{F}, P), \ldots, L_-^\infty(\Omega, \mathcal{F}, P)]$;
$M_1 \triangleq K[\Theta_1, \ldots, \Theta_1]$, *where* $\Theta_1 \triangleq \{H \in L^\infty(\Omega, \mathcal{F}, P) : \|H\|_\infty \leq 1\}$.
Similar definitions apply for each component (for example we set $M_+^j \triangleq K[X^j, L_+^\infty(\Omega, \mathcal{F}, P)]$).

Lemma 1 $\mathcal{Q}^{M-}, \mathcal{Q}^{M+}, \mathcal{Q}^M, \mathcal{Q}^{M_1}$ *represent respectively the sets of submartingale, supermartingale, martingale and quasimartingale laws (for X) absolutely continuous w.r.t. P.*

Proof First note that the integrability requirements are satisfied for $Q \in \mathcal{L}^K$ (note also that if $\mathcal{F}_0$ was not suppose to be trivial, an integrability condition on X_0 should have been assumed). A probability $Q \in \mathcal{L}^K$ is a quasimartingale law for X^j if and only if $\sup_{f \in M_1^j} E_Q[f] < +\infty$; indeed let

$$\text{Var}_Q[X^j] \triangleq \sup_\pi \left\{ E_Q \left[\sum_{i=0}^n \left| E_Q[X_{t_{i+1}}^j - X_{t_i}^j | \mathcal{F}_{t_i}] \right| \right] \right\},$$

where the sup is taken over all finite partitions π of $[0, T]$. By definition X^j is a Q-quasimartingale if $\text{Var}_Q[X^j] < +\infty$. Note that if $f \in M_1^j$, then:

$$E_Q[f] = E_Q \left[\sum_{i=0}^n H_i^j (X_{t_{i+1}}^j - X_{t_i}^j) \right]$$
$$= E_Q \left[\sum_{i=0}^n H_i^j E_Q[X_{t_{i+1}}^j - X_{t_i}^j | \mathcal{F}_{t_i}] \right] \leq \text{Var}_Q[X^j]$$

and one gets the equality $\sup_{f \in M_1^j} E_Q[f] = \mathrm{Var}_Q[X^j]$ from considering $H_i^j = \mathrm{sgn}\,(E_Q[X_{t_{i+1}}^j - X_{t_i}^j | \mathcal{F}_{t_i}])$.

Hence $\sup_{f \in M_1} E_Q[f] = \sum_{j=1}^N \mathrm{Var}_Q[X^j]$, so that $Q \in \mathcal{L}^K$ is a quasimartingale measure for X if and only if $\sup_{f \in M_1} E_Q[f] < +\infty$.

$Q \in \mathcal{L}^K$ is a submartingale (resp. supermartingale) law for X^j if and only if $E_Q[f] = E_Q[\sum_{i=0}^n H_i^j(X_{t_{i+1}}^j - X_{t_i}^j)] \leq 0 \; \forall f \in M_-^j$ (resp. M_+^j) and hence if and only if $Q \in \mathcal{Q}^{M-}$ (resp. Q^{M+}) since M_- and M_+ are positively homogeneous.

$Q \in \mathcal{L}^K$ is martingale law for X^j if and only if

$$E_Q[f] = E_Q \left[\sum_{i=0}^n H_i^j(X_{t_{i+1}}^j - X_{t_i}^j) \right] = 0 \; \forall f \in M^j$$

and hence if and only if and $Q \in \mathcal{Q}^M$, since M is a linear space.   □

2.2   Tamed families

We briefly introduce some definitions and state theorem 2 which was proved in a previous paper (see Frittelli [11]); we will apply it several times in the sequel since it is crucial for the proof of the corollaries in the next section.

We recall that two families $\mathcal{R}, \mathcal{Q} \subset \mathcal{L}$ of probabilities are *equivalent* ($\mathcal{R} \sim \mathcal{Q}$) iff $\forall A \in \mathcal{F} \; (R(A) = 0 \; \forall R \in \mathcal{R}) \Leftrightarrow (Q(A) = 0 \; \forall Q \in \mathcal{Q})$ and that a family of probabilities $\mathcal{Q}$ is *dominated* by a probability $\hat{Q}$ if $Q << \hat{Q} \; \forall Q \in \mathcal{Q}$. Clearly a single probability $\hat{Q} \in \mathcal{Q}$ dominates the family $\mathcal{Q}$ if and only if $\hat{Q} \sim \mathcal{Q}$.

Definition 3 *A family $\mathcal{Q}$ is* tamed *if it is dominated by a single measure $\hat{Q} \in \mathcal{Q}$.*

The Halmos–Savage lemma [14] guarantees that any dominated family $\mathcal{Q}$ of probabilities contains a countable equivalent family. Thus, to show that a dominated family $\mathcal{Q}$ is tamed it is sufficient to prove that any given countable family $\{Q_n\}_{n \in \mathbb{N}} \in \mathcal{Q}$ is dominated by some measure in $\mathcal{Q}$. Note that this certainly occurs whenever it is possible to select, for any given countable family $\{Q_n\}_{n \in \mathbb{N}} \in \mathcal{Q}$, a sequence $\{a_n\}_{n \in \mathbb{N}}$ of positive real numbers such that $\sum_{n=1}^\infty a_n Q_n \in \mathcal{Q}$.

In this way one may easily show, for example, that if $K \subseteq L^0(\Omega, \mathcal{F}, P)$ is countable, then $\mathcal{Q}^K$ is tamed, and also that if $K \subseteq L^\infty(\Omega, \mathcal{F}, P)$ is such that $\mathcal{Q}^K$ is not empty, then $\mathcal{Q}^K$ is tamed.

For the families of quasi/sub/super/martingale measures we have:

Theorem 2 *Each (not empty) family $\mathcal{Q}^{M_1}, \mathcal{Q}^{M-}, \mathcal{Q}^{M+}, \mathcal{Q}^M$ is tamed.*

2.3 The topology

As already mentioned, we do not want to impose a priori any further restrictions on X, but, in order to define the notion of extreme arbitrage opportunities we need to introduce a topology on the space

$$S_K \triangleq \mathrm{Lin}\,\{K \cup L^\infty(\Omega, \mathcal{F}, P)\}.$$

To this end define:

$$S'_K \triangleq \left\{ \mu \in \Sigma(\Omega, \mathcal{F}) : \mu << P \text{ and } \int_\Omega |s| d|\mu| < +\infty \ \forall s \in S_K \text{ where} \right.$$

$$\left. \Sigma(\Omega, \mathcal{F}) \text{ is the space of finite signed measures on } (\Omega, \mathcal{F}) \right\}$$

$$= \left\{ s' \in L^1(\Omega, \mathcal{F}, P) : E_P[|s's|] < +\infty \ \forall s \in S_K \right\}.$$

The intuition behind this definition is that S'_K is formed with all (finite and signed) measures under which all terminal payoffs have finite 'expected value'. Note that S'_K is a subspace of the algebraic dual of S_K, and so (S_K, S'_K) defines a dual system.

Definition 4 *Let τ_K be any topology on S_K compatible with the dual system (S_K, S'_K).*

(S_K, τ_K) is (see Grothendieck, [13], Ch. II, Sec. 8 and 13) a locally convex topological vector space that, in general, is not metrizable (see [12] and [19] for further details).

One example of a topology compatible with (S_K, S'_K) is the weakest topology $\hat{\tau}_K$ on S_K which is stronger than each $L^1(\Omega, \mathcal{F}, Q)$-norm topology for all $Q \in \mathcal{L}^K_e$; in other words $\hat{\tau}_K$ is the upper bound of these topologies (again see [12] and [19] for more details).

Let us call *admissible* the agents whose beliefs are represented by a (subjective) probability measure $Q \in \mathcal{L}^K_e$. Then each admissible agent would regard his subjective $L^1(\Omega, \mathcal{F}, Q)$-norm topology as the 'right' one for determining the notion of a free lunch. However, we do not want to assign any special role to one specific agent versus the others.

By definition of $\hat{\tau}_K$, two random variables in $(S_K, \hat{\tau}_K)$ are *close* if they are considered close in the norm topology by all admissible agents so that $\hat{\tau}_K$ may be interpreted as the topology of common agreement among all admissible agents.

We stress that the use of this topology is needed for the sake of generality. If one prefers to impose restrictions on the class X, then one ends up with more familiar topologies. For example, if one requires $K \subseteq L^\infty(\Omega, \mathcal{F}, P)$, then $L^\infty(\Omega, \mathcal{F}, P) = S_K$, $S_K' = L^1(\Omega, \mathcal{F}, P)$, and one may, for instance, choose as τ_K the weak* topology $\sigma(L^\infty, L^1)$ on $L^\infty(\Omega, \mathcal{F}, P)$.

Note that S_K, S_K' and τ_K are invariant under substitution of P with an equivalent probability measure; this implies that also the notion of no extreme arbitrage opportunity (Definition 5) will have this remarkable property.

3   Extreme arbitrage opportunities

In this section we show the equivalence between the existence of an equivalent quasi / super / sub / martingale measure and the appropriate notion of no arbitrage opportunity. We adopt the same notations and assumptions as in Section 2; L^0 (and similarly L^∞, L^1) is the abbreviation for $L^0(\Omega, \mathcal{F}, P)$.

The (convex) set $K = K[\Theta] \subseteq L^0$ is to be interpreted as the set of terminal cumulative gains as explained in the introduction.

We say that a random variable $k \in K$ is an *arbitrage opportunity* if

$$k \geq 0 \ P - a.s. \quad \text{and} \quad P(k > 0) > 0.$$

The interpretation is that one may find an admissible trading strategy $H = (H^j, \ j = 1, \dots, N)$ which provides a non negative terminal cumulative gain $k = \sum_{j=1}^N \left(\sum_{i=0}^n H_i^j (X_{t_{i+1}}^j - X_{t_i}^j) \right)$ that is positive with positive probability. The absence of arbitrage may therefore be written as

$$K \cap L_+^0 = \{0\}. \tag{No A}$$

In frictionless markets $(K = M)$ with a finite number of trading dates Harrison and Pliska [15] and Dalang and Morton and Willinger [5] proved the equivalence between no arbitrage and the existence of an Equivalent Martingale Measure (EMM).

However, many examples (see for instance Back and Pliska [1], Dalang and Morton and Willinger [5], McBeth [20], Schachermayer [22]) show that, if one considers the possibility of trading in a countable number of dates, the absence of arbitrage is not any more sufficient to imply the existence of an EMM, so that a condition stronger than No A is needed.

To this end the notion of a *free lunch* was introduced by Kreps (1979, [18]); it involves an appropriate topological closure of the set of replicable gains by admissible trading strategies. Since then, many other notions of free lunches appeared in literature ([5], [6], [7], [9], [12], [15], [18], [19], [20], [22], [23]), and their various definitions reflect different assumptions made on the stochastic security market model.

In general, the notion of a free-lunch requires the existence of a sequence (or generalized sequence) of terminal cumulative gains $k_n \in K$ (or contingent claims replicable at zero initial cost with self financing strategies) which dominates a sequence of r.v. f_n converging (in some topology) to a non negative non zero random variable f: there exist

$$(k_n, f_n, f)_{n \in \mathbb{N}} \in K \times L^0 \times L^0_{++} \text{ such that } k_n \geq f_n \to f.$$

Letting $h_n = k_n - f_n \in L^0_+$ so that $f_n = k_n - h_n \in K - L^0_+$, one may formulate the no free-lunch condition as:

$$\overline{K - L^0_+} \cap L^0_+ = \{0\}, \tag{No FL}$$

where the closure is taken in some specific topology.

This form of geometric condition of no free lunch was initially introduced by Stricker [23], who discovered the pertinence of a theorem of Yan [24] to this subject. Yan's theorem has become an important tool in most subsequent work in the area, and indeed the following definition of extreme arbitrage opportunity is inspired by Yan's work.

We now study these notions in the framework of our model.

From the previous section it is evident that the condition of the existence of a martingale (resp. submartingale, supermartingale, quasimartingale) law Q equivalent to P is:

$$\mathcal{Q}^K_e \neq \emptyset, \tag{A}$$

when K is replaced with M (resp. with M_-, M_+, M_1). Suppose that (A) holds and $\hat{Q} \in \mathcal{Q}^K_e$; then clearly the following condition (B) holds with Q_A replaced by $\hat{Q}$:

$$\forall A \in \mathcal{F} : P(A) > 0 \ \exists Q_A \in \mathcal{Q}^K : Q_A(A) > 0. \tag{B}$$

Note that whenever the set K is positively homogeneous, condition (B) implies the absence of arbitrage opportunities. Indeed, suppose k is an arbitrage opportunity and let $A \triangleq \{k > 0\}$; then there exists a $Q_A \in \mathcal{Q}^K$ such that $Q_A(k > 0) > 0$ and $Q_A(k < 0) = 0$ (since $P(k < 0) = 0$ and $Q_A << P$). Therefore $E_{Q_A}[k] > 0$, which contradicts the requirement that $Q_A \in \mathcal{Q}^K$ because this implies $\sup_K E_{Q_A}[k] = 0$.

However, (B) is not equivalent to the no arbitrage condition. Indeed we will show that (B) is a preliminary version of a no free lunch type condition which we will call *No Extreme Arbitrage* condition; to this end we first show that (A) and (B) are, in some cases, equivalent:

Proposition 3 *If the family $\mathcal{Q}^K$ is tamed then* (A) *and* (B) *are equivalent.*

Proof (B) $\Rightarrow$(A): Since $\mathcal{Q}^K$ is tamed, there exists a $\hat{Q} \in \mathcal{Q}^K$ such that $Q << \hat{Q} \ \forall Q \in \mathcal{Q}^K$. We need to show that $P << \hat{Q}$. Let $A \in \mathcal{F}$ be such that $P(A) > 0$; then there exists $Q_A \in \mathcal{Q}^K$ such that $Q_A(A) > 0$. But $Q_A << \hat{Q}$ and so $\hat{Q}(A) > 0$. $\qquad \square$

Definition 5 *We say that there exists in (K, τ_K) an extreme arbitrage opportunity if:*

$$\exists f \in L^\infty_{++}, \ \exists \{\alpha_m\}_{m \in \mathbb{N}} \in \mathbb{R}_+ \ s.t. \ \alpha_m \to +\infty$$
$$and \ (\alpha_m f) \in \overline{K - L^\infty_+}^{\tau_K} \ \forall m \in \mathbb{N}.$$

We say that (K, τ_K) admits no extreme arbitrage *if:*

$$\forall f \in L^\infty_{++}, \ \exists \hat{r} \geq 0 : \forall r > \hat{r} \ \ rf \notin \overline{K - L^\infty_+}^{\tau_K}. \tag{No EA}$$

Note that when K is positively homogeneous then the No EA condition may be simplified and rewritten as:

$$\overline{K - L^\infty_+}^{\tau_K} \cap L^\infty_+ = \{0\}, \tag{No GFL}$$

which we label as a *No Global Free Lunch* condition in analogy with the previous No FL type condition and with the result in Frittelli and Lakner [12].

The following theorem is quite general since we do not need to assume any particular form of the convex set K; it is the basic result that we will apply several times to cover the different constraint cases.

Theorem 4 *If $K \subseteq L^0(\Omega, \mathcal{F}, P)$ is a convex set with $0 \in K$ and if $\mathcal{Q}^K$ is tamed, then conditions* (A), (No EA), *and* (B) *are equivalent.*

Proof (A) $\Rightarrow$ (No EA): Let $\hat{Q} \in \mathcal{Q}^K_e$ and $\hat{c} = \sup_K \hat{E}[k] < +\infty$, where $\hat{E}[k] \triangleq E_P[\hat{y}k]$ and $\hat{y} \triangleq \frac{d\hat{Q}}{dP} \in S'_K$. Define $G = \left\{ s \in S_K : \hat{E}[s] \leq \hat{c} \right\}$. G is closed in (S_K, τ_K) since $\hat{y} \in S'_K$. If $g \in (K - L^\infty_+)$ then $\hat{E}[g] \leq \hat{c}$ and therefore $\overline{(K - L^\infty_+)}^{\tau_K} \subseteq G$. Let $f \in L^\infty_{++}$; then $\hat{E}[f] > 0$ (since $P \sim \hat{Q}$), and we may select $\hat{r} = \frac{\hat{c}}{\hat{E}[f]}$. For $r > \hat{r}$ we have $\hat{E}[rf] > \hat{r}\hat{E}[f] = \hat{c}$; hence $rf \notin G$.

(No EA) $\Rightarrow$ (B): This part of the proof, which is inspired by Yan [24], is a simple application of the separation theorem in locally convex T.V.S. Fix $A \in \mathcal{F}$ such that $P(A) > 0$ so that $1_A \in L^\infty_{++}$. Fix also $r > \hat{r}$. Since the compact set $\{r1_A\}$ is disjoint from the closed convex set $\overline{(K - L^\infty_+)}^{\tau_K}$, the separation theorem (Grothendieck, [13], Ch. II, Sec. 7, Prop. 10) guarantees the existence of a non trivial functional $s' \in S'_K$ such that

$$\sup_{k \in K, h \in L^\infty_+} E_P[s'(k - h)] < E_P[s'r1_A] < +\infty .$$

Take $k = 0$; then $E_P[-s'h] < +\infty$, $\forall h \in L^\infty_+$, which implies $s' \geq 0 \ P-a.s..$ Define Q_A by $dQ_A/dP = s'/\|s'\|_{L^1}$; so then $Q_A << P$.

Take $h = 0$; then $\sup_{k \in K} E_{Q_A}[k] < E_{Q_A}[r1_A] = rQ_A(A)$. Hence $Q_A \in Q^K$, so that Q^K is not empty. Moreover $0 \in K$ implies $Q_A(A) > 0$, so (B) is satisfied. Proposition 3 concludes the proof. $\qquad \square$

If K is equal to the linear space M, then the notion of No GFL corresponds exactly to the notion described in Frittelli and Lakner [12]; in this case Theorems 2 and 4 provide a new formulation of the fundamental theorem of asset pricing:

Corollary 5 *There exists a martingale law Q (for X) equivalent to P if and only if*

$$\overline{M - L_+^\infty}^{\tau_M} \cap L_+^\infty = \{0\}.$$

Consider now the case where K is positively homogeneous; again the No EA condition may be formulated as a no global free lunch condition. Substituting M_- or M_+ for K in (A) and in (No GFL) Theorems 2 and 4 now yield:

Corollary 6 *There exists a submartingale (resp. supermartingale) law Q (for X) equivalent to P if and only if (M_-, τ_{M_-}) (resp. (M_+, τ_{M_+})) admits no global free lunch.*

A *Global free lunch* (for M_+) consists of an a.s. non negative and non zero random variable f which may be 'approximated' by a random variable

$$\hat{f} = g - h \in M_+ - L_+^\infty, \text{ where } g = \sum_{j=1}^N \left(\sum_{i=0}^n H_i^j (X_{t_{i+1}}^j - X_{t_i}^j) \right) \in M_+$$

is replicable by a bounded trading strategy $H = (H_i^j, i = 0, \ldots, n, \ j = 1, \ldots, N)$ s.t. $H_i^j \in L^\infty(\Omega, \mathcal{F}_{t_i}, P)$ and $H_i^j \geq 0$ $P - a.s.$ (prohibition of short-selling). Note that, even though each H_i^j is bounded, the approximation of f may eventually require *arbitrarily large* investments.

We note that Corollary 6 agrees with a previous result obtained by Jouini and Kallal [16], although they assumed that $X_t \in L^2(\Omega, \mathcal{F}, P)$ and used the $L^2(\Omega, \mathcal{F}, P)$ norm topology to define free lunches.

The above definition of global free lunch may be compared with the notion of extreme arbitrage opportunity. Suppose that K is the convex set M_1. An *extreme arbitrage opportunity* (for M_1) consists of a $P - a.s.$ non negative and non zero random variable f and a diverging sequence α_m of positive real numbers such that each (*eventually large*) $(\alpha_m f)$ is arbitrarily close to some random variable

$$f_m = g_m - h_m \in M_1 - L_+^\infty, \text{ where } g_m = \sum_{j=1}^N \left(\sum_{i=0}^n H_i^{j,m} (X_{t_{i+1}} - X_{t_i}) \right) \in M_1$$

is replicable by a bounded trading strategy $H^m = (H_i^{j,m}, i = 0, \ldots, n, j = 1, \ldots, N)$ such that $H_i^{j,m} \in L^\infty(\Omega, \mathcal{F}_{t_i}, P)$ and $\|H^m\|_\infty \leq 1$. This means that in order to approximate each $\alpha_m f$, only investments that do not exceed a *limited amount* are admitted.

Substituting M_1 for K in (A) and in (No EA) from Theorems 2 and 4 we get:

Corollary 7 *There exists a quasimartingale law Q for X equivalent to P if and only if (M_1, τ_{M_1}) admits no extreme arbitrage.*

It is well known that the process X is a semimartingale if and only if there exists a probability measure Q equivalent to P under which X is a quasimartingale (see for example Dellacherie and Meyer [8] Ch. VII, Th. 58), so we get

Theorem 8 *X is a semimartingale if and only if (M_1, τ_{M_1}) admits no extreme arbitrage.*

Given the generality of the assumptions on X and the fact that the selected topology is invariant under equivalent change of probability measures, this theorem also provides a new characterization of semimartingales.

The existence of an EA clearly implies the existence of a GFL. If X is not a semimartingale, then it will never become a martingale under any equivalent change of probability measure, and it is quite evident that a global free lunch has to exist. This fact is of no surprise; it is also in accordance with the findings described by Cutland, Kopp and Willinger [2] and by Rogers [21] when X is a fractional Brownian Motion. However, note that theorem 8 establishes the existence of the *stronger* notion of an extreme arbitrage opportunity.

This adds one more reason to be cautious when modeling asset prices with non semimartingale processes (as for example fractional Brownian Motions): an extreme arbitrage opportunity shouldn't appear in any reasonable market model allowing the type of frictions here described.

Acknowledgement

I am very grateful to the organizers of the Financial Mathematics Programme at the Isaac Newton Institute for Mathematical Sciences, University of Cambridge, UK, for their invitation to present a preliminary version of this paper and to the Newton Institute for their hospitality during the first semester 1995.

References

[1] K. Back and S. Pliska, 'On the Fundamental Theorem of Asset Pricing with an Infinite State Space', *J. of Math. Econ.* **20**, (1991), 1–18.

[2] N.J. Cutland, P.E. Kopp and W. Willinger, 'Stock price returns and the Joseph Effect: A fractional version of the Black-Scholes model', to appear in *Proceeding of the Monte Verità Conference*, Ascona, Switzerland, June 1993.

[3] J. Cvitanic and I. Karatzas, 'Convex duality in constrained portfolio optimization', *Ann. Appl. Prob.* **2**, (1992), 767–818.

[4] J. Cvitanic and I. Karatzas, 'Hedging contingent claims with constrained portfolios', *Ann. Appl. Prob.* **3**, (1993), 652–681.

[5] R.C. Dalang, A. Morton and W. Willinger, 'Equivalent martingale measures and no arbitrage in stochastic securities market models', *Stoc. and Stoc. Rep.* **29**, (1990), 185–201.

[6] F. Delbaen, 'Representing martingale measures when asset prices are continuous and bounded', *Math. Fin.* **2**, (1992), 107–130.

[7] F. Delbaen and W. Schachermayer, 'A general version of the fundamental theorem of asset pricing', *Math. Ann.* **300**, (1994), 463–520.

[8] C. Dellacherie and P.A. Meyer, *Probabilities and Potential*, North-Holland, (1978).

[9] D. Duffie and C. Huang, 'Multiperiod security markets with differential information', *J.M.E.* **15**, (1986), 283–303.

[10] N. El Karoui and M.C. Quenez, 'Dynamic programming and pricing of contingent claims in an incomplete market', *SIAM J. Control Optim.* **33**, (1995), 29–66.

[11] M. Frittelli, 'Dominated families of martingale, supermartingale and quasimartingale laws', Working Paper #89, Università degli Studi di Brescia, Brescia, Italy (1995).

[12] M. Frittelli and P. Lakner, A'lmost sure characterization of martingales', *Stoc. and Stoc. Rep.* **49**, (1994), 181–190.

[13] A. Grothendieck, *Topological Vector Spaces*, Gordon and Breach Science Publishers, (1973).

[14] P.R. Halmos and L.J. Savage, 'Application of the Radon-Nikodym theorem to the theory of sufficient statistics', *Ann. Math. Stat.* **20**, (1947), 225–241.

[15] J.M. Harrison and S.R. Pliska, 'Martingales and stochastic integrals in the theory of continuous trading', *Stoc. Proc. Appl.* **11**, (1981), 215–260.

[16] E. Jouini and H. Kallal, 'Arbitrage in securities markets with short-sales constraints', *Math. Fin.* **5**, (1991), 197–231.

[17] I. Karatzas and S. Kou, 'On the pricing of contingent claims under constraints', to appear in *Ann. Appl. Prob.*

[18] D.M. Kreps, 'Arbitrage and equilibrium in economics with infinitely many commodities', *J.M.E.* **8**, (1981) 15–35.

[19] P. Lakner, 'Martingale measures for a class of right-continuous processes', *Math. Fin.* **3**, (1993), 43–53.

[20] D.W. McBeth, 'On the existence of equivalent local martingale measures', Technical Report No 980, Cornell, 1991.

[21] L.C.G. Rogers, 'Arbitrage with fractional Brownian motion', Working Paper, 1995.

[22] W. Schachermayer, 'Martingale measures for discrete time processes with infinite horizon', *Math Fin.* **4**, (1994), 25–55.

[23] C. Stricker, 'Arbitrage et lois de martingale', *Ann. Inst. H. Poincaré* **26**, (1990), 451–460.

[24] J.A. Yan, 'Caracterisation d'une classe d'ensembles convexex de L^{1} ou H^{1}', Sem Prob. XIV, LN (1978/1979).

Part IV
Term Structure and
Interest Rate Derivatives

Bond and Bond Option Pricing Based on the Current Term Structure

Philip H. Dybvig

Abstract

Ho and Lee derive a term structure model that is based on the term structure at a point in time. The Ho and Lee model is shown to be a a binomial version of Vasicek's model without mean reversion, but with an added deterministic drift as a function of time. From this perspective, we derive a whole class of term structure models based on existing bond and bond option pricing models and the term structure at a point of time. Because the Ho and Lee model has an unreasonable implicit short rate process (that has a larger and larger drift over time), these alternative models are theoretically more realistic. Like the Ho and Lee model, all of these models are *ad hoc* in the sense that most shifts and twists in the yield curve are captured not by changes of state variables but rather by changing the model as a whole period by period. Exploratory data analysis indicates that a one-factor model with slight mean-reversion explains nearly all of the variability of interest rates. Therefore, only the one factor is needed for analyzing the variance, and the Ho and Lee model should fit well for ordinary options on bonds maturing in less than a year. For options on bonds maturing in more than a year, an analogous model based on the Vasicek one-factor mean-reverting model should fit well. Future research should focus on the time series properties of the common factor in bond returns and in particular its variance process.

1   Introduction

The desire to price interest rate options has placed new demands on our modeling of interest rates and the term structure. An earlier literature was content to ask about the relation between today's term structure and the expectation of future interest rates. Now, we require a complete model of co-movements of bond prices so we can compute the dynamic hedges used to price a universe of interest-sensitive securities. Ho & Lee (1986) have combined the two traditions, by building a dynamic hedging model that reflects the information contained in the current term structure. This combination comes at the expense of an *ad hoc* structure in which many types of interest rate

271

movements are taken as shifts in the model rather than shifts in the state variables (as in a correctly specified model). This paper has two primary purposes. One primary purpose is to synthesize existing theoretical results to show how to derive a whole class of models in the spirit of Ho and Lee's model. The class allows us to take advantage of existing models for which closed-form bond and bond option pricing formulas exist. The other primary purpose is to examine theoretically and empirically the features of interest rate co-movements that have made the Ho and Lee model useful for practitioners in spite of its *ad hoc* features.

The analysis of the Ho and Lee Model in this paper has some common ground with the discussion of Ho and Lee in Heath, Jarrow, & Morton (1990, 1992). This paper's perspective on Vasicek models was previously discussed in Section 3.2 of Dybvig (1988). The empirical results here are based on that perspective. This perspective could be viewed as consistent with leading cases of the Heath, Jarrow, & Morton (1992) approach of treating the term structure itself as the set of state variables (but without the complicated notation and exposition). However, the philosophy here is much different, with the emphasis being on having a model in which all sorts of local moves in the shape of the observed discrete points on the yield curve are possible. This need not be inconsistent with the Heath, Jarrow, and Morton model (if finitely many maturities are observed), but it is inconsistent with most implementations of the model.

Acknowledgements

I am grateful for useful conversations with Jim Bodurtha, Jon Ingersoll, Chris Lamoureux, Jean Masson, Ehud Ronn, and Greg Willard. This is a minor revision (primarily to update the statistics) of a paper written in 1988 and revised slightly in 1989.

2   The Ho and Lee Model

We will start with the primitive assumption that prices are given by discounted expected values, where discounting uses the rolled-over short rate. For example, in continuous time the price P_s at s of an asset paying P_t at $t > s$ is given by

$$P_s = E_s \left[P_t \exp \left(- \int_{\tau=s}^t r_\tau d\tau \right) \right], \tag{1}$$

where r_τ is the instantaneous riskless rate at time τ and E_s is expectation conditional on information available at time s. There are two possible interpretations of this assumption. The most general is that we are working in the *risk-neutral probabilities* of Cox & Ross (1976) (referred to by Harrison & Kreps (1979) as the *equivalent martingale measure* and by Cox, Ross, &

Rubinstein (1979) as the *artificial probabilities*). Provided we are concerned only about option pricing theories (and not about optimal portfolio choice), using risk-neutral probabilities is without loss of generality, although any empirical work must keep in mind that these are not the same as the actual probabilities. A more specific interpretation is that the local expectations hypothesis (Cox, Ingersoll, & Ross (1981)) holds, in which case the actual probabilities are the risk-neutral probabilities. Except as noted, it does not matter which interpretation is taken. When we refer to expectations, they will be the expectations as in (1), whichever interpretation is taken.

The discrete-time version of (1) is similar, except with a sum in place of the integral. In particular, if each time interval has length Δt,

$$P_{m\Delta t} = E_{m\Delta t}\left[P_{n\Delta t}\exp\left(-\sum_{i=m}^{n-1} r_i\Delta t\right)\right], \tag{2}$$

where r_τ is now the logarithmic short rate. Our analysis of the Ho and Lee (1986) model can specialize (2) to

$$P_s = E_s\left[P_t\exp\left(-\sum_{\tau=s}^{t-1} r_\tau\right)\right], \tag{3}$$

because they take $\Delta t = 1$. (We will want to look at the general form (2) later when we want to take limits as Δt goes to 0.)

Unlike most of the literature on the term structure, the primitive in Ho and Lee's model is the vector of discount bond prices. Without repeating their motivation and derivation, here are the essential elements of their model. Let the discount bond price $D_{i,n}$ be the value at time i of receiving 1 for sure at time n (which is $n - i$ periods out). The assumed process has as parameters a risk-neutral probability π of the 'upstate', the complementary risk-neutral probability $1 - \pi$ of the 'downstate', and positive functions $h(j)$ and $h^*(j)$ specifying the up and down shifts of the yield curves. Specifically,

$$D_{i,n} = h(n - i)\,D_{i-1,n}/D_{i-1,i} \tag{4}$$

if the upstate occurs at time i, and

$$D_{i,n} = h^*(n - i)\,D_{i-1,n}/D_{i-1,i} \tag{5}$$

if the downstate occurs at time i. There is a draw of an upstate and a downstate at each time i, and the draws are i.i.d. over time. The functions h and h^* are assumed to satisfy

$$\pi\,h(j) + (1 - \pi)\,h^*(j) \equiv 1, \tag{6}$$

which ensures that π and $1 - \pi$ are correctly interpreted as risk-neutral probabilities, and to satisfy

$$h(0) = h^*(0) = 1. \tag{7}$$

Additional structure is also assumed (they assume that up-down and down-up give the same final term structure) so that h and h^* are determined up to a parameter δ to be

$$h(t) = \frac{1}{(\pi + (1 - \pi)\,\delta^t)} \tag{8}$$

and

$$h^*(t) = \frac{\delta^t}{(\pi + (1 - \pi)\,\delta^t)}. \tag{9}$$

Which of the upstate and downstate corresponds to an upward shift in the yield curve depends on whether δ is larger or smaller than 1.

Now we want to restate the Ho and Lee model in more familiar terms. (This is really just elaborating on the analysis in Section III of Ho and Lee's paper.) The short rate at time i is

$$r_i = -\log(D_{i,i+1}), \tag{10}$$

and the 1-period forward rate quoted at i for money borrowed or lent from n to $n+1$ is

$$f_{i,n} = -\log(D_{i,n+1}/D_{i,n}). \tag{11}$$

Note that $D_{i,i} = 1$ and therefore

$$r_i \equiv f_{i,i}. \tag{12}$$

By (4), (8), and (11), we have that

$$
\begin{aligned}
f_{i,n} &= -\log\!\left(\frac{D_{i,n+1}}{D_{i,n}}\right) \\
&= -\log\left(\frac{h(n+1-i)\,D_{i-1,n+1}/D_{i-1,i}}{h(n-i)\,D_{i-1,n}/D_{i-1,i}}\right) \\
&= f_{i-1,n} - \log\left(\frac{h(n+1-i)}{h(n-i)}\right) \\
&= f_{i-1,n} + \log\left(\frac{\pi + (1-\pi)\,\delta^{n+1-i}}{\pi + (1-\pi)\,\delta^{n-i}}\right)
\end{aligned} \tag{13}
$$

in the upstate. Similarly, (5), (9), and (11) imply that

$$
\begin{aligned}
f_{i,n} &= f_{i-1,n} - \log\left(\frac{h^*(n+1-i)}{h^*(n-i)}\right) \\
&= f_{i-1,n} + \log\left(\frac{\pi + (1-\pi)\,\delta^{n+1-i}}{\pi + (1-\pi)\,\delta^{n-i}}\right) - \log(\delta)
\end{aligned} \tag{14}
$$

in the downstate.

To make better sense of the formulas in (13) and (14), it is useful to separate the change in $f_{.,n}$ into a constant and a mean-zero noise term. Define ε_i by

$$\varepsilon_i = \begin{cases} (1 - \pi)\log(\delta) & \text{if upstate at time i} \\ -\pi\log(\delta) & \text{if downstate at time i.} \end{cases} \tag{15}$$

Then $E[\varepsilon_i] = 0$ and we can rewrite (13) and (14) as

$$f_{i,n} = f_{i-1,n} + \log\left(\frac{\pi + (1 - \pi)\,\delta^{n+1-i}}{\pi + (1 - \pi)\,\delta^{n-i}}\right) - (1 - \pi)\log(\delta) + \varepsilon_i. \tag{16}$$

This says that $f_{i,n}$ is the sum of $f_{i-1,n}$, a constant depending on $n - i$ and the parameters π and δ (the two middle terms of (16)), and a random noise term that is the same for all maturities n and i.i.d. across time.

In (16), we have the expression for movements of the forward rates. Now we consider the analogous expression for movements in the short rate r. First, note that we can aggregate (16) over time to obtain the following expression for $f_{i,n}$ as it depends on $f_{0,n}$, the ε's, and the parameters π and δ. Applying (16) recursively we have that

$$\begin{aligned} f_{i,n} &= f_{0,n} + \sum_{j=1}^{i}\left(\log\left(\frac{\pi + (1 - \pi)\delta^{n+1-j}}{\pi + (1 - \pi)\,\delta^{n-j}}\right) - (1 - \pi)\log(\delta) + \varepsilon_j\right) \\ &= f_{0,n} + \log\left(\frac{\pi + (1 - \pi)\,\delta^{n}}{\pi + (1 - \pi)\,\delta^{n-i}}\right) - i(1 - \pi)\log(\delta) + \sum_{j=1}^{i}\varepsilon_j. \end{aligned} \tag{17}$$

From (17) and (12), we can specialize (17) to an expression for r_i.

$$\begin{aligned} r_i &= f_{0,i} + \log\left(\frac{\pi + (1 - \pi)\,\delta^{i}}{\pi + (1 - \pi)\,\delta^{0}}\right) - i(1 - \pi)\log(\delta) + \sum_{j=1}^{i}\varepsilon_j \\ &= f_{0,i} + \log\left(\pi + (1 - \pi)\,\delta^{i}\right) - i(1 - \pi)\log(\delta) + \sum_{j=1}^{i}\varepsilon_j. \end{aligned} \tag{18}$$

Differencing (18) gives us that

$$r_i = r_{i-1} + (f_{0,i} - f_{0,i-1}) + \log\left(\frac{\pi + (1 - \pi)\,\delta^{i}}{\pi + (1 - \pi)\,\delta^{i-1}}\right) - (1 - \pi)\log(\delta) + \varepsilon_i. \tag{19}$$

In other words, today's interest rate is the sum of yesterday's interest rate, the relevant slope of the initial yield curve, a constant depending on time and the parameters π and δ, and i.i.d. noise. Note further from (16) and (19) that the innovation to each forward rate is the same as the innovation to the spot rate, so that different possible yield curves at a point in time are all parallel shifts of each other.

The stochastic part of the Ho and Lee interest rate process (19) is a binomial version of the Vasicek (1977) model without mean reversion in interest rates.[1] In this degenerate model, changes in interest rates are i.i.d. normal random variables with mean zero (in the risk-neutral probabilities). (In general, the Vasicek model admits a large class of joint normally distributed interest rate processes.) This is referred to as the *driftless absolute interest rate process* – absolute refers to the fact that the variance of the spot rate is the same absolute number independent of the interest rate and everything else known before the realization. It is well-known that the driftless Vasicek model is not a useful model of the term structure (except perhaps as an approximation at short maturities), because the model implies that $\lim_{n \uparrow \infty} f_{i,n} = -\infty$ and $\lim_{n \uparrow \infty} D_{i,n} = \infty$.[2] Intuitively, the reason is that although the mean interest rate at future dates is not unreasonable (it is equal to today's rate), its variance is too large and allows long runs of negative interest rates in some realizations, and these realizations dominate the expectation in (3) for large t.

Given that the Ho and Lee model has the same unreasonable variance process as the driftless absolute Vasicek model, how can it be consistent with reasonable initial yield curves? The answer is that the Ho and Lee model assumes an equally unreasonable mean interest rate process which ultimately is tending towards $+\infty$ at a rate linear in time. Assuming the forward rate tends to a constant at large times to maturity, the expected change in interest rates from time $i - 1$ to i is

$$\lim_{i \uparrow \infty} E[r_i - r_{i-1}] = \begin{cases} -(1 - \pi) \log(\delta) & \text{if } \delta < 1 \\ \pi \log(\delta) & \text{if } \delta > 1. \end{cases} \tag{20}$$

The surprising feature of (20) is that the drift in both cases is equal to minus the smaller possible value of epsilon. In other words, for large i the drift plus noise takes on two values, one of which is approximately zero and the other of which is positive and not approximately zero.[3]

To summarize the results of this section, the Ho and Lee model starts with an unreasonable implicit assumption about innovations in interest rates, but can obtain a sensible initial yield curve by making an unreasonable assumption about expected interest rates. Unfortunately, while this patch using the initial yield curve is designed to give correct pricing of discount bonds at

---

[1] It may seem from (19) that the Ho and Lee model agrees essentially with the Vasicek model without mean reversion when $\delta = 1$ and the initial term structure is flat. However, when $\delta = 1$, $\varepsilon_i \equiv 0$ (by (15)).

[2] In general, suppose $r_t = r_{t-1} + \varepsilon_t$ where the ε_t's are nonconstant i.i.d. shocks with $E[\varepsilon_t] \leq 0$ under the risk-neutral probabilities. Then, letting ε be drawn from the common distribution of the ε_t's, (3) and (11) imply that $D_{i,n} = \exp(-r_i(n-i)) \prod_{t=i+1}^{n} E[\exp(-\varepsilon(n-i))]$ and $f_{i,n} = r_i - \log(E[\exp(-\varepsilon(n-i))])$. Now, ε nonconstant and $E[\varepsilon] \leq 0$ imply that $\lim_{n \uparrow \infty} E[\exp(-\varepsilon(n-i))] = \infty$, and therefore $\lim_{n \uparrow \infty} f_{i,n} = -\infty$ and $\lim_{n \uparrow \infty} D_{i,n} = \infty$.

[3] An interesting corollary of (20) is that the limiting forward rate can never fall, that is, $\lim_{n \uparrow \infty} f_{i,n} \geq \lim_{n \uparrow \infty} f_{i-1,n}$ with probability one. This is a special case of a very general result proven in Dybvig, Ingersoll, & Ross (1996). The result proven there (that the limiting forward rate can never fall) relies only on the existence of limiting forward rates and absence of arbitrage.

time 0, there is every reason to believe it will give incorrect pricing of interest rate options. After all, changing means and changing variances cannot have offsetting effects simultaneously on all types of options. Therefore, we should not rely on the Ho and Lee model for pricing options except at very short maturities for which many one-factor models give similar option prices.

The next section derives a class of models in the spirit of Ho and Lee's model (in the sense of using initial term structure data), but with two important advantages. First, the models permit interest rate processes with reasonable means and variances. Second, the class of models allows us to obtain a closed form solution for the yield curve and bond option pricing using any existing model of the term structure for which closed form solutions exist.

3 Generalizing Ho and Lee

The Ho and Lee model and its successors seem to be based on two premises. First, it is important for a pricing model to fit today's term structure. Second, it is difficult to work with a multifactor model of the term structure. These two premises lead us to using a model with a single source of noise but a very flexible parameter space to fit today's term structure. In this section we start to explore these two premises. We find that it is actually easy to fit today's term structure and that it it is easy to build multifactor models. A later section shows that it is also easy to estimate multifactor models.

First we turn to the question of building multifactor models. Cox, Ingersoll, & Ross (1985) implicitly used a clever trick to generate multifactor models of the term structure from single-factor models. Using continuous-time notation (as in (1)), if $\{r_t^a\}$ and $\{r_t^b\}$ are two independent[4] single-factor term structure models with discount bond pricing given by $D_{s,t}^a = E_s[\exp(-\int_{\tau=s}^t r_\tau^a d\tau)]$ and $D_{s,t}^b = E_s[\exp(-\int_{\tau=s}^t r_\tau^b d\tau)]$ respectively, then the interest rate process $\{r_t^c\}$ defined by $r_t^c \equiv r_t^a + r_t^b$ has discount bond pricing given by the product $D_{s,t}^c = D_{s,t}^a D_{s,t}^b$.

Obviously, by repeating this procedure (using the derived multi-factor interest-rate processes in place of the single-factor interest rate processes), one can obtain a term structure process with as many factors as one desires. As Cox, Ingersoll, & Ross pointed out, we can interpret the factors as a real rate and an inflation rate, but of course none of the analysis relies on this interpretation.

Ho & Lee (1986) can be interpreted in this context. Specifically, one of the interest rate processes is taken to be the driftless absolute version of the Vasicek model, and the other interest rate process is a deterministic function of time (a 'fudge factor'). In general, the decomposition of an interest rate

---

[4]Not all of the two-factor models developed by Cox, Ingersoll, & Ross use independent factor processes.

process into a stochastic interest rate process plus a constant function of time is ambiguous up to a constant function of time. For the Ho and Lee model, one choice is to decompose (18) as follows: the Vasicek process is

$$r_i^a = r_0 + \sum_{j=1}^{i} \varepsilon_j \tag{21}$$

(where r_0 is the observed short rate) and the deterministic process is

$$r_i^b = f_{0,i} - r_0 + \log\left(\pi + (1-\pi)\,\delta^i\right) - i(1-\pi)\log(\delta), \tag{22}$$

where r_0 is again the observed short rate and the $f_{0,i}$'s are the observed forward rates.

This observation motivates a whole class of term-structure models. Each model in the class uses a known term structure model and a deterministic model. Given the known term structure model, the deterministic model can be chosen to fit the initial yield curve. The variance assumption is as reasonable as it is in the chosen term structure model. The following Theorem shows that any bond or option pricing results available in the known term structure model can be extended to the new model.

Theorem 1 *Let $\{r_t^a\}$ be any interest rate process, let $\{r_t^b\}$ be an interest rate process that is a function of time alone, and let the interest rate process $\{r_t^c\}$ be defined by $r_t^c \equiv r_t^a + r_t^b$. Then if we let discount bond pricing for processes a and b be given by $D_{s,t}^a = E_s[\exp(-\int_{\tau=s}^{t} r_\tau^a d\tau)]$ and $D_{s,t}^b = \exp(-\int_{\tau=s}^{t} r_\tau^b d\tau)$ respectively, then the discount bond price for process c is given by the product $D_{s,t}^c = D_{s,t}^a D_{s,t}^b$. Furthermore, if interest rates follow the c process, consider an option that pays at t some function $f(r_t^c, D_{t,T_1}^c, \ldots, D_{t,T_n}^c)$ of the short rate r_t^c at t and various bond prices $D_{t,T_1}^c, \ldots, D_{t,T_n}^c$. At time $s < t$, this option has the same price as the option paying $D_{s,t}^b f(r_t^a + r_t^b, D_{t,T_1}^a D_{t,T_1}^b, \ldots, D_{t,T_n}^a D_{t,T_n}^b)$ at t under process a.*

Proof By (1),

$$
\begin{aligned}
D_{s,t}^c &= E_s\left[\exp\left(-\int_{\tau=s}^{t} r_\tau^c d\tau\right)\right] = E_s\left[\exp\left(-\int_{\tau=s}^{t} (r_\tau^a + r_\tau^b)d\tau\right)\right] \\
&= E_s\left[\exp\left(-\int_{\tau=s}^{t} r_\tau^a d\tau\right)\right]\exp\left(-\int_{\tau=s}^{t} r_\tau^b d\tau\right) \\
&= D_{s,t}^a D_{s,t}^b,
\end{aligned}
$$

which is the first result we are to prove.

By (1) and the first result, the price at s of receiving $f(r_t^c, D_{t,T_1}^c, \ldots, D_{t,T_n}^c)$ at t under the process c is equal to

$$E_s\left[f(r_t^c, D_{t,T_1}^c, \ldots, D_{t,T_n}^c)\exp\left(-\int_{\tau=s}^{t} r_\tau^c d\tau\right)\right]$$

$$= E_s\left[f(r_t^a + r_t^b, D_{t,T_1}^a D_{t,T_1}^b, \ldots, D_{t,T_n}^a D_{t,T_n}^b)\exp\left(-\int_{\tau=s}^t (r_\tau^a + r_\tau^b)d\tau\right)\right]$$

$$= E_s\left[\exp\left(-\int_{\tau=s}^t r_\tau^b d\tau\right)f(r_t^a + r_t^b, D_{t,T_1}^a D_{t,T_1}^b, \ldots, D_{t,T_n}^a D_{t,T_n}^b)\times \right.$$
$$\left. \exp\left(-\int_{\tau=s}^t r_\tau^a d\tau\right)\right]$$

$$= E_s\left[D_{s,t}^b f(r_t^a + r_t^b, D_{t,T_1}^a D_{t,T_1}^b, \ldots, D_{t,T_n}^a D_{t,T_n}^b)\exp\left(-\int_{\tau=s}^t r_\tau^a d\tau\right)\right],$$

which is the price of the option paying $D_{s,t}^b f(r_t^a + r_b, D_{t,T_1}^a D_{t,T_1}^b, \ldots, D_{t,T_n}^a D_{t,T_n}^b)$ at t under process a, and we are done. ∎

Because the interest rate process b is known and deterministic in Theorem 1, the option pricing result says that any standard option we can always price in model a can also be priced in model c. Often, the option payoff we are pricing will depend only on one of the arguments. For example, suppose that we want to know the price at s in model c of a call option with exercise price X maturing at t on a unit discount bond maturing at T. In this case,

$$f(r_t^c, D_{t,T_1}^c, \ldots, D_{t,T_n}^c) = \max(D_{t,T}^c - X, 0)$$

and

$$\begin{aligned}
D_{s,t}^b f(r_t^a + r_b, D_{t,T_1}^a D_{t,T_1}^b, \ldots, D_{t,T_n}^a D_{t,T_n}^b) &= D_{s,t}^b \max(D_{t,T}^a D_{t,T}^b - X, 0)\\
&= D_{s,t}^b D_{t,T}^b \max(D_{t,T}^a - X/D_{t,T}^b, 0)\\
&= D_{s,T}^b \max(D_{t,T}^a - X/D_{t,T}^b, 0)
\end{aligned}$$

which is $D_{s,T}^b$ times the price at s in model a of a call option with exercise price $X/D_{t,T}^b$ maturing at t on a unit discount bond maturing at T.

In other words, Theorem 1 allows us to perform the Ho and Lee type of analysis using a perturbation (by b) of any term structure model a that is convenient and reasonable to use. Furthermore, the option pricing in the original model a extends to the new model c. This means that we can generate the interest rate process based on the initial yield curve and still use the closed-form bond and bond option pricing models including those derived by Vasicek (1977) and Cox, Ingersoll, & Ross (1985).

In a certain sense, all of these models in this section (including the Ho and Lee model) are *ad hoc* and should be treated as temporary fixes to be used only until we are successful at building and testing more realistic multifactor models of the term structure. The unreasonable feature of these models is that in each period we change our whole model (the parameters) in response to moves in the term structure. This is like comparative statics results that assume each period's change is interpreted as a once-and-for-all change of the sort that can never happen again. This is not rational and is not reasonable. Imperfectly correlated moves in all forward rates should be a feature of the stochastic structure of the model, and ideally we should have enough flexibility from the factor structure of a to eliminate entirely the need for a 'fudge factor' b. On the other hand, as a practical matter, if a model with a few factors explains almost all of the variance of interest rate movements, the Ho and Lee model or a model developed in this section will give a good approximation for pricing many bond options even if it ignores small factors.[5]

It is worth noting at this point that while we have focused on the question of fitting today's term structure of interest rates using a translation of interest rates, it is also easy to fit today's term structure of interest rate volatilities using a change of time. The change of time would suffer from the same conceptual problems we have discussed for the translation of interest rates.

The next section discusses a class of multi-factor Vasicek models that overcome the somewhat *ad hoc* nature of the Ho and Lee approach. Using that general model as a starting point, it is explored whether the Ho and Lee approach, a generalization discussed in this section, or a more general model best fits the data.

4   Multifactor Models: Empirical Exploration

In the Ho and Lee model and the extensions discussed in Section 3, we have the *ad hoc* feature that many changes in the yield curve over time are interpreted as parameter changes rather than changes in state variables. In other words, what happens again and again each period is a realization that is completely outside the model, and yet we retain the model. Having an assumed term structure model that is so completely inconsistent with the data every period (because the realization is outside of the support of the probability distribution) makes it hard to put much faith in these models.

There are different approaches one might take to resolving this problem. One is to assume that the observed term structure is not the correct term structure, and that any deviations from our model are the result of measurement error. This interpretation is not really consistent with the Ho and Lee approach, because it says that rather than accomodate today's term structure (as in Ho and Lee's approach) we ought to use it as a guide and base our pricing on a nearby term structure conforming to the model instead.

In general, basing a pricing model primarily on today's term structure cannot be a very good approach if measurement error (for example from not knowing how to interpret the bid-ask spread, from non-contemporaneous quotes, or from the difficulty of inferring discount bond prices from the prices

---

[5]It is an interesting intellectual question, and given widespread use of these models in practice an interesting practical question, to understand what options are badly priced by these models.

of coupon bonds in the presence of tax effects) makes it difficult to measure today's term structure.

A more palatable approach, which is the approach taken in this section, is to use a term structure model that is rich enough to admit all possible movements in the term structure. To illustrate this approach, we will use a many-factor model in the style of Vasicek (1977). Basically, Vasicek showed that we can do bond and bond option pricing whenever the interest rate process is Gaussian (in the risk-neutral probabilities).[6] This is because joint normally distributed interest rates imply that the exponent in the valuation formula ((1), (2), or (3)) is normally distributed and therefore discount bond prices can be computed using the normal moment generating function. Also, the joint lognormality of discount bond prices and the discount factor implies that pricing European puts and calls on discount bonds involves what is basically a transformation of the Black-Scholes model. Depending on how one does this derivation, it can be viewed as arising from Merton (1973), Rubinstein (1976), or Vasicek (1977).

A simple implication of the assumption of a Gaussian interest rate process is that discount bond yields also follow a jointly Gaussian process. Predictions in a Gaussian world are linear with homoskedastic Gaussian errors: joint normality of yields across times and maturities therefore follows from the valuation formula. We use the simple discrete-time valuation formula (3), and we assume that there is a finite vector of Markov state variables. More specifically, we assume that we can take the state variables at time i to be the yields to maturity of the discount bonds maturing at times $i + 1$ through $i + n$. (Or, equivalently, these yields together are an invertible function of the state variables.)

The *log discount* at s on a discount bond maturing at a later time t is given by

$$d_{s,t} \equiv -\log(D_{s,t}). \tag{23}$$

We will assume that innovations in the vector $d_s \equiv (d_{s,s+1}, \ldots, d_{s,s+n})$ have a covariance matrix Σ. If we apply (3) for $t = s + 1$ and price a discount bond maturing at $T \geq t$, then dividing by $D_{s,s+1}$ we have that

$$E_s[D_{s+1,T}] = \frac{D_{s,T}}{D_{s,s+1}} \tag{24}$$

$$= \exp\left(-d_{s,T} + d_{s,s+1}\right). \tag{25}$$

By joint normality of the log discounts, the value of (24) is given by the normal moment generating function as $\exp(-E_s[d_{s+1,T}] + \mathrm{var}_s[d_{s+1,T}]/2)$, where $\mathrm{var}_s[\cdot]$ indicates variance conditional on information available at s. Consequently,

$$E_s[d_{s+1,T}] = d_{s,T} - d_{s,s+1} + \frac{1}{2}\mathrm{var}_s[d_{s+1,T}]. \tag{26}$$

---

[6]This is discussed in Dybvig (1988).

Because the discount factors are joint Gaussian and Markov, $\text{var}_s[d_{s+1,T}]$ is a constant that depends only on $T - (s + 1)$. Of course, $E_s[d_{s+1,T}]$ varies depending on the value of the state variables at t.

What we have learned in the previous paragraph is that we do not have to estimate $E_s[d_{s+1,T}]$ and $\text{var}_s[d_{s+1,T}]$ separately. In fact, this is essential (in the absence of a strong assumption such as the local expectations hypothesis), because what we observe empirically is the mean and variance under actual probabilities, not under the risk-neutral probabilities. For short time periods, variances of prediction errors under the risk-neutral probabilities are the same as under the actual probabilities–in other words, the change of measure affects the local mean, not the local variance.[7] This is a generalization of the fact that the actual mean does not affect asset pricing in the Black-Scholes model. We will take this result as accurate for the actual periods we use, which seems reasonable for monthly data and perhaps less reasonable for annual data. Roughly speaking, this means that when we run regressions on the data, we will ignore the estimated regression coefficients and intercept, and that we will base our asset pricing model on the covariance matrix of the residuals.

If our data always go n periods out, the only indeterminacy will be for $E_s[d_{s+1,s+1+n}]$, corresponding to the last bond traded at $s + 1$, which is also the shortest bond not traded at s. However, this is not an issue provided we are finding the price at s of bonds maturing at or before $s + n$, or options on those bonds. Those bonds are all priced at s: by assumption, their prices are our state variables (up to conversion from yields to prices). Furthermore, (26) implies that the distribution of their movement over time does not depend on knowing the process for $d_{s,s+n+1}$, and by (3) neither does valuation of their options. Therefore, valuation of options on assets maturing within the horizon of observed short rates does not require us to pin down the process on $d_{s,s+n+1}$. If we want to extrapolate and price assets and options beyond this horizon, there are reasonable ways to choose $E_s[d_{s+1,T}]$ (for example by using the estimated process in actual probabilities or making some assumption about risk premia), but within this paper it will be assumed that we are concerned with pricing options on bonds maturing within our observed maturity structure.

Now, the strategy is to estimate the covariance matrix of the innovations in log discounts. Then, (26) will tell us the means under the risk-neutral probabilities. This analysis is carried out using two data sets provided by the Center for Research in Security Prices (CRSP) at the University of Chicago. One data set contains yields from the 12-month version of the Fama Treasury Bill Term Structure file. The file used here generates the term structure for 1 to 12 months out using the average-of-bid-and-ask price of T-Bills of those approximate maturities. To avoid missing data, we use maturities from 1

---

[7]See Harrison & Pliska (1981). Dybvig & Huang (1988) is an example of a more modern treatment.

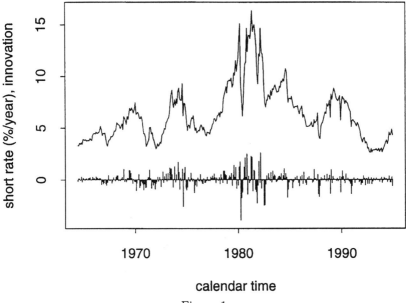

Short Rate and Innovations
monthly data 6407-9412

Figure 1

through 9 months over the period June, 1964 through December, 1994. The other data set is the Fama-Bliss file of derived prices for discount bonds with maturities 1 through 5 years. This file is based on an elaborate selection procedure using bonds near par that do not have any special tax treatment. The second data set has monthly prices; from them was derived an annual series for December, 1952 through December, 1994, and a monthly series for the same dates as the Fama Treasury Bill Term Structure file. The time series of one-month interest rates and their innovations (as measured in the vector auto-regression below) are shown in Figure 1.

The first step in using the data was computation of the discount factors $d_{s,t}$, which was straightforward given the form of the data. In doing so, most of the Fama-Bliss data was discarded to convert it to a non-overlapping annual series. In each data set, the innovations in the discount factors were computed as the residuals of first-order vector autoregressions. These regressions are shown in Tables 1 and 4. While individual regression coefficients are not significant in these regressions (because interest rates tend to move together), the large R^2 and F values indicate that the regressions are very significant overall.[8] Including all the lagged discount factors should tend to minimize

---

[8] Chris Lamoureux has pointed out that adding a second lag to these regressions seems to

$d_{s,s+i} = b_0 + \sum_{j=1}^{9} b_j d_{s-1,s-1+j} + \varepsilon_{s,s+i}$ monthly 6406–9412									
Independent	Dependent variables								
variables	$d_{s,s+1}$	$d_{s,s+2}$	$d_{s,s+3}$	$d_{s,s+4}$	$d_{s,s+5}$	$d_{s,s+6}$	$d_{s,s+7}$	$d_{s,s+8}$	$d_{s,s+9}$
intercept	0.00	0.02	0.02	0.03	0.05	0.06	0.08	0.11	0.14
×100	(−0.08)	(0.95)	(0.63)	(0.92)	(1.23)	(1.19)	(1.44)	(1.80)	(1.99)
$d_{s-1,s}$	0.03	0.05	0.29	0.59	0.88	0.98	1.18	1.36	1.49
	(0.24)	(0.24)	(0.90)	(1.40)	(1.68)	(1.61)	(1.66)	(1.68)	(1.61)
$d_{s-1,s+1}$	0.61	0.42	0.51	0.45	0.40	0.53	0.49	0.62	0.81
	(4.73)	(1.80)	(1.45)	(0.97)	(0.70)	(0.79)	(0.64)	(0.70)	(0.80)
$d_{s-1,s+2}$	−0.33	0.16	−0.29	−0.35	−0.59	−0.92	−1.12	−1.37	−1.67
	(−2.28)	(0.63)	(−0.73)	(−0.68)	(−0.92)	(−1.23)	(−1.30)	(−1.39)	(−1.49)
$d_{s-1,s+3}$	0.22	0.26	0.68	0.42	0.59	0.89	0.87	1.05	0.99
	(1.89)	(1.23)	(2.13)	(1.00)	(1.15)	(1.47)	(1.23)	(1.31)	(1.09)
$d_{s-1,s+4}$	−0.13	−0.23	−0.39	−0.10	−0.82	−1.08	−1.19	−1.40	−1.53
	(−1.36)	(−1.40)	(−1.54)	(−0.30)	(−2.00)	(−2.24)	(−2.14)	(−2.19)	(−2.11)
$d_{s-1,s+5}$	−0.11	−0.18	−0.17	−0.16	0.56	0.33	0.59	0.68	0.92
	(−1.23)	(−1.10)	(−0.69)	(−0.48)	(1.40)	(0.69)	(1.07)	(1.08)	(1.28)
$d_{s-1,s+6}$	0.04	0.02	0.03	−0.09	−0.21	−0.15	−0.68	−0.95	−1.23
	(0.42)	(0.11)	(0.11)	(−0.24)	(−0.46)	(−0.29)	(−1.12)	(−1.37)	(−1.56)
$d_{s-1,s+7}$	0.33	0.65	0.98	1.44	1.76	2.17	2.84	2.86	3.07
	(3.59)	(3.93)	(3.94)	(4.37)	(4.35)	(4.57)	(5.16)	(4.55)	(4.28)
$d_{s-1,s+8}$	−0.20	−0.39	−0.59	−0.84	−1.02	−1.09	−1.21	−0.90	−0.77
	(−3.43)	(−3.76)	(−3.71)	(−4.03)	(−3.98)	(−3.60)	(−3.46)	(−2.25)	(−1.69)
$\overline{R}^2$	0.94	0.95	0.95	0.95	0.95	0.95	0.95	0.95	0.95
$F_{(9,357)}$	596.59	754.07	765.00	782.14	810.08	840.88	843.42	836.59	816.62

Table 1: First-order vector autoregression of monthly log discounts. This table reports the first-order vector autoregressions of the first nine monthly log discounts for the sample period 6406–9412. The regression uses the Fed version of the Fama T-Bill data provided by CRSP. The numbers in parentheses are t-statistics; these numbers are low because the discount rates are nearly multicollinear. This is no problem for the paper because the regressions are very significant overall (see the F-statistics) and because we are interested primarily in the residuals.

cov$[\varepsilon_{s,s+1}, \ldots, \varepsilon_{s,s+9}]$ monthly 6406–9412									
×1e5	$\varepsilon_{s,s+1}$	$\varepsilon_{s,s+2}$	$\varepsilon_{s,s+3}$	$\varepsilon_{s,s+4}$	$\varepsilon_{s,s+5}$	$\varepsilon_{s,s+6}$	$\varepsilon_{s,s+7}$	$\varepsilon_{s,s+8}$	$\varepsilon_{s,s+9}$
$\varepsilon_{s,s+1}$	0.03	0.05	0.07	0.09	0.11	0.12	0.14	0.16	0.18
$\varepsilon_{s,s+2}$	0.05	0.10	0.15	0.19	0.23	0.26	0.30	0.34	0.38
$\varepsilon_{s,s+3}$	0.07	0.15	0.23	0.30	0.36	0.42	0.48	0.54	0.61
$\varepsilon_{s,s+4}$	0.09	0.19	0.30	0.40	0.49	0.57	0.65	0.73	0.82
$\varepsilon_{s,s+5}$	0.11	0.23	0.36	0.49	0.61	0.70	0.81	0.92	1.04
$\varepsilon_{s,s+6}$	0.12	0.26	0.42	0.57	0.70	0.84	0.96	1.09	1.24
$\varepsilon_{s,s+7}$	0.14	0.30	0.48	0.65	0.81	0.96	1.12	1.28	1.45
$\varepsilon_{s,s+8}$	0.16	0.34	0.54	0.73	0.92	1.09	1.28	1.47	1.66
$\varepsilon_{s,s+9}$	0.18	0.38	0.61	0.82	1.04	1.24	1.45	1.66	1.90

Table 2: Covariance matrix of residuals from the regression reported in Table 1. Each entry is ×1e5; for example the first entry 0.03 really represents a variance of 3×10^{-7}. The diagonal elements of this matrix are used in the bond and bond option pricing. Small numbers imply that the effect on bond pricing is very small at these maturities, but of course the effect on bond option pricing is significant because the numbers here represent the only source of variance.

the effect of any errors-in-variables on the residuals without a significant loss in degrees of freedom. (Some possible sources of errors-in-variables include quote errors, the handling of the bid-ask spread, yields taken from bonds not exactly one month out, tax effects, and the selection procedure in the Fama–Bliss data.)

The estimated variances and covariances of innovations in log discounts are reported in Tables 2 and 5. To better understand these covariance matrices, each was decomposed into principal components. The principal components model takes each eigenvector (normalized to Euclidian length 1) of the covariance matrix to be an independent component of the covariance matrix, and its corresponding eigenvalue measures the amount of variance explained by the component. (Essentially, principal components analysis is like factor analysis without the idiosyncratic noise. See Gnanadesikan (1977) for a simple description of the principal components technique and a discussion of its relation to factor analysis.) The results of the principal components analysis are striking. For both monthly and annual data, there is a dominant component

---

add significantly to the explanatory power, which suggests that the particular log discounts being used may not span a full set of state variables, depending on whether the additional explanatory power is due to a factor or to a conditional risk premium. This merits additional attention.

Principal Components of $\varepsilon_{s,s+1}, \ldots, \varepsilon_{s,s+9}$ monthly 6406–9412									
eigenvalues of cov$[\varepsilon_{s,s+1}, \ldots, \varepsilon_{s,s+9}]$ ×1e5									
	6.56	0.09	0.02	0.01	0.01	0.01	0.00	0.00	0.00
maturity	corresponding normalized eigenvectors (as columns)								
1 month	0.05	−0.24	0.47	0.17	0.40	−0.39	0.00	−0.62	0.05
2 months	0.11	−0.36	0.45	0.26	0.18	−0.01	−0.11	0.66	−0.33
3 months	0.18	−0.44	0.20	0.04	−0.21	0.41	0.13	0.02	0.71
4 months	0.24	−0.45	−0.02	−0.22	−0.41	0.21	0.14	−0.34	−0.58
5 months	0.30	−0.31	−0.26	−0.38	−0.14	−0.68	−0.19	0.21	0.20
6 months	0.36	−0.16	−0.42	−0.09	0.64	0.37	−0.33	−0.07	−0.03
7 months	0.41	0.05	−0.30	0.39	0.12	−0.16	0.73	0.07	−0.02
8 months	0.47	0.25	0.00	0.55	−0.38	−0.03	−0.50	−0.14	0.01
9 months	0.53	0.47	0.46	−0.50	0.07	0.09	0.12	0.06	0.01

Table 3: Principal Components decomposition of the covariance matrix of residuals from the regression reported in Table 1. The dominant eigenvector corresponds to roughly parallel shifts in the yield curve (exactly parallel shifts would give an eigenvector proportional to $(1, 2, \ldots, 9)$). The second eigenvector corresponds to changes the overall slope of the yield curve. The third eigenvector corresponds to changes of curvature in the yield curve. Subsequent eigenvectors seem hard to interpret. The first eigenvalue is dominant, at 70 times the size of the next eigenvalue. Furthermore, the third and subsequent eigenvectors are even less important because their associated eigenvalues are so small: the largest in this group is less than $1/300^{\text{th}}$ of the largest eigenvalue.

corresponding to co-movements throughout the yield curve. In the monthly data, the movement corresponds roughly to parallel movements in the yield curve, while the annual data shows less response in the forward rates at higher maturities (note that the forward rates are differences in the log discounts). Given that the first component explains almost the entire variability in the data, this factor will contribute the dominant variance effect in bond and bond option prices.[9]

---

[9]It is possible (and equivalent theoretically) to study the covariance matrix of some linear combination of the log discounts instead of the log discounts themselves. In theory, this would change somewhat the principal components decomposition. In fact, the alternatives still yield a dominant component, with an eigenvector corresponding to comovements. Furthermore, using the log discounts is probably less sensitive to various sources of errors-in-variables. Using forward rates puts lots of weight in little bumps in the yield curve, and using yields puts too much weight on the shortest maturity (whose yield is most affected

$d_{s,s+i} = b_0 + \sum_{j=1}^{9} b_j d_{s-1,s-1+j} + \varepsilon_{s,s+i}$ annual 52–94					
Independent	Dependent variables				
variables	$d_{s,s+1}$	$d_{s,s+2}$	$d_{s,s+3}$	$d_{s,s+4}$	$d_{s,s+5}$
intercept	1.24	2.26	3.64	4.51	5.79
$\times 100$	(1.90)	(1.92)	(2.28)	(2.25)	(2.40)
$d_{s-1,s}$	1.50	2.52	3.34	4.48	4.12
	(1.08)	(1.01)	(0.99)	(1.06)	(0.81)
$d_{s-1,s+1}$	1.20	2.22	3.21	3.78	5.23
	(0.86)	(0.89)	(0.95)	(0.89)	(1.02)
$d_{s-1,s+2}$	-1.66	-3.32	-5.35	-7.00	-8.56
	(-1.83)	(-2.04)	(-2.42)	(-2.53)	(-2.57)
$d_{s-1,s+3}$	-0.65	-0.93	-0.52	-0.69	-0.64
	(-0.82)	(-0.65)	(-0.27)	(-0.28)	(-0.22)
$d_{s-1,s+4}$	0.90	1.69	2.18	3.04	3.60
	(1.85)	(1.93)	(1.85)	(2.05)	(2.02)
$\overline{R}^2$	0.73	0.78	0.81	0.83	0.84
$F(5,37)$	24.08	30.23	37.07	43.00	46.09

Table 4: First-order vector autoregression of annual log discounts. These results are similar to the monthly autoregression in Table 1. As in that regression, near multicollinearity makes the individual t-statistics insignificant, but each regression is significant overall as indicated by the F-statistics.

To test whether the same factor is being measured in both data sets, a combined monthly data set is formed for the same period as the monthly regression (June, 1964 through December, 1994). The vector autoregression now regresses each maturity's log discount on all 14 log discounts from the previous month. The actual regression is similar in flavor to the other two and the actual coefficients are not reported. Table 7 reports the dominant components of the principal components decomposition of the residuals. This decomposition, which again has a totally dominant component, confirms that the dominant component is the same in both cases.

The shape of the principal component in Table 7 indicates that it may be possible to explain almost all the variation in the data using a simple one-factor model. Next to the driftless model, the simplest Vasicek model is the

---

by data problems such as inaccurate quotes, the bid-ask spread, or deviation of settlement or bond payments from the nominal date.

$\times 1e5$	cov$[\varepsilon_{s,s+1}, \ldots, \varepsilon_{s,s+9}]$ monthly 6406–9412								
	$\varepsilon_{s,s+1}$	$\varepsilon_{s,s+2}$	$\varepsilon_{s,s+3}$	$\varepsilon_{s,s+4}$	$\varepsilon_{s,s+5}$	$\varepsilon_{s,s+6}$	$\varepsilon_{s,s+7}$	$\varepsilon_{s,s+8}$	$\varepsilon_{s,s+9}$
$\varepsilon_{s,s+1}$	0.03	0.05	0.07	0.09	0.11	0.12	0.14	0.16	0.18
$\varepsilon_{s,s+2}$	0.05	0.10	0.15	0.19	0.23	0.26	0.30	0.34	0.38
$\varepsilon_{s,s+3}$	0.07	0.15	0.23	0.30	0.36	0.42	0.48	0.54	0.61
$\varepsilon_{s,s+4}$	0.09	0.19	0.30	0.40	0.49	0.57	0.65	0.73	0.82
$\varepsilon_{s,s+5}$	0.11	0.23	0.36	0.49	0.61	0.70	0.81	0.92	1.04
$\varepsilon_{s,s+6}$	0.12	0.26	0.42	0.57	0.70	0.84	0.96	1.09	1.24
$\varepsilon_{s,s+7}$	0.14	0.30	0.48	0.65	0.81	0.96	1.12	1.28	1.45
$\varepsilon_{s,s+8}$	0.16	0.34	0.54	0.73	0.92	1.09	1.28	1.47	1.66
$\varepsilon_{s,s+9}$	0.18	0.38	0.61	0.82	1.04	1.24	1.45	1.66	1.90

Table 5: Covariance matrix of residuals from the regression reported in Table 4. This is similar to Table 2.

	Principal Components of $\varepsilon_{s,s+1}, \ldots, \varepsilon_{s,s+5}$ annual 52–94				
	eigenvalues of cov$[\varepsilon_{s,s+1}, \ldots, \varepsilon_{s,s+5}]$ $\times 1e5$				
	628.84	5.98	0.63	0.43	0.11
maturity	corresponding normalized eigenvectors (as columns)				
1 year	0.17	−0.53	−0.48	−0.30	0.61
2 year	0.31	−0.54	−0.25	0.04	−0.74
3 year	0.42	−0.36	0.55	0.56	0.29
4 year	0.53	0.12	0.43	−0.72	−0.06
5 year	0.64	0.53	−0.47	0.28	0.06

Table 6: Principal Components decomposition of the covariance matrix of residuals from the regression reported in Table 4. The dominant eigenvector corresponds to co- movement of all interest rates, although this movement no longer looks parallel as in the monthly case. This shows evidence of mean reversion in risk-neutral probabilities (exactly parallel shifts would give an eigenvector proportional to $(1, 2, \ldots, 5)$). As in the monthly data, the second eigenvector corresponds to changes the overall slope of the yield curve, the third eigenvector corresponds to changes of curvature in the yield curve, and subsequent eigenvectors seem hard to interpret. In this data, the largest eigenvector is even more dominant than in the monthly data: it is more than 100 times larger than the next largest eigenvalue.

simple mean-reverting model for which

$$r_t = r_{t-1} + \kappa(\bar{r} - r_{t-1}) + \sigma \eta_t. \tag{27}$$

In this model,

$$r_t = \bar{r} + (1 - \kappa)^t(r_0 - \bar{r}) + \sigma \sum_{\tau=1}^{t}(1 - \kappa)^{t-\tau}\eta_\tau, \tag{28}$$

and it is easy to show that

$$
\begin{aligned}
d_{s,t} - E_{s-1}[d_{s,t}] &= \sigma \sum_{\tau=s+1}^{t}(1 - \kappa)^{t-(s+1)}\eta_s \\
&= \sigma \frac{1 - (1 - \kappa)^{t-s}}{\kappa}\eta_s
\end{aligned}
\tag{29}
$$

is the innovation in the log discount.

In terms of the empirics, this model would have a single component for which the eigenvector times the square root of the eigenvalue would equal

$$\frac{1 - (1 - \kappa)^{t-s}}{\kappa}\sigma\sqrt{\Delta}, \tag{30}$$

where $t - s$ is the maturity (1–9 months and 1–5 years in our sample) and Δ is the time interval over which we measure the change. For estimation, we use the continuous version of the model (as described for example in Dybvig (1988)), for which largest eigenvalue can be written as[10]

$$\frac{1 - \exp(-\kappa(t - s))}{\kappa}\sigma\sqrt{\Delta}. \tag{31}$$

Figure 2 shows the nonlinear-least-squares fit of the dominant eigenvalue times the square root of the eigenvector to (30), and the line with the same variance but no mean reversion. These fitted equations have parameters $\kappa = 0.0126$ (standard error 0.016) and $\sigma = 0.01848$ (standard error 0.00087).[11] Apparently, this parsimonious representation has a very good fit at all maturities, and the even more parsimonious representation with $\kappa = 0$ fits very well up to a year. (Remember that this means it should do well for options on instruments maturing in less than a year.) This says that the original Ho and Lee procedure should be a good fit for options on very short assets, and the extension of Ho and Lee should be a good fit for options on longer assets.

---

[10]Note that this is the same as (30) up to a change in parameter definitions.

[11]The estimated standard errors are based on sampling variation in the error term but not in the parameter estimates of the underlying regression, which might be unstable due to a unit root in sample.

Largest Principal Components of $\varepsilon_{s,s+i}'s$ combined 6406–9412					
eigenvalues of cov$[\varepsilon_{s,s+i}'s]$ $\times 1e5$					
101.85	3.16	0.95	0.49	0.24	0.10
maturity	corresponding normalized eigenvectors (as columns)				
1 month	0.01	−0.05	−0.01	0.05	0.05	0.04
2 month	0.02	−0.10	−0.02	0.09	0.10	0.13
3 month	0.04	−0.15	−0.04	0.12	0.16	0.20
4 month	0.05	−0.19	−0.06	0.15	0.17	0.28
5 month	0.06	−0.23	−0.07	0.16	0.15	0.23
6 month	0.08	−0.26	−0.06	0.15	0.15	0.17
7 month	0.09	−0.28	−0.06	0.16	0.13	0.12
8 month	0.11	−0.31	−0.05	0.17	0.12	0.06
9 month	0.12	−0.34	−0.07	0.15	0.11	−0.05
12 month	0.16	−0.38	−0.07	0.10	0.01	−0.84
24 month	0.30	−0.34	−0.03	−0.13	−0.85	0.21
36 month	0.42	−0.17	0.17	−0.80	0.35	0.06
48 month	0.53	0.22	0.72	0.38	0.01	−0.01
60 month	0.62	0.43	−0.64	0.12	0.05	−0.01

Table 7: Principal Components decomposition of the covariance matrix of residuals from combined monthly and annual data. This is the decomposition (as in Table 3 or 6) of the residuals from the log discount autoregression (as in Table 1 or 4) using monthly observations of combined monthly and annual data. This table offers evidence that the dominant factor in the annual and monthly regressions are in fact the same, because essentially the same factor shows up here. In fact, the second and third factors (which are again much smaller) again can be interpreted as slope and curvature of the yield curve.

The consistency and simplicity of the empirical results are evidence in favor of the integrity of the data sets. If the data sets contained significant errors in variables, these errors would show up as significant components in the principal components analysis. Furthermore, the nice fit in Figure 2 would be unlikely to show up.

One implication of the empirical work is that conditional on the information at a given point in time, the next period's innovations in log discounts are almost perfectly correlated. This suggests that any second factor will not be an additive second factor (such as those in Cox, Ingersoll, & Ross (1985)

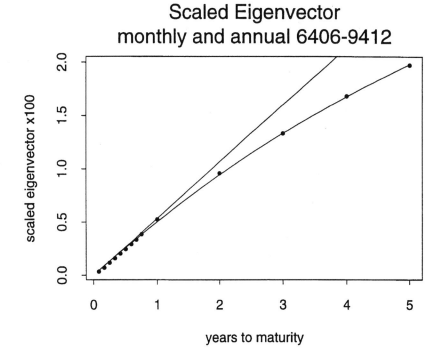

Figure 2. The points indicate the computed values of the largest eigenvector. The straight line is the fitted theoretical curve without mean reversion, and the curved line is the fitted theoretical curve with mean reversion. The estimated value of kappa corresponds to mean reversion of 12.8% per year, with sampling error of at least 1.66% per year.

or Brennan & Schwartz (1979)). Rather, the second factor is more likely to be a variance factor which has a small effect on bond pricing (and almost none at short maturities) but perhaps a significant effect on bond option pricing.

5 Conclusion

The Ho and Lee model and its descendents (such as popular implementations of Heath, Jarrow, & Morton (1992)) are conceptually flawed because they treat all but one dimension of local shocks as complete surprises. This conceptual problem is corrected in a general model with many factors, so that all possible local movements in interest rates are possible.

Empirically, this problem may be less severe than one might expect, at least for most common options not specifically designed to exploit this weakness. The principal components analysis in this paper, although exploratory and not backed up by formal tests, indicates that there is a single factor that explains almost all variability in the term structure. This suggests that treating all but a one-dimensional space of price movements as complete surprises may do reasonably well in pricing most options (since the amount of volatility being neglected is small and is probably less important than estimation error and other sources of misspecification). Nonetheless it is undoubtedly possible to identify options (e.g. on spreads) for which the error is large.

The analysis here assumes but does not imply that the dominant factor has a constant variance. The analysis of this paper suggests (as did Brown & Dybvig (1986)) that the second factor (if any) in a term structure model should be related to the variance or other distributional features of interest rates, not additive in levels of interest rates as is usually assumed. Subsequent research should integrate a study of the variance process.

References

Brennan, Michael J., & Eduardo S. Schwartz, 1979, 'A Continuous Time Approach to the Pricing of Bonds', *Journal of Banking and Finance* **3**, 133–155.

Brown, Stephen & Philip Dybvig, 1986, 'The Empirical Implications of the Cox, Ingersoll, Ross Theory of the Term Structure of Interest Rates', *Journal of Finance* **41**, 617–630.

CRSP (Center for Research in Security Prices), 1995, 1994 Bond Files Guide, Graduate School of Business, University of Chicago.

Cox, John C., Jonathan E. Ingersoll, Jr., & Stephen A. Ross, 1981, 'A Re-examination of Traditional Hypotheses about the Term Structure of Interest Rates', *Journal of Finance* **36**, 769–799.

Cox, John C., Jonathan E. Ingersoll, Jr., & Stephen A. Ross, 1985, 'A Theory of the Term Structure of Interest Rates', *Econometrica* **53**, 385–407.

Cox, John C., & Stephen Ross, 1976, 'A Survey of Some New Results in Financial Option Pricing Theory', *Journal of Finance* **21**, 383–402.

Cox, John, Stephen Ross, & Mark Rubinstein, 1979, 'Option Pricing: A Simplified Approach', *Journal of Financial Economics* **7**, 229–263.

Dybvig, Philip H., 1988, 'Inefficient Dynamic Portfolio Strategies, or How to Throw Away a Million Dollars in the Stock Market', *Review of Financial Studies* **1**, 67–88.

Dybvig, Philip H., & Chi-fu Huang, 1988, 'Nonnegative Wealth, Absence of Arbitrage, and Feasible Consumption Streams', *Review of Financial Studies* **1**, 377–401.

Dybvig, Philip H., Jonathan E. Ingersoll, Jr., & Stephen A. Ross, 1996, 'Long Forward and Zero-Coupon Rates Can Never Fall', *Journal of Business* **69**, 1–25.

Gnanadesikan, 1977, *Methods for Multivariate Data Analysis of Multivariate Observations*, Wiley.

Harrison, Michael, & David Kreps, 1979, 'Martingales and multiperiod securities markets', *Journal of Economic Theory* **20**, 381–408.

Harrison, Michael, & Stanley Pliska, 1981, 'Martingales and stochastic integrals in the theory of continuous trading', *Stochastic Processes and Their Applications* **11**, 215–260.

Heath, David, Robert Jarrow, & Andrew Morton, 1990, 'Bond Pricing and the Term Structure of Interest Rates: A Discrete Time Approximation', *Journal of Financial and Quantitative Analysis* **25**, 419–440.

Heath, David, Robert Jarrow, & Andrew Morton, 1992, 'Bond Pricing and the Term Structure of Interest Rates: A New Methodology for Contingent Claims Valuation', *Econometrica* **60**, 77–105.

Ho, Thomas S.Y., & Sang-Bin Lee, 1986, 'Term Structure Movements and Pricing Interest Rate Contingent Claims', *Journal of Finance* **41**, 1011–1029.

Merton, Robert C., 1973, 'Theory of Rational Option Pricing', *Bell Journal of Economics and Management Science* **4**, 141–183.

Rubinstein, Mark, 1976, 'The Valuation of Uncertain Income Streams and the Pricing of Options', *Bell Journal of Economics* **7**, 407–425.

Vasicek, Oldrich, 1977, 'An Equilibrium Characterization of the Term Structure', *Journal of Financial Economics* **5**, 177–188.

Dynamic Models for Yield Curve Evolution

Bjorn Flesaker and Lane Hughston

Abstract

This chapter presents a concise 'working' formulation of the general arbitrage-free framework for the valuation of interest rate derivatives. Our approach is cast in terms of 'financial observables', notably the initial yield curve and the zero coupon bond price volatility term structure, which is given by a one-parameter family of vector processes adapted to the filtration of an n-dimensional Brownian motion. Simple exact representations for the money market account process and the zero coupon bond price processes are derived by the use of martingale arguments, and we are able to demonstrate succinctly and elegantly the relation between conventional 'short rate models' and Heath-Jarrow-Morton style 'forward rate models'. A novel geometric interpretation of the interest rate volatility term structure is outlined, and we show that market completeness in an n-factor interest rate model is equivalent to the condition that the random n-dimensional 'volatility curve' generated at any instant of time by the one-parameter vector volatility process is in fact a 'space curve'. In conclusion we present a new framework for interest rate dynamics based on a general characterisation for the condition that interest rates should be positive.

1   Introduction

Our purpose in this article is to present a general framework for the dynamics of yield curve evolution, with a view specifically to the valuation of interest rate derivatives. The object is not so much to develop the theory from first principles, but rather to assemble it in a workable form, stripped as far as possible of technicalities. A further objective here is to supply a notation sufficiently flexible that it will readily sustain the arguments of the general theory and the deduction of qualitative results, while at the same time allowing for a specialisation to models for specific deal structures in complex interest rate environments.

We begin with a brief introduction to the general theory of interest rate processes. We do not propose to give a mathematically rigorous or systematic development of the subject, but rather to present it from a slightly unusual point of view that is well-suited for our purposes and is directed at the problems associated with the valuation of contingent claims. The underlying interest rate methodology is essentially equivalent to the theory of Heath, Jarrow and Morton (1992). The intention here, however, is to improve in various significant ways on the usual presentation, and to be much more directly applications oriented.

The HJM theory in its conventional formulation is presented as a model for the stochastic evolution of the instantaneous forward rates. A recurrent theme in our investigations, however, will be the exploration of alternative starting points. Here we begin by taking the *short* term interest rate as the principal object of study. This turns out to be essentially equivalent to the HJM theory as it is usually put forward, and generally speaking nothing is different; but one gains a great deal in economy of thought, and the relationship to other interest rate models is clarified.

Historically, finance theorists tended to give the short term interest rate r_t a preferred significance, building models in which it appeared in essence as the sole or primary economic state variable. Such constructions include, for example, the well-known models investigated by Merton (1973), Vasicek (1977), Cox, Ingersoll and Ross (1985), Jamshidian (1990), and Hull and White (1990). Many other interest rate models have been developed in essentially the same spirit.

The HJM model appears at first consideration to represent a complete departure from this point of view, treating the instantaneous forward rates as fundamental. This is deceptive however, since the HJM framework can also be developed as a theory of the short rate, without even the need to mention the instantaneous forward rates.

When seen in this light the distinctive feature of the HJM framework is the facility with which it incorporates a generally non-Markovian process for the short rate, thus diminishing or eliminating the status of the value of the short rate itself as a sole or preferred state variable.

There is another more subtle reason for emphasising the role of the short rate, as opposed to the instantaneous forward rates. This relates to the special status of the money market process in interest rate derivatives pricing. The money market process represents the value of a unit-initialised money market account that accumulates interest at the short rate on a continuously compounding basis. At any future time t the value of the money market process is given by one unit of currency times the exponential of the integral of the short rate from the present to time t. The point is that the short rate never enters directly into the valuation of an interest rate derivative, but always via such an integral. This means that in a working version of the HJM framework one can always formulate matters 'upstairs', in terms of variables associated with the integrated short rate. We call such variables 'financial observables'. These have the property of always being representable directly in terms of the price levels, volatilities, and correlations of zero coupon bonds.

This point will be touched on several occasions in this article and will also be further developed elsewhere (cf. Hughston 1994, Flesaker and Hughston 1996a,b, 1997).

2   Probabilistic Characterisation of the Short Rate

Let us begin with a few philosophical points that will indicate the nature of our strategy. As much as possible we would like to emphasise the merits of financial intuition, probabilistic reasoning, and geometric insight. We find it especially useful, as is often customary in derivatives pricing arguments, to leap directly to the standard risk neutral measure, thus simply assuming the main brunt of the HJM argument. One of the primary goals of their analysis was to eliminate the interest rate risk premium variables from the valuation problem for derivative securities, and we shall find it convenient to regard that job as already having been accomplished. We also avoid, where possible, the specification of stochastic differential equations for interest rate processes; instead we work directly with the processes themselves.

Now let $\mathbf{W}_s$ be a standard n-dimensional Wiener process under the risk neutral measure. In our picture all interest rate risk will be adapted to the filtration generated by $\mathbf{W}_s$ in the sense that the trajectory of $\mathbf{W}_s$ between the present (time 0) and time s represents the entire source of interest rate randomness in the domestic economy up to that time.

Let σ_{st} ($s \leq t$) be a t-parametrised n-vector of adapted processes, where s is the 'process' (or 'evolutionary') index, and the parameter t will be called the 'maturity' (or 'asset') index. Thus for each value of t between zero and some essentially arbitrarily specified sufficiently large value (the 'trading horizon') we have a set of n adapted processes defined on the time interval $s \in [0, t]$. We shall assume that σ_{st} is sufficiently well-behaved to ensure the existence of all the various stochastic integrals and related expressions that follow. It is not our intention to go into the relevant technical details of stochastic analysis here, but we refer the interested reader for example to Heath, Jarrow and Morton (1992) and Carverhill (1995) for discussions of some of the non-trivial points arising in this connection. We shall also require that σ_{st} satisfies a certain non-degeneracy condition, which will be indicated below. It should be observed that the non-degeneracy requirement is not merely a 'technical condition', but rather has a substantive economic content (market completeness).

Instead of working directly with the yield curve we shall study the price processes of simple discount bonds (zero coupon bonds). Let us denote by P_{0t} the value at time zero of a discount bond that pays out one unit of currency at time t. Now let f_{0t} be the instantaneous forward rate determined by the present discount factor P_{0t} according to the standard formula

$$f_{0t} = -\frac{\partial \ln P_{0t}}{\partial t}. \tag{2.1}$$

Then the general interest rate model is completely characterised by the following *ansatz* for the short term interest rate r_t in the risk neutral measure:

$$r_t = f_{0t} + \int_{s=0}^{t} \sigma_{st} \cdot \left(\int_{u=s}^{t} \sigma_{su} du \right) ds + \int_{s=0}^{t} \sigma_{st} \cdot d\mathbf{W}_s. \tag{2.2}$$

Since here σ_{st} and $\mathbf{W}_t$ are both 'vector' processes, each with n components, we say we have an 'n-factor interest rate model'. It should be noted that this does not imply that the model can be necessarily characterised in terms of a finite number of economic state variables, since additional rather strong assumptions are required for that to be the case. Even in the single factor models the number of state variables is in general effectively infinite.

The important, widely studied 'Gaussian' interest rate models arise when the process σ_{st} is deterministic. In that case we see that σ_{st} acts as a kind of linear filter through which the 'noise' of the multi-variate Brownian motion is passed before it impinges on the short rate.

As a further specialisation, we note that in the single factor case we obtain the basic mean-reverting Gaussian models with term volatility if we choose σ_{st} to be of the 'separable' form $\sigma_{st} = C_s B_t$ where C_s and B_t are deterministic. In that situation the model has a single state variable, which we can take to be the short term interest rate itself if we wish. Alternatively by the prescription just specified we are led to a natural martingale characterisation of the model, for which the primary state variable is given by the stochastic integral $\int_0^t C_s dW_s$. By extending σ_{st} to be a sum of expressions of the general separable form $C_s B_t$ we obtain examples of multiple state variable models in a one-factor setting.

Returning now to the main course of the development, we remark that in expression (2.2) for the short rate the bracketed term involves the following integrated volatility process:

$$\Omega_{st} = \int_{u=s}^{t} \sigma_{su} du, \tag{2.3}$$

where the integration is taken over the maturity variable, ranging from s (the value of the process index) to t (the value of the maturity index of the integrated process). By use of the fundamental process Ω_{st} we can write the formula for the short rate more succinctly in the form:

$$r_t = f_{0t} + \int_{s=0}^{t} \sigma_{st} \cdot \Omega_{st} ds + \int_{s=0}^{t} \sigma_{st} \cdot d\mathbf{W}_s. \tag{2.4}$$

Furthermore, on account of the relations $\frac{\partial}{\partial t}\Omega_{st} = \sigma_{st}$ and $\Omega_{ss} = 0$ we find that the short rate can be written as follows as an exact differential:

$$r_t dt = d\left[-\ln P_{0t} + \frac{1}{2}\int_{s=0}^{t} \Omega_{st}^2 ds + \int_{s=0}^{t} \Omega_{st} \cdot d\mathbf{W}_s \right], \tag{2.5}$$

where Ω_{st}^2 denotes the squared vector norm. This relation then allows us easily to deduce an expression for the integral of the short rate over any given time interval, given by

$$\int_a^b r_t dt = -\ln\left[\frac{P_{0b}}{P_{0a}}\right] + \left[\frac{1}{2}\int_{s=0}^t \Omega_{st}^2 ds + \int_{s=0}^t \Omega_{st} \cdot d\mathbf{W}_s\right]_a^b \qquad (2.6)$$

from which we infer after some elementary rearrangement that:

$$\int_a^b r_t dt = -\ln\left[\frac{P_{0b}}{P_{0a}}\right] + \frac{1}{2}\int_{s=0}^a (\Omega_{sb}^2 - \Omega_{sa}^2) ds + \int_{s=0}^a (\Omega_{sb} - \Omega_{sa}) \cdot d\mathbf{W}_s$$

$$+\frac{1}{2}\int_{s=a}^b \Omega_{sb}^2 ds + \int_{s=a}^b \Omega_{sb} \cdot d\mathbf{W}_s. \qquad (2.7)$$

This basic formula will be used in the derivation of the general bond price process (3.2). As a special case of (2.7) we have the following important expression for the integrated short rate process:

$$\int_0^a r_t dt = -\ln[P_{0a}] + \int_{s=0}^a \Omega_{sa} \cdot d\mathbf{W}_s + \frac{1}{2}\int_{s=0}^a \Omega_{sa}^2 ds. \qquad (2.8)$$

This expression enters into formula (7.3) below for the random future value of a money market account initiated today with a unit of cash.

Note that in formula (2.8) above for the integrated short rate it is only the integrated volatility process Ω_{st} that appears, not its progenitor σ_{st}. We shall see later that Ω_{st} represents the volatility structure at time s of a t-maturity zero coupon bond, whereas σ_{st} is the corresponding volatility structure for the associated instantaneous forward rate.

Now we can return to the matter of the non-degeneracy condition for the volatility process. We shall require that Ω_{st} is of rank n for each value of s. The non-degeneracy assumption is essentially a 'market completeness' condition for interest rate related securities. See Harrison and Pliska (1981) and Heath, Jarrow and Morton (1992) for some of the measure-theoretic technicalities we have suppressed here.

3   The Zero Coupon Bond Price Process

The valuation of interest rate contingent claims requires appropriate formulae for two fundamental price processes: these are the zero coupon bond process P_{ab}, and the money market process I_a. The bond price process (which actually represents a *family* of processes indexed by the maturity date b) is defined as follows:

$$P_{ab} = \mathbb{E}_a\left[\exp\left(-\int_a^b r_s ds\right)\right] \qquad (3.1)$$

where $\mathbb{E}_a$ is the conditional expectation based on the filtration of the Brownian motion up to time a.

This is the random value at time a of a contingent claim that pays off one unit of currency at time b. If we insert (2.7) into (3.1) and calculate the conditional expectation (the details of this calculation will be given in Section 6) we get

$$P_{ab} = \tilde{P}_{ab} \exp\left[\int_{s=0}^{a} (\Omega_{sa} - \Omega_{sb}) \cdot d\mathbf{W}_s + \frac{1}{2}\int_{s=0}^{a} (\Omega_{sa}^2 - \Omega_{sb}^2) ds\right] \qquad (3.2)$$

for the general discount bond process in the risk neutral measure, where for convenience we have written

$$\tilde{P}_{ab} = \frac{P_{0b}}{P_{0a}} \qquad (3.3)$$

for the (deterministic) forward zero coupon bond price, the 'tilde' being there to signify the deterministic property. For convenience later we shall also sometimes write $P(a, b)$ for P_{ab}.

Note that the value of the random bond price P_{ab} only depends (as it should) on random events up to time a, and that the dependency of P_{ab} on the vector σ_{st} is entirely via the principal integrated volatility vector process Ω_{st} (the infinitesimal forward rate volatility σ_{st} does not itself enter directly into the discount bond process).

4 Geometric Interpretation of the Bond Volatility Structure

Once we have the random discount bond process (3.2) we can begin to make a systematic interpretation of its features. Our goal here is to develop a geometric picture of the stochastic properties of zero coupon bonds. First if we take the stochastic differential of (3.2) a short calculation making use of (2.5) gives

$$dP_{ab} = P_{ab}(r_a da - \Omega_{ab} \cdot d\mathbf{W}_a) \qquad (4.1)$$

which confirms that the instantaneous expected rate of return on the risk-neutral bond price process is indeed the short rate. Furthermore we see at once that the squared instantaneous (or 'local') volatility at time a of a b-maturity discount bond is Ω_{ab}^2:

$$\text{var}_a\left[\frac{dP_{ab}}{P_{ab}}\right] = \Omega_{ab}^2 da \qquad (4.2)$$

Equivalently, we note that the quadratic variation of the logarithm of the bond price process is given by the cumulative integral (over the process index) of the squared volatility process, a relation which expression (4.2) represents in differential form. This result gives us a useful start at a kind of 'geometro-economic' understanding of the vector process Ω_{ab}, since it says that the length of this vector is the volatility of the corresponding zero coupon bond.

More generally for any pair of bonds of differing maturities b and b' we have
the instantaneous covariance relation

$$\text{cov}_a \left[\frac{dP_{ab}}{P_{ab}}, \frac{dP_{ab'}}{P_{ab'}} \right] = \Omega_{ab} \cdot \Omega_{ab'} da, \qquad (4.3)$$

which can be interpreted as a formula for the quadratic covariation of the
logarithms of the two bond price processes. For their instantaneous correlation
of motion we then get

$$\rho_a \left[\frac{dP_{ab}}{P_{ab}}, \frac{dP_{ab'}}{P_{ab'}} \right] = \frac{\Omega_{ab} \cdot \Omega_{ab'}}{\sqrt{\Omega_{ab}^2 \Omega_{ab'}^2}}. \qquad (4.4)$$

These relations can be given a striking geometrical interpretation. In the case
of an n-factor model we consider an n-dimensional Euclidean space $\mathbb{R}^n$. The
elements of this space correspond to possible values of the volatility vector.

For a fixed value of the evolutionary time a we have a one-parameter
family of bonds labelled by the maturity date b. Each bond in this family is
represented by a volatility vector emanating from the origin in $\mathbb{R}^n$, and the
endpoints of these vectors form a curve, parametrised by the maturity date
b. We shall call this the 'volatility curve'.

The length of the vector joining the point on the curve corresponding to
a particular maturity date gives the volatility of that bond. The cosine of
the angle between a pair of vectors corresponding to different maturity dates
determines the correlation between the percentage price movements of those
two bonds.

Thus at each instant of time the volatility curve summarises the full information of the volatilities and correlations of discount bonds of all maturities,
i.e. the 'volatility term structure' of the discount bond spectrum.

We can fix one end of the curve to the origin by requiring that point always
to correspond to a bond of maturity a, for which the volatility is zero. By
construction the curve is at least once-differentiable, i.e. it has a well-defined
tangent vector. The evolution of the bond volatility can then be pictured as
a random motion of this curve. The 'normal' or 'Gaussian' interest models
are those for which the motion is deterministic.

It is worth taking this line of argument a step further and considering the
relationship at a given instant of time between two such curves only differing
from one another by an overall rotation of the n-dimensional space about
the origin. Two such apparently distinct volatility processes give rise to one
and the same interest rate model. More generally we can multiply the volatility vector process by any adapted orthogonal matrix process (corresponding
to a continuous random rigid rotation) and this will leave all economically
significant quantities invariant.

It follows as a corollary of this observation that in a multi-factor model
price data are in principle never sufficient to completely specify the volatility

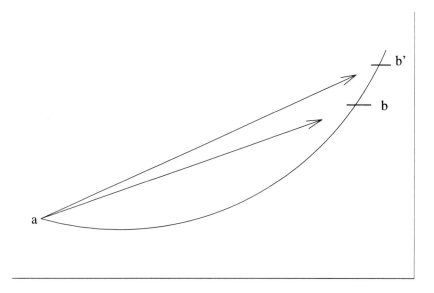

Figure 1. Volatility term structure of a zero coupon bond. The instantaneous volatility at time a of a bond that matures at time b is given by the length of the vector which joins the origin, corresponding to time a, to the point b on the volatility curve representing the maturity date. The cosine of the angle between two such vectors gives the instantaneous correlation between the percentage price movements of the corresponding bonds. In an n-factor interest rate model market completeness implies that at each instant the volatility curve does not lie in a hyperplane of dimension less than n.

vector process; the best one can obtain (and all one ever needs) are the invariants (inner products) represented by the zero coupon bond volatilities and correlations.

These quantities are invariant under random orthogonal transformations of the volatility process and determine the volatility process completely up to such transformations. More precisely, the invariants can be represented by a doubly parametrised process $g_{abb'}$ defined by:

$$g_{abb'} = \Omega_{ab} \cdot \Omega_{ab'} \qquad (4.5)$$

Here a is the 'process' index, and the two 'maturity' parameters are b and b'. For each value of a, the infinite matrix $g_{abb'}$ has rank n. Note that $g_{abb'}$ is essentially the instantaneous covariance matrix at time a for bonds of maturities b and b' (cf. formula 4.3 above).

Now suppose F_T represents the payoff at time T of a contingent claim. Then invariance considerations show that its value F_0 at time zero can always be expressed as a functional of the matrix $g_{abb'}$ for values of a in the range $0 \le a \le T$. It follows that Ω_{ab} itself is not directly required for the valuation of interest rate derivatives; we only require its equivalence class modulo adapted orthogonal transformations. Another way of looking at this is to note that the probability law of the discount bond process (3.2) only depends on the equivalence class of the volatility structure. This is because the volatility structure enters into (3.2) in only the following two ways: (1) via integrals of the squared volatility invariants, and (2) via a stochastic integral with respect to Brownian motion. However, the law of Brownian motion is itself invariant under orthogonal transformations of the Brownian motion; thus any function of the stochastic integral of the vector volatility process is necessarily invariant in law under orthogonal transformations of the volatility process.

With all this in mind we can say a little more about the non-degeneracy condition on the volatility structure. This is essentially the condition that for each instant of time the volatility curve (as described above) should be a 'space-curve', that is to say with probability one it should not lie entirely within any hyperplane of co-dimension one. This gives us a convenient visual embodiment of the complete markets condition on interest rate related securities. The volatility curve is a space-curve if and only if there exist n distinct points along the curve such that the vectors joining these points to the origin are linearly independent.

This general line of reasoning can be carried further, and suggests that geometric methods, particularly some of the techniques of stochastic differential geometry, may ultimately play a useful role in dynamic asset pricing models (cf. Hughston 1993, 1994) and possibly other areas of finance theory.

5   Relation to the HJM Framework

Now let us turn to consider the process followed by the instantaneous forward rate. This is where we connect up with the HJM model in its original and more usual formulation. Ito's lemma applied to the logarithm of the bond price process gives

$$d \ln P_{ab} = \frac{dP_{ab}}{P_{ab}} - \frac{1}{2}\left[\frac{dP_{ab}}{P_{ab}}\right]^2 \tag{5.1}$$

and thus by use of (4.1) and (4.2),

$$d \ln P_{ab} = (r_a - \frac{1}{2}\Omega_{ab}^2)da - \Omega_{ab} \cdot d\mathbf{W}_a. \tag{5.2}$$

Taking a derivative of (5.2) with respect to the maturity index b we obtain

$$df_{ab} = \sigma_{ab} \cdot \Omega_{ab}da + \sigma_{ab}.d\mathbf{W}_a, \tag{5.3}$$

by use of $\frac{\partial}{\partial t}\Omega_{st} = \sigma_{st}$ and the definition of the instantaneous forward rates

$$f_{ab} = -\frac{\partial \ln P_{ab}}{\partial b}. \tag{5.4}$$

We thus see that the length of σ_{ab} can be interpreted as the absolute volatility of the forward rate f_{ab}. Instead of formula (5.3) one can use the equivalent integral representation.

$$f_{ab} = f_{0b} + \int_{s=0}^{a} \sigma_{sb} \cdot \Omega_{sb} ds + \int_{s=0}^{a} \sigma_{sb} \cdot d\mathbf{W}_s, \tag{5.5}$$

which by virtue of the relation $r_a = f_{aa}$ reduces to (2.4) if we set $a = b$. Formula (5.5) is in fact essentially the starting point in the standard presentation of the Heath-Jarrow-Morton theory, though interestingly in our formulation it appears almost as an afterthought. For the purposes of derivatives price modelling and risk management the arguments put forward here suggest that the instantaneous forward rates should be viewed as purely subsidiary processes. In the case of interest rate contingent claims pricing attention thus needs to be focused on: (a) the bond price process, and (b) the money market process.

6   Remarks on Exponential Martingales

Let us return now to the derivation of formula (3.2) for the bond price process. This is an item of considerable significance since the probabilistic arguments involved lie at the heart of the study of interest rate processes. To see in more detail that (3.2) follows from (3.1) we observe that the first three terms of the right hand side of (2.7) are known as at time a, and if we exponentiate the sum of these three terms (with a minus sign) we are led directly to expression (3.2). It remains therefore to see that

$$\mathbb{E}_a \exp\left[-\int_{s=a}^{b} \Omega_{sb} \cdot d\mathbf{W}_s - \frac{1}{2}\int_{s=a}^{b} \Omega_{sb}^2 ds\right] = 1. \tag{6.1}$$

The reasoning behind this is of some interest in its own right, and it is worth digressing to go into the matter in more depth. Formula (6.1) follows from the fact that for any (suitably well-behaved) adapted vector process Θ_s we have

$$\mathbb{E}\left[\exp\left(\int_{s=0}^{b} \Theta_s \cdot d\mathbf{W}_s - \frac{1}{2}\int_{s=0}^{b} \Theta_s^2 ds\right)\right] = 1 \tag{6.2}$$

for each value of b, and more generally for each value of a and b

$$\mathbb{E}_a\left[\exp\left(\int_{s=a}^{b} \Theta_s \cdot d\mathbf{W}_s - \frac{1}{2}\int_{s=a}^{b} \Theta_s^2 ds\right)\right] = 1. \tag{6.3}$$

These formulae can be interpreted from two different perspectives, and interestingly both come into play in what follows. In the first interpretation we are given the process Θ_s, and from it we create a new process

$$\Psi_b = \exp\left(\int_{s=0}^{b} \Theta_s \cdot d\mathbf{W}_s - \frac{1}{2}\int_{s=0}^{b} \Theta_s^2 ds\right) \tag{6.4}$$

which by construction has the martingale property

$$\mathbb{E}_a\left[\Psi_b\right] = \Psi_a, \tag{6.5}$$

which can be seen to be equivalent to (6.3). The fact that Ψ_b as defined by (6.4) satisfies the martingale condition follows more visibly from the standard integral representation

$$\Psi_b = 1 + \int_{s=0}^b \Psi_s \Theta_s \cdot d\mathbf{W}_s \tag{6.6}$$

which by Ito's lemma turns out to be equivalent to (6.4). A process Ψ_b of the form (6.4) is called an 'exponential martingale'.

On the other hand, if we are given a family of processes Θ_{sb} depending explicitly on the parameter b, and form a process Ψ_b according to formula (6.4), then it can be verified that Ψ_b does not (generally) satisfy the martingale condition (6.5). Nevertheless if we construct the integral

$$\Psi_{ab} = \exp\left(\int_{s=a}^b \Theta_{sb}.d\mathbf{W}_s - \frac{1}{2}\int_{s=a}^b \Theta_{sb}^2 ds\right) \tag{6.7}$$

in that circumstance, we find that the weaker relation

$$\mathbb{E}_a\left[\Psi_{ab}\right] = 1 \tag{6.8}$$

still holds for any value of a and b. This 'weak' relation can be thought of as generalising the much-used identity

$$\mathbb{E}\left[\exp\left(X - \frac{1}{2}\mathbf{V}[X]\right)\right] = 1, \tag{6.9}$$

that holds for any normally distributed random variable X with mean zero and variance $\mathbf{V}[X]$, a result that is handy in the analysis of log-normally distributed asset prices.

In particular we can think of (6.8) as providing a kind of natural link between the Gaussian interest rate models, and the general interest rate framework. Given (6.8), if we set $\Theta_{sb} = -\Omega_{sb}$ in (6.7), then (6.1) follows at once, which is what we set out to demonstrate in this section, thereby justifying formula (3.2).

7 Money Market Accounts

A useful alternative representation of the bond price process P_{ab} is given by

$$P_{ab} = I_a P_{0b} \exp\left[-\int_{s=0}^a \Omega_{sb} \cdot d\mathbf{W}_s - \frac{1}{2}\int_{s=0}^a \Omega_{sb}^2 ds\right], \tag{7.1}$$

where here we have introduced the so-called 'money market process' (inflator process) I_a defined by

$$I_a = \exp\left[\int_0^a r_t dt\right],$$ (7.2)

which represents the (random) value at time a of a money market account initialised at time 0 with one unit of currency. In the literature I_a is sometimes said to represent the price of a 'riskless' security, but this can be misleading and is true only in a specific local sense. From equation (2.8) it follows that the explicit formula for I_a is given by:

$$I_a = (P_{0a})^{-1} \exp\left[\int_{s=0}^a \Omega_{sa} \cdot d\mathbf{W}_s + \frac{1}{2}\int_{s=0}^a \Omega_{sa}^2 ds\right],$$ (7.3)

and it is straightforward to verify that any two of formulae (3.2), (7.1), and (7.3) imply the third. Now let us turn to the stochastic discount factor D_a, defined by

$$D_a = \exp\left[-\int_0^a r_t dt\right]$$ (7.4)

and given explicitly by:

$$D_a = P_{0a} \exp\left[-\int_{s=0}^a \Omega_{sa} \cdot d\mathbf{W}_s - \frac{1}{2}\int_{s=0}^a \Omega_{sa}^2 ds\right],$$ (7.5)

It is a principle of contingent claims analysis that for any random flow of cash F_a occurring at time a we can compute its present value F_0 by formation of the risk neutral expectation of its product with the stochastic discount factor: $F_0 = \mathbb{E}[D_a F_a]$. Three very special cases of this are:

(1) $F_a = 1$, corresponding to receipt at time a of a unit of cash;

(2) $F_a = P_{ab}$, corresponding to receipt at time a of the value of a b-maturity zero coupon bond

(3) $F_a = I_a$, corresponding to receipt at time a of the contents of a unit-initialised money market account.

From economic reasoning we know that the present value of case (1) must be P_{0a}; whereas the value of (2) has to be P_{0b}, and the value of (3) clearly has to be unity. It follows therefore that

$$\mathbb{E}[D_a] = P_{0a},$$ (7.6)

and

$$\mathbb{E}[D_a P_{ab}] = P_{0b}.$$ (7.7)

Relation (7.6) follows directly as a consequence of (6.2). Relation (7.7) is obtained by inspection of the formula

$$D_a P_{ab} = P_{0b} \exp\left[-\int_{s=0}^a \Omega_{sb} \cdot d\mathbf{W}_s - \frac{1}{2}\int_{s=0}^a \Omega_{sb}^2 ds\right],$$ (7.8)

which shows us that $D_a P_{ab}$ is an exponential martingale. Equivalently for any time $a' \leq a$ we can write

$$\mathbb{E}_{a'}\left[\frac{P_{ab}}{I_a}\right] = \frac{P_{a'b}}{I_{a'}}, \tag{7.9}$$

which demonstrates that the ratio of the value of the discount bond to the value of the money market account is a martingale, which must of course be the case in an arbitrage-free model. That is to say, if we use units of the money market account as a currency ('numeraire'), the value of the discount bond is a martingale when expressed in such units. Equation (7.7) then follows if we set $a' = 0$.

8 Valuation of Interest Rate Derivatives

Suppose we are interested in calculating the present value of an interest rate contingent claim with a single terminal payoff that depends on the values of various zero coupon bonds (e.g. via Libor and swap rates) on a series of dates preceding and including the terminal date T. The various 'decision dates' on which the zero coupon bond values are to be noted will be denoted a_i with $i = 1, \ldots, m$. For convenience we shall sometimes also use the notation $T = a_m$ for the terminal date. Let us write $P(a_i, b_{ij})$ for the values of a set of $m \times q$ zero coupon bonds at times a_i $(i = 1, \ldots, m)$ with maturity dates b_{ij} $(j = 1, \ldots, q)$. Then for the terminal payoff we have some given function $F_T[P(a_i, b_{ij})]$ of these $m \times q$ zero coupon bonds. For convenience we sometimes refer to the various a_i and b_{ij} values as the 'a-dates' and the 'b-dates'. The 'a-dates' are the times at which the relevant random events actually occur.

So far we have had in mind the situation where there is a single payment. Suppose however we have a succession of payments (as in a cap or an index amortising rate swap), such that each payment is structured according to the prescription given above. Then we can still treat the problem (for the purposes of valuation) as though all the money is delivered on the terminal date. If one of the random payments is scheduled to occur at time t, say, and the terminal date is T, then the earlier random payment can be treated as effectively occurring at time T if we divide it by the discount factor $P(t, T)$. In this way we can 'reschedule' all the earlier payments to create a single combined effective payment on the terminal date T.

Henceforth we shall assume the rescheduling already to have been taken into account, and it suffices to consider contingent claims with a single payment made on the terminal date. Our problem then is to calculate the present value F_0 of the contingent claim by taking the following mean:

$$F_0 = \mathbb{E}\left[D_T F_T \left[P(a_i, b_{ij})\right]\right], \tag{8.1}$$

where $\mathbb{E}$ as before denotes the expectation operator in the risk neutral measure, for which $\mathbf{W}_t$ is an n-dimensional standard Wiener process. Here for each zero coupon bond price $P(a, b)$ appearing in the argument of the payoff function F_T we substitute the appropriate expression of the form (3.2) above, and for the terminal stochastic discount factor D_T we write

$$D_T = P_{0T} \exp\left[-\int_{s=0}^{T} \Omega_{sT} \cdot d\mathbf{W}_s - \frac{1}{2}\int_{s=0}^{T} \Omega_{sT}^2 ds\right]. \tag{8.2}$$

Now we highlight a few interesting points that have practical implications. Suppose first we scrutinise more closely formula (8.1). This is the formula for the present value F_0 of a general interest rate derivative. To calculate F_0 we need to know D_T, as given by (8.2), and $F_T[P(a_i, b_{ij})]$ which is given by some prescribed function of one or more zero coupon bonds, and these in turn are given according to the basic formula

$$P_{ab} = \tilde{P}_{ab} \exp\left[\int_{s=0}^{a} (\Omega_{sa} - \Omega_{sb}) \cdot d\mathbf{W}_s + \frac{1}{2}\int_{s=0}^{a} (\Omega_{sa}^2 - \Omega_{sb}^2) ds\right], \tag{8.3}$$

derived earlier in Section 3. It should be apparent that to value any interest rate contingent claim there are only two ingredients that we need to know.

These are the initial discount function P_{0t}, and the bond volatility process Ω_{st}. These quantities both go directly into (8.2) and (8.3), which in turn go into (8.1) to give the value of F_0.

Equivalently, we can say that the value of any interest rate derivative can be expressed as a functional of the initial discount function and the bond volatility structure.

In particular, we note that the derivatives of P_{0t} and Ω_{st} with respect to their maturity indices do not enter directly into the formula for F_0. This is good news since it implies that providing Ω_{st} does not itself depend directly on the derivative of P_{0t} with respect to t, we only require a continuous (but not necessarily differentiable) specification of the yield curve to proceed in practical terms for the valuation of interest rate contingent claims.

In short, the valuation of yield curve derivatives doesn't require the evaluation of yield curve derivatives.

This situation can be contrasted vividly with the more usual approach to interest rate modelling, which typically begins with a stochastic differential equation for the short rate or the instantaneous forward rates, in which both the first and second derivatives of P_{0t} make an appearance at the very outset. But once the model is assembled into full working form these derivatives drop out of the picture (they all get 'integrated up'), and we are left merely with the specification of the initial discount function P_{0t} and the volatility structure Ω_{st}.

Conversely, given the discount function P_{0t} we can examine price data for contingent claims, and from these data we can infer estimates for the statistical properties of the Ω_{st} process, in the form of implied bond volatilities and correlations.

9   Forward Measures

We remark also that the form of (8.2) suggests the use of D_T to define a new measure on the underlying probability space. For any random variable X defined on the filtration of the probability space up to time T we define a new probability measure by setting

$$\mathbb{E}^T[X] = (P_{0T})^{-1}\mathbb{E}[D_T X].\tag{9.1}$$

Then under the new measure corresponding to $\mathbb{E}^T$, called the 'T-measure' or 'forward adjusted risk measure' (cf. Geman 1989, Jamshidian 1989, El Karoui, Myneni and Viswanathan 1992, El Karoui, Geman and Lacoste 1995), one finds that the process W_t^T defined by

$$W_t^T = W_t + \int_0^t \Omega_{sT}ds\tag{9.2}$$

is a Brownian motion. It turns out that there are a number of significant practical advantages in the use of this measure for actual implementations. This is on account of the fact that in the T-measure the valuation of a derivative proceeds via the formula

$$F_0 = P_{0T}\mathbb{E}^T\left[F_T[P(a_i, b_{ij})]\right]\tag{9.3}$$

which is simpler than (9.1) due to the absence of the stochastic discount factor. The idea here is to use a probability measure that is naturally tailored to the payout structure of the particular derivative being valued.

One exciting conclusion that can be drawn in a general way from the foregoing analysis is that we have just about reached the stage now where it becomes a valid goal to formulate a compelling dynamical framework for the determination of the process Ω_{st} itself, and a precise account of the extent to which this process is determined by suitable initial conditions.

An outstanding problem in connection with this, to be addressed in the next section, is the specification of a general tractable ansatz for the discount bond process volatility Ω_{ab} sufficient to ensure that interest rates are always positive.

10   Positive Interest Rate Processes

Now we turn to consider the problem of interest rate positivity. The general interest rate model, when formulated in the risk neutral measure, is curiously refractory when it comes to the matter of ensuring, in a satisfying way, that interest rates are to remain positive.

The HJM framework entails as special cases various examples in which interest rates are positive, but these examples are typically either highly restricted in character, such as models of the CIR class, or else severely artificial and intractable, such as the examples based on truncated forward rate volatilities studied by HJM and by Miltersen (1994). In any case it cannot be said that the approaches offered hitherto show any promise of generality.

Fortunately, however, there does exist a natural characterisation of the positivity condition within the general framework. This will be outlined briefly, without technicalities, in the material following. The key idea is to work with the T-measure associated with a bond of given maturity, and take the limit as the maturity date recedes to infinity. Let us fix therefore a date T and consider the family of bond price processes P_{ab} for which $0 \le a \le T$. For any bond P_{ab} in this family we form the ratio P_{ab}/P_{aT}, where P_{aT} is the T-maturity bond.

Since the system of bond prices is by assumption complete and arbitrage-free, there exists a unique equivalent martingale measure $\mathbb{E}^T$ (the T-measure, described in 9) with respect to which the ratio $N_{ab} = P_{ab}/P_{aT}$ is a martingale for any bond in the given family, subject to the boundary condition $N_{aT} = 1$. We can then solve for P_{ab} in terms of the martingale family N_{ab} to get

$$P_{ab} = \frac{N_{ab}}{N_{aa}}. \tag{10.1}$$

Now suppose c is such that $0 \le a \le b \le c$. Then according to (9.1) we have

$$\frac{P_{ac}}{P_{ab}} = \frac{N_{ac}}{N_{ab}}. \tag{10.2}$$

The positive interest rate property is the condition that the forward bond price process $P_{abc} = P_{ab}/P_{ac}$ should satisfy $P_{abc} < 1$ for $b < c$. This says that all the forward rates determined at any time a are positive. From (10.2) it follows that under this condition we have $N_{ac}/N_{ab} < 1$, which implies that the derivative of N_{ab} with respect to the maturity index is negative: $\partial_b N_{ab} < 0$, where ∂_b stands for differentiation with respect to b. If N_{ab} is a martingale for all b such that $0 \le a \le b$, then (given certain reasonable technical conditions) so is its derivative with respect to b. The point is that if $N_{ab} = P_{ab}/P_{aT}$ is a martingale for two given values of the parameter b, then by linearity so is the difference of these two processes, and we want this to remain the case in the limit that the two given values of b approach one another. Thus $\partial_b N_{ab}$ represents a family of negative martingales.

The condition that interest rates are initially positive, however, is that $\partial_b N_{0b}$ is negative. It follows therefore that there exists a family of positive martingales M_{ab}, subject to the unit initialisation $M_{0b} = 1$, such that $\partial_b N_{ab} = \partial_b N_{0b} M_{ab}$. It is not difficult to see that the solution of this differential equation, subject to the boundary condition $N_{aT} = 1$, is given by:

$$N_{ab} = 1 - (P_{0T})^{-1} \int_b^T \partial_s P_{0s} M_{as} ds. \tag{10.3}$$

Note that (10.3) automatically incorporates the initial condition $N_{0b} = P_{0b}/P_{0T}$. Inserting the solution (10.3) into (10.1) we obtain the following formula for the discount bond prices:

$$P_{ab} = \frac{P_{0T} - \int_b^T \partial_s P_{0s} M_{as} ds}{P_{0T} - \int_a^T \partial_s P_{0s} M_{as} ds}. \tag{10.4}$$

This expression represents the general bond price process, over the indicated range, for which interest rates remain positive. Note that P_{ab} appears to depend directly on the derivative $\partial_s P_{0s}$, but this dependence can be removed by an integration by parts.

Suppose we consider a random interest rate derivative payoff H_a at time a. Thus H_a can be any measurable function of discount bond prices, over the interval $[0, a]$, as indicated in Section 8. Since the ratio H_a/P_{aT} is a martingale under $\mathbb{E}^T$, the present value H_0 of the derivative is given by

$$\frac{H_0}{P_{0T}} = \mathbb{E}^T \left[\frac{H_a}{P_{aT}} \right], \tag{10.5}$$

where $\mathbb{E}^T$ is the equivalent martingale measure associated with T-maturity bond, described in the previous section. On account of (10.4) we have

$$P_{aT} = \frac{P_{0T}}{P_{0T} - \int_a^T \partial_s P_{0s} M_{as} ds} \tag{10.6}$$

for the T-maturity bond price process, and it follows after a simple calculation that formula (10.5) can be re-expressed in the form:

$$H_0 = \mathbb{E}^T \left[\left(P_{0T} - \int_a^T \partial_s P_{0s} M_{as} ds \right) H_a \right]. \tag{10.7}$$

So far the discussion has been based on the choice of a particular discount bond as numeraire, such that other bond prices are martingales when expressed in units of it. It is remarkable, however, that both expression (10.4) for the bond price process and formula (10.7) for contingent claims evaluation have natural limiting expressions as we let T get large.

Let us therefore take the limit as T goes to infinity, for which P_{0T} goes to zero. This leads us to a new equivalent martingale measure that we call the terminal measure. By inspection of (10.4) and (10.7) we obtain

$$P_{ab} = \frac{\int_b^\infty \partial_s P_{0s} M_{as} ds}{\int_a^\infty \partial_s P_{0s} M_{as} ds} \tag{10.8}$$

for the bond price process and

$$H_0 = \mathbb{E}^\infty \left[\left(-\int_a^\infty \partial_s P_{0s} M_{as} ds \right) H_a \right] \tag{10.9}$$

as the contingent claims valuation formula, where $\mathbb{E}^\infty$ signifies the terminal measure.

These two formulae are the main structural equations of what we like to call the 'positive interest rate framework'. It follows from (10.8) by differentiation that in this framework the instantaneous forward rates are given by:

$$f_{ab} = \frac{\partial_b P_{0b} M_{ab}}{\int_b^\infty \partial_s P_{0s} M_{as} ds} \tag{10.10}$$

Since the numerator and denominator of this expression are both by construction negative, we see that the instantaneous forward rates are manifestly positive. The short term interest rate $r_a = f_{aa}$ is then given by

$$r_a = \frac{\partial_a P_{0a} M_{aa}}{\int_a^\infty \partial_s P_{0s} M_{as} ds}, \tag{10.11}$$

which is also strictly positive. We note, incidentally, that by virtue of the martingale property (whereby the expectation of the modulus of M_{ab} is guaranteed to exist) the interest rate processes under consideration here are necessarily 'non-explosive'.

Let us consider the stochastic equation satisfied by the bond price process. Our goal is to construct the analogue of formula (4.1). Since M_{ab} represents a family of positive martingales it follows that there exists a family of proportional volatility processes ω_{ab} such that

$$dM_{as} = M_{as} \omega_{as} \cdot d\mathbf{W}_a^\infty, \tag{10.12}$$

where $\mathbf{W}_a^\infty$ is an n-dimensional Brownian motion with respect to the terminal measure. To proceed further we define a family of vector processes $\mathbf{V}_{ab}$ in terms of ω_{ab} by the following formula:

$$\mathbf{V}_{ab} = \frac{\int_b^\infty \partial_s P_{0s} M_{as} \omega_{as} ds}{\int_b^\infty \partial_s P_{0s} M_{as} ds}. \tag{10.13}$$

We shall require that ω_{ab} goes to zero for large values of b, which implies that $\mathbf{V}_{ab}$ shares the same property, and that M_{ab} is the constant process, with value unity, in the limit of large b. A short calculation shows that the stochastic equation satisfied by the bond price process is given by:

$$dP_{ab} = P_{ab}[r_a da - \Omega_{ab} \cdot (d\mathbf{W}_a^\infty - \mathbf{V}_{aa} da)], \tag{10.14}$$

where the short rate r_a is given as in formula (10.11) above, and the discount bond vector volatility Ω_{ab} is given by

$$\Omega_{ab} = \mathbf{V}_{aa} - \mathbf{V}_{ab}. \tag{10.15}$$

This solves the problem raised in the previous section, namely, the specification of a general *ansatz* for Ω_{ab} necessary and sufficient to ensure interest rate positivity. Note that in deriving (10.14) we have made use of the following elementary Ito quotient identity, here expressed for convenience in 'proportional' form:

$$d\left(\frac{X}{Y}\right) = \left(\frac{X}{Y}\right)\left[\frac{dX}{X} - \frac{dY}{Y} + \frac{dY^2}{Y^2} - \frac{dXdY}{XY}\right]. \tag{10.16}$$

The form of the 'correction term' in the expression involving the Brownian motion in the stochastic equation (10.14) suggests the appropriate change of measure to move from the terminal measure back to the risk neutral measure, for which $\mathbf{W}_t = \mathbf{W}_t^\infty - \int_0^t \mathbf{V}_{ss}ds$ is a Brownian motion, consistent with (9.2) in the limit $T \to \infty$.

This concludes our brief sketch of the positivity property for interest rates. A similar analysis can be carried through in a much broader context, e.g. with application to the world of foreign exchange and international interest rate markets, which we pursue in greater detail elsewhere (Flesaker and Hughston 1996a,b), along with a fuller discussion of the material presented here. The positive interest rate framework represents a significant step forward in interest rate derivatives pricing methodology since it allows for the isolation, in the form of a fundamental family of positive martingale processes, of the true dynamical degrees of freedom in yield curve evolution.

Glossary

For convenience we summarise some of the notation used here, with an indication of the relevant economic interpretation.

$\mathbf{W}_t$ standard n-dimensional Wiener process in the risk neutral measure, where n is the number of factors driving the economy

$\tilde{P}_{ab}$ forward zero coupon bond price (definition: $\tilde{P}_{ab} = P_{0b}/P_{0a}$)

P_{ab} zero coupon bond price process [alternative notation: $P(a,b)$]

F_0 present value of an interest rate derivative

F_T payoff value of a derivative at time T

Ω_{st} volatility vector for discount bonds

I_t money market account

D_t stochastic discount factor

$\mathbb{E}[-]$ risk neutral expectation

$\mathbb{E}_a[-]$ conditional expectation up to time a, in risk neutral measure

$\mathbb{E}^T[-]$ T-measure (forward adjusted risk measure)

$\mathbf{W}_t^T$ standard n-dimensional Brownian motion under the T-measure

$\mathbb{E}^\infty[-]$ terminal measure

$\mathbf{W}_t^\infty$ standard n-dimensional Brownian motion under the terminal measure

M_{ab} fundamental family of positive martingales

Acknowledgements

The authors would like to thank R. Brenner, R. Jarrow, O. Jonsson, D. Madan, S. Pliska, N. Rabeau, L. Sankarasubramanian, and J. Tigg for stimulating discussions.

References

A. Carverhill (1995) 'A simplified exposition of the Heath, Jarrow, and Morton model', *Stochastics Reports* **53**, 227–240.

J. Cox, J. Ingersoll and S. Ross (1985) 'A theory of the term structure of interest rates', *Econometrica* **53**, 385–408.

N. El Karoui, R. Myneni and R. Viswanathan (1992) 'Arbitrage pricing and hedging of interest rate claims with state variables: I Theory, II Applications'. Université de Paris VI and Stanford University working paper.

N. El Karoui, H. Geman, V. Lacoste (1995) 'On the role of state variables in interest rate models', Université de Paris VI and ESSEC Départment Finance working paper.

B. Flesaker and L.P. Hughston (1996a) 'Positive interest', *Risk Magazine* **9**, 46–49.

B. Flesaker and L.P. Hughston (1996b) 'Positive interest: foreign exchange', Chapter 22 in *Vasicek and Beyond: Approaches to Building and Applying Interest Rate Models*, L.P. Hughston (ed.), Risk Publications.

B. Flesaker and L.P. Hughston (1997) 'Exotic interest rate options', Chapter 6 in *Exotic Options: the State of the Art*, L. Clewlow and C. Strickland, (eds.) Chapman and Hall.

H. Geman (1989) 'The importance of the forward neutral probability measure in a stochastic approach to interest rates', ESSEC working paper.

J. M. Harrison and S.R. Pliska (1981) 'Martingales and stochastic integrals in the theory of continuous trading', *Stochastic Processes and their Applications* **11**, 215–260.

D. Heath, R. Jarrow and A. Morton (1992) 'Bond pricing and the term structure of interest rates: a new methodology for contingent claims valuation', *Econometrica* **60**, (1), 77–105.

J. Hull and A. White (1990) 'Pricing interest rate derivative securities', *Review of Financial Studies* **3**, 573–592.

L.P. Hughston (1993) 'Financial geometry: a new angle on risk', *International Derivative Review*, September 1993, 11–14 (SunGard Capital Markets).

L.P. Hughston (1994) 'Financial observables', *International Derivative Review*, December 1994, 11–14 (SunGard Capital Markets).

F. Jamshidian (1989) 'An exact bond option formula', *J. Finance* **44**, 205–209. (See also numerous Merrill Lynch working papers dating from this period.)

F. Jamshidian (1990) 'The preference-free determination of bond and option prices from the spot interest rate', *Advances in Futures and Options Research* **4**, 51–67.

R.C. Merton (1973) 'Theory of rational option pricing', *Bell Journal of Economics and Management Science* **4**, 141–183. Reprinted as Chapter 8 in R. C. Merton (1990) *Continuous-Time Finance*, Blackwell.

K. Miltersen (1994)'An arbitrage theory of the term structure of interest rates', *Ann. Appl. Prob.* **4**, 953–967.

General Interest-Rate Models and the Universality of HJM

Martin W. Baxter

There are now many models of interest-rate markets available. Many are based on the powerful HJM model of Heath, Jarrow and Morton (1992). Others, surveyed by Rogers (1995), are not (explicitly) set within the HJM framework, but are driven by, say, the short-term interest rate. In this paper we shall describe the appearance of a general interest-rate market. We shall also show, under some additional technical restrictions, that the general model is the short-rate model, and that the short-rate model is the HJM model. In other words, every 'sufficiently nice' model is simultaneously an HJM model and a short-rate model.

1 Introduction

We choose to work within the framework where all our processes are adapted to the filtration of an n-dimensional Brownian motion.

We do this for three important reasons. Firstly, there are great technical simplifications, such as the continuity of all martingales, which allow stronger results. Secondly, the Brownian case is recognizably distinct from other frameworks such as Poisson processes and Markov chains, and worthy of consideration. And thirdly, much market practice and interest is focused in this direction.

But, this framework assumption aside, we will try to be as general as possible.

Many approaches to interest-rate modelling can be divided into two types: short rate (SR) models and forward rate (HJM) models. A market of discount bonds $P(t, T)$ can be defined as

(SR)
$$P(t, T) = \mathbb{E}\left(\exp(-\int_t^T r_u \, du) \mid \mathcal{F}_t\right),$$
for some adapted short-rate process r_t,

(HJM)
$$P(t, T) = \exp(-\int_t^T f(t, u) \, du),$$
for some family of forward rate processes $f(t, T)$.

In each case, the defining equation is augmented by some conditions or constraints on the driving process(es). In SR, the process r_t must be such that the integral $\int_0^T |r_s|\, ds$ exists and that the expectation of the reciprocal of the bank account process $B_t = \exp(\int_0^t r_s\, ds)$ is finite. In HJM, it is firstly assumed that the bonds are differentiable in T to give the forward rates $f(t,T)$, and further that $f(t,T)$ is a continuous semimartingale in t whose volatility and drift satisfy certain conditions.

It is immediate that an HJM model is also an SR model. This is because if we take the short rate process r_t to be $f(t,t)$, then the SR-equation is just equation (19) of Heath, Jarrow and Morton (1992).

It is not so obvious that any SR model is also an HJM model. The bond prices might not even be T-differentiable, and the other HJM conditions are awkward to prove directly. We shall address this question and show that:

Theorem. *An SR model, satisfying the regularity condition* $\mathbb{E} \int_0^T |r_u| B_u^{-1} du < \infty$, *is also an HJM model.*

These models are both prescriptive of the interest-rate market – its form is determined by these definitions. It would be interesting to have a descriptive picture of the market, which would tell us whether there are any other models we could use. We will answer this too and show, in fact, that there are not. Before we can state the theorem, we need to say what a general interest-rate market adapted to the Brownian filtration looks like.

We will take as axioms that there is a market of discount bonds $P(t,T)$ such that for each maturity T,

(BM)
$$\left\{\begin{array}{l} \bullet \quad P(t,T) \text{ is positive for all } t \leqslant T \\[2ex] \bullet \quad P(T,T) = 1 \\[2ex] \bullet \quad \text{there is no arbitrage in the market} \\[2ex] \bullet \quad P(t,T) \text{ is a continuous semimartingale in } t. \end{array}\right.$$

We might call these the BM or bond market axioms, and it is immediate that all models of either SR or HJM type satisfy them. We shall further justify the axioms momentarily, but we can now state a theorem about such a market:

Theorem. *A model satisfying BM will have a short rate process r_t and, if the corresponding bank account process B_t is tradable or hedgeable, it is also an SR model.*

Some justification of the BM axioms is necessary to see that any useful model must have these characteristics. The positivity of $P(t,T)$ follows from the economic assumptions of the bond holder's limited liability, which keeps things non-negative, and of a utility function which is not completely indifferent to future rewards, which gives the strict positivity. Also the bond cannot default, which implies that $P(T,T)$ is exactly 1. The no-arbitrage condition speaks for itself.

If there is no arbitrage, work by Delbaen and Schachermayer (1994) proves that if a security is adapted (you can't see into the future), is right-continuous with left-limits (the right way round to avoid arbitrage profits from discontinuous shocks), and is locally bounded (doesn't explode) then firstly it is a semimartingale (their theorem 7.2) and secondly there is a measure under which it is a local martingale (their corollary 1.2). The security is then continuous, as are all Brownian local martingales. We can thus replace the continuous semimartingale condition by the equivalent requirement that

- $P(t, T)$ is locally bounded, adapted and right-continuous with left-limits.

Although there is a martingale measure for each bond individually, there is not yet a proof that there is a measure under which simultaneously all bonds are martingales (though work in preparation by Lowther will go a good way towards this). Until then, we shall assume that

- there is a measure which makes the (discounted) bond prices into martingales.

Ab initio, neither SR nor HJM models can claim to be truly general. The SR model assumes the existence of a short-rate process r_t, and the HJM model assumes the existence of the forward rates $f(t, T)$. We shall show that these assumptions are warranted and that any market of bonds satisfying BM, plus two regularity conditions, is both an SR model and an HJM model. This includes the burden of showing that a short-rate process exists, and that the bond prices are jointly measurable and T-differentiable.

In the next section we will lay out our main theorems. To show that a model is HJM we need to prove the joint measurability of some families of processes parameterised by maturity as well as a stochastic version of Fubini's theorem. The existing stochastic calculus literature does not seem to address our precise problem, and Section 3 contains the technical details of how they may be solved. The subsequent Section then uses those results to prove the original theorems.

Once we have understood the full generality of the interest-rate market it is possible to explore its wilder shores. Section 5 contains two examples of pathological behaviour of BM models. One is an example for which one of the regularity conditions fails, and the model is neither SR nor fully HJM. The other is a model which is both SR and HJM, some of whose bond prices have non-vanishing volatility as time approaches maturity.

2   Main results

As stated earlier, we shall show that any model adapted to a finite Brownian motion filtration and satisfying some technical constraints is both an SR and an HJM model. Our generality allows us to separate out properties of an interest-rate model which are absolutely essential for the mathematics to operate, and those which are just desirable for modelling or econometric purposes.

Notation note: to ease the proliferation of subscripts, it may be assumed that a summation sign without limits is being summed over the range 1 to n. For example, we will write $\sum_i \sigma_i \, dW_i$ for $\sum_{i=1}^n \sigma_i \, dW_i$.

For ease of proof, we shall also assume that all the semimartingales involved have absolutely continuous drifts.

We have three theorems.

Theorem 1. *Let $P(t,T)$ be a market of pure discount bonds under a measure $\mathbb{P}$, with the boundary condition that $P(T,T) = 1$ for every maturity date T. We assume that*

- *(BM1)for each maturity date T the process $P(t,T)$ is a positive-valued continuous semimartingale in t and is adapted to the filtration $\mathcal{F}_t$ of n-dimensional Brownian motion.*

- *(BM2)the market is 'arbitrage-free', in the sense that for any particular bond there is a measure equivalent to $\mathbb{P}$ under which the bonds (discounted by the chosen bond) are martingales.*

Then

- *(i)for each maturity date T, there exist $\mathcal{F}$-previsible processes $\Sigma_i(t,T)$ $(i = 1,\ldots,n)$ and $\alpha(t,T)$ such that $\int_0^T (|\Sigma(t,T)|^2 + |\alpha(t,T)|)\, dt < \infty$ and*

$$d_t P(t,T) = P(t,T)\Big(\sum_i \Sigma_i(t,T)\, dW_i(t) + \alpha(t,T)\, dt\Big).$$

- *(ii)fixing a maturity horizon date τ and choosing the bond $P(t,\tau)$ to be numeraire, there is a measure $\mathbb{P}^\tau$ equivalent to $\mathbb{P}$ and a $\mathcal{F}$-previsible n-vector $\gamma_i(t)$ $(i = 1,\ldots,n)$ such that*

$$\frac{d\mathbb{P}^\tau}{d\mathbb{P}}\Big|_{\mathcal{F}_t} = \exp\left(-\sum_i \int_0^t \gamma_i(s)\, dW_i(s) - \tfrac{1}{2}\int_0^t |\gamma(s)|^2\, ds\right),$$

and $W_i^\tau(t) = W_i(t) + \int_0^t \gamma_i(s)\, ds$ is $\mathbb{P}^\tau$-Brownian motion. Additionally, for every maturity date T

$$d_t P(t,T) = P(t,T)\Big(\sum_i \Sigma_i(t,T)\, dW_i^\tau(t) + \tilde{\alpha}(t,T)\, dt\Big),$$

where $\tilde{\alpha}(t,T) = \alpha(t,T) - \sum_i \gamma_i(t)\Sigma_i(t,T)$ is also a $\mathcal{F}$-previsible process whose integral $\int_0^T |\tilde{\alpha}(t,T)|\, dt$ is finite. The τ-bond discounted bond price

$$\frac{P(t,T)}{P(t,\tau)} \quad \text{is a } \mathbb{P}^\tau\text{-martingale for every } T.$$

- (iii)there is a $\mathcal{F}$-previsible process r_t such that $\int_0^T |r_t|\, dt < \infty$ and

$$d_t P(t,T) = P(t,T)\left(\sum_i \Sigma_i(t,T)\, dW_i^\tau(t) + \left(\sum_i \Sigma_i(t,T)\Sigma_i(t,\tau) + r_t\right) dt\right).$$

Also the process $B_t = \exp(\int_0^t r_s\, ds)$ exists and is absolutely continuous.

- (iv)there is a version of $P(t,T)$ which is jointly measurable and t-continuous, and such that

$$P(t,T) = P(t,\tau)\, \mathbb{E}_{\mathbb{P}^\tau}(P^{-1}(T,\tau) \mid \mathcal{F}_t).$$

Further there are jointly measurable versions of $\Sigma_i(t,T)$ and $\tilde{\alpha}(t,T)$.

It should be noted that while the chosen numeraire τ does provide a local time horizon for the market, it need not be an ultimate limit. The market can extend beyond τ, even up to infinity, but at or before τ a new numeraire must be picked to allow further progress.

Theorem 2. *Suppose that BM holds, then the market is an SR model if and only if the additional regularity condition holds; that*

(A1) $\qquad \zeta_t = \exp\left(-\sum_i \int_0^t \Sigma_i(s,\tau)\, dW_i^\tau(s) - \tfrac{1}{2}\int_0^t |\Sigma(s,\tau)|^2\, ds\right)$

is a uniformly integrable $\mathbb{P}^\tau$-martingale up to time τ (that is that $\mathbb{E}_{\mathbb{P}^\tau}(\zeta_\tau) = 1$, or it is sufficient that $\mathbb{E}_{\mathbb{P}^\tau} \exp \tfrac{1}{2}\int_0^\tau |\Sigma(t,\tau)|^2\, dt < \infty$).
This condition is also equivalent to each of the following

- (v)the τ-bond discounted value of the process B_t, $P^{-1}(t,\tau)B_t$ is a $\mathbb{P}^\tau$-martingale.

- (vi)there exists a measure $\mathbb{Q}$ equivalent to $\mathbb{P}$ and $\mathbb{P}^\tau$ such that $\tilde{W}_i(t)$ defined to be $W_i^\tau(t) + \int_0^t \Sigma_i(s,\tau)\, ds$ is $\mathbb{Q}$-Brownian motion and

$$d_t P(t,T) = P(t,T)\left(\sum_i \Sigma_i(t,T)\, d\tilde{W}_i(t) + r_t\, dt\right).$$

- (vii)there exists a measure $\mathbb{Q}$ equivalent to $\mathbb{P}$ such that the B-discounted bond prices $B_t^{-1}P(t,T)$ are $\mathbb{Q}$-martingales and the bond prices themselves can be written as the expectation

$$P(t,T) = \mathbb{E}_{\mathbb{Q}}\left(\exp(-\int_t^T r_u\, du) \,\Big|\, \mathcal{F}_t\right).$$

Theorem 3. *Suppose that BM and (A1) all hold and further that*

- (A2) the expectation $\mathbb{E}_{\mathbb{Q}}(\int_0^\tau |r_u| B_u^{-1}\, du)$ is finite, *then the market is an HJM model, in that*

- *(viii)the bond price $P(t,T)$ is absolutely continuous in T with $-\frac{\partial}{\partial T}P(t,t) = r_t$, and*

$$-\frac{\partial}{\partial T}\log P(t,T) = f(t,T) := \frac{\mathbb{E}_{\mathbb{Q}}(r_T \exp(-\int_t^T r_u\,du) \mid \mathcal{F}_t)}{P(t,T)}.$$

- *(ix)the process $f(t,T)$ is a semimartingale in t with SDE*

$$d_t f(t,T) = \sum_i \sigma_i(t,T)\,d\tilde{W}_i(t) - \sum_i \sigma_i(t,T)\Sigma_i(t,T)\,dt,$$

  where $\sigma_i(t,T)$ $(i=1,\ldots,n)$ is a $\mathcal{F}$-previsible process in t with

$$\int_0^T |\sigma(t,T)|^2\,dt < \infty.$$

- *(x) the bond volatilities $\Sigma_i(t,T)$ are absolutely continuous in T with $\Sigma_i(t,t) = 0$ and*

$$\Sigma_i(t,T) = -\int_t^T \sigma_i(t,u)\,du.$$

Theorem 1 describes what we say about the basic BM model with no additional regularity conditions. The conditions (BM1) and (BM2) are our formulation of the BM axioms. Parts (i) and (ii) recall familiar material. There is a link between a market being arbitrage-free and the existence of an equivalent measure, under which the discounted securities are martingales. Quite what this link is in the case of an infinite number of tradable securities is an open question, but the finite case has been explored widely from Harrison and Pliska (1981) to Delbaen and Schachermayer (1994). Here the market has a risk-neutral τ-forward measure, under which bond prices (discounted by the numeraire τ-bond) are martingales.

The theorem continues by showing that bond drift is a function only of the volatilities and a process (suggestively) labelled r_t. If the model is SR then r_t will be the short rate, but we do not know this for sure yet. At this stage the putative bank account process (or cash bond) B_t exists as a mathematical process, but may not be a tradable security. The burden of result (iv) is less the formula for $P(t,T)$ in terms of the τ-bond, than in the statements of the joint measurability of P, Σ_i and $\tilde{\alpha}$. These (or equivalent) measurabilities were assumed by HJM, but we have them as results.

Theorem 2 contains nothing new technically, but is included to make clear the role of r_t and the bond B_t. Mathematically, Theorem 1 posits the existence of the bonds $P(t,T)$ and doesn't specify the existence or tradability of B_t. In fact we showed in part (iii) that B_t did exist, but that didn't prove it was tradable. By the equivalence of lack of arbitrage and existence of a martingale measure, the bond B_t can be traded only if it is a martingale under the forward measure $\mathbb{P}^\tau$ (when discounted). As stated, the regularity condition (A1) is actually equivalent to results (v), (vi), (vii) and the tradability of B_t without arbitrage. Although almost all models do assume this condition, it needn't happen and a completely general model need not satisfy condition (A1). An example of a market where the cash bond is not a $\mathbb{P}^\tau$-martingale is given in Section five. If the condition does hold, then the bonds themselves can now be expressed as expectations of discount factors involving r_t, and the market is seen to be an SR model. The bonds discounted by the cash bond are $\mathbb{Q}$-martingales, but the new measure $\mathbb{Q}$ may depend on the τ originally chosen if the market is incomplete.

The main results of Theorem 3 are (viii) and (x) which prove that $P(t, T)$ and $\Sigma_i(t, T)$ respectively are absolutely continuous in T. In their paper, HJM assume the differentiability of $P(t, T)$ and that the resulting forward rates are semimartingales. Here we show that those assumptions are unnecessary. The condition (A2) makes the bond term-structure differentiable (at almost all maturities) and ensures that the forward rate process is well-behaved. It is this that actually lets us see the cash bond as the limit of holding very short-dated bonds and makes it tradable. What we have by the end of item (x) is an HJM model, as defined by Heath, Jarrow and Morton (1992). It should be noted that any model of the bond market without continuous yield curves, differentiable at almost all maturities, cannot be satisfying assumptions BM and (A1–2). We also note that it is sufficient that the interest rate r_t be bounded below by some constant, that is $r_t \geqslant -K$ for all $t \leqslant \tau$, for condition (A2) to hold.

We can now address the first theorem we stated in Section 1.

Corollary 4. *Suppose that r_t is a previsible process adapted to the Brownian motion filtered space $(\Omega, \mathcal{F}, \mathbb{P})$ such that $\int_0^\tau |r_t|\, dt$ is finite and that $\mathbb{E}\int_0^\tau |r_u| B_u^{-1}\, du$ is also finite, where $B_t = \exp(\int_0^t r_s\, ds)$. Then the market of bond prices $P(t, T)$, defined by*

$$P(t, T) = \mathbb{E}\left(\exp(-\textstyle\int_t^T r_u\, du)\,\big|\, \mathcal{F}_t\right),$$

is an arbitrage-free market satisfying the conditions BM and (A1–2) of the theorems and results (i) to (x).

In other words, an SR model satisfying (A2) is an HJM model.

3   Preliminary results

To prove the three theorems, we do require some technical results on the measurability and existence of some random processes, as well as a stochastic variant of Fubini's theorem. In all of what follows, we work with a one-dimensional Brownian motion W_t, its filtration $\mathcal{F}_t$, and its probability space

$(\Omega, \mathcal{F}, \mathbb{P})$. The one-factor case is presented purely for simplicity, and the obvious multi-factor versions of these results also hold.

Lemma 5. If $(A, \mathcal{A})$ is a measurable space and $X : A \rightarrow L^1(\Omega, \mathcal{F}_\tau)$, $a \mapsto X_a$, is a measurable function (giving L^1 the Borel σ-algebra induced by its norm), then there is a jointly measurable function F

$$F : A \times \Omega \rightarrow \mathbb{R},$$

such that $F(a, \omega)$ is a version of X_a for every a.

Proof of Lemma. We recall firstly that $L^1(\Omega, \mathcal{F}_\tau)$ is separable, because $L^2(\Omega, \mathcal{F}_\tau)$ is both separable itself (the filtration $\mathcal{F}_\tau$ has a countable basis, such as $\{W_q \leqslant q'\}$, $q \in \mathbb{Q} \cap [0, \tau]$, $q' \in \mathbb{Q}$) and is also a dense subspace of $L^1(\Omega, \mathcal{F}_\tau)$. Let (Y_n) denote a dense sequence in L^1, choosing a version $Y_n(\omega)$ of each one.

Then for any positive ϵ, we define the measurable index $n_\epsilon : L^1(\Omega, \mathcal{F}_\tau) \rightarrow \mathbb{N}$ by

$$n_\epsilon(X) = \inf\{n : \|X - Y_n\|_1 < \epsilon\}.$$

This lets us define an approximation to F as

$$F_\epsilon(a, \omega) = Y_{n_\epsilon(X_a)}(\omega),$$

which is certainly jointly measurable for every ϵ. For any a, there is a version $F_0(a, \omega)$ of X_a, which is ω-measurable, but F_0 is not necessarily a-measurable.

We can now use Markov's inequality to see that the set

$$A_\epsilon^a = \{\omega : |F_\epsilon(a, \omega) - F_0(a, \omega)| \geqslant \sqrt{\epsilon}\}$$

has size $\mathbb{P}(A_\epsilon^a) < \sqrt{\epsilon}$. Moving along the fast subsequence $\epsilon_n = 2^{-2n}$, we have that the set $A_0^a = \limsup_{n \rightarrow \infty} A_{\epsilon_n}^a$ is $\mathbb{P}$-null. For all ω not in A_0^a, $F_{\epsilon_n}(a, \omega)$ tends to $F_0(a, \omega)$. If we define

$$F(a, \omega) = \begin{cases} \lim_{n \rightarrow \infty} F_{\epsilon_n}(a, \omega) & \text{if this limit exists,} \\ 0 & \text{otherwise.} \end{cases}$$

Then $F(a, \omega)$ is jointly measurable, and is a version of X_a for every a.   □

Proposition 6. If $X : [0, \tau] \rightarrow L^1(\Omega, \mathcal{F}_\tau)$, $T \mapsto X_T$, is a measurable function, then there is a jointly measurable function

$$N : [0, \tau] \times [0, \tau] \times \Omega \rightarrow \mathbb{R},$$

such that $N(t, T, \omega)$ is a t-continuous version of the martingale

$$N(t, T) = \mathbb{E}(X_T \mid \mathcal{F}_t).$$

Proof of Proposition. The function

$$E : [0, \tau] \times L^1(\Omega, \mathcal{F}_\tau) \to L^1(\Omega, \mathcal{F}_\tau),$$

which takes (t, X) to $\mathbb{E}(X|\mathcal{F}_t)$ is continuous. This is because

$$\|E(t, X) - E(s, Y)\|_1 \leqslant \|X - Y\|_1 + \|E(t, X) - E(s, X)\|_1,$$

and the second term of the right-hand side tends to zero as s tends to t for any X because the process $E(t, X)$ is uniformly integrable and (almost surely) continuous in t. Thus the continuous function E is jointly measurable, and so too must be the function $\tilde{X}$,

$$\tilde{X} : [0, \tau] \times [0, \tau] \to L^1(\Omega, \mathcal{F}_\tau),$$

which takes (t, T) to $E(t, X_T)$. The Lemma 5 now applies to give a jointly measurable (but not necessarily t-continuous) function $\tilde{N}(t, T, \omega)$ which is a version of $\mathbb{E}(X_T|\mathcal{F}_t)$.

Following the notation of II.61 of Rogers and Williams (1994), the set A defined to be the set

$$\{(\omega, T) : \forall \epsilon > 0, \ \exists \delta > 0, \ \forall q, q' \in \mathbb{Q} \cap [0, \tau], \ |q - q'| < \delta \Rightarrow$$
$$|\tilde{N}(q, T, \omega) - \tilde{N}(q', T, \omega)| < \epsilon\}$$

is measurable and has sections $A^T = \{\omega : (\omega, T) \in A\}$ of size $\mathbb{P}(A^T) = 1$ for every T. If we define

$$N(t, T, \omega) = \begin{cases} \lim_{q \downarrow \downarrow t} \tilde{N}(q, T, \omega) & \text{if } (\omega, T) \in A, \\ 0 & \text{if } (\omega, T) \notin A, \end{cases}$$

then this is a measurable t-continuous modification of $\tilde{N}$. $\qquad\square$

Lemma 7. *Let $H^0 = H^0(\Omega, \mathcal{F}_\tau, \mathbb{P})$ be the space of $\mathcal{F}$-previsible processes ψ up to time τ such that $\int_0^\tau \psi_t^2 \, dt$ is finite almost surely, and give H^0 the (metric) topology under which ψ_n is defined to converge to ψ if*

$$\int_0^\tau (\psi_n(t) - \psi_t)^2 \, dt \to 0 \quad \text{in probability.}$$

Define the map $\Phi : L^1(\Omega, \mathcal{F}_\tau, \mathbb{P}) \to H^0(\Omega, \mathcal{F}_\tau, \mathbb{P})$ which takes the random variable X to the process $\Phi_t(X)$ which is the Brownian motion representation of $\mathbb{E}(X|\mathcal{F}_t)$, that is

$$\mathbb{E}(X|\mathcal{F}_t) = \mathbb{E}(X) + \int_0^t \Phi_s(X) \, dW_s.$$

Then the map Φ is continuous.

Proof of Lemma. Let M_t^X be the martingale $\mathbb{E}(X|\mathcal{F}_t)$. By Doob's submartingale inequality, the maximum process $M_*^X = \sup_{t \leqslant \tau} |M_t^X|$ satisfies

$$\mathbb{P}(M_*^X \geqslant a) \leqslant \frac{\|X\|_1}{a},$$

and thus $M_*^X \to 0$ in probability as X tends to 0 in L^1. Define F to be the moderate (and bounded) function $F(x) = x/(x+1)$, so that the Burkholder-Davis-Gundy inequality (see IV.42 of Rogers and Williams (1987)) holds:

$$\mathbb{E}F([M^X]_\tau^{\frac{1}{2}}) \leqslant C_F \, \mathbb{E}F(M_*^X),$$

for some constant C_F. Now $F(M_*^X)$ tends to zero in probability and is bounded by 1, so converges in L^1. So $F([M^X]_\tau^{\frac{1}{2}})$ converges in L^1 and hence in probability. As F is continuous, we deduce that

$$[M^X]_\tau \to 0 \quad \text{in probability},$$

which is the desired result. $\square$

Proposition 8. *Under the conditions of Proposition 6, there is a jointly measurable function*

$$\phi : [0, \tau] \times [0, \tau] \times \Omega \to \mathbb{R}$$

such that $\phi(t, T, \omega)$ is a version of the $\mathcal{F}$-previsible process $\Phi_t(X_T)$, and that

$$N(t, T) = N(0, T) + \int_0^t \phi(u, T) \, dW_u.$$

Proof of Proposition. Let $H^0(\Omega, \mathcal{F}_\tau, \mathbb{P})$ be as in Lemma 7. Recall our dense sequence Y_n in $L^1(\Omega, \mathcal{F}_\tau, \mathbb{P})$ and let $\psi_n(t, \omega)$ be a version of the process $\Phi_t(Y_n)$. Then for every positive ϵ the function $n_\epsilon : [0, \tau] \to \mathbb{N}$, defined by

$$n_\epsilon(T) = \min\{n : \|X_T - Y_n\| < \epsilon\}$$

is measurable. For each T, let $\phi_0(t, T)$ be the $\mathcal{F}$-previsible process $\Phi_t(X_T)$, which will not necessarily be T-measurable. Define our measurable approximation to ϕ as ϕ_ϵ, where

$$\phi_\epsilon(t, T, \omega) = \psi_{n_\epsilon(T)}(t, \omega).$$

For any T, the process $\phi_\epsilon(t, T)$ is just a version of $\Phi_t(Y_{n_\epsilon(X_T)})$ which tends in the H^0-topology to $\Phi_t(X_T)$ because Φ is continuous (by Lemma 7). As in Lemma 5, along a fast subsequence ϵ_n there will be almost sure convergence and we can define ϕ to be the limit

$$\phi(t, T, \omega) = \begin{cases} \lim_{n \to \infty} \phi_{\epsilon_n}(t, T, \omega) & \text{if this limit exists,} \\ 0 & \text{otherwise.} \end{cases}$$

This completes the proof. $\square$

Proposition 9 (Stochastic Fubini). *If* $X : [0, \tau] \to L^1(\Omega, \mathcal{F}_\tau)$, $u \mapsto X_u$, *is a measurable function such that* $\int_0^\tau |X_u| \, du$ *is also in* $L^1(\Omega, \mathcal{F}_\tau)$, *then firstly conditional expectation commutes with integration in that*

$$\mathbb{E}\left(\int_0^\tau X_u \, du \mid \mathcal{F}_t\right) = \int_0^\tau \mathbb{E}(X_u | \mathcal{F}_t) \, du,$$

and secondly the map Φ *of Lemma 7 also commutes with integration:*

$$\Phi_t\left(\int_0^\tau X_u \, du\right) = \int_0^\tau \Phi_t(X_u) \, du.$$

In other words, writing $\phi(t, u)$ *for a measurable version of* $\Phi_t(X_u)$, *and* Y *for the integral* $\int_0^\tau X_u \, du$, *then*

$$\mathbb{E}(Y|\mathcal{F}_t) - \mathbb{E}(Y) = \int_0^t \left(\int_0^\tau \phi(s, u) \, du\right) dW_s = \int_0^\tau \left(\int_0^t \phi(s, u) \, dW_s\right) du.$$

Proof of Proposition. Let us set $f(u)$ to be $\mathbb{E}(|X_u|)$. We know that $\int_0^\tau f(u) \, du$ is finite, so $f(u)$ must be finite for almost all times u. We can set X_u to be zero on the set of 'bad' u without loss of generality, and thus assume that $f(u)$ is finite for all times u. By Proposition 6, there is a jointly measurable function $N(t, u)$, which is a version of the martingale

$$N(t, u) = \mathbb{E}(X_u | \mathcal{F}_t).$$

We want to show that

$$\int_0^\tau N(t, u) \, du = \mathbb{E}\left(\int_0^\tau X_u \, du \mid \mathcal{F}_t\right).$$

The left-hand side above is $\mathcal{F}_t$-measurable and L^1-integrable. For any event A in $\mathcal{F}_t$

$$\mathbb{E}\left(I_A \int_0^\tau N(t, u) \, du\right) = \int_0^\tau \mathbb{E}(I_A N(t, u)) \, du = \int_0^\tau \mathbb{E}(I_A X_u) \, du.$$

The standard version of Fubini's theorem allows us to rewrite this as

$$\mathbb{E}\left(I_A \int_0^\tau X_u \, du\right).$$

which proves the first part of the result.

For the second part, we can define the stopping times

$$T_u^n = \inf\{t : |N(t, u)| \geqslant n\} \wedge \tau,$$

which are measurable in u. Then there is an approximation X^n to X given by

$$X_u^n = N(T_u^n, u),$$

which is jointly measurable and bounded. Because X_u^n is just X_u stopped at time T_u^n, it follows that

$$\Phi_t(X_u^n) = \Phi_t(X_u), \qquad t \leqslant T_u^n.$$

By the L^2-version of the stochastic Fubini theorem in Ikeda and Watanabe (1981), we can deduce that

$$\Phi_t\left(\int_0^\tau X_u^n\, du\right) = \int_0^\tau \Phi_t(X_u^n)\, du.$$

In addition, for each u the random variable X_u^n tends to X_u in $L^1(\Omega, \mathcal{F}_\tau)$ as n tends to infinity, because $\|X_u^n - X_u\|_1 \leqslant 2\mathbb{E}(|X_u|; T_u^n < \tau)$. As this upper bound converges (to zero) monotonically, we can also see that

$$\int_0^\tau X_u^n\, du \to \int_0^\tau X_u\, du \quad \text{in } L^1(\Omega, \mathcal{F}_\tau).$$

By Lemma 7, Φ is continuous, so

$$\Phi\left(\int_0^\tau X_u^n\, du\right) \to \Phi\left(\int_0^\tau X_u\, du\right) \quad \text{in } H^0(\Omega, \mathcal{F}_\tau).$$

Let $\Psi_n(t) = \Phi_t(\int_0^\tau X_u^n\, du) = \int_0^\tau \Phi_t(X_u^n)\, du = \int_{A_n(t)} \Phi_t(X_u)\, du$, where $A_n(t)$ is the set $\{u \in [0, \tau] : t < T_u^n\}$. Because $A_n(t)$ tends upwards to the whole of the interval $[0, \tau]$ as n tends to infinity, so $\Psi_n(t)$ converges to $\int_0^\tau \Phi_t(X_u)\, du$. (We actually have dominated convergence here because $\int_0^\tau |\Phi_t(X_u)|\, du$ is finite, seen by considering the related system $\tilde{X}$ which has $\Phi_t(\tilde{X}_u) = |\Phi_t(X_u)|$ and automatic (monotone) convergence.) Hence $\int_0^\tau \Phi_t(X_u)\, du$ is also equal to $\Phi_t(\int_0^\tau X_u\, du)$. $\qquad\square$

4   Proofs of the Theorems

We can now use our preliminary results from Section three to prove the results stated as theorems in the Section two.

Proof of Theorem 1.   Result (i). Because $P(t, T)$ is a semimartingale, the drift $\alpha(t, T)$ exists by our assumption that the drift is absolutely continuous. The volatilities $\Sigma_i(t, T)$ exist by the Brownian martingale representation theorem. See, for instance, IV.36.5 of Rogers and Williams (1987).

Result (ii) is a re-statement of assumption (BM2). The change of measure comes from the Cameron-Martin-Girsanov theorem. See IV.38.5(i) of Rogers and Williams (1987).

Result (iii). Using the SDE for $P(t, T)$ in part (ii) above, the SDE for $Z(t, T) = P(t, T)/P(t, \tau)$ is

$$d_t Z(t, T) = Z(t, T)\Big(\sum_i (\Sigma_i(t, T) - \Sigma_i(t, \tau))\, dW_i^\tau(t) + (r(t, T) - r(t, \tau))\, dt\Big),$$

where $r(t,T)$ is the $\mathcal{F}$-previsible process $\tilde{\alpha}(t,T) - \sum_i \Sigma_i(t,T)\Sigma_i(t,\tau)$, such that the integral $\int_0^T |r(t,T)|\,dt$ is finite. For this to be a $\mathbb{P}^\tau$-local martingale for every T, the drift term must be zero — in other words, every $r(t,T)$ must equal $r(t,\tau)$. Suppressing its dependence on the fixed maturity τ, we shall call this common value r_t. In other words, to be arbitrage free the drift must have the form

$$\tilde{\alpha}(t,T) = \sum_i \Sigma_i(t,T)\Sigma_i(t,\tau) + r_t,$$

from which the stated SDE follows. (The process r_t will only be fully independent of τ if the market is complete, which property we are not concerned with here.)

Result (iv). The formula follows from the assumption (BM2) that the discounted bond prices are $\mathbb{P}^\tau$-martingales. Joint measurability is harder, and follows from Proposition 6 applied to the function $T \mapsto P^{-1}(T,\tau)$. To show that this function is measurable it is enough to remark that it is the monotone limit of the functions $T \mapsto \max\{P^{-1}(T,\tau), K\}$ as K goes to infinity, which itself is continuous (and hence measurable) as a function from $[0,\tau]$ into $L^1(\Omega, \mathcal{F}_\tau)$. The Proposition then gives a jointly measurable and t-continuous function $N(t,T,\omega)$ which is a version of

$$N(t,T) = \mathbb{E}_{\mathbb{P}^\tau}(P^{-1}(T,\tau) \mid \mathcal{F}_t).$$

We then choose our version of $P(t,T)$ to be $P(t,\tau)N(t,T)$. The measurability of $\Sigma_i(t,T)$ and $\tilde{\alpha}(t,T)$ follows from Proposition 8. $\qquad\square$

Proof of Theorem 2. Result (v). The process $Z_t = P^{-1}(t,\tau)B_t$ has SDE

$$dZ_t = -Z_t \sum_i \Sigma_i(t,\tau)\,dW_i^\tau(t),$$

so that Z_t can be seen to just be a normalisation of ζ_t. (In fact $Z_t = \zeta_t/P(0,\tau)$.) Thus is (v) equivalent to (A1).

Result (vi) follows from (A1) as an application of the converse of the Cameron-Martin-Girsanov theorem for changing measure. For more details, see IV.38.5(ii) of Rogers and Williams (1987). Condition (A1) follows from (vi) by the C-M-G theorem proper.

Result (vii). It is immediate that SR and (vii) are equivalent, so all that remains is to link (vii) with (A1). To prove that (A1) is sufficient for (vii), we consider ζ_t to be the Radon-Nikodym derivative up to time t of $\mathbb{Q}$ with respect to $\mathbb{P}^\tau$. Then for any X in $L^1(\Omega, \mathcal{F}_T, \mathbb{P}^\tau)$

$$\mathbb{E}_{\mathbb{P}^\tau}(X|\mathcal{F}_t) = \zeta_t \mathbb{E}_{\mathbb{Q}}(\zeta_T^{-1}X|\mathcal{F}_t).$$

In particular, for X equal to $P^{-1}(T,\tau)$, then

$$\frac{P(t,T)}{P(t,\tau)} = \frac{B_t}{P(t,\tau)}\mathbb{E}_{\mathbb{Q}}(B_T^{-1} \mid \mathcal{F}_t),$$

which is the desired result (vii).

Conversely, if $B_t^{-1}P(t,\tau)$ is a $\mathbb{Q}$-martingale, for some $\mathbb{Q}$ equivalent to $\mathbb{P}$, then we can perform similar calculations to show that there is a forward measure $\tilde{\mathbb{P}}^\tau$ under which $P^{-1}(t,\tau)B_t$ is a martingale, and so result (v) holds which is equivalent to (A1). (We might have to go back and pick a different forward measure at result (ii) if the market is incomplete, but that doesn't present any serious problems.)   □

Proof of Theorem 3. Result (viii). Our basic approach will be to define the forward rates via expectations and show that their integral is a bond price, rather than trying to differentiate the bond prices directly.

By Proposition 6, there is a jointly measurable function $F(t,u)$ which, for each u, is a t-continuous version of the martingale

$$F(t,u) = \mathbb{E}_{\mathbb{Q}}(r_u B_u^{-1} \mid \mathcal{F}_t).$$

By the first part of Proposition 9 (Stochastic Fubini), we can integrate this with respect to u on the interval $[0,T]$

$$\int_0^T F(t,u)\,du = \mathbb{E}_{\mathbb{Q}}\left(\int_0^T r_u B_u^{-1}\,du \;\middle|\; \mathcal{F}_t\right) = 1 - \mathbb{E}_{\mathbb{Q}}(B_T^{-1} \mid \mathcal{F}_t).$$

Thus

$$P(t,T) = B_t\left(1 - \int_0^T F(t,u)\,du\right).$$

So that $P(t,T)$ is absolutely continuous in T and

$$-\frac{\partial}{\partial T}P(t,T) = B_t F(t,T).$$

Clearly, $F(t,t) = r_t B_t^{-1}$, so that $-\frac{\partial}{\partial T}P(t,t)$ is r_t. The forward rate $f(t,T)$ is just $B_t F(t,T)/P(t,T)$. Writing $Z(t,T)$ for $B_t^{-1}P(t,T)$, then $f(t,T) = F(t,T)/Z(t,T)$.

Result (ix). Obviously $F(t,T)$ is a $\mathbb{Q}$-martingale, so that f is a semimartingale. Let $\Lambda_i(t,T)$ be the volatility of $F(t,T)$ with respect to $\tilde{W}_i$, so that

$$d_t F(t,T) = \sum_i \Lambda_i(t,T)\,d\tilde{W}_i(t),$$

$$\text{and}\quad d_t Z(t,T) = Z(t,T)\sum_i \Sigma_i(t,T)\,d\tilde{W}_i(t).$$

By Proposition 8, we can choose a jointly measurable version of Λ_i, which will allow us to integrate it against T later. It can be deduced that

$$d_t f(t,T) = \sum_i \sigma_i(t,T)\,d\tilde{W}_i(t) - \sum_i \sigma_i(t,T)\Sigma_i(t,T)\,dt,$$

where $\sigma_i(t,T)$ is the $\mathcal{F}$-previsible process

$$\sigma_i(t,T) = Z^{-1}(t,T)(\Lambda_i(t,T) - F(t,T)\Sigma_i(t,T)).$$

Result (x). By the second part of Proposition 9 (Stochastic Fubini),

$$\int_0^T F(t,u)\,du - \int_0^T F(0,u)\,du = \sum_i \int_0^t \left(\int_0^T \Lambda_i(s,u)\,du \right) d\tilde{W}_i(s).$$

But by the proof of part (viii) the left-hand side above is $Z(0,T) - Z(t,T)$, where $Z(t,T) = B_t^{-1} P(t,T)$. This has SDE

$$d_t Z(t,T) = Z(t,T) \sum_i \Sigma_i(t,T)\,d\tilde{W}_i(t).$$

Hence

$$Z(t,T) \Sigma_i(t,T) = -\int_0^T \Lambda_i(t,u)\,du.$$

Now $Z(t,T)$ is absolutely continuous with derivative $-\frac{\partial}{\partial T} Z(t,T) = F(t,T)$, so $\Sigma_i(t,T)$ is absolutely continuous with derivative

$$-\frac{\partial}{\partial T} \Sigma_i(t,T) = Z^{-1}(t,T)(\Lambda_i(t,T) - F(t,T)\Sigma_i(t,T)),$$

which is just $\sigma_i(t,T)$. $\qquad\square$

The condition (A2), that $\mathbb{E}_{\mathbb{Q}}(\int_0^\tau |r_u| B_u^{-1}\,du) < \infty$, is required to enable application of the stochastic Fubini theorem. In the Heath, Jarrow and Morton (1992) paper, they use a condition (the last part of their C3) which is equivalent to requiring merely that the integral $\int_0^\tau |r_u| B_u^{-1}\,du$ is finite. It is an open question as to whether this is enough, but it is obvious that (A2) is more than sufficient, as there are some weaker pathwise conditions which can be proved to be sufficient.

Additionally, our models may fail the strict meaning of the HJM paper in another way. In particular, a model under Theorems 1–3 will not necessarily satisfy the second inequality of the HJM condition C2 in Heath, Jarrow and Morton (1992), though it will satisfy their inequality (7) which that condition is used to prove.

5 Two Examples

A market with a non-martingale cash bond

Pick a maturity τ, and let W_t be a $\mathbb{P}$-Brownian motion. (Here $\mathbb{P}$ will actually be the forward measure $\mathbb{P}^\tau$ as well.) Then we shall choose for our $P(t,\tau)$ to follow a Bessel(3) process up to some time $\tau_0 < \tau$. Details of this Bessel process can be found in, for example, VI.3 of Revuz and Yor (1994). (How $P(t,\tau)$ evolves after τ_0 will be immaterial.) Then $\mathbb{P}(t,\tau)$ has SDE

$$d_t P(t,\tau) = P(t,\tau)(P^{-1}(t,\tau)\,dW_t + P^{-2}(t,\tau)\,dt), \qquad t \leqslant \tau_0.$$

This gives the volatility $\Sigma(t, \tau) = P^{-1}(t, \tau)$ and drift $\tilde{\alpha}(t, \tau) = P^{-2}(t, \tau)$. We can define the rest of the market via result (iv) as

$$P(t, T) = P(t, \tau)\mathbb{E}(P^{-1}(T, \tau) \mid \mathcal{F}_t),$$

which is well-defined as $P^m(t, \tau)$ has finite expectation for all powers $m > -3$. This model satisfies the BM conditions, and has the r_t process equal to $r_t = \Sigma^2(t, \tau) - \tilde{\alpha}(t, \tau) = 0$. This is the nicest interest rate process we could hope for — it is non-negative, bounded and deterministic. But the discounted bond B_t is not a martingale. As $B_t = 1$, the process $P^{-1}(t, \tau)B_t$ is just the reciprocal of a Bessel(3) which by VI.33 of Rogers and Williams (1987) is only a local martingale and not a full martingale. The condition (A1) fails to hold.

Bond prices as Brownian bridges

We can show that even an SR/HJM-style model can have unusual properties. We will need to recall the notation that the left-limit and right-limit of any function f at x can be denoted by $f(x-)$ and $f(x+)$ respectively, or equivalently by $\lim_{y \uparrow x} f(y)$ and $\lim_{y \downarrow x} f(y)$ respectively.

Returning to the HJM model, for example, it is true that the volatility $\Sigma(t, T)$ is absolutely continuous in T and

$$\Sigma(t, t+) = 0,$$

that does not mean that the limit in the other direction

$$\Sigma(t-, t)$$

either exists or is zero. In other words, there is no mathematical reason (as opposed to economic reasons) that a bond's volatility should get smaller as it approaches maturity. If the volatility stays away from zero, we might expect the bond price to follow some sort of Brownian bridge path, but existing models tend to discount this possibility. The fact that non-vanishing volatility is possible has not been well appreciated in the literature. Hull and White (1993) remark that it would imply unbounded drifts, and don't make it clear that $\Sigma(T, T) = 0$ is insufficient to avoid this. Cheng (1991) does show that an exponential Brownian bridge cannot be a bond price in an arbitrage-free market, but in a setting which places restrictions on the interest rate r_t.

We can construct an arbitrage-free complete bond market in which some of the log bond prices behave as Brownian bridges with constant volatility.

Our measure throughout will be the martingale measure $\mathbb{Q}$. Let a be a positive constant, and fix a date τ. The τ-bond will be made to be a 'log-Brownian bridge', that is, $P(t, \tau) = \exp(X_t)$, where X_t $(0 \leqslant t \leqslant \tau)$ is a Brownian bridge from $-a\tau$ to 0. The process X has SDE

$$dX_t = \sigma \, dW_t - \frac{X_t}{\tau - t} \, dt,$$

whose solution simultaneously satisfies the two integral equations:

$$X_t + a\tau = \sigma W_t - \int_0^t \frac{X_s}{\tau - s}\, ds,$$

and

$$X_t + a\tau = at + \sigma(\tau - t) \int_0^t \frac{dW_s}{\tau - s}.$$

Setting r_t to be the previsible interest-rate process

$$r_t = \tfrac{1}{2}\sigma^2 - \frac{X_t}{\tau - t}, \qquad t < \tau,$$
$$r_\tau = 0,$$

our aim is to define a market via Corollary 4. We can re-express r_t, for $t < \tau$, as

$$r_t = \tfrac{1}{2}\sigma^2 + a - \sigma \int_0^t \frac{dW_s}{\tau - s}.$$

We can use this expression to calculate the variance of r_t. The reason for doing that is to be able to discover the asymptotic size of r_t as t nears τ. In fact $\|r_t\|_2 \sim \sigma(\tau - t)^{-\frac{1}{2}}$, which is t-integrable. Thus the first integral condition of Corollary 4 is satisfied. Then we can integrate r between the limits of t and T to get

$$\int_t^T r_u\, du = \tfrac{1}{2}\sigma^2(T - t) - \frac{T - t}{\tau - t} X_t - \sigma \int_t^T \frac{T - u}{\tau - u}\, dW_u, \quad t \leqslant T \leqslant \tau.$$

Let Z be the normal random variable $\int_t^T (T - u)/(\tau - u)\, dW_u$, which is the last term on the right-hand side above (without the factor of σ). This has variance

$$\mathrm{Var}(Z) = \int_t^T \left(\frac{T - u}{\tau - u}\right)^2 du = (T - t) + (\tau - T)\left(\frac{T - t}{\tau - t} + 2\log\frac{\tau - T}{\tau - t}\right).$$

In the case where $t = 0$, this variance is bounded by τ, and the mean of $\int_0^T r_s\, ds$ is bounded by $(\tfrac{1}{2}\sigma^2 + a)\tau$. Hence $\|B_t^{-1}\|_2$ is bounded, and so $\|r_t B_t^{-1}\|_1 \leqslant c(\tau - t)^{-\frac{1}{2}}$, which is t-integrable. So the other integral condition of Corollary 4 is satisfied.

Let us define the bond prices $P(t, T)$ to be $P(t, T) = \mathbb{E}(\exp - \int_t^T r_u\, du | \mathcal{F}_t)$ which gives

$$P(t, T) = \exp\left\{\frac{T - t}{\tau - t} X_t + \tfrac{1}{2}\sigma^2(\tau - T)\left(\frac{T - t}{\tau - t} + 2\log\frac{\tau - T}{\tau - t}\right)\right\}, \quad t \leqslant T < \tau,$$
$$P(t, \tau) = \exp(X_t), \qquad t \leqslant \tau.$$

We note that the price of the τ-bond is a log-Brownian bridge process. Extending the r_t process to the interval $(\tau, 2\tau]$ by taking an independent copy of the distribution of the process r on $[0, \tau]$, then

$$P(t, T) = \exp\left\{X_t - a(T - \tau) + \tfrac{1}{2}\sigma^2(2\tau - T)\left(\frac{T - \tau}{\tau} + 2\log\frac{2\tau - T}{\tau}\right)\right\},$$

$$t \leqslant \tau < T < 2\tau.$$

Taking the SDE of these bond price equations, we have that

$$d_t P(t, T) = P(t, T)(\Sigma(t, T)\, dW_t + r_t\, dt),$$

where Σ is the non-vanishing volatility function

$$\Sigma(t, T) = \begin{cases} \frac{T-t}{\tau-t}\sigma & t \leqslant T < \tau \\ \sigma & t < \tau \leqslant T \\ 0 & t = \tau \leqslant T. \end{cases}$$

Indeed the Σ tends to zero as maturity decreases, that is $\Sigma(\tau, \tau+) = \Sigma(\tau, \tau) = 0$, but not as time increases, because $\Sigma(\tau-, \tau) = \sigma$.

The forward rates also exist with $f(t, T) = -\frac{\partial}{\partial T}\log P(t, T)$, which is

$$f(t, T) = -\frac{X_t}{\tau - t} + \tfrac{1}{2}\sigma^2 + \sigma^2\left(\frac{T - t}{\tau - t} + \log\frac{\tau - T}{\tau - t}\right), \quad t \leqslant T < \tau$$

$$f(t, T) = a + \tfrac{1}{2}\sigma^2\frac{T}{\tau}, \quad t \leqslant \tau \leqslant T < 2\tau.$$

Their SDEs are

$$d_t f(t, T) = -\frac{\sigma}{\tau - t}\, dW_t + \sigma^2\frac{T - t}{(\tau - t)^2}\, dt, \quad t \leqslant T < \tau$$

$$d_t f(t, T) = 0, \quad t \leqslant \tau \leqslant T < 2\tau.$$

This is an HJM-style model with

$$\sigma(t, T) = -\frac{\sigma}{\tau - t}, \quad t \leqslant T < \tau$$

$$\sigma(t, T) = 0, \quad t \leqslant \tau \leqslant T < 2\tau,$$

We can confirm that the equations of results (ix) and (x) hold.

We notice that the behaviour of the process r_t, which is thought of as the short-term interest rate, is rather erratic in this case. In particular, $\sup_{t<\tau}|r_t|$ is infinite almost surely. We should remember that the bond prices and the cash bond however are quite well-behaved, and we never did posit even the existence of r_t initially.

In this case interest rates also go very negative with $\inf_{t<\tau} r_t = -\infty$. This need not happen though. For instance, if we define X_t to be a driftless bridge with non-vanishing volatility, which has SDE

$$dX_t = \max\left\{\epsilon, \frac{|X_t|}{\sqrt{\tau - t}}\right\}dW_t,$$

where ϵ is a positive constant, then $X_\tau = 0$ a.s. Then, with $P(t,\tau) = \exp(X_t)$, $\sigma_t = \max\{\epsilon, |X_t|/\sqrt{\tau-t}\}$ and $r_t = \frac{1}{2}\sigma_t^2$, we can create an SR/HJM market in a similar way, but with positive interest rates, though these too are unbounded.

We can even make some progress towards a general result linking non-vanishing volatilities with unbounded interest rates.

Proposition 10. *In an SR model (or equivalently, a general model satisfying BM and (A1)), for any maturity T satisfying $P(0,T) < 1$, it is impossible that both the volatility be bounded below and the short rate be bounded above.*

Proof of Proposition. (For simplicity, we take a one-factor model, but this makes no difference, as each bond in isolation is equivalent to single-factor model.) For a fixed maturity T, let σ_t be the bond volatility $\Sigma(t,T)$. Then rewriting result (vi) of Theorem 2, under the martingale measure, we have that

$$d_t P(t,T) = P(t,T)(\sigma_t\, dW_t + r_t\, dt).$$

To try and derive a contradiction, let us suppose that $\sigma_t \geq \epsilon$ for some positive ϵ and $|r_t| \leq K$ for some constant K, for all $t \leq T$. Then we set $\gamma_t = r_t/\sigma_t$, which is absolutely bounded as $|\gamma_t| \leq K/\epsilon$. The Cameron-Martin-Girsanov theorem applies to give a measure $\mathbb{Q}$ equivalent to $\mathbb{P}$ under which

$$\tilde{W}_t = W_t + \int_0^t \gamma_s\, ds \quad \text{is } \mathbb{Q}\text{-Brownian motion.}$$

Thus $d_t P(t,T) = P(t,T)\sigma_t\, d\tilde{W}_t$, and so $P(t,T)$ is a local $\mathbb{Q}$-martingale. As it is non-negative it is also a $\mathbb{Q}$-supermartingale in that

$$\mathbb{E}_{\mathbb{Q}}(P(t,T) \mid \mathcal{F}_s) \leq P(s,T), \qquad s \leq t \leq T.$$

Evaluating this inequality at $s = 0$, $t = T$ gives the contradiction $1 \leq P(0,T)$. $\square$

Note. This does not mean that a shrinking Σ and an unbounded $|r_t|$ are mutually exclusive. It is possible that both can happen at once. For instance, a market based on a log-bond price X_t, with SDE

$$dX_t = \sigma(T-t)\, dW_t - \frac{X_t}{(T-t)^\alpha}\, dt,$$

for any $\alpha > 2$, has $\Sigma(t,T) \downarrow 0$ as $t \uparrow T$, but $\limsup_{t \uparrow T} |r_t| = \infty$.

6   Conclusions

What we have done in our three principal theorems is to describe a model with as few conditions as possible. In doing so, two things became apparent.

Firstly, that (essentially) all models adapted to a finite-dimensional Brownian motion are in fact short-rate/Heath–Jarrow–Morton models, whether or not that was intended. This particularly underlines the generality and rigorous approach of the Heath, Jarrow and Morton paper (1992), and suggests that the model cannot be ignored by model makers, who *de facto* are working within it. At a rarified level, all such interest-rate models are just restrictions of the general HJM model and/or changes of its notation. At a practical level, however, a convenient notation for a sub-model may reveal the wood which the HJM trees obscure.

What we have not been about here is creating new models. The BM framework resembles the philosopher's ladder that we climb up only in order to be able to throw it away. The BM model, the most general we could think of, turns out just to be both the SR and the HJM models. This means not that we should phrase things in terms of the BM model, but rather that it is pointless to look outside SR/HJM for new Brownian based models. Model-makers should focus on restricting the SR/HJM model to sub-models which suit their particular needs.

Secondly, we have seen that many of the 'conditions' of the HJM model can be taken as proved theorems of the generalised model. In particular, the assumptions about the measurability and integrability of the forward rate volatilities and drifts; the existence of the interest rate process r_t and the regularity of the cash bond B_t; and the T-differentiability of the bond prices are not necessary conditions. Instead they follow from the arbitrage-free nature of the market and the two integrability conditions (A1–2) in the preambles to Theorems 2 and 3.

It is tempting and not entirely unjustified to deduce that, as the HJM model is no better than the SR model, it is pointless to work within its notation rather than the simpler framework of the short-rate. In the end, this is a question that others must decide, but one should beware one thing. In the multi-factor setting that we have worked with throughout, it is true that bond and option prices can be written in terms of the short rate. The price at time t of a claim X maturing at time T is

$$\mathbb{E}_{\mathbb{Q}}\Big(\exp(-\textstyle\int_t^T r_s\,ds)X \;\Big|\; \mathcal{F}_t\Big).$$

We must remember that although the discount factor is a function of the short-rate process, the claim X might not be and the filtration $\mathcal{F}_t$ almost certainly will not be. In the multi-factor setting, it is necessary to keep track of all the factors, and the HJM notation is set up to help with that, whereas the SR notation conceals it.

Further generalisations will come in weakening some of the conditions of the theorem. Most notably, in a market comprising an infinite number of securities we need some version of the results of Delbaen and Schacher-mayer (1994) giving the equivalence of a no-arbitrage condition and the existence of a martingale measure (BM2). There is also the question of completeness which we have not tried to address here.

Acknowledgements

I would like to record the debt I owe to Stanley Pliska for his support and the referee for his comments, which have both made the final version of this paper a great improvement on the former. I must also thank David Heath for his help with the stochastic Fubini's theorem, and Gareth Roberts for the example of the driftless bridge.

References

Cheng, S.T. (1991). On the Feasibility of Arbitrage-Based Option Pricing when stochastic bond price processes are involved. *J. Econ. Theory.* **53** 185–198.

Delbaen, F. and Schachermayer, W. (1994) The fundamental theorem of asset pricing. *Math. Ann.* **300** 463–520.

Harrison, J.M. and Pliska, S.R. (1981). Martingales and stochastic integrals in the theory of continuous trading. *Stoch. Procs. and their Applns.* **11** 215–260.

Heath, D., Jarrow, R. and Morton, A. (1992). Bond pricing and the term structure of interest rates: a new methodology for contingent claims valuation. *Econometrica.* **60** 77–105.

Hull, J. and White, A. (1993). One-factor interest-rate models and the valuation of interest-rate derivative securities. *J. Financial and Quantitative Analysis.* **28** 235–254.

Ikeda, N. and Watanabe, S. (1981). *Stochastic Differential Equations and Diffusion Processes.* North Holland.

Revuz, D. and Yor, M. (1994). *Continuous Martingales and Brownian Motion.* Springer Verlag.

Rogers, L.C.G. (1995). Which model for the term-structure of interest rates should one use ? In *Mathematical Finance* (ed. M.H.A.Davis, D.Duffie *et al*), IMA Volume 65. Springer Verlag, 93–116.

Rogers, L.C.G. and Williams, D. (1987). *Diffusions, Markov Processes and Martingales, Volume 2: Itô Calculus.* Wiley.

Rogers, L.C.G. and Williams, D. (1994). *Diffusions, Markov Processes and Martingales, Volume 1: Foundations.* Wiley.

Swap Derivatives in a Gaussian HJM Framework

Alan Brace and Marek Musiela

Abstract

The paper reviews a range of practical techniques to calculate prices of at the money European options, both domestic and cross-economy, within the Gaussian HJM framework.

1   Introduction

A lot is already known about Gaussian HJM models, and we warn readers that they will find little that is particularly new or original in this review chapter. Rather, we draw together information and results sufficient to calculate reliable prices of at-the money European options in a dealing environment.

The word 'reliable' needs some qualification. In this article, volatility is specified by a deterministic function, and we assume the user can fit that function to liquid market instruments. For example, the cap and swaption markets in the US, and historical correlation of the yield curve (the yield curve itself is calculated from cash, futures and swap rates). We assume also that the model is well-engineered in the sense that the chosen volatility function leads to robust hedging – see Brace and Musiela (1994) for ideas on this topic.

Under those circumstances, our judgement is that at-the-money prices, generated by formulae like the ones presented here, are commercially quotable. That seems to be the main advantage of the Gaussian HJM model: reliable at-the-money prices via formulae that can be mathematically derived and programmed reasonably quickly (i.e. within one or two days).

Most deals are, of course, done at- or near-the-money. As time passes they move either well in- or well out-of-the-money, and our comfort levels with Gaussian HJM models decline accordingly. A lognormal model, like that of Brace et al (1995), then becomes attractive for portfolio revaluation and risk management.

A potential objection to the supposedly practical nature of this paper is that a minority of over-the-counter deals are European, the majority have American or path-dependent features. We agree, but note that European options give bounds and analytic insight, and can be used as building blocks to simulate more complex options. Moreover, instruments that might be used for parametrisation, like options on swap rate spreads (see Subsection 2.4) to input correlation, will probably be European.

Furthermore, the Gaussian HJM model can often substitute for more complex models. For example, the price of a stock option (an easy exercise using the results of Section 2) is determined far more by the lognormality of its stock index, than by the normality or lognormality of the underlying interest rate model.

A standard technique used throughout this paper is to calculate expectations under the forward measure at the maturity of the option. We refer the reader to El Karoui et al (1995) for a comprehensive survey of this method which applied in Section 2 yields the price in the domestic economy of any option (stock, FX or interest rate) as an integral. Later sections examine vanilla, compound and exotic caps and swaptions and also options on swap spreads.

Note the affine approximation to the boundary function 'h' in Subsection 2.6 on compound swaptions. Gaussian densities effectively restrict integration to the interval [-5,5]; our experience is that many boundaries appearing in interest rate option integrals can be linearly approximated within that range with no significant loss of accuracy.

Forward measure techniques are extended to foreign economies in Section 3, and applied to differential swaps, basket caps and cross-currency swaptions in later sections.

The usual mathematical assumptions are made: all processes are defined on the probability space $(\Omega, \{\mathcal{F}_t; t \geq 0\}, \mathbf{P})$; the filtration $\{\mathcal{F}_t; t \geq 0\}$ is the $\mathbf{P}$−augmentation of the natural filtration generated by a d-dimensional Brownian motion $W = \{W(t); t \geq 0\}$; the discounted price processes are martingales under $\mathbf{P}$; we trade the domestic and foreign zero coupon bonds of all maturities and other domestic and foreign assets or indices of asset values; the market is frictionless.

2 Domestic Economy

We denote by $f(t, T)$ the **spot forward rate** at time t for time $T(t \leq T)$ and by $P(t, T)$ the time t price of a T maturity zero coupon bond. In terms of the forward rates $f(t, u), t \leq u \leq T$,

$$P(t, T) = \exp\left(-\int_t^T f(t, u) du\right) .$$

The instantaneous continuously compounded **spot rate** at date t is $r(t) = f(t, t)$. The price process of the **savings account** is

$$\beta(t) = \exp\left(\int_0^t r(s) ds\right) .$$

Under the measure **P**

$$
\begin{aligned}
df(t, T) &= \sigma(t, T)^* \int_t^T \sigma(t, u) du\ dt + \sigma(t, T)^* dW(t) \\
dP(t, T) &= P(t, T)\left(r(t) dt - \int_t^T \sigma(t, u)^* du\ dW(t)\right),
\end{aligned}
$$

where $\sigma(t, T), t \leq T$, is a **deterministic** locally bounded function with values in $\mathbb{R}^d$.

Let for $t \leq T \leq T_1$

$$F_T(t, T_1) = \frac{P(t, T_1)}{P(t, T)}$$

denote the **forward price** at time t for settlement at time T on a T_1 maturity zero coupon bond, then

$$dF_T(t, T_1) = -F_T(t, T_1) \int_T^{T_1} \sigma(t, u)^* du\left(\int_t^T \sigma(t, u) du\ dt + dW(t)\right) .$$

A T maturity **forward measure** $\mathbf{P}_T$ makes the forward price process $\{F_T(t, T_1);\ 0 \leq t \leq T\}$ a martingale. It is defined by

$$\mathbf{P}_T = \mathcal{E}(M_T(\cdot))(T)\mathbf{P} ,$$

where for $0 \leq t \leq T$

$$
\begin{aligned}
M_T(t) &= -\int_0^t \int_s^T \sigma(s, u)^* du\ dW(s), \\
\mathcal{E}(M_T(\cdot))(t) &= \exp\left(M_T(t) - \frac{1}{2} < M_T(\cdot) > (t)\right) .
\end{aligned}
$$

Under the measure $\mathbf{P}_T$ the process

$$W_T(t) = W(t) + \int_0^t \int_s^T \sigma(s, u) du\ ds$$

is a Brownian motion and

$$dF_T(t, T_1) = -F_T(t, T_1) \int_T^{T_1} \sigma(t, u)^* du\ dW_T(t) .$$

Therefore

$$P(T, T_1) = F_T(T, T_1)$$
$$= F_T(t, T_1) \exp\left(-\int_t^T \int_T^{T_1} \sigma(s, u)^* du \, dW_T(s) - \frac{1}{2} \int_t^T \left|\int_T^{T_1} \sigma(s, u) du\right|^2 ds\right).$$

Hence

$$\text{Var}\left(\log P(T, T_1)|\mathcal{F}_t\right) = \text{Var}_T(\log P(T, T_1)|\mathcal{F}_t)$$
$$= \int_t^T \left|\int_T^{T_1} \sigma(s, u) du\right|^2 ds,$$

where Var_T stands for the variance under the measure $\mathbf{P}_T$.

The **futures price** at time t in the contract with expiry date $T(t \leq T)$ on a zero coupon bond with maturity $T_1(T \leq T_1)$ is

$$G_T(t, T_1) = E(P(T, T_1)|\mathcal{F}_t).$$

It follows (cf. Jamshidian 1993) that

$$G_T(t, T_1) = F_T(t, T_1) \exp\left(-\int_t^T \int_T^{T_1} \sigma(s, u)^* du \int_s^T \sigma(s, u) du \, ds\right)$$

and hence also that

$$dG_T(t, T_1) = -G_T(t, T_1) \int_T^{T_1} \sigma(t, u)^* du \, dW(t).$$

But (cf. Brace and Musiela 1993, p. 6)

$$\int_0^T r(t) dt = \int_0^T f(0, u) du + \frac{1}{2} \int_0^T \left|\int_s^T \sigma(s, u) du\right|^2 ds$$
$$+ \int_0^T \int_s^T \sigma(s, u)^* du \, dW(s)$$

and consequently

$$\text{Cov}\left(\log P(T, T_1), \log \beta(T)|\mathcal{F}_t\right) = -\int_t^T \int_T^{T_1} \sigma(s, u)^* du \int_s^T \sigma(s, u) du \, ds.$$

In general, if the price process $\{X(t); t \geq 0\}$ of an asset follows

$$dX(t) = X(t)\left(r(t) dt + \eta(t)^* dW(t)\right)$$

then the corresponding processes of **forward and futures** prices, denoted by $F_T^X(t)$ and $G_T^X(t)$, $0 \le t \le T$ respectively, satisfy $F_T^X(t) = \frac{X(t)}{P(t,T)}$, $G_T^X(t) = E(X(T)|\mathcal{F}_t)$ and

$$dF_T^X(t) = F_T^X(t) \left(\int_t^T \sigma(t,u)du + \eta(t) \right)^* \left(\int_t^T \sigma(t,u)du\ dt + dW(t) \right),$$

$$dG_T^X(t) = G_T^X(t) \left(\int_t^T \sigma(t,u)du + \eta(t) \right)^* dW(t).$$

Using the Ito formula it follows that

$$d\log(F_T^X(t)/G_T^X(t)) = \left(\int_t^T \sigma(t,u)du + \eta(t) \right)^* \int_t^T \sigma(t,u)du\ dt$$

and therefore because $F_T^X(T) = G_T^X(T) = X(T)$ also

$$\log(F_T^X(t)/G_T^X(t)) = - \int_t^T \left(\int_s^T \sigma(s,u)du + \eta(s) \right)^* \int_s^T \sigma(s,u)du\ ds.$$

Finally

$$X(T) = F_T^X(T) = F_T^X(t) \exp \left(\int_t^T \left(\int_s^T \sigma(s,u)du + \eta(s) \right)^* dW_T(s) \right.$$
$$\left. - \frac{1}{2} \int_t^T \left| \int_s^T \sigma(s,u)du + \eta(s) \right|^2 ds \right)$$

and hence

$$\mathrm{Cov}\ (\log X(T),\ \log \beta(T)|\mathcal{F}_t)$$
$$= \int_t^T \left(\int_s^T \sigma(s,u)du + \eta(s) \right)^* \int_s^T \sigma(s,u)du\ ds$$

or equivalently

$$G_T^X(t) = F_T^X(t) \exp \Big(\mathrm{Cov}\ (\log X(T), \log \beta(T)|\mathcal{F}_t) \Big).$$

Pricing of derivatives involves calculations of the expectation

$$E \left(\frac{\beta(t)}{\beta(T)} \xi | \mathcal{F}_t \right),$$

where ξ is a $\mathcal{F}_T$ measurable random variable. If $E|\xi|^p < \infty$ for some $p > 1$ then (cf. Brace and Musiela 1993)

$$E \left(\frac{\beta(t)}{\beta(T)} \xi | \mathcal{F}_t \right) = P(t,T) E_T(\xi | \mathcal{F}_t),$$

where E_T stands for the expectation under the measure $\mathbf{P}_T$. In particular, if

$$\xi = g(X_1(T), \ldots, X_n(T)) \, ,$$

where $g : \mathbb{R}^n \to \mathbb{R}$ and the price processes $\{X_i(t); \ t \geq 0\}$, $i = 1, \ldots, n$ satisfy

$$dX_i(t) = X_i(t)\Big(r(t)dt + \eta_i(t)^*dW(t)\Big)$$

then

$$
\begin{aligned}
X_i(T) = F_T^{X_i}(T) \;\; = \;\; & F_T^{X_i}(t)\exp\left(\int_t^T \left(\int_s^T \sigma(s,u)du + \eta_i(s)\right)^* dW_T(s) \right. \\
& \left. -\frac{1}{2}\int_t^T \left|\int_s^T \sigma(s,u)du + \eta_i(s)\right|^2 ds\right) \, .
\end{aligned}
$$

But the $\mathcal{F}_t$ conditional distribution under the measure $\mathbf{P}_T$ of the vector

$$
\left(\int_t^T \left(\int_s^T \sigma(s,u)du + \eta_1(s)\right)^* dW_T(s), \ldots, \right.
$$
$$
\left. \int_t^T \left(\int_s^T \sigma(s,u)du + \eta_n(s)\right)^* dW_T(s)\right)^*
$$

is $N(0, \triangle)$ with $\triangle = ((\triangle_{ij}))$, where

$$
\begin{aligned}
\triangle_{ij} \;\; = \;\; & \mathrm{Cov}\left(\log X_i(T), \log X_j(T)|\mathcal{F}_t\right) \\
= \;\; & \int_t^T \left(\int_s^T \sigma(s,u)du + \eta_i(s)\right)^* \left(\int_s^T \sigma(s,u)du + \eta_j(s)\right) ds \, .
\end{aligned}
$$

Changing variables and identifying a $k \times n$ matrix $\Gamma = [\gamma_1, \ldots, \gamma_n]$ such that $\triangle = \Gamma^*\Gamma$ leads to

$$
\begin{aligned}
& E\left(\frac{\beta(t)}{\beta(T)}\xi|\mathcal{F}_t\right) \\
& = E\left(\frac{\beta(t)}{\beta(T)}g(X_1(T), \ldots, X_n(T))|\mathcal{F}_t\right) \\
& = P(t,T)E_T\left(g(F_T^{X_1}(T), \ldots, F_T^{X_n}(T))|\mathcal{F}_t\right) \\
& = P(t,T)\int_{\mathbb{R}^k} g\left(F_T^{X_1}(t)e^{\gamma_1^* x - \frac{1}{2}|\gamma_1|^2}, \ldots, F_T^{X_n}(t)e^{\gamma_n^* x - \frac{1}{2}|\gamma_n|^2}\right) n_k(x)dx \\
& = P(t,T)\int_{\mathbb{R}^k} g\left(\frac{X_1(t)n_k(x+\gamma_1)}{P(t,T)n_k(x)}, \ldots, \frac{X_n(t)n_k(x+\gamma_n)}{P(t,T)n_k(x)}\right) n_k(x)dx,
\end{aligned}
$$

where

$$n_k(x) = (2\pi)^{-\frac{k}{2}}\exp\left(-\frac{1}{2}|x|^2\right), \quad |x|^2 = \sum_{i=1}^k x_i^2 \, .$$

The **decomposition** $\triangle = \Gamma^*\Gamma$, where $\Gamma = [\gamma_1, \ldots, \gamma_n]$ is a $k \times n$ matrix can be obtained from the eigenvalues and the eigenvectors of the matrix $\triangle$. Namely, if $\delta_1, \ldots, \delta_n$ are the **eigenvalues** and $v_1, \ldots, v_n$ the corresponding **orthonormal eigenvectors** of $\triangle$ then with

$$D = \text{diag}\, (\delta_1, \ldots, \delta_n), \ V = [v_1, \ldots, v_n]$$

we have

$$\triangle = VDV^* = VD^{\frac{1}{2}}(VD^{\frac{1}{2}})^* = \sum_{i=1}^n \delta_i v_i v_i^* \ .$$

Assume that $\delta_1 \geq \delta_2 \geq \ldots \geq \delta_k$, $k \leq n$ are positive and $\delta_{k+1} = \ldots = \delta_n = 0$. Then

$$VD^{\frac{1}{2}} = [\sqrt{\delta_1} v_1, \ldots, \sqrt{\delta_k} v_k, 0, \ldots, 0]$$

and with

$$\Gamma^* = [\sqrt{\delta_1} v_1, \ldots, \sqrt{\delta_k} v_k]$$

we have

$$\triangle = \Gamma^*\Gamma$$

Hence, we can define γ_i as the ith row of Γ^*, i.e.

$$\gamma_i = \begin{pmatrix} \sqrt{\delta_1} v_{i1} \\ \vdots \\ \sqrt{\delta_k} v_{ik} \end{pmatrix} \ .$$

2.1   Forward swaps and forward swap rates

In a **payer** (receiver) **forward start swap** the fixed rate is paid (received) and the floating rate is received (paid). Swaps can be **settled in arrears** or in advance.

Consider a payer forward swap on principal 1 settled quarterly in arrears. The floating rate $L(T_j)$ received at time $T_{j+1}, j = 0, \ldots, n-1$, is set at time T_j by the reference to the zero coupon over that period i.e.,

$$P(T_j, T_{j+1})^{-1} = 1 + L(T_j)(T_{j+1} - T_j) \ .$$

The swap cash flows at times $T_j, j = 1, \ldots, n$, where $T_j - T_{j-1} = \delta$ ($\delta = .25$), $T_0 = T$, are $L(T_{j-1})\delta$ and $-\kappa\delta$. Therefore the time $t(t \leq T)$ **value of the forward swap** is

$$E\left(\sum_{j=1}^n \frac{\beta(t)}{\beta(T_j)}(L(T_{j-1}) - \kappa)\delta|\mathcal{F}_t\right)$$

$$= \sum_{j=1}^n E\left(\frac{\beta(t)}{\beta(T_j)}(P(T_{j-1}, T_j)^{-1} - (1 + \kappa\delta))|\mathcal{F}_t\right)$$

$$= \sum_{j=1}^{n} E\left(\frac{\beta(t)}{\beta(T_{j-1})} P(T_{j-1}, T_j)^{-1} E\left(\frac{\beta(T_{j-1})}{\beta(T_j)}|\mathcal{F}_{T_{j-1}}\right)|\mathcal{F}_t\right)$$

$$- \sum_{j=1}^{n}(1+\kappa\delta)E\left(\frac{\beta(t)}{\beta(T_j)}|\mathcal{F}_t\right)$$

$$= \sum_{j=1}^{n}\left(P(t, T_{j-1}) - (1+\kappa\delta)P(t, T_j)\right)$$

$$= P(t, T_0) - \sum_{j=1}^{n} C_j P(t, T_j) ,$$

where $C_j = \kappa\delta$ for $j = 1, \ldots, n-1$ and $C_n = 1 + \kappa\delta$.

In the equivalent **swap settled in advance** the discounting method vary from country to country. In Australia the cash flows at times $T_j, j = 0, \ldots, n-1$ are $L(T_j)\delta(1+L(T_j)\delta)^{-1}$ and $-\kappa\delta(1+\kappa\delta)^{-1}$. Hence the time $t(t \le T = T_0)$ value of the swap is

$$E\left(\sum_{j=0}^{n-1} \frac{\beta(t)}{\beta(T_j)}\left(L(T_j)\delta(1+L(T_j)\delta)^{-1} - \kappa\delta(1+\kappa\delta)^{-1}\right)|\mathcal{F}_t\right)$$

$$= E\left(\sum_{j=0}^{n-1} \frac{\beta(t)}{\beta(T_j)}\left((1+\kappa\delta)^{-1} - (1+L(T_j)\delta)^{-1}\right)|\mathcal{F}_t\right)$$

$$= (1+\kappa\delta)^{-1} \sum_{j=0}^{n-1}\left(P(t, T_j) - (1+\kappa\delta)E\left(\frac{\beta(t)}{\beta(T_j)} P(T_j, T_{j+1})|\mathcal{F}_t\right)\right)$$

$$= (1+\kappa\delta)^{-1} \sum_{j=1}^{n}\left(P(t, T_{j-1}) - (1+\kappa\delta)P(t, T_j)\right)$$

which is the discounted with the fixed rate κ value of the swap settled in arrears. In the US the cash flows at times $T_j, j = 0, \ldots, n-1$ are $L(T_j)\delta(1+L(T_j)\delta)^{-1}$ and $-\kappa\delta(1+L(T_j)\delta)^{-1}$. Therefore the time t value of the swap is

$$E\left(\sum_{j=0}^{n-1} \frac{\beta(t)}{\beta(T_j)}(L(T_j) - \kappa)\delta(1+L(T_j)\delta)^{-1}|\mathcal{F}_t\right)$$

$$= E\left(\sum_{j=0}^{n-1} \frac{\beta(t)}{\beta(T_j)}(L(T_j) - \kappa)\delta P(T_j, T_{j+1})|\mathcal{F}_t\right)$$

$$= E\left(\sum_{j=0}^{n-1} \frac{\beta(t)}{\beta(T_{j+1})}(L(T_j) - \kappa)\delta|\mathcal{F}_t\right)$$

which is the value of the swap settled in arrears. The same approach is also used in many European markets.

The **forward swap rate** $\omega_T(t, n)$ at time t for maturity T is that value of the fixed rate κ which makes the value of the forward swap zero, i.e.,

$$\omega_T(t, n) = \left(\delta \sum_{j=1}^{n} P(t, T_j) \right)^{-1} (P(t, T) - P(t, T_n))$$

A **swap** (**swap rate**) is the forward swap (forward swap rate) with $t = T$.

2.2   Forward caps and floors

A **forward start cap** (floor) is a strip of caplets, each of which is a call (put) option on a forward rate. Similarly to swaps, caps and floors may be settled in arrears or in advance.

In a forward cap and floor on principal $R = 1$ settled in arrears at times $T_j, j = 1, \ldots, n$, where $T_j - T_{j-1} = \delta, T_0 = T$, the cash flows at times T_j are

$$(L(T_{j-1}) - \kappa)^+ \delta \quad \text{and} \quad (\kappa - L(T_{j-1}))^+ \delta,$$

respectively. The forward rate $L(T_{j-1})$ is set at time T_{j-1} and satisfies

$$P(T_{j-1}, T_j)^{-1} = 1 + L(T_{j-1})(T_j - T_{j-1}) .$$

The **cap price** at time $t \leq T_0$ is (cf. Brace and Musiela 1994)

$$
\begin{aligned}
\text{Cap}(t) &= \sum_{j=1}^{n} E\left(\frac{\beta(t)}{\beta(T_j)} (L(T_{j-1}) - \kappa)^+ \delta | \mathcal{F}_t \right) \\
&= \sum_{j=0}^{n-1} \left(P(t, T_j) N(-h_j(t)) - (1 + \kappa\delta) P(t, T_{j+1}) N(-h_j(t) - \zeta_j(t)) \right) ,
\end{aligned}
$$

where

$$h_j(t) = \left(\log \frac{(1 + \kappa\delta) P(t, T_{j+1})}{P(t, T_j)} - \frac{1}{2} \zeta_j(t)^2 \right) / \zeta_j(t)$$

and

$$\zeta_j(t)^2 = \text{Var} \left(\log P(T_j, T_{j+1}) | \mathcal{F}_t \right) = \int_t^{T_j} \left| \int_{T_j}^{T_{j+1}} \sigma(s, u) du \right|^2 ds .$$

The **floor price** at time $t \leq T_0$ is

$$\sum_{j=1}^{n} E\left(\frac{\beta(t)}{\beta(T_j)} (\kappa - L(T_{j-1}))^+ \delta | \mathcal{F}_t \right)$$

and because

$$(\kappa - L(T_{j-1}))^+ \delta = (L(T_{j-1}) - \kappa)^+ \delta - (L(T_{j-1}) - \kappa)\delta$$

it follows from the formulae for caps and swaps that

$$
\begin{aligned}
\text{Floor}\,(t) &= \sum_{j=1}^{n} E\left(\frac{\beta(t)}{\beta(T_j)}(\kappa - L(T_{j-1}))^+\delta|\mathcal{F}_t\right) \\
&= \sum_{j=0}^{n-1}\left(P(t,T_j)N(-h_j(t)) - (1+\kappa\delta)P(t,T_{j+1})N(-h_j(t) - \zeta_j(t))\right) \\
&\quad - \sum_{j=0}^{n-1}\left(P(t,T_j) - (1+\kappa\delta)P(t,T_{j+1})\right) \\
&= \sum_{j=0}^{n-1}\left((1+\kappa\delta)P(t,T_{j+1})N(h_j(t) + \zeta_j(t)) - P(t,T_j)N(h_j(t))\right)
\end{aligned}
$$

The above explains the well known parity between caps and floors which says that

$$\textbf{Forward Cap} - \textbf{Forward Floor} = \textbf{Forward Swap.}$$

In Australia the price of the equivalent forward cap (floor) settled in advance is equal to the discounted with the strike rate κ price of the cap (floor) settled in arrears. In the US the prices do not depend on the settlement method. A **cap (floor)** is the forward cap (forward floor) with $t = T$.

2.3 Swaptions

The owner of a **payer** (receiver) **swaption** at strike κ maturing at time $T = T_0$ has the right to receive at time T the cash flows of the corresponding forward payer (receiver) swap settled in arrears. Hence the time $t \leq T$ **price of the swaption** is

$$
E\left(\frac{\beta(t)}{\beta(T)}\left(E\left(\sum_{j=1}^{n}\frac{\beta(T)}{\beta(T_j)}(L(T_{j-1}) - \kappa)\delta \mid \mathcal{F}_T\right)\right)^+ |\mathcal{F}_t\right)
$$

and

$$
E\left(\frac{\beta(t)}{\beta(T)}\left(E\left(\sum_{j=1}^{n}\frac{\beta(T)}{\beta(T_j)}(\kappa - L(T_{j-1}))\delta|\mathcal{F}_T\right)\right)^+ |\mathcal{F}_t\right)
$$

for the **payer** and the **receiver**, respectively. It follows easily that

$$\textbf{Payer swaption} - \textbf{Receiver swaption} = \textbf{Forward swap.}$$

A **payer** (receiver) **swaption** can also be viewed as a **portfolio of call** (put) **options on a swap rate**, which can not be exercised separately. The owner receives at time T the discounted from time $T_j, j = 1, \ldots, n$ to T value of the cash flows defined by

$$
(\omega_T(T, n) - \kappa)^+\delta
$$

for the payer and by

$$(\kappa - \omega_T(T, n))^+ \delta$$

for the receiver option, where

$$\omega_T(T, n) = \left(\delta \sum_{j=1}^{n} P(T, T_j)\right)^{-1} (1 - P(T, T_n))$$

is the corresponding swap rate. The time $t \leq T$ **price of the payer option** is

$$E\left(\frac{\beta(t)}{\beta(T)} E\left(\sum_{j=1}^{n} \frac{\beta(T)}{\beta(T_j)}(\omega_T(T, n) - \kappa)^+ \delta|\mathcal{F}_T\right)|\mathcal{F}_t\right)$$

$$= E\left(\frac{\beta(t)}{\beta(T)}\left(1 - \sum_{j=1}^{n} C_j P(T, T_j)\right)^+ |\mathcal{F}_t\right)$$

$$= E\left(\frac{\beta(t)}{\beta(T)}\left(E\left(\sum_{j=1}^{n} \frac{\beta(T)}{\beta(T_j)}(L(T_{j-1}) - \kappa)\delta|\mathcal{F}_T\right)\right)^+ |\mathcal{F}_t\right)$$

which is the payer swaption price. Similarly, for all $t \leq T$, the receiver option price is

$$E\left(\frac{\beta(t)}{\beta(T)} E\left(\sum_{j=1}^{n} \frac{\beta(T)}{\beta(T_j)}(\kappa - \omega_T(T, n))^+ \delta|\mathcal{F}_T\right)|\mathcal{F}_t\right)$$

$$E\left(\frac{\beta(t)}{\beta(T)}\left(\sum_{j=1}^{n} C_j P(T, T_j) - 1\right)^+ |\mathcal{F}_t\right)$$

$$E\left(\frac{\beta(t)}{\beta(T)}\left(E\left(\sum_{j=1}^{n} \frac{\beta(T)}{\beta(T_j)}(\kappa - L(T_{j-1}))\delta|\mathcal{F}_T\right)\right)^+ |\mathcal{F}_t\right)$$

which equals the price of the receiver swaption. As before $C_j = \kappa\delta$, $j = 1, \ldots, n-1$, $C_n = 1 + \kappa\delta$.

The above shows that a **payer** (receiver) **swaption is a put** (call) **option on a coupon bond** with the coupon rate equal to the strike κ. To compute the price we can use the general formula for pricing of derivatives with the functions

$$g(x_1, \ldots, x_n) = \left(1 - \sum_{j=1}^{n} C_j x_j\right)^+, C_j = \kappa\delta, j = 1, \ldots, n-1, \ C_n = 1 + \kappa\delta\,,$$

and

$$g(x_1, \ldots, x_n) = \left(\sum_{j=1}^{n} C_j x_j - 1\right)^+.$$

for the payer and the receiver swaptions, respectively. The time $t \leq T$ price of the payer (receiver) swaption is

$$\int_{\mathbb{R}^k} \left(\lambda \left(P(t,T) n_k(x) - \sum_{i=1}^{n} C_i P(t,T_i) n_k(x + \gamma_i) \right) \right)^+ dx$$

with $\lambda = 1$ for the payer and $\lambda = -1$ for the receiver, respectively, and

$$n_k(x) = (2\pi)^{-\frac{k}{2}} \exp\left(-\frac{1}{2}|x|^2\right), \quad |x|^2 = \sum_{i=1}^{k} x_i^2; \quad n_1(x) = n(x) .$$

If the corresponding covariance matrix $\triangle$ is of rank 2 and $s = s(y)$ satisfies

$$\sum_{i=1}^{n} C_i P(t,T_i) n(s(y) + \gamma_{i1}) n(y + \gamma_{i2}) = P(t,T) n(s(y)) n(y) ,$$

then the payer and receiver swaption prices are, respectively,

$$
\begin{aligned}
\text{Payer}(t) \;=\; & P(t,T) \int_{-\infty}^{\infty} N(-s(y)) n(y) dy \\
& - \sum_{i=1}^{n} C_i P(t,T_i) \int_{-\infty}^{\infty} N(-s(y - \gamma_{i2}) - \gamma_{i1}) n(y) dy
\end{aligned}
$$

and

$$
\begin{aligned}
\text{Receiver}(t) \;=\; & \sum_{i=1}^{n} C_i P(t,T_i) \int_{-\infty}^{\infty} N(s(y - \gamma_{i2}) + \gamma_{i1}) n(y) dy \\
& - P(t,T) \int_{-\infty}^{\infty} N(s(y)) n(y) dy .
\end{aligned}
$$

If the boundary $s(y)$ can be approximated by a line going through points $(y_1, s(y_1))$ and $(y_2, s(y_2))$, then

$$s(y) \simeq ay + b$$

with

$$a = \frac{s(y_2) - s(y_1)}{y_2 - y_1}, \quad b = \frac{s(y_1) y_2 - s(y_2) y_1}{y_2 - y_1}$$

and because

$$
\begin{aligned}
& \int_{-\infty}^{\infty} N(s(y - \alpha) + \beta) n(y) dy \\
& \simeq \int_{-\infty}^{\infty} N(ay + b + \beta - a\alpha) n(y) dy = N\left(\frac{b + \beta - a\alpha}{\sqrt{1 + a^2}}\right)
\end{aligned}
$$

we obtain

$$
\begin{aligned}
\text{Payer}(t) \;=\; & P(t,T) N(-h_0) - \sum_{i=1}^{n} C_i P(t,T_i) N(-h_i) , \\
\text{Receiver}(t) \;=\; & \sum_{i=1}^{n} C_i P(t,T_i) N(h_i) - P(t,T) N(h_0) ,
\end{aligned}
$$

where, for $i = 1, \ldots, n$,

$$h_i = \frac{s(y_1)y_2 - s(y_2)y_1 + \gamma_{i1}(y_2 - y_1) - \gamma_{i2}(s(y_2) - s(y_1))}{\sqrt{(y_2 - y_1)^2 + (s(y_2) - s(y_1))^2}}$$

and

$$h_0 = \frac{s(y_1)y_2 - s(y_2)y_1}{\sqrt{(y_2 - y_1)^2 + (s(y_2) - s(y_1))^2}} \ .$$

Taking the limit as $y_2 - y_1$ tends to zero we obtain

$$a = -\frac{\sum_{i=1}^{n} \gamma_{i2} C_i P(t, T_i) n(b + \gamma_{i1}) n(\gamma_{i2})}{\sum_{i=1}^{n} \gamma_{i1} C_i P(t, T_i) n(b + \gamma_{i1}) n(\gamma_{i2})}$$

where b satisfies

$$\sum_{i=1}^{n} C_i P(t, T_i) n(b + \gamma_{i1}) n(\gamma_{i2}) \sqrt{2\pi} - P(t, T) n(b) = 0 \ .$$

Therefore in the above formulae we can also use

$$h_i = \frac{b + \gamma_{i1} - \gamma_{i2} a}{\sqrt{1 + a^2}}$$

for $i = 1, \ldots, n$ and

$$h_0 = \frac{b}{\sqrt{1 + a^2}} \ .$$

If the covariance matrix $\triangle$ is of rank 1 then for all $i = 1, \ldots, n$ we have $\gamma_{i2} = 0$,

$$\gamma_{i1}^2 = \int_t^T \left| \int_T^{T_i} \sigma(s, u) du \right|^2 ds$$

and $a = 0$.

2.4   Options on swap rate spreads

The owner of a **call** option **on a spread** between the swap rate

$$\omega_T(T, n_1) = \left(\delta \sum_{j=1}^{n_1} P(T, T_j) \right)^{-1} (1 - P(T, T_{n_1}))$$

and the swap rate

$$\omega_T(T, n_2) = \left(\delta \sum_{j=1}^{n_2} P(T, T_j) \right)^{-1} (1 - P(T, T_{n_2}))$$

at strike κ maturing at time $T = T_0$ on principal 1 receives at time T the amount

$$(\omega_T(T, n_1) - \omega_T(T, n_2) - \kappa)^+ .$$

The time $t \leq T$ price can be computed using the general formula with the function

$$g(x_1, \ldots, x_{n_1 \vee n_2}) = \left(\frac{1 - x_{n_1}}{\delta \sum_{j=1}^{n_1} x_j} - \frac{1 - x_{n_2}}{\delta \sum_{j=1}^{n_2} x_j} - \kappa \right)^+ ,$$

where $n_1 \vee n_2 = \max(n_1, n_2)$ and $X_i(t) = P(t, T_i)$, $i = 1, \ldots, n_1 \vee n_2$. We obtain

$$E\left(\frac{\beta(t)}{\beta(T)} (\omega_T(T, n_1) - \omega_T(T, n_2) - \kappa)^+ | \mathcal{F}_t \right)$$

$$= P(t, T) \int_{\mathbb{R}^k} \left(\frac{P(t, T)n_k(x) - P(t, T_{n_1})n_k(x + \gamma_{n_1})}{\delta \sum_{j=1}^{n_1} P(t, T_j)n_k(x + \gamma_j)} \right.$$

$$\left. - \frac{P(t, T)n_k(x) - P(t, T_{n_2})n_k(x + \gamma_{n_2})}{\delta \sum_{j=1}^{n_2} P(t, T_j)n_k(x + \gamma_j)} - \kappa \right)^+ n_k(x)dx .$$

The vectors $\gamma_1, \ldots, \gamma_{n_1 \vee n_2}$ are calculated from the decomposition $\triangle = \Gamma^* \Gamma$, where $\Gamma = [\gamma_1, \ldots, \gamma_{n_1 \vee n_2}]$ and $\triangle = ((\triangle_{ij}))$ with

$$\triangle_{ij} = \int_t^T \int_T^{T_i} \sigma(s, u)^* du \int_T^{T_j} \sigma(s, u) du \, ds .$$

The corresponding **put** option pays the amount

$$(\kappa - \omega_T(T, n_1) + \omega_T(T, n_2))^+$$

and has the price at time $t \leq T$ given by

$$E\left(\frac{\beta(t)}{\beta(T)} (\kappa - \omega_T(T, n_1) + \omega_T(T, n_2))^+ | \mathcal{F}_t \right)$$

$$= P(t, T) \int_{\mathbb{R}^k} \left(\kappa - \frac{P(t, T)n_k(x) - P(t, T_{n_1})n_k(x + \gamma_{n_1})}{\delta \sum_{j=1}^{n_1} P(t, T_j)n_k(x + \gamma_j)} \right.$$

$$\left. + \frac{P(t, T)n_k(x) - P(t, T_{n_2})n_k(x + \gamma_{n_2})}{\delta \sum_{j=1}^{n_2} P(t, T_j)n_k(x + \gamma_j)} \right)^+ n_k(x)dx$$

A **one period yield curve swap** is a contract in which at time T one side pays $\omega_T(T, n_2) + \kappa$ and the other pays $\omega_T(T, n_1)$. Obviously, because

$$\omega_T(T, n_1) - \omega_T(T, n_2) - \kappa = (\omega_T(T, n_1) - \omega_T(T, n_2) - \kappa)^+$$
$$-(\kappa - \omega_T(T, n_1) + \omega_T(T, n_2))^+$$

we have

$$\textbf{Call} - \textbf{Put} = \textbf{One period yield curve swap.}$$

The margin rate is that value of the constant κ which makes the value of the swap zero, i.e.,

$$\kappa = \int_{\mathbb{R}^k} \frac{P(t,T)n_k(x) - P(t,T_{n_1})n_k(x+\gamma_{n_1})}{\delta \sum_{j=1}^{n_1} P(t,T_j)n_k(x+\gamma_j)} n_k(x)dx$$
$$- \int_{\mathbb{R}^k} \frac{P(t,T)n_k(x) - P(t,T_{n_2})n_k(x+\gamma_{n_2})}{\delta \sum_{j=1}^{n_2} P(t,T_j)n_k(x+\gamma_j)} n_k(x)dx \ .$$

A multi-period **yield curve swap** consists of n payments at times $T_i, i = 1, \ldots, n$. The swap rate

$$\omega_{T_{i-1}}(T_{i-1}, n_1) = \left(\delta \sum_{j=1}^{n_1} P(T_{i-1}, T_{i-1+j})\right)^{-1} (1 - P(T_{i-1}, T_{i-1+n_1}))$$

is received and the swap rate $\omega_{T_{i-1}}(T_{i-1}, n_2)$ plus the margin rate m is paid. The yield curve swap value is

Yield Curve Swap (t)

$$= E\left(\sum_{i=1}^n \frac{\beta(t)}{\beta(T_i)} \left(\omega_{T_{i-1}}(T_{i-1}, n_1) - \omega_{T_{i-1}}(T_{i-1}, n_2) - m\right) | \mathcal{F}_t\right)$$
$$= \sum_{i=1}^n P(t,T_i) \left(\int_{\mathbb{R}^k} \frac{P(t,T_{i-1})n_k(x) - P(t,T_{i-1+n_1})n_k(x+\gamma_{i-1+n_1})}{\delta \sum_{j=1}^{n_1} P(t,T_{i-1+j})n_k(x+\gamma_{i-1+j})} n_k(x)dx\right.$$
$$\left. - \int_{\mathbb{R}^k} \frac{P(t,T_{i-1})n_k(x) - P(t,T_{i-1+n_2})n_k(x+\gamma_{i-1+n_2})}{\delta \sum_{j=1}^{n_2} P(t,T_{i-1+j})n_k(x+\gamma_{i-1+j})} n_k(x)dx - m\right).$$

2.5   Compound caps and floors

A **compound cap** (floor) is a call or put option on the underlying forward cap (floor) with strike κ settled in arrears at times $T_j, j = 1, \ldots, n$, where $T_j - T_{j-1} = \delta$, $T_0 = T$. The compound option has strike κ^c and maturity T^c.

We use a rank 1 approximation to the covariance matrix Δ^c obtained from the zero coupons $P(T^c, T_0), \ldots, P(T^c, T_n)$ defining the cap (floor) price at time T^c. Hence the compound option gammas are given by

$$(\gamma_{i1}^c)^2 = \int_t^{T^c} \left|\int_{T^c}^{T_i} \sigma(s,u)du\right|^2 ds$$

for $i = 0, 1, \ldots, n$.

Consider a **call on floor** option. The price at time $t \leq T^c$ is

$$E\left(\frac{\beta(t)}{\beta(T^c)} (\text{Floor}(T^c) - \kappa^c)^+ | \mathcal{F}_t\right) \ ,$$

where the floor price at time T^c is given in Section 2.2. It follows, using the general pricing formula, that

$$\text{Call on Floor}(t) = \int_{\mathbb{R}} (f(x) - \kappa^c P(t, T^c))^+ n(x) dx ,$$

where

$$f(x) = \sum_{j=0}^{n-1} P(t, T_j) \frac{n(x + \gamma_{j1}^c)}{n(x)} \left(\frac{(1 + \kappa\delta) P(t, T_{j+1}) n(x + \gamma_{j+1,1}^c)}{P(t, T_j) n(x + \gamma_{j1}^c)} N(\alpha_j x \right.$$
$$\left. + \beta_j + \zeta_j(T^c)) - N(\alpha_j x + \beta_j) \right) ,$$

while for $j = 0, 1, \ldots, n-1$

$$\alpha_j = -\zeta_j(T^c)^{-1}(\gamma_{j+1,1}^c - \gamma_{j1}^c)$$

$$\gamma_{j+1,1}^c - \gamma_{j1}^c = \left(\int_t^{T^c} \left| \int_{T_j}^{T_{j+1}} \sigma(s, u) du \right|^2 ds \right)^{1/2} ,$$

$$\beta_j = \zeta_j(T^c)^{-1} \left(\log \frac{(1 + \kappa\delta) P(t, T_{j+1})}{P(t, T_j)} - \frac{1}{2}(\gamma_{j+1,1}^c)^2 \right.$$
$$\left. + \frac{1}{2}(\gamma_{j1}^c)^2 - \frac{1}{2}\zeta_j(T^c)^2 \right).$$

It is easy to see that

$$f'(x) =$$
$$-\sum_{j=0}^{n-1} P(t, T_j) \gamma_{j1}^c \frac{n(x + \gamma_{j1}^c)}{n(x)} \left(\frac{(1 + \kappa\delta) P(t, T_{j+1}) n(x + \gamma_{j+1,1}^c)}{P(t, T_j) n(x + \gamma_{j1}^c)} N(\alpha_j x + \right.$$
$$\left. \beta_j + \zeta_j(T^c)) - N(\alpha_j x + \beta_j) \right)$$
$$-\sum_{j=0}^{n-1} P(t, T_j) \frac{n(x + \gamma_{j1}^c)}{n(x)} N(\alpha_j x + \beta_j + \zeta_j(T^c))(\gamma_{j+1,1}^c - \gamma_{j1}^c) \times$$
$$\frac{(1 + \kappa\delta) P(t, T_{j+1}) n(x + \gamma_{j+1,1}^c)}{P(t, T_j) n(x + \gamma_{j1}^c)}$$

and hence f is monotonically decreasing. Integrating over the interval $(-\infty, m^c)$, where constant m^c is given by

$$\sum_{j=0}^{n-1} \left((1 + \kappa\delta) P(t, T_{j+1}) n(m^c + \gamma_{j+1,1}^c) N(\alpha_j m^c + \beta_j + \zeta_j(T^c)) \right.$$
$$\left. - P(t, T_j) n(m^c + \gamma_{j1}^c) N(\alpha_j m^c + \beta_j) \right)$$
$$= \kappa^c P(t, T^c) n(m^c),$$

leads to

Call on Floor (t)

$$
= \sum_{j=0}^{n-1} (1 + \kappa\delta) P(t, T_{j+1}) N_2(m^c + \gamma_{j+1,1}^c, h_j(t) + \zeta_j(t); \rho_j(t))
$$

$$
- \sum_{j=0}^{n-1} P(t, T_j) N_2(m^c + \gamma_{j1}^c, h_j(t); \rho_j(t)) - \kappa^c P(t, T^c) N(m^c) ,
$$

where

$$
\rho_j(t) = \zeta_j(t)^{-1} \left(\int_t^{T^c} \left| \int_{T_j}^{T_{j+1}} \sigma(s, u) du \right|^2 ds \right)^{1/2} .
$$

for $j = 0, 1, \ldots, n - 1$. Similarly for a **put on floor** option we have

$$
\begin{aligned}
\text{Put on Floor} (t) &= E\left(\frac{\beta(t)}{\beta(T^c)} (\kappa^c - \text{Floor}(T^c))^+ | \mathcal{F}_t \right) \\
&= E\left(\frac{\beta(t)}{\beta(T^c)} (Floor(T^c) - \kappa^c)^+ | \mathcal{F}_t \right) \\
&\quad - E\left(\frac{\beta(t)}{\beta(T^c)} (\text{Floor} (T^c) - \kappa^c) | \mathcal{F}_t \right) \\
&= \text{Call on Floor} (t) - \text{Floor} (t) + \kappa^c P(t, T^c) .
\end{aligned}
$$

Formulae for **call and put options on caps** can be written as integrals over $\mathbb{R}$. We have

$$
\begin{aligned}
\text{Call on Cap} (t) &= \int_{\mathbb{R}} (g(x) - \kappa^c P(t, T^c)^+ n(x) dx \\
\text{Put on Cap} (t) &= \text{Call on Cap} (t) - \text{Cap} (t) + \kappa^c P(t, T^c) ,
\end{aligned}
$$

where

$$
\begin{aligned}
g(x) &= \sum_{j=0}^{n-1} \left(P(t, T_j) \frac{n(x + \gamma_{j1}^c)}{n(x)} N(-\alpha_j x - \beta_j) \right. \\
&\quad \left. - (1 + \kappa\delta) P(t, T_{j+1}) \frac{n(x + \gamma_{j+1,1}^c)}{n(x)} N(-\alpha_j x - \beta_j - \zeta_j(T^c)) \right) .
\end{aligned}
$$

In this case the function $\{g(x); x \geq 0\}$ is not monotonic and hence there is no unique solution to the equation

$$
g(x) - \kappa^c P(t, T^c) = 0
$$

in terms of x. Therefore it seems easier to approximate the integral by a sum rather than write the formula in terms of bivariate normal distribution. Our result differs from the formula given in Hull and White (1994).

2.6 Compound swaptions

A **compound swaption** is a call or put option on the underlying receiver or payer swaption with strike κ settled in arrears at times $T_j, j = 1, \ldots, n$, where $T_j - T_{j-1} = \delta, T_0 = T$. The compound option has strike κ^c and maturity T^c. As in Section 2.5 we use a rank 1 approximation to the covariance matrix $\triangle^c = ((\triangle_{ij}^c))$, where for $i, j = 0, 1, \ldots, n$

$$\triangle_{ij}^c = \text{Cov}\left(\log\ P(T^c, T_i), \log\ P(T^c, T_j) | \mathcal{F}_t\right) .$$

The compound option gammas can be calculated as follows

$$(\gamma_{i1}^c)^2 = \int_t^{T^c} \left| \int_{T^c}^{T_i} \sigma(s, u) du \right|^2 ds, \ i = 0, 1, \ldots, n .$$

The underlying swaption gammas are also calculated assuming a rank 1 approximation to the corresponding covariance matrix. Hence we have

$$\gamma_{i1}(t)^2 = \int_t^{T_0} \left| \int_{T_0}^{T_i} \sigma(s, u) du \right|^2 ds, \ i = 1, \ldots, n .$$

Constants b, a, s^c and $s(t)$ are given by

$$b: \quad \sum_{i=1}^n C_i P(t, T_i) n(\gamma_{i1}^c) n(b + \gamma_{i1}(T^c)) = P(t, T_0) n(\gamma_{01}^c) n(b) ,$$

$$a: \quad a = -\frac{\sum_{i=1}^n C_i P(t, T_i) \gamma_{i1}^c n(\gamma_{i1}^c) n(b + \gamma_{i1}(T^c)) - P(t, T_0) \gamma_{01}^c n(\gamma_{01}^c) n(b)}{\sum_{i=1}^n C_i P(t, T_i) \gamma_{i1}(T^c) n(\gamma_{i1}^c) n(b + \gamma_{i1}(T^c))} ,$$

$$s^c: \quad \sum_{i=1}^n C_i P(t, T_i) n(s^c + \gamma_{i1}^c) N(a s^c + b + \gamma_{i1}(T^c))$$

$$= P(t, T_0) n(s^c + \gamma_{01}^c) N(a s^c + b) + \kappa^c P(t, T^c) n(s^c) ,$$

$$s(t): \quad \sum_{i=1}^n C_i P(t, T_i) n(s(t) + \gamma_{i1}(t)) = P(t, T_0) n(s(t)) .$$

As before $C_j = \kappa \delta, j = 1, \ldots, n - 1, C_n = 1 + \kappa \delta$.

Consider a **call on receiver** option. The underlying swaption price at time T^c is

$$\text{Receiver}\,(T^c) = \sum_{i=1}^n C_i P(T^c, T_i) N(s(T^c) + \gamma_{i1}(T^c)) - P(T^c, T_0) N(s(T^c))$$

and hence the compound option price at time $t \leq T^c$ is

Call on Receiver (t)

$$= E\left(\frac{\beta(t)}{\beta(T^c)} \left(\text{Receiver}\,(T^c) - \kappa^c \right)^+ | \mathcal{F}_t \right)$$

$$= \int_{\mathbb{R}} \left(\sum_{i=1}^n C_i P(t, T_i) n(x + \gamma_{i1}^c) N(h(x) + \gamma_{i1}(T^c)) \right.$$

$$\left. - P(t, T_0) n(x + \gamma_{01}^c) N(h(x)) - \kappa^c P(t, T^c) n(x) \right)^+ dx ,$$

where $h(x)$ is given by

$$\sum_{i=1}^{n} C_i P(t, T_i) n(x + \gamma_{i1}^c) n(h(x) + \gamma_{i1}(T^c)) = P(t, T_0) n(x + \gamma_{01}^c) n(h(x)) .$$

The function

$$f(x) = \sum_{i=1}^{n} C_i P(t, T_i) \frac{n(x + \gamma_{i1}^c)}{n(x)} N(h(x) + \gamma_{i1}(T^c))$$
$$- P(t, T_0) \frac{n(x + \gamma_{01}^c)}{n(x)} N(h(x))$$

is positive and monotonically decreasing from infinity at minus infinity to zero at infinity (we assume that for all $i = 1, \ldots, n$ $0 \leq \gamma_{01}^c \leq \gamma_{i1}^c \leq \gamma_{i+1,1}^c$). Moreover the function $h(x)$ can be approximated by a line $h(x) = ax + b$, and hence

Cal on Receiver (t)

$$= \sum_{i=1}^{n} C_i P(t, T_i) N_2 \left(s^c + \gamma_{i1}^c, \; \frac{b + \gamma_{i1}(T^c) - a\gamma_{i1}^c}{\sqrt{1 + a^2}}; \; -\frac{a}{\sqrt{1 + a^2}} \right)$$
$$- P(t, T_0) N_2 \left(s^c + \gamma_{01}^c, \; \frac{b - a\gamma_{01}^c}{\sqrt{1 + a^2}}; \; -\frac{a}{\sqrt{1 + a^2}} \right) - \kappa^c P(t, T^c) N(s^c) ,$$

and

Put on Receiver(t) = Call on Receiver (t) − Receiver (t) + $\kappa^c P(t, T^c)$.

Formulae for **options on payer swaptions** can be written as integrals over $\mathbb{R}$. We have

Call on Payer (t)

$$= \int_{\mathbb{R}} \left(P(t, T_o) n(x + \gamma_{01}^c) N(-ax - b) \right.$$
$$\left. - \sum_{i=1}^{n} C_i P(t, T_i) n(x + \gamma_{i1}^c) N(-ax - b - \gamma_{i1}(T^c)) - \kappa^c P(t, T^c) n(x) \right)^{+} dx$$

and

Put on Payer (t) = Call on Payer (t) − Payer (t) + $\kappa^c P(t, T^c)$.

Exotic caps and swaptions

There is a large number of exotic caps and swaptions. In this section we develop pricing formulae for some of them

The **N-Cap** is an interest rate cap that has a lower strike κ_1, an upper strike κ_2 ($\kappa_1 \leq \kappa_2$), and a trigger ℓ. As long as the forward rate L is below the level ℓ the owner enjoys protection at the lower strike κ_1. For periods when L is at or above ℓ, the $N-$Cap owner has protection at the upper strike κ_2.

Consider the $N-$Cap on principal 1 settled in arrears at times $T_j, j = 1, \ldots, n$, where $T_j - T_{j-1} = \delta, T_0 = T$. The cash flows at times T_j are

$$(L(T_{j-1}) - \kappa_1)^+ \delta I_{\{L(T_{j-1}) < \ell\}} + (L(T_{j-1}) - \kappa_2)^+ \delta I_{\{L(T_{j-1}) \geq \ell\}}.$$

It follows easily that the $N-$Cap price at time $t \leq T$ is

$$N - \text{Cap}\,(t)$$
$$= \sum_{j=0}^{n-1} P(t, T_j)\Big(N(h_j(t, \ell)) - N(h_j(t, \kappa_1 \wedge \ell)) + N(-h_j(t, \kappa_2 \vee \ell)) \Big)$$
$$- \sum_{j=0}^{n-1} (1 + \kappa_1 \delta) P(t, T_{j+1})\Big(N(h_j(t, \ell) + \zeta_j(t)) - N(h_j(t, \kappa_1 \wedge \ell) + \zeta_j(t)) \Big)$$
$$- \sum_{j=0}^{n-1} (1 + \kappa_2 \delta) P(t, T_{j+1}) N(-h_j(t, \kappa_2 \vee \ell) - \zeta_j(t)),$$

where

$$h_j(t, \kappa) = \left(\log \frac{(1 + \kappa\delta) P(t, T_{j+1})}{P(t, T_j)} - \frac{1}{2}\zeta_j(t)^2 \right) / \zeta_j(t).$$

A **B-Cap** (Bounded Cap) consists of a sequence of caplets in which the difference between the fixed and floating is paid only if total payments to date are less than some fixed amount.

Consider a **B-Caplet** maturing at reset date T_{j-1} on which the cash flows in arrears at time T_j only if the accumulated cash flow at time T_{j-1} due to resets at times T_k and paid in arrears at $T_{k+1}, k = 0, 1, \ldots, j-1$, is less than b. The time $t \leq T = T_0$ value of the $B-$Caplet is

$$B - \text{Caplet}\,(t)$$
$$= E\left(\frac{\beta(t)}{\beta(T_j)} (L(T_{j-1}) - \kappa)^+ \delta I_{\{\sum_{k=0}^{j-1}(L(T_k) - \kappa)^+ \delta \leq b\}} \Big| \mathcal{F}_t \right)$$
$$= E\left(\frac{\beta(t)}{\beta(T_{j-1})} \left(1 - (1 + \kappa\delta) P(T_{j-1}, T_j) \right)^+ I_{\{\sum_{k=0}^{j-1}(P(T_k, T_{k+1})^{-1} - (1+\kappa\delta))^+ \leq b\}} \Big| \mathcal{F}_t \right)$$
$$= P(t, T_{j-1}) E_{T_{j-1}}\left(\left(1 - (1 + \kappa\delta) P(T_{j-1}, T_j) \right)^+ \times \right.$$
$$\left. I_{\{\sum_{k=0}^{j-1}(P(T_k, T_{k+1})^{-1} - (t+\kappa\delta))^+ \leq b\}} \Big| \mathcal{F}_t \right),$$

where $E_{T_{j-1}}$ is the expectation under the forward measure $\mathbf{P}_{T_{j-1}}$. Each of the zero coupons $P(T_k, T_{k+1})$, $k = 0, \ldots, j-1$, can be expressed as follows

$$P(T_k, T_{k+1})$$
$$= \frac{P(t, T_{k+1})}{P(t, T_k)} \exp\left(-\int_t^{T_k} \int_{T_k}^{T_{k+1}} \sigma(s, u)^* du dW_{T_{j-1}}(s)\right.$$
$$\left. -\frac{1}{2} \int_t^{T_k} \left|\int_{T_k}^{T_{k+1}} \sigma(s, u) du\right|^2 ds + \int_t^{T_k} \int_{T_k}^{T_{k+1}} \sigma(s, u)^* du \int_{T_k}^{T_{j-1}} \sigma(s, u) du ds\right)$$

and the $\mathcal{F}_t$ conditional distribution under $\mathbf{P}_{T_{j-1}}$ of the vector $(X_k, k = 0, 1, \ldots, j-1)$, where

$$X_k = \int_t^{T_k} \int_{T_k}^{T_{k+1}} \sigma(s, u)^* du dW_{T_{j-1}}(s)$$

is $N(0, \triangle)$ with $\triangle = ((\triangle_{k\ell}(t)))$ and

$$\triangle_{k\ell}(t) = \int_t^{T_k \wedge T_\ell} \int_{T_k}^{T_{k+1}} \sigma(s, u)^* du \int_{T_\ell}^{T_{\ell+1}} \sigma(s, u) du ds ,$$
$$\zeta_k(t)^2 = \triangle_{kk}(t) .$$

Hence

$$B - \text{Caplet}\,(t) =$$
$$P(t, T_{j-1}) E_{T_{j-1}}\left(\left(1 - (1 + \kappa\delta)\frac{P(t, T_j)}{P(t, T_{j-1})} \exp(-X_{j-1} - \frac{1}{2}\zeta_{j-1}(t)^2)\right)^+ \times\right.$$
$$\left. I_{B_{j-1}}|\mathcal{F}_t\right),$$

where

$$B_{j-1} = \left\{\sum_{k=0}^{j-1} \left(\frac{P(t, T_k)}{P(t, T_{k+1})} \exp\left(X_k - \frac{1}{2}\zeta_k(t)^2 - \alpha_{kj-1}(t)\right) - (1 + \kappa\delta)\right)^+ \leq b\right\},$$
$$\alpha_{k\ell}(t) = \int_t^{T_k} \int_{T_k}^{T_{k+1}} \sigma(s, u)^* du \int_{T_{k+1}}^{T_\ell} \sigma(s, u) du ds .$$

Letting

$$A_{j-1} = \left\{X_{j-1} \geq \log \frac{(1 + \kappa\delta)P(t, T_j)}{P(t, T_{j-1})} - \frac{1}{2}\zeta_{j-1}(t)^2\right\}$$

we get

$$B - \text{Caplet}\,(t) = P(t, T_{j-1})\mathbf{P}(A_{j-1} \cap B_{j-1})$$
$$-(1 + \kappa\delta)P(t, T_j)E \exp\left(-X_{j-1} - \frac{1}{2}\zeta_{j-1}(t)^2\right) I_{A_{j-1} \cap B_{j-1}},$$

where the vector $X = (X_k, k = 0, \ldots, j-1)$ has $N(0, \triangle)$ distribution under the measure $\mathbf{P}$. To calculate the second expectation recall that

$$E \exp(z^* X - \frac{1}{2} z^* \triangle z) f(X) = Ef(X + \triangle z)$$

what gives

$$B - \text{Caplet}(t) = P(t, T_{j-1})\mathbf{P}(A_{j-1} \cap B_{j-1}) - (1 + \kappa\delta)P(t, T_j)\mathbf{P}(A_{j-1}^+ \cap B_{j-1}^+),$$

where

$$A_{j-1}^+ = \left\{ X_{j-1} \geq \log \frac{(1 + \kappa\delta)P(t, T_j)}{P(t, T_{j-1})} + \frac{1}{2}\zeta_{j-1}(t)^2 \right\},$$

$$B_{j-1}^+ = \left\{ \sum_{k=0}^{j-1} \left(\frac{P(t, T_k)}{P(t, T_{k+1})} \exp\left(X_k - \frac{1}{2}\zeta_k(t)^2 - \alpha_{kj}(t) \right) - (1 + \kappa\delta) \right)^+ \leq b \right\}.$$

Finally, rearranging the above formula we get that

$$B - \text{Caplet}(t) = \text{Caplet}(t) - \text{Discount}(t),$$

where $\text{Caplet}(t)$ is a standard caplet and

$$\text{Discount}(t)$$
$$= P(t, T_{j-1})\mathbf{P}(A_{j-1} \cap B_{j-1}^c) - (1 + \kappa\delta)P(t, T_j)\mathbf{P}(A_{j-1}^+ \cap (B_{j-1}^+)^c).$$

Probabilities of the events $A_{j-1} \cap B_{j-1}^c$ and $A_{j-1} \cap (B_{j-1}^+)^c$ can be calculated by simulation. We can also approximate the sets B_{j-1}^c and $(B_{j-1}^+)^c$ by the set

$$\{ (\mu \exp(\sigma Z - \frac{1}{2}\sigma^2) - \nu)^+ \geq b \}$$

where Z is $N(0, 1)$ and (X, Z) is jointly normal and then derive an analytic approximation to the discount.

The **Q–Cap** (Cumulative Cap) provides protection against increases of interest costs as opposed to interest rates for a conventional cap. The interest rate cost is accumulated over the accumulation period with consists of m time intervals of length δ. The cash flows at times $T_j, j = 1, \ldots, m$ of a **Q–Caplet** are

$$\left(\sum_{k=0}^{j-1} L(T_k) - \kappa m \right)^+ \delta$$

and hence its price at time $t \leq T = T_0$ is

$$Q - \text{Caplet}(t) = E\left(\frac{\beta(t)}{\beta(T_j)} \left(\sum_{k=0}^{j-1} L(T_k) - \kappa m \right)^+ \delta \Big| \mathcal{F}_t \right).$$

Using the method and notation used in pricing a *B–Cap* one can show that

$$Q - \text{Caplet}(t) =$$
$$P(t, T_j) \sum_{k=0}^{j-1} \frac{P(t, T_k)}{P(t, T_{k+1})} \exp(-\alpha_{kj}(t))\mathbf{P}(C_{j-1}^k) - (j + \kappa m\delta)P(t, T_j)\mathbf{P}(C_{j-1})$$

where

$$C_{j-1}^k = \left\{ \sum_{\ell=0}^{j-1} \frac{P(t,T_\ell)}{P(t,T_{\ell+1})} \exp\left(X_\ell - \frac{1}{2}\zeta_\ell(t)^2 - \alpha_{\ell j}(t) + \triangle_{\ell k}(t) \right) \geq j + \kappa m\delta \right\}$$

and

$$C_{j-1} = \left\{ \sum_{\ell=0}^{j-1} \frac{P(t,T_\ell)}{P(t,T_{\ell+1})} \exp\left(X_\ell - \frac{1}{2}\zeta_\ell(t)^2 - \alpha_{\ell j}(t) \right) \geq j + \kappa m\delta \right\} .$$

Probabilities of the events C_{j-1}^k and C_{j-1} can be calculated by simulation. We can also approximate sums of the corresponding lognormal variables by one lognormal variable with moments calculated from the sum. This leads to an analytic approximation of the price.

Trigger swaptions are options on swaps for which the terminal payoff depends on whether or not the trigger crosses over some boundary ℓ at the maturity T. If the trigger is the underlying swap rate and we consider a **down and out receiver swaption** with strike κ, the owner receives at time T the discounted from time $T_j, j = 1, \ldots, n$ to T value of the cash flows defined by

$$(\kappa - \omega_T(T,n))^+ \delta I_{\{\omega_T(T,n) \geq \ell\}}.$$

If $\ell \leq \kappa$ and a rank one approximation is used, then it follows easily that the time $t \leq T$ price of the option is given by

$$\sum_{j=1}^n C_j(\kappa) P(t,T_j)\Big(N(s_\kappa + \gamma_{j1}) - N(s_\ell + \gamma_{j1}) \Big) - P(t,T)\Big(N(s_\kappa) - N(s_\ell) \Big),$$

where s_κ satisfies

$$\sum_{j=1}^n C_j(\kappa) P(t,T_j) n(s_\kappa + \gamma_{j1}) = P(t,T) n(s_\kappa),$$

and $C_j(\kappa) = \kappa\delta, \ j = 1, \ldots, n-1, \ C_n = 1 + \kappa\delta$.

In general, the trigger could be defined on any swap rate, in particular it could be set on LIBOR.

3   Foreign economies

To analyse cross economy products we need to expand our model to include in it foreign assets and indices. The superscript i on the quantities defined in the first part of the paper indicates that they represent the corresponding quantiles in the economy i. The domestic economy has $i = 0$ (or no superscript), foreign economies have $i = 1, 2, \ldots, N$.

The **exchange rate** $S^i(t)$ of the currency $i, i \geq 1$, denominated in the domestic currency per unit of the currency i establishes the link between the two economies. Under the measure **P**

$$dS^i(t) = S^i(t)\Big((r^0(t) - r^i(t))dt + v^i(t)^* dW(t)\Big)$$

where $r^0(t) = r(t)$ and $r^i(t)$ stand for the spot rates in the domestic and foreign economies, respectively.

The dynamics under **P** of the **foreign forward rate** $f^i(t, T)$, **foreign zero coupon bond** $P^i(t, T)$ denominated in the foreign currency, or more generally of a **foreign asset** $X^i(t)$ which pays no dividend are, respectively

$$df^i(t, T) = \sigma^i(t, T)^* \left(\int_t^T \sigma^i(t, u)du - v^i(t)\right) dt + \sigma^i(t, T)^* dW(t),$$

$$dP^i(t, T) = P^i(t, T)\left((r^i(t) + \int_t^T \sigma^i(t, u)^* du v^i(t))dt\right.$$
$$\left. - \int_t^T \sigma^i(t, u)^* du dW(t)\right),$$

$$dX^i(t) = X^i(t)\Big((r^i(t) - \eta^i(t)^* v^i(t))dt + \eta^i(t)^* dW(t)\Big).$$

Indeed the processes $P^i(t, T)S^i(t)$ and $X^i(t)S^i(t)$ satisfy, respectively,

$$d(P^i(t, T)S^i(t)) = P^i(t, T)S^i(t)\left(r(t)dt + (v^i(t) - \int_t^T \sigma^i(t, u)du)^* dW(t)\right),$$

$$d(X^i(t)S^i(t)) = X^i(t)S^i(t)\left(r(t)dt + (v^i(t) + \eta^i(t))^* dW(t)\right),$$

and hence the processes

$$\frac{P^i(t, T)S^i(t)}{\beta^0(t)} \quad \text{and} \quad \frac{X^i(t)S^i(t)}{\beta^0(t)},$$

where $\beta^0(t) = \beta(t)$, are martingales under **P**. They represent the discounted (in the domestic economy) prices (in the domestic currency) of foreign assets.

The time t **forward price** for settlement at time T on a T_1 maturity foreign zero coupon bond is

$$F_T^i(t, T_1) = \frac{P^i(t, T_1)S^i(t)}{P(t, T)}.$$

If $T = T_1$ then

$$F_T^i(t, T) = \frac{P^i(t, T)S^i(t)}{P(t, T)}$$

is the **forward exchange rate**. The dynamics of $F_T^i(t, T_1)$ under $\mathbf{P}$ is

$$dF_T^i(t, T_1) = F_T^i(t, T_1) \times$$
$$\left(\int_t^T \sigma(t, u)du + \nu^i(t) - \int_t^{T_1} \sigma^i(t, u)du \right)^* \left(\int_t^T \sigma(t, u)dudt + dW(t) \right).$$

The time t forward price for settlement at time T on X^i is

$$F_T^{X^i}(t) = \frac{X^i(t)S^i(t)}{P(t, T)}$$

and its dynamics satisfies

$$dF_T^{X^i}(t) = F_T^{X^i}(t) \left(\int_t^T \sigma(t, u)du + \nu^i(t) + \eta^i(t) \right)^* \left(\int_t^T \sigma(t, u)dudt + dW(t) \right).$$

The **futures price** $G_T^{X^i}(t)$ at time t in the contract expiring at T on the foreign asset X^i satisfies

$$dG_T^{X^i}(t) = G_T^{X^i}(t) \left(\int_t^T \sigma(t, u)du + \nu^i(t) + \eta^i(t) \right)^* dW(t) ,$$
$$G_T^{X^i}(T) = X^i(T)S^i(T) .$$

This follows from the semimartingale representation of the process $X^i(t)S^i(t)$ and the dynamics of the futures price on a domestic asset.

Let $\mathbf{P}^i$ be the **probability measure** defined by

$$\mathbf{P}^i = \mathcal{E}(N^i)\mathbf{P} ,$$

where

$$N^i(t) = \int_0^t \nu^i(s)^* dW(s) .$$

¿From the Girsanov theorem it follows that the process

$$W^i(t) = W(t) - \int_0^t \nu^i(s)ds$$

is a Brownian motion under $\mathbf{P}^i$. Also because

$$df^i(t, T) = \sigma^i(t, T)^* \int_t^T \sigma^i(t, u)dudt + \sigma^i(t, T)^* dW^i(t) ,$$
$$dP^i(t, T) = P^i(t, T) \left(r^i(t)dt - \int_t^T \sigma^i(t, u)^* dudW^i(t) \right) ,$$
$$dX^i(t) = X^i(t) \left(r^i(t)dt + \eta^i(t)^* dW(t) \right) ,$$

we conclude that $\mathbf{P}^i$ is the **arbitrage free measure of the foreign economy** i.

Let us recall that the forward measure in the domestic economy was defined by

$$\mathbf{P}_T = \mathcal{E}(M_T(\cdot))(T)\mathbf{P},$$

where

$$M_T(t) = -\int_0^t \int_s^T \sigma(s,u)^* du\, dW(s)\ ,$$

and that under $\mathbf{P}_T$ the process

$$W_T(t) = W(t) + \int_0^t \int_s^T \sigma(s,u)du\, ds$$

is a Brownian motion. Analogously

$$\mathbf{P}_T^i = \mathcal{E}(M_T^i(\cdot))(T)\mathbf{P}^i\ ,$$

with

$$M_T^i(t) = -\int_0^t \int_s^T \sigma^i(s,u)^* du\, dW^i(s)$$

defines the **forward measure for the foreign economy**. Under $\mathbf{P}_T^i$ the process

$$W_T^i(t) = W^i(t) + \int_0^t \int_s^T \sigma^i(s,u)du\, ds = W(t) - \int_0^t \left(\nu^i(s) - \int_s^T \sigma^i(s,u)du\right)ds$$

is a Brownian motion and hence the process $f^i(\cdot,T)$ is a martingale. Obviously, because $\mathcal{E}(X)\mathcal{E}(Y) = \mathcal{E}(X+Y+\langle X,Y\rangle)$ for continuous semimartingales X and Y, we also have

$$
\begin{aligned}
\mathbf{P}_T^i &= \mathcal{E}(M_T^i(\cdot))\mathcal{E}(N^i)(T)\mathbf{P}\\
&= \mathcal{E}\left(-\int_0^\cdot \int_s^T \sigma^i(s,u)^* du\, dW(s) + \int_0^\cdot \int_s^T \sigma^i(s,u)^* du\, \nu^i(s)ds\right) \times\\
&\qquad \mathcal{E}\left(\int_0^\cdot \nu^i(s)^* dW(s)\right)(T)\mathbf{P}\\
&= \mathcal{E}\left(\int_0^\cdot \left(\nu^i(s) - \int_s^T \sigma^i(s,u)du\right)^* dW(s)\right)(T)\mathbf{P}\ .
\end{aligned}
$$

Moreover

$$
\begin{aligned}
\mathbf{P}_T^i &= \mathcal{E}(M_T^i(\cdot))\mathcal{E}(N^i)\mathcal{E}(-M_T(\cdot) + \langle M_T(\cdot)\rangle)(T)\mathbf{P}_T\\
&= \mathcal{E}\left(\int_0^\cdot \left(\int_s^T (\sigma(s,u) - \sigma^i(s,u))du + \nu^i(s)\right)^* dW(s)\right.\\
&\qquad + \int_0^\cdot |\int_s^T \sigma(s,u)du|^2 ds\\
&\qquad \left.+ \int_0^\cdot \left(\nu^i(s) - \int_s^T \sigma^i(s,u)du\right)^* \int_s^T \sigma(s,u)du\, ds\right)(T)\mathbf{P}_T
\end{aligned}
$$

$$= \mathcal{E}\left(\int_0^{\cdot}\left(\int_s^T (\sigma(s,u) - \sigma^i(s,u))\, du + \nu^i(s)\right)^*\right.$$
$$\left.\left(dW(s) + \int_s^T \sigma(s,u)\, du\, ds\right)(T)\mathbf{P}_T\right.$$
$$= \mathcal{E}\left(\int_0^{\cdot}\left(\int_s^T (\sigma(s,u) - \sigma^i(s,u))\, du + \nu^i(s)\right)^* dW_T(s)\right)(T)\mathbf{P}_T \ .$$

Consider a **contingent claim** ξ^i **in the economy** i (i.e, denominated in the currency of economy i). Assume that ξ^i is measurable with respect to $\mathcal{F}_T$. Under the appropriate integrability conditions the time t price of ξ^i in the domestic currency is

$$E\left(\frac{\beta(t)}{\beta(T)}\xi^i S^i(T)|\mathcal{F}_t\right) = P(t,T)E_T(\xi^i S^i(T)|\mathcal{F}_t)$$

However, an alternative way of computing this price is to evaluate first

$$E^i\left(\frac{\beta^i(t)}{\beta^i(T)}\xi^i|\mathcal{F}_t\right) = P^i(t,T)E_T^i(\xi^i|\mathcal{F}_t)$$

and then convert the result to the domestic currency, that is calculate

$$S^i(t)P^i(t,T)E_T^i(\xi^i|\mathcal{F}_t) \ .$$

Obviously the arbitrage free price is uniquely defined and hence both methods must give the same answer. This follows from the following formula

$$S^i(t)E^i\left(\frac{\beta^i(t)}{\beta^i(T)}\xi^i|\mathcal{F}_t\right) = E\left(\frac{\beta(t)}{\beta(T)}\xi^i S^i(T)|\mathcal{F}_t\right)$$

which can be proved using standard arguments. Namely

$$S^i(t) = S^i(0)\frac{\beta(t)}{\beta^i(t)}\mathcal{E}(N^i)(t),$$

for all $t \geq 0$ and we have for all $A \in \mathcal{F}_t$

$$\int_A E\left(\frac{\beta(t)}{\beta(T)}\xi^i S^i(T)|\mathcal{F}_t\right) d\mathbf{P}^i$$
$$= \int_A E\left(\frac{\beta(t)}{\beta(T)}\xi^i S^i(T)|\mathcal{F}_t\right)\mathcal{E}(N^i)(T)d\mathbf{P}$$
$$= \int_A E\left(E\left(\frac{\beta(t)}{\beta(T)}\xi^i S^i(T)|\mathcal{F}_t\right)\mathcal{E}(N^i)(T)|\mathcal{F}_t\right)d\mathbf{P}$$
$$= \int_A E\left(\frac{\beta(t)}{\beta(T)}\xi^i S^i(T)|\mathcal{F}_t\right)\mathcal{E}(N^i)(t)d\mathbf{P}$$

$$= \int_A \frac{\beta(t)}{\beta(T)} \xi^i S^i(T) \mathcal{E}(N^i)(t) d\mathbf{P}$$

$$= \int_A S^i(t) \frac{\beta^i(t)}{\beta^i(T)} \xi^i \mathcal{E}(N^i)(T) d\mathbf{P} = S^i(t) \int_A \frac{\beta^i(t)}{\beta^i(T)} \xi^i d\mathbf{P}^i$$

$$= S^i(t) \int_A E^i \left(\frac{\beta^i(t)}{\beta^i(T)} \xi^i | \mathcal{F}_t \right) d\mathbf{P}^i$$

Consider two foreign economies, say ℓ and m, and denote by $S^{m/\ell}(t)$ the exchange rate of the currency ℓ denominated in the currency m per unit of the currency ℓ. In terms of our previous notation we can write

$$S^{m/\ell}(t) = \frac{S^\ell(t)}{S^m(t)}$$

and hence, by the Ito formula

$$
\begin{aligned}
dS^{m/\ell}(t) &= S^{m/\ell}(t) \left(\left(r^m(t) - r^\ell(t) - \nu^m(t)^* \left(\nu^\ell(t) - \nu^m(t) \right) \right) dt \right. \\
&\quad \left. + \left(\nu^\ell(t) - \nu^m(t) \right)^* dW(t) \right) \\
&= S^{m/\ell}(t) \left(\left(r^m(t) - r^\ell(t) \right) dt + \left(\nu^\ell(t) - \nu^m(t) \right)^* dW^m(t) \right) ,
\end{aligned}
$$

where W^m is a Brownian motion under the arbitrage free measure $\mathbf{P}^m$ of the economy m. Therefore we can identify the volatility $\nu^{m/\ell}$ of the exchange rate $S^{m/\ell}$ in terms of the volatilities ν^ℓ and ν^m of the exchange rates S^ℓ and S^m, respectively, as

$$\nu^{m/\ell} = \nu^\ell - \nu^m .$$

3.1 Differential forward swaps

A differential forward swap or a **diffswap** per unit of the currency m consists of swaping floating rates of another two currencies. At each of the payment dates T_{j+1}, $j = 0, \ldots, n-1$ the floating rate $L^k(T_j)$ of the currency k is received and the corresponding floating rate $L^\ell(T_j)$ of the currency ℓ is paid. The rates $L^i(T_j)$, $i = k, \ell$; $j = 0, \ldots, n-1$ satisfy

$$P^i(T_j, T_{j+1})^{-1} = 1 + L^i(T_j)(T_{j+1} - T_j) ,$$

where $T_j = T + j\delta$. The time $t(t \le T)$ value, in the domestic currency, of the differential forward swap is

$$E\left(\sum_{j=1}^{n}\frac{\beta(t)}{\beta(T_j)}\Big(L^k(T_{j-1})-L^\ell(T_{j-1})\Big)\delta S^m(T_j)|\mathcal{F}_t\right)$$

$$= E\left(\sum_{j=1}^{n}\frac{\beta(t)}{\beta(T_j)}\Big(P^k(T_{j-1},T_j)^{-1}-P^\ell(T_{j-1},T_j)^{-1}\Big)S^m(T_j)|\mathcal{F}_t\right)$$

$$= E\left(\sum_{j=1}^{n}\frac{\beta(t)}{\beta(T_{j-1})}\Big(P^k(T_{j-1},T_j)^{-1}-P^\ell(T_{j-1},T_j)^{-1}\Big)\times\right.$$
$$\left. E\left(\frac{\beta(T_{j-1})}{\beta(T_j)}S^m(T_j)|\mathcal{F}_{T_{j-1}}\right)|\mathcal{F}_t\right)$$

$$= E\left(\sum_{j=1}^{n}\frac{\beta(t)}{\beta(T_{j-1})}\Big(P^k(T_{j-1},T_j)^{-1}-P^\ell(T_{j-1},T_j)^{-1}\Big)\times\right.$$
$$\left. P^m(T_{j-1},T_j)S^m(T_{j-1})|\mathcal{F}_t\right)$$

$$= \sum_{j=1}^{n}S^m(t)E^m\left(\frac{\beta^m(t)}{\beta^m(T_{j-1})}\Big(P^k(T_{j-1},T_j)^{-1}-P^\ell(T_{j-1},T_j)^{-1}\Big)\times\right.$$
$$\left. P^m(T_{j-1},T_j)|\mathcal{F}_t\right)$$

$$= \sum_{j=1}^{n}S^m(t)P^m(t,T_{j-1})E^m_{T_{j-1}}\left(\frac{P^m(T_{j-1},T_j)}{P^k(T_{j-1},T_j)}-\frac{P^m(T_{j-1},T_j)}{P^\ell(T_{j-1},T_j)}|\mathcal{F}_t\right).$$

In the economy m the dynamics of $\{P^m(t,T_j);\ t\le T_j\}$ and $\{P^i(t,T_j);\ t\le T_j\}$, $i\ne m$, are

$$dP^m(t,T_j) = P^m(t,T_j)\left(r^m(t)dt-\int_t^{T_j}\sigma^m(t,u)^*du\,dW^m(t)\right),$$

$$dP^i(t,T_j) = P^i(t,T_j)\left(\Big(r^i(t)+\int_t^{T_j}\sigma^i(t,u)^*du\,\nu^{m/i}(t)\Big)dt\right.$$
$$\left. -\int_t^{T_j}\sigma^i(t,u)^*du\,dW^m(t)\right).$$

Hence by the Ito formula

$$d\frac{P^m(t,T_j)}{P^i(t,T_j)} = \frac{P^m(t,T_j)}{P^i(t,T_j)}\left(\Big(r^m(t)-r^i(t)-\int_t^{T_j}\sigma^i(t,u)^*du\Big(\nu^{m/i}(t)+\right.$$
$$\int_t^{T_j}(\sigma^m(t,u)-\sigma^i(t,u))du\Big)\Big)dt$$
$$\left. -\int_t^{T_j}\Big(\sigma^m(t,u)-\sigma^i(t,u)\Big)^*du\,dW^m(t)\right),$$

$$d\frac{P^i(t, T_{j-1})}{P^m(t, T_{j-1})} = \frac{P^i(t, T_{j-1})}{P^m(t, T_{j-1})}\left(\left(r^i(t) - r^m(t) + \int_t^{T_{j-1}} \sigma^i(t, u)^* du \nu^{m/i}(t)\right.\right.$$
$$+ \int_t^{T_{j-1}} \sigma^m(t, u)^* du \int_t^{T_{j-1}} (\sigma^m(t, u) - \sigma^i(t, u)) du\right) dt$$
$$\left.+ \int_t^{T_{j-1}} (\sigma^m(t, u) - \sigma^i(t, u))^* du dW^m(t)\right)$$

and

$$d\frac{P^m(t, T_j)P^i(t, T_{j-1})}{P^i(t, T_j)P^m(t, T_{j-1})} = -\frac{P^m(t, T_j)P^i(t, T_{j-1})}{P^i(t, T_j)P^m(t, T_{j-1})}\left(\alpha_{ijm}(t)dt\right.$$
$$\left.+ \int_{T_{j-1}}^{T_j} (\sigma^m(t, u) - \sigma^i(t, u))^* du dW_{T_{j-1}}^m(t)\right) ,$$

where

$$\alpha_{ijm}(t) = \int_{T_{j-1}}^{T_j} \sigma^i(t, u)^* du \left(\nu^{m/i}(t) + \int_t^{T_j} (\sigma^m(t, u) - \sigma^i(t, u)) du\right) .$$

Therefore

$$E_{T_{j-1}}^m\left(\frac{P^m(T_{j-1}, T_j)}{P^i(T_{j-1}, T_j)}|\mathcal{F}_t\right) = E_{T_{j-1}}^m\left(\frac{P^m(T_{j-1}, T_j)P^i(T_{j-1}, T_{j-1})}{P^i(T_{j-1}, T_j)P^m(T_{j-1}, T_{j-1})}|\mathcal{F}_t\right)$$
$$= \frac{P^m(t, T_j)P^i(t, T_{j-1})}{P^i(t, T_j)P^m(t, T_{j-1})}e^{-\int_t^{T_{j-1}} \alpha_{ijm}(s)ds}$$

and the time t value of the differential forward swap is

$$\text{Diffswap}(t) = \sum_{j=1}^n S^m(t)P^m(t, T_{j-1})\left(\frac{P^m(t, T_j)P^k(t, T_{j-1})}{P^k(t, T_j)P^m(t, T_{j-1})}e^{-\int_t^{T_{j-1}} \alpha_{kjm}(s)ds}\right.$$
$$\left.-\frac{P^m(t, T_j)P^\ell(t, T_{j-1})}{P^\ell(t, T_j)P^m(t, T_{j-1})}e^{-\int_t^{T_{j-1}} \alpha_{\ell jm}(s)ds}\right)$$
$$= \sum_{j=1}^n S^m(t)P^m(t, T_j)\left(\frac{P^k(t, T_{j-1})}{P^k(t, T_j)}e^{-\int_t^{T_{j-1}} \alpha_{kjm}(s)ds}\right.$$
$$\left.-\frac{P^\ell(t, T_{j-1})}{P^\ell(t, T_j)}e^{-\int_t^{T_{j-1}} \alpha_{\ell jm}(s)ds}\right) .$$

3.2 Basket caps

In a settled in arrears cap on a basket of libors the cash flow at times $T_j, j = 1, \ldots, n$ is

$$\left(\sum_{i=0}^{b-1} \lambda_i L^i(T_{j-1}) - \kappa\right)^+ \delta,$$

where $\lambda_i, i = 0, \ldots, b-1$ are constants. The time t price of each caplet is

Basket Caplet (t)

$$= E\left(\frac{\beta(t)}{\beta(T_j)}\left(\sum_{i=0}^{b-1}\lambda_i L^i(T_{j-1}) - \kappa\right)^+ \delta | \mathcal{F}_t\right)$$

$$= E\left(\frac{\beta(t)}{\beta(T_{j-1})}\left(\sum_{i=0}^{b-1}\lambda_i \frac{P(T_{j-1},T_j)}{P(T_{j-1},T_j)} - \left(\kappa\delta + \sum_{i=0}^{b-1}\lambda_i\right)P(T_{j-1},T_j)\right)^+ | \mathcal{F}_t\right)$$

$$= P(t,T_{j-1})E_{T_{j-1}}\left(\sum_{i=0}^{b-1}\lambda_i\frac{P(t,T_j)P^i(t,T_{j-1})}{P^i(t,T_j)P(t,T_{j-1})}\exp\left(-\int_t^{T_{j-1}}\alpha_{ij0}(s)ds\right.\right.$$

$$-\int_t^{T_{j-1}}\int_{T_{j-1}}^{T_j}(\sigma(s,u) - \sigma^i(s,u))^*dudW_{T_{j-1}}(s)$$

$$\left.-\frac{1}{2}\int_t^{T_{j-1}}\left|\int_{T_{j-1}}^{T_j}(\sigma(s,u) - \sigma^i(s,u))du\right|^2 ds\right)$$

$$-(\kappa\delta + \sum_{i=0}^{b-1}\lambda_i)\frac{P(t,T_j)}{P(t,T_{j-1})}\exp\left(-\int_t^{T_{j-1}}\int_{T_{j-1}}^{T_j}\sigma(s,u)^*dudW_{T_{j-1}}(s)\right.$$

$$\left.\left.-\frac{1}{2}\int_t^{T_{j-1}}\left|\int_{T_{j-1}}^{T_j}\sigma(s,u)du\right|^2\right)^+ | \mathcal{F}_t\right),$$

where

$$\alpha_{ij0}(t) = \int_{T_{j-1}}^{T_j}\sigma^i(s,u)^*du\left(\nu^i(t) + \int_t^{T_j}\left(\sigma(t,u) - \sigma^i(t,u)\right)du\right).$$

Hence

Basket Caplet (t)

$$= P(t,T_j)\int_{\mathbb{R}^k}\left(\sum_{i=0}^{b-1}\lambda_i\frac{P^i(t,T_{j-1})}{P^i(t,T_j)}\exp\left(-\int_t^{T_{j-1}}\alpha_{ij0}(s)ds\right)n_k(x+\eta_i)\right.$$

$$\left.-(\kappa\delta + \sum_{i=0}^{b-1}\lambda_i)n_k(x+\eta_b)\right)^+ dx,$$

where the vectors $\eta_0, \ldots, \eta_{b-1}, \eta_b$ are such that

$$\eta_m^*\eta_n = \int_t^{T_{j-1}}\int_{T_{j-1}}^{T_j}\left(\sigma(s,u) - \sigma^m(s,u)\right)^*du\int_{T_{j-1}}^{T_j}\left(\sigma(s,u) - \sigma^n(s,u)\right)duds$$

for $m, n = 0, \ldots, b-1$,

$$\eta_m^*\eta_b = \int_t^{T_{j-1}}\int_{T_{j-1}}^{T_j}\left(\sigma(s,u) - \sigma^m(s,u)\right)^*du\int_{T_{j-1}}^{T_j}\sigma(s,u)duds$$

for $m = 0, \ldots, b-1$ and

$$\eta_b^*\eta_b = \int_t^{T_{j-1}}\left|\int_{T_{j-1}}^{T_j}\sigma(s,u)du\right|^2 ds.$$

3.3 Cross-currency swaps

A cross-currency fixed-floating swaption is an option to pay fixed rate κ of currency k and receive floating rate L^ℓ of currency ℓ plus margin m. The option price at time $t \leq T$ is

$$
E\left(\frac{\beta(t)}{\beta(T)}\left(E\left(\sum_{j=1}^{n}\frac{\beta(T)}{\beta(T_j)}(L^\ell(T_{j-1})+m-\kappa)\delta|\mathcal{F}_T\right)\right)^+|\mathcal{F}_t\right)
$$

$$
= E\left(\frac{\beta(t)}{\beta(T)}\left(\sum_{j=1}^{n}P(T,T_j)\left(\frac{P^\ell(T,T_{j-1})}{P^\ell(T,T_j)}\times\right.\right.\right.
$$

$$
\left.\left.\left.\exp\left(-\int_{T}^{T_{j-1}}\alpha_{\ell j0}(s)ds\right)-(\kappa-m)\delta\right)\right)^+|\mathcal{F}_t\right)
$$

$$
= P(t,T)E_T\left(\left(\sum_{j=1}^{n}\left(\frac{P(t,T_j)P^\ell(t,T_{j-1})}{P(t,T)P^\ell(t,T_j)}\right)\exp\left(-\int_{t}^{T}\left(\int_{T}^{T_j}\sigma(s,u)du\right.\right.\right.\right.
$$

$$
\left.-\int_{T_{j-1}}^{T_j}\sigma^\ell(s,u)du\right)^* dW_T(s)
$$

$$
-\frac{1}{2}\int_{t}^{T}\left|\int_{T}^{T_j}\sigma(s,u)du-\int_{T_{j-1}}^{T_j}\sigma^\ell(s,u)du\right|^2 ds\right)\exp\left(-\int_{t}^{T_{j-1}}\alpha_{\ell j0}(s)ds\right)
$$

$$
-(\kappa-m)\delta\frac{P(t,T_j)}{P(t,T)}\exp\left(-\int_{t}^{T}\int_{T}^{T_j}\sigma(s,u)^* du dW_T(s)\right.
$$

$$
\left.\left.\left.-\frac{1}{2}\int_{t}^{T}\left|\int_{T}^{T_j}\sigma(s,u)du\right|^2 ds\right)\right)^+|\mathcal{F}_t\right)
$$

$$
= \int_{\mathbb{R}^d}\left(\sum_{j=1}^{n}P(t,T_j)\left(\frac{P^\ell(t,T_{j-1})}{P^\ell(t,T_j)}\right)\exp\left(-\int_{t}^{T_{j-1}}\alpha_{\ell j0}(s)ds\right)n_d(x+\eta_j^\ell)\right.
$$

$$
\left.-(\kappa-m)\delta n_d(x+\gamma_j)\right)^+ dx,
$$

where the vectors $\eta_1^\ell,\ldots,\eta_n^\ell,\gamma_1,\ldots,\gamma_n$ satisfy

$$
(\eta_i^\ell)^*\eta_j^\ell = \int_{t}^{T}\left(\int_{T}^{T_i}\sigma(s,u)du-\int_{T_{i-1}}^{T_i}\sigma^\ell(s,u)du\right)^*\times
$$

$$
\left(\int_{T}^{T_i}\sigma(s,u)du-\int_{T_{j-1}}^{T_j}\sigma^\ell(s,u)du\right)ds ,
$$

$$
(\eta_i^\ell)^*\gamma_j = \int_{t}^{T}\left(\int_{T}^{T_i}\sigma(s,u)du-\int_{T_{i-1}}^{T_i}\sigma^\ell(s,u)du\right)^*\int_{T}^{T_j}\sigma(s,u)du ds ,
$$

$$
\gamma_i^*\gamma_j = \int_{t}^{T}\int_{T}^{T_i}\sigma(s,u)^* du\int_{T}^{T_j}\sigma(s,u)du ds
$$

for $i,j=1,\ldots,n$.

A cross currency floating-floating swaption is an option to pay floating rate L^k of currency k and receive floating rate L^ℓ of currency ℓ plus margin m. The time t value of the option is

$$E\left(\frac{\beta(t)}{\beta(T)}\left(E\left(\sum_{j=1}^{n}\frac{\beta(T)}{\beta(T_j)}\left(L^\ell(T_{j-1})+m-L^k(T_{j-1})\right)\delta|\mathcal{F}_T\right)\right)^+|\mathcal{F}_t\right)$$

$$=\int_{\mathbb{R}^d}\left(\sum_{j=1}^{n}P(t,T_j)\left(\frac{P^\ell(t,T_{j-1})}{P^\ell(t,T_j)}\exp\left(-\int_t^{T_{j-1}}\alpha_{\ell j0}(s)ds\right)n_d(x+\eta_j^\ell)\right.\right.$$

$$\left.\left.+mn_d(x+\gamma_j)-\frac{P^k(t,T_{j-1})}{P^k(t,T_j)}\exp\left(-\int_t^{T_{j-1}}\alpha_{kj0}(s)ds\right)n_d(x+\eta_j^k)\right)\right)^+dx,$$

where the vectors $\eta_1^\ell,\ldots,\eta_n^\ell,\eta_1^k,\ldots,\eta_n^k,\gamma_1,\ldots,\gamma_n$ are defined in the standard way.

References

Brace, A. and Musiela, M. (1994) 'A multifactor Gauss Markov implementation of Heath, Jarrow, and Morton', *Mathematical Finance* **4**, 259–283.

Brace, A., Gatarek, D. and Musiela, M. (1995) 'The market model of interest rate dynamics'. Preprint, University of New South Wales.

El Karoui, N., Geman, H. and Rochet, J.C. (1995) 'Changes of numeraire, changes of probability measures and pricing of options', *J. Appl. Prob.* **32**, 443–458.

Heath, D., Jarrow, R. and Morton, A. (1992) 'Bond pricing and the term structure of interest rates: a new methodology', *Econometrica* **60**, 77–105.

Hull, J. and White, A. (1994) 'The pricing of options on interest rate caps and floors using the Hull–White model', *J. Financial Eng.* **2**, 287–296.

Jamshidian, F. (1993) 'Options and futures evaluation with deterministic volatilities', *Mathematical Finance* **3**, 149–159.

Modelling Bonds and Derivatives with Default Risk

David Lando

Abstract

An overview of different approaches to modelling credit risk in bonds and derivatives is given. The classical Merton model and some extensions are outlined which use first passage times of the value of the firm to model defaults. After this outline the remainder of the paper considers intensity based models. These are models in which defaults have a Poisson-like behavior but where the default intensities are random. We show that in this framework modelling of bonds, and defaultable securities in general, becomes very similar to ordinary term structure modelling.

1 Introduction

This chapter gives an overview of some approaches to valuing bonds and derivatives in which default risk (or, synonymously, credit risk) is a significant factor. Default risk may be important either because the issuer of the security could default or because an underlying quantity of a derivative, such as a yield spread, may be closely related to default risk.

The fundamental example of a defaultable bond is a corporate bond, i.e. a bond issued by a firm which promises the holder a nominally fixed payment stream but which due to the possibility of financial distress of the issuer may default on its promise. Municipal bonds and bonds issued by heavily indebted governments are examples of credit risky bonds as well, whereas usually we fell safe ignoring credit risk in Treasury bonds issued by the US government[1].

The typical examples of derivatives with default risk are swaps and over-the-counter instruments in general. Options with risk of default by the issuer are often referred to as *vulnerable* options, see Johnson and Stulz (1987).

Default is to be thought of in a strict sense: A delay in the promised payment would qualify as a default and this is one of many default scenarios which is captured in our models where default implies a partial repayment of contractually specified payments.

---

[1] Recently this assumption has become less innocuous as Moody placed 387 billion dollars worth of short maturity US Treasury bonds on a credit watch list because of the impasse in the federal budget negotiations.

Although it is of course a matter of judgement when a bond or a derivative security has a likelihood of default which is significant enough to make it 'credit risky' there will be a clear cut definition in our probabilistic models: The payments of a default-free security are made with probability one whereas payments of credit risky securities are not.

The quantities which determine prices of credit risky bonds and derivatives are the default probabilities of the issuers and the recovery rates, i.e. the fraction of promised payments which the defaulting entities are able to pay. The problem is to find analytically tractable ways of modelling this.

This article only considers pricing issues and may be seen as an overview of the financial engineering of credit risky instruments. As markets for over-the-counter derivatives grow, an understanding of how to price credit risky instruments and derivatives used to hedge credit risk has become an essential part of financial institutions' risk management.

The goal is to motivate and survey the methods used in a large number of these models and to facilitate the reader's own attack on the literature. The survey emphasizes intensity based models (to be defined below) since these models have been drawing some attention lately.

Virtually no attention is paid to institutional detail which in the area of corporate bonds is overwhelming. Other important issues related to default-able securities, such as general equilibrium implications of default risk and collateral, optimal (re)negotiation of debt contracts, models for financial re-structuring of firms and macroeconomic causes and consequences of changes in default risk are not considered either.

Acknowledgements

This paper is based on a survey lecture given at the term structure week at the Isaac Newton Institute, June 1995. Participants at this meeting, an anonymous reviewer and participants at the Odense University Symposium on Fixed Income Derivatives have provided helpful comments. Carsten Sørensen has provided important corrections and special thanks are due to Darrell Duffie for organizing the term structure week, for a helpful discussion at the meeting and for access to his (and co-authors') unpublished working papers on default risk. All errors are of course my own.

2   Some Basic Terminology and Organization of the Paper

We will be working in an arbitrage-free setting and we will be considering the behavior of the involved processes directly under an equivalent martingale measure Q, i.e. a filtered probability space $(\Omega, \mathcal{F}, (\mathcal{F}_t), Q)$ is given and all processes are assumed to be defined on this space and adapted to the filtration $(\mathcal{F}_t)$. The short-hand notation $E_t(\cdot)$ will be used throughout instead of $E(\cdot | \mathcal{F}_t)$. Unless otherwise indicated, all expectations are with respect to the measure Q.

In all of contingent claims pricing we are interested in the model having nice mathematical properties under possible equivalent martingale measures, and most if not all successful models have been defined under the real world measure P with a view towards obtaining a particular structure under an equivalent martingale measure. In this paper we choose to focus directly on this structure and will only occasionally refer to the probability measure P. Of course, for purposes of statistical estimation and for choosing appropriate martingale measures in incomplete models, the specification of P is important.

Throughout the paper some model of default-free bonds is given in which the money market account is given as

$$B(t) = \exp\left(\int_0^t r_s ds\right)$$

where r_s is a spot rate process, and the price at time t of a default free zero coupon bond with maturity date T is given as

$$p(t, T) = E_t \exp\left(-\int_t^T r_s ds\right).$$

How exactly the spot rate is modelled is not important at this point: The reader may choose to think of any arbitrage-free term structure model for treasuries. In some approaches to modelling of derivatives this will not be needed at all.

To fix some terminology, consider the price $v^i(t, T)$ at time t of a credit risky zero coupon bond maturing at T issued by a party i. In all models discussed here it will have the form

$$v^i(t, T) = E_t\left(\frac{B(t)}{B(T)}(1_{\{\tau^i > T\}} + \delta^i(T, \omega)1_{\{\tau^i \leq T\}})\right) \tag{1}$$

or

$$v^i(t, T) = E_t\left(\frac{B(t)}{B(T)}1_{\{\tau^i > T\}} + \frac{B(t)}{B(\tau^i)}\delta^i(\tau^i, \omega)1_{\{\tau^i \leq T\}}\right). \tag{2}$$

In both expressions τ^i is the default time of the issuer and δ^i is the recovery rate which specifies the fraction of the promised payment that the bondholders receive in the event of default. In the first expression this recovery is thought of as a payout received at maturity whereas in the second expression we think of a payment made at the time of default. Clearly, given the existence of the money market account we can easily translate from one representation to the other by appropriately rolling payments backward or forward using the money market account:

$$\delta^i(T,\omega) = \delta^i(\tau^i,\omega) \exp\left(\int_{\tau^i}^T r_s ds\right).$$

The problem now is to produce, in addition to the chosen term structure model, good specifications of the firm specific *default time* τ^i and the *recovery rate* δ^i. A 'good specification' should of course provide a good compromise between analytical tractability, realism and flexibility.

The models we will consider can broadly be placed in two categories: Models based on the value of the firm (where 'firm' is used as a generic term for the issuer of the bond), and intensity based models. Another terminology, used by Duffie and Singleton (1995b), denotes value-of-firm based models *structural* models and denotes intensity based models *reduced form* models. In models based on the value of the firm the default time is determined by an underlying process describing the value of the firm. Default will occur when this process hits a certain boundary. This class of models is treated in Section 3. Intensity based models use a Poisson process-like environment to describe default capturing the idea that the timing of a default takes the bondholders by surprise. Much of the newer work in the area is of this category and it is treated in Section 4 which will take up a fairly large part of this paper.

It should be noted, that the distinction between the two approaches is not clear cut. Models which use the value of the firm could easily be intensity based by describing the value of the firm as a jump process, and intensity based models could easily incorporate the value of the firm by using it as a variable affecting the default intensity. So perhaps the distinction is better portrayed as follows: The value of firm based models typically result in problems similar to those encountered when extending the Black-Scholes model to include American options, dividends, stochastic interest rates and exotic features, whereas the intensity based models result in setups which more closely resemble term structure modelling. But again, the classification is not precise.

Recovery rates can be modelled in roughly two ways: Either they are modelled as a function of the value of the firm using the option features of corporate debt or they are modelled as an essentially exogenous quantity which must be estimated from historical data or implied out from observed prices. As a rule, exogenous recovery rates are typically applied in the intensity based framework, whereas value-of-firm based models use one or both of these approaches to recovery rates. Examples will be given in Sections 3 and 4. Section 5 concludes with references to some applications of credit risk models and some possible directions of future research.

3 Models Based on Firm Value

In the classical approach of Merton (1974) a firm is considered which is financed by equity and zero coupon debt with a face value of D maturing at time T. The value of the firm is given by V which is modelled as a geometric Brownian motion which under the martingale measure has the form

$$dV_t = rV_t dt + \sigma V_t dB_t$$

where B is a standard Brownian motion, and the interest rate r is assumed constant.

At the maturity date T the bondholders receive the amount

$$\min(V_T, D) = D - \max(D - V_T, 0)$$

which we recognize as the payoff of a riskless zero coupon bond minus a put option on the value of the firm with exercise price D. Now using the Black–Scholes formula on this option we see that

$$D \cdot v(t, T) = D \exp(-r(T - t)) - P(V_t, D, \sigma, r, T - t)$$

where P is the Black–Scholes European put option pricing function.

The risk structure of interest rates can then be obtained by varying T. We note that formally this setup corresponds to the formula 1 with face value D instead of 1, and $\{\tau = T\} = \{V_T < D\}, \{\tau > T\} = \{V_T \geq D\}$ and with $\delta(T) = \frac{V_T}{D}$. There are several generalizations of the Merton model which retain the option interpretation of corporate debt but where some of the fairly restrictive assumptions are relaxed.

Geske (1977) uses a compound option approach to modelling bonds with coupons in the Merton model. To see the basic idea we need only consider a bond which promises to pay D at the terminal date T and D_1 at date $t_1 < T$. If the firm makes the coupon payment at time t_1 we are back in the classical model. Given the value of the firm is V_{t_1} the equity value immediately after the payment is given as

$$S_{t_1} = C(V_{t_1}, D)$$

where C denotes the value of a call option with the relevant parameters and hence the bond value is given by $V_{t_1} - S_{t_1}$. Turning now to the value of the equity before the coupon payment it is important to note that in Geske's model it is assumed that the coupon payment is made by issuing new equity. It may be easiest to imagine that the existing shareholders decide right before the coupon date whether to pay the coupon out of their own pockets or give up the firm to the bondholders. The shareholders will pay the coupon as long as the value of their equity after the coupon payment is larger than D_1. In other words, the value of the equity an instant before t_1 is

$$S_{t_1-} = \max(S_{t_1} - D_1, 0)$$

and since S_{t_1} is already an option we see that before the coupon payment the value of equity is a compound option and hence the value of the debt has compound option features as well. Note that in this model we may have $V_{t_1} > D_1$ and still have a default at that date. If D is much greater than V_{t_1} and V_{t_1} is only slightly greater than D_1 the value of the equity will be very close to zero after the coupon payment and the shareholders will be better off not paying the coupon. The firm's assets will then be sufficient to compensate for the coupon payment in full but will of course have limited hopes of compensating for the promised final payment as well.

If the coupon could be made by liquidating assets of the firm, the situation would be different: In this case the shareholders would have an interest in the firm making the coupon payment precisely when $V_{t_1} > D_1$ because if the payment is not made they lose everything whereas if the payment is made at least the equity will have non-zero value.

Geske obtains closed form solutions whose only numerical challenge is computation of integrals of multivariate normal distributions.

Although coupons are of course lump sum payments in real life, it is practical in the diffusion based models to view coupons as continuous payment streams since this more naturally fits within the analytical framework. Other extensions of the classical model contain one or both of the following two features:

First, the default-free term structure is built on a model with stochastic interest rates and this includes a description of the correlation between the evolution of the default-free term structure and the value of the firm. It is easy to think of both assets and liabilities on a firm's balance sheet whose values depend on the level of interest rates and therefore we would expect this correlation to exist. Empirical findings in for example Duffee (1995) and Longstaff and Schwartz (1995) indicate a negative correlation between the level of treasury yields and credit spreads.

Second, in practice bondholders often have some sort of safety covenant linked to the terms of the loan which in case the condition of the firm deteriorates in some measurable fashion allows the bondholders to call the firm into paying the debt immediately or alternatively have its assets taken over by the bondholders. Note that in the Merton model, default never occurs before the maturity date. Bondholders could in theory watch the firm value nearly evaporate but be unable to act before maturity. Several models operate with a boundary which – if crossed by the value of the firm – allows the bondholders to act and force a liquidation or restructuring of the firm.

The typical way of capturing some of these features can be summarized as follows:

Under an equivalent martingale measure the interest rate on default-free bonds and the value of the firm evolve as

$$dr_t = \mu_r dt + \sigma_r dB_t^1 \tag{3}$$

$$dV_t = r_t V_t dt + \sigma_V (\rho dB_t^1 + \sqrt{1 - \rho^2} dB_t^2) \tag{4}$$

where μ_r, σ_V and σ_r are well-behaved functions of (V_t, r_t, t), ρ is a constant between -1 and 1 and B_t^1, B_t^2 are independent standard Brownian motions.

In addition, a lower boundary process K_t – which in most cases is merely a function of t – is specified which describes the level V_t has to cross for the firm to be forced into bankruptcy, and a pay-off function $\delta(K_t, t)$ giving the pay-off to bondholders in bankruptcy completes the model.

At this point there are two paths one can follow. One is the classical PDE approach in which the value of the bond at time t is assumed to be a twice continuously differentiable function F of V_t, r_t and t. Applying Ito's lemma to obtain the stochastic differential equation for F and requiring that under the martingale measure the drift in this equation equals $r_t F_t$, one obtains a partial differential equation that his claim must satisfy:

$$F_t + \mu_r F_r + rV F_V + \frac{1}{2}\sigma_r^2 F_{rr} + \frac{1}{2}\sigma_V^2 F_{VV} + \frac{1}{2}\rho\sigma_r\sigma_V F_{rV} - rF = 0.$$

The boundary conditions for this PDE are then supplied by the payoff at maturity of the bond in the event of no default, the payoff when defaulting, i.e. when hitting the boundary K_t. Additional conditions may also be required stating that the value of the bond approaches that of a riskless bond when the firm value is large, or in the form of smooth contact conditions when the boundary K is derived from optimal exercise conditions (as in Titman and Torous (1989)). Examples of this approach are found in Black and Cox (1976); Brennan and Schwartz (1980); Titman and Torous (1989); Kim, Ramaswamy and Sundaresan (1993); and Shimko, Tejima and Van Deventer (1993).

A second approach is to try and evaluate (2) directly. One faces two problems in this approach: First, one has to be able to compute the distribution of the default time, i.e. a first passage time distribution for the diffusion V. Also, since in some models the firm may survive to maturity of the bond while still having insufficient value to cover a full repayment, one has to be able to compute distributions of V_T conditionally on V not having hit the boundary $K(t)$ before T. Such distributions are known for Brownian motion with constant drift and boundaries which are affine functions of time (see for example Harrison (1990)) but not for much more than that. Models with boundary crossing which obtain closed form solutions all use first hitting times based on Brownian motion. In Black and Cox (1976) the boundary is an exponentially affine function of time, but taking logs of firm value and the boundary we are back to the above mentioned case. Nielsen, Saa-Requejo and Santa Clara (1993) also reduces to this case except that an extra complication arises here compared to the Black and Cox model: Since interest rates are stochastic the discount factor and the first passage time distributions are correlated, but closed form solutions are obtained nevertheless. Longstaff and Schwartz (1995) propose a model in which a first hitting time problem for the Ornstein–Uhlenbeck process arises, and they use numerical techniques based on Buonocore, Nobile and Ricciardi (1987).

Shimko, Tejima and Van Deventer (1993) are able to obtain closed form solutions in a Merton type model where bankruptcy only can occur at maturity but where a Vasicek model and a geometric Brownian motion are used in (3) and (4), respectively.

The classical Merton model is an indispensable tool for discussing the distribution of the firm's value between shareholders and bondholders (possibly with several layers of seniority). In addition to the references given above, see for example the study of bond indenture provisions of Ho and Singer (1982), (1984), and the works of Cooper and Mello (1991) and Rendleman (1993) which demonstrate the clear insight provided by the Merton model into the valuation and analysis of risk sharing in swaps with default risk. Finally, we mention Johnson and Stulz (1987) who consider vulnerable options and Leland (1994) who extends the model to incorporate bankruptcy costs and taxes (which in turn makes it possible to work with notions of optimal capital structure.)

The main problem with the classical approach is that firm value is a somewhat abstract quantity which typically is not observable. For some types of issuers, such as municipal authorities, it is not clear what 'firm value' to use.

Even if many quantities are observable which could lead to sensible estimates of firm value, it would seem to be a formidable task for a financial institution dealing with thousands of default-prone counterparties to work out and update these estimates for each counterparty. Some sort of easily observed proxy for firm value might be preferable.

Also, as noted in Kim, Ramaswamy and Sundaresan (1993) realistic values of leverage and volatility of the value of the firm seem incapable of producing the yield spreads that are actually observed in the market. Of course, one could then replace the diffusion model of firm value with a jump-diffusion and use option pricing for jump-diffusions which is fairly tractable. However, again there is a problem with observability, and this is an even larger problem for jump-diffusions in which we need to estimate jump rates and jump size distributions in addition to the diffusion parameters.

Finally, it seems that the framework for modelling default risk which we are about to present offers a high degree of tractability not only for credit risky bonds but for derivative securities with credit risk in general.

4 Intensity Based Models

In newer approaches to modelling default risk the focus is on models in which the default time is specified in terms of a hazard rate. Formally, it is assumed in these models that the one-jump process $N_t = 1_{\{\tau \leq t\}}$ has an absolutely continuous compensator A, i.e. A is a predictable and increasing process such that $N - A$ is a martingale, and

$$A_t = \int_0^t \lambda_s ds \qquad (5)$$

for some non-negative, adapted[2] process λ which we will denote the intensity process or the hazard rate. The canonical example of such a process would be the process whose only jump is the first jump of a Poisson process with intensity λ. This process would have an intensity of the form $\lambda_t = \lambda 1_{\{N_t=0\}}$ and the term hazard rate captures the idea that

$$\lim_{h \to 0} \frac{1}{h} P(N_{t+h} - N_t = 1 \,|\, N_t = 0) = \lambda,$$

i.e. the probability of a jump occurring over a very small time interval is proportional to the hazard rate. The models we consider in this section are all based on this structure with the key modification being that the models of the hazard rates are gradually made more sophisticated. Whenever we condition on information available at time t, it will be understood that this information contains the natural filtration generated by the process N, i.e. we observe the default as it occurs.

4.1 Models using an independence assumption

Consider again the general valuation expression for a zero coupon bond with maturity date T :

$$v(t,T) = E_t \left(\left(\exp - \int_t^T r_s ds \right) \left(1_{\{\tau > T\}} + \delta(T, \omega) 1_{\{\tau \leq T\}} \right) \right).$$

If we assume that the spot rate process r, the stopping time τ and the recovery rate function δ are mutually independent under the martingale measure, and recalling that we assume that the conditioning information at time t contains the event $1_{\{\tau \leq t\}}$ then the expression for the bond price simplifies and becomes

$$
\begin{aligned}
v(t,T) &= E_t \exp \left(-\int_\tau^T r_s ds \right) E_t (1_{\{\tau > T\}} + \delta(T, \omega) 1_{\{\tau \leq T\}}) \\
&= \delta_t p(t,T) + 1_{\{\tau > t\}} (1 - \delta_t) p(t,T) S(t,T)
\end{aligned}
$$

---

[2] Actually, it may be chosen to be predictable. For more on this and the uniqueness of predictable versions of the intensity process, see Bremaud (1981).

where

$$S(t,T) = Q(\tau > T \,|\, \tau > t)$$
$$\delta_t = E_t \delta(T, \omega)$$

are the conditional survival probability and the conditional recovery rate respectively. This independence assumption of course simplifies modelling considerably: The default-free term structure of interest rates becomes separated from the modelling of defaultable bonds and only recovery rates and intensity functions have to be modelled. Thus the default-free zero coupon prices $p(t,T)$ in this section can arise from any consistent term structure model of ones choice.

The simplest case would be one where the recovery rate is constant and the default time is exponentially distributed with parameter λ. This example is considered in Jarrow and Turnbull (1995). Here, the authors model the dollar value of a corporate bond as a product

$$v(t,T) = e_1(t)p_1(t,T)$$

where

$$e_1(t) = \begin{cases} 1 & \text{when } t < \tau \\ \delta & \text{when } t \geq \tau \end{cases}$$

and where $p_1(t,T)$ is modelled through a forward rate curve with a jump component driven by the one-jump process $N_t = 1_{\{\tau \leq t\}}$:

$$df_1(t,T) = (\alpha_1(t,T) - \theta_1(t,T)\lambda)dt + \sigma(t,T)dW_t + \theta_1(t,T)dN_t.$$

The goal of this representation is to use a foreign currency analogy: If we were modelling a firm issuing corporate bonds in Japanese yen, it is reasonable to model the firms' forward rate structure denominated in yen and then convert to dollars by the exchange rate. Jarrow and Turnbull propose viewing 'promised dollars' of corporate bonds as a foreign currency.

The equation for the forward rate gives the evolution of the process $p_1(t,T)$ and after multiplying $e_1(t)$ one may show that

$$\frac{1}{v(t-,T)}dv(t,T) = (r_1(t) + \beta_1(t,T) - \Theta_1(t,T)\lambda)\,dt + a(t,T)dW_t$$
$$+ (\delta \exp(\Theta_1(t,T)) - 1)\,dN_t$$

where $r_1(t) = f_1(t,t)$ and $\beta_1(t,T)$, $\Theta_1(t,T)$ and $a(t,T)$ are integral processes of the type arising when modelling in the HJM framework (see Jarrow and Turnbull (1995) for the definition). The key property to note from this expression is the jump component. At the date of default τ the price of the defaultable bond is affected by two jumps: The jump in the forward rate causes $p_1(\tau-,T)$ to jump to a new value $p_1(\tau-,T)\exp(\Theta_1(\tau,T))$ and the simultaneous jump in the process $e_1(t)$ causes the extra contribution δ in the jump of v which then jumps from $v(\tau-, T)$ to the post default value $v_1(\tau-, T)\delta \exp(\Theta_1(\tau, T))$.

The assumption $\delta \exp(\Theta_1(t, T)) \neq 1$ for all $t \leq \tau$ and all T ensures that the jump in the forward rate is not exactly offset by the jump in the 'exchange rate' in which case the corporate bond would have no default risk. Conditions are then derived which ensure that the market for default-free bonds and corporate bonds is arbitrage-free.

After the default time the process of the corporate bond in this model becomes identical to that of a default-free bond. Clearly, the assumption of a constant default intensity is too simple. The default intensity of a highly rated firm would be expected to be low in the near future, but increase in the distant future as the effects of possible worsening of the firm's condition come into play. This is one motivation behind the work of Jarrow, Lando and Turnbull (1997).

There, the default time is modelled as the first time a continuous time Markov chain η with state space $\{1, ..., K\}$ hits the absorbing state K. The idea is to let the state 1 represent the highest rating (Aaa in Moody's terminology), let $K - 1$ represent the lowest rating before default, and let K be default[3]. The evolution of a continuous time Markov chain with finite state space is conveniently specified in terms of its generator matrix

$$\Lambda = \begin{pmatrix} -\lambda_1 & \lambda_{12} & \cdots & \cdots & \lambda_{1K} \\ \lambda_{21} & -\lambda_2 & \lambda_{23} & \cdots & \lambda_{2K} \\ \vdots & \ddots & \ddots & & \vdots \\ \lambda_{K-1,1} & \cdots & & -\lambda_{K-1} & \lambda_{K-1,K} \\ \lambda_{K,1} & \cdots & \cdots & \lambda_{K,K-1} & -\lambda_{KK} \end{pmatrix}$$

where

$$\lambda_{ij} \geq 0 \text{ all } i, j; \ i \neq j$$
$$\lambda_i = \sum_{j \neq i} \lambda_{ij}.$$

With state K absorbing the last row has all zeros. Recall that the generator matrix specifies the evolution of the Markov chain in the following sense: Given a starting state $\eta_0 = i$, the chain stays in this state a length of time which is exponentially distributed with parameter λ_i.[4] The chain then jumps to state $j \neq i$ with probability

$$p_{ij} = \frac{\lambda_j}{\lambda_i}.$$

---

[3]But of course the chain could also be a discrete state space approximation to firm value, and then we would have an intensity based approach using firm value!

[4]If $\lambda_i = 0$ the chain never leaves this state.

It then stays in state j an exponentially distributed length of time and so forth. From the generator matrix one easily obtains the transition probabilities by computing the matrix exponential

$$P(t) = \exp(\Lambda t)$$

where $P(t)$ is a $K \times K$ matrix whose ijth entry specifies the probability of η being in state j at time t given that the chain is in state i at time 0. Since state K is absorbing it is easy to see that the iKth entry is the probability that there is a default before time t, and this is precisely what we need for our model. To see how the model is related to (5), note that the intensity process of default here is given as $\lambda_s = \lambda_{\eta(s),K}$.

With enough securities trading one could in principle imply out the relevant parameters of this Markov chain from market prices, but in reality this will be difficult. Furthermore, since there is a lot of empirical data on the transition matrices it would seem natural to use this information. Jarrow, Lando and Turnbull (1997) propose using the following structure of the generator matrix Λ under the equivalent martingale measure:

$$\Lambda = \widehat{\Lambda} U$$

where $\widehat{\Lambda}$ is an empirically estimated generator matrix (i.e. the estimate of the transition intensities under P) and U is a diagonal matrix of possibly time dependent risk adjustments (all except the last diagonal element are positive). A method for recovering Λ from market prices is then presented in the paper. A natural way to estimate the components of $\widehat{\Lambda}$ from observations of transitions over a time interval of length T is to use the estimator

$$\widehat{\lambda}_{ij} = \frac{N_{ij}(T)}{\int_0^T Y_i(s)ds}$$

where $N_{ij}(T)$ is the number of transitions from state i to state j observed in the period and $Y_i(s)$ is the number of firms in state i at time s.

Jarrow, Lando and Turnbull (1997) did not have these detailed data available but relied instead of transition probabilities reported in Standard and Poor's Credit Review (1993) and then used a method for approximating the generator matrix from the one-year transition probabilities. Recovery rates may be found in Moody's Special report of January 1992, and risk premia were estimated using yield-to-worst data for coupon bonds with varying maturities and credit ratings.

Using ratings is a controversial issue since it is debatable to what extent prices react to ratings changes. The overall conclusion of Hand, Holthausen and Leftwich (1992) (where additional references may be found related to this problem) is that there are bond and stock price effects associated with additions to Credit Watch lists and to up-and downgrading, but the evidence is mixed. Clearly, if ratings are slow to react to information about firms it poses a problem for pricing, whereas if the adjustment occurs within a period of a few weeks the effect for valuation purposes may not be significant. A comforting property of the credit ratings is the property of stochastic monotonicity: Default probabilities do seem to decrease with increased ratings. For a study of ratings changes effect on outstanding volume in the commercial paper market, see Crabbe and Post (1994).

Despite all this, there is no doubt that a formulation involving credit ratings is necessary for the pricing of instruments whose contractual terms explicitly involve ratings changes of the issuer. An application of the Markov model above, modified to have random recovery rates, can be found in Das and Tufano (1995). Furthermore, the intensity based formulation is the natural framework from a statistical viewpoint for testing a number of hypotheses related to credit rating.

A problem in the model of Jarrow, Lando and Turnbull is that credit spreads are deterministic (holding the time to maturity fixed) as long as there are no ratings changes. An extension of the model where this is remedied may be found in Lando (1994b).

Another early use of stochastic intensity is in Madan and Unal (1995). Here the intensity is modelled as a function of the excess return on the issuer's equity s:

$$\lambda(s) = \frac{c}{\log(s/\delta)^2}$$

where c, δ are positive constants, and where s under the martingale measure satisfies

$$ds_t = \sigma s_t dB_t.$$

Note that the intensity is an increasing function of s for $s < \delta$ and a decreasing function of s for $s > \delta$. Although default here is totally inaccessible there is still a boundary $s = \delta$ at which the excess return on equity cannot cross without default occurring.

Madan and Unal (1995) also use a random recovery rate represented by letting $\delta(T, \omega)$ in (1) be a random variable independent of the processes r and $1_{\{\tau \geq t\}}$. Hence the expression for the price of a defaultable bond becomes

$$v(t, T) = E(\delta)p(t, T) + 1_{\{\tau > t\}}(1 - E(\delta))p(t, T)S(t, T).$$

A key question then is whether it is possible to imply out the relevant intensities and the recovery rate from observed market prices. If all one observes is the market prices of v it is impossible in general to imply out the hazard and the recovery rate distribution separately: It is easy to show examples of two equivalent measures under which both recovery rates and intensities are different but in such a way that the differences offset when prices are computed on bonds. (Note that if the recovery rate is 0, intensities can be implied out from market prices under technical conditions, see Artzner and Delbaen (1995)). But by combining a parametric restriction on the distribution of the recovery rate under the martingale measure with the existence of bonds of different priority, it is possible. In a slight deviation from Madan and Unal (1995) we show how this can be done if δ arises from a one-parameter family of distributions. Let δ as usual denote the fraction of the total outstanding zero coupon debt which is recovered in the event of default. Given that the fraction of the total debt held by senior bondholders is p_S, then if the recovery on all debt is δ, the senior bondholders will have a recovery rate on their debt given by

$$S(\delta) = \min\left(\frac{\delta}{p_S}, 1\right)$$

and the junior debtholders will have a recovery rate of

$$J(\delta) = \frac{(\delta - p_S)^+}{1 - p_S}.$$

Now let the density of the recovery δ to bondholders be denoted q_θ. Then the expected recovery rates as a function of θ for senior and junior debt are given as

$$\delta_S(\theta) = \int_0^1 S(y)q_\theta(y)dy$$
$$\delta_J(\theta) = \int_0^1 J(y)q_\theta(y)dy$$

respectively and therefore the values of senior and junior debt can be expressed as

$$v_S(t,T) = p(t,T)\left(S(t,T) + (1 - S(t,T))\delta_S(\theta)\right)$$
$$v_J(t,T) = p(t,T)\left(S(t,T) + (1 - S(t,T))\delta_J(\theta)\right)$$

if default has not occurred at time t. Since we may observe the market values of senior debt, junior debt and treasury bonds, we have here two equations whose unknowns are the parameter θ and the survival probabilities.

In Madan and Unal (1995) the parameter θ is two dimensional (the parameters of a beta-distribution) and they therefore use time variation in p_S to imply out θ and the survival probabilities separately.

For other works in which the independence assumption appears in some form, see Johnson and Stulz (1987); Hull and White (1995); and Litterman and Iben (1991).

4.2   Relaxing the independence assumption

Intensity based approaches must of course address the empirical evidence of correlation between default risk and yields on default-free bonds that we mentioned earlier. A framework for handling this in an intensity based setting has been laid out in Duffie and Singleton (1995b) and Duffie, Schroder and Skiadas (1995) using recursive methods and backward stochastic differential equations and in Lando (1994b) (see also Lando (1994a)) using Cox processes. We start by outlining the approach used by Lando (1994b), and then proceed to the more general approach laid out in Duffie, Schroder and Skiadas (1995).

Our goal is to produce a model where the intensity of default is adapted to a filtration $\mathcal{G}$ (which may contain information related to the default-free term structure and other state variables), i.e. we want

$$Q(\text{default in } (t, t + \Delta t] \,|\mathcal{G}_t) \simeq \lambda_t \Delta t \qquad (6)$$

for Δt small and on the set $\{\tau > t\}$. Let $\mathcal{N}$ denote the filtration generated by the one-jump process $1_{\{\tau \leq t\}}$ and let

$$\mathcal{F}_t = \mathcal{N}_t \vee \mathcal{G}_t.$$

A way to achieve this setup is as follows: Let λ be a non-negative process adapted to $\mathcal{G}$ and let E_1 be an exponentially distributed random variable with mean 1 which is independent of $\mathcal{G}$. Now define the default time τ as

$$\tau = \inf\{t : \int_0^t \lambda_s ds > E_1\}.$$

This is simply a way of constructing a model where (6) holds. With this construction the default time, conditionally on $\mathcal{G}$, is distributed as the first jump of a non-homogeneous Poisson process with intensity function equal to the given sample path of the intensity process and the unconditional survival probability is given as

$$Q(\text{default after } t) = E \exp\left(-\int_0^t \lambda_s ds\right).$$

Note how this functional is exactly of the type which maps the spot rate process into a bond price in ordinary term structure modelling. Now consider evaluating the following expressions which all have clear interpretations in terms of defaultable securities:

Let X be a $\mathcal{G}_T$−measurable random variable, let Y, Z be $\mathcal{G}_t$−adapted processes. Consider the following default related claims:

1. $X1_{\{\tau > T\}}$ - i.e. a claim which pays X if there has been no default at time T. Particular cases include a corporate zero coupon bond and vulnerable European options.

2. $Y1_{\{\tau > s\}}$ - a claim which pays at a rate of Y as long as there is no default. An example could be the stream of coupon payments on a variable-rate loan.

3. $Z_\tau 1_{\{\tau \le T\}}$ - i.e. a 'settlement' payment Z_τ paid at the time of default if default occurs before time T.

It is then straightforward to show (see Lando 1994a) that the values of these contingent claims are given by the expressions

$$E_t \left(\exp \left(- \int_t^T r_s ds \right) X 1_{\{\tau > T\}} \right) \qquad (7)$$

$$= E_t \left(\exp \left(- \int_t^T (r_s + \lambda_s) ds \right) X \right)$$

$$E_t \left(\int_t^T Y_s 1_{\{\tau > s\}} \exp \left(- \int_t^s r_u du \right) ds \right) \qquad (8)$$

$$= E_t \left(\int_t^T Y_s \exp \left(- \int_t^s (r_u + \lambda_u) du \right) ds \right)$$

$$E_t \left(\exp \left(- \int_t^\tau r_s ds \right) Z_\tau \right) \qquad (9)$$

$$= E_t \left(\int_t^T Z_s \lambda_s \exp \left(- \int_t^s (r_u + \lambda_u) du \right) ds \right)$$

There are a couple of things to note in these expressions: First, for valuation purposes the spot rate is replaced by the default adjusted spot rate given by the sum of the default free spot rate and the hazard rate. Second, nowhere in the expressions does the default time τ occur. It has been replaced by the intensity λ.

This means that when modelling a bond with (say) a constant recovery rate, all that matters for pricing is the sum of the spot rate and the hazard rate – not the individual components. So one may for example use an affine model for this default-adjusted spot rate, and this would bring us back to the framework of ordinary term structure modelling. This point may be carried even further as we shall see below.

With Markovian state variables we find the usual interpretation in terms of a PDE. Let X be an $\mathbb{R}^n$-valued diffusion with

$$dX_t = \mu(X_t, t)dt + \sigma(X_t, t)dW_t \qquad (10)$$

where W is a d-dimensional standard Brownian motion, $\mu : \mathbb{R}^n \times [0, \infty] \longrightarrow \mathbb{R}^n$ and $\sigma : \mathbb{R}^n \times [0, \infty] \longrightarrow \mathbb{R}^{n \times d}$ are measurable functions. With suitable regularity conditions the SDE defines an n-dimensional diffusion, which in particular is a (strong) Markov process. The Feynman–Kac framework allows us, for given functions $r, h : \mathbb{R}^n \times [0, T] \longrightarrow \mathbb{R}$ and $g : \mathbb{R}^n \longrightarrow \mathbb{R}$, to translate expectations of the form

$$f(t, x) = E^{t,x} \left(\int_t^T \exp\left(- \int_t^s r(X_u, u) du \right) h(X_s, s) ds \right. \tag{11}$$
$$\left. + \exp\left(- \int_t^T r(X_s, s) ds \right) g(X_T) \right)$$

into solutions of the PDE

$$\mathcal{D}f(x, t) - r(x, t)f(x, t) + h(x, t) = 0 \tag{12}$$

with the boundary condition

$$f(x, T) = g(x) \text{ for } x \in \mathbb{R}^n$$

where

$$\mathcal{D}f(x, t) = f_t(x, t) + f_x(x, t)\mu(x, t) + \frac{1}{2}\text{tr}\left[\sigma(x, t)\sigma(x, t)^T f_{xx}(x, t) \right].$$

For regularity conditions ensuring that the SDE (10) has a solution and which ensure that solutions to the PDE (12) and the functional (11) are in correspondence, see Duffie (1992). All we wish to note here is that the same technique which we use in ordinary term structure modelling for valuing contingent claims applies to defaultable claims in the intensity based framework: Each of the expressions (7), (8) and (9) are of the form (11) with the spot rate adjusted by the default intensity. As the easiest possible illustration, note that the price $v(t, T)$ of a zero coupon bond with zero recovery in the event of default, is given by

$$v(t, T) = 1_{\{\tau > t\}} E\left(\exp\left(- \int_t^T r(X_s, s) + \lambda(X_s, s) ds \right) | \mathcal{G}_t \right)$$

and this type of functional is exactly the one which is our key concern in ordinary term structure modelling.

Below we shall see that this framework can be extended into handling more complicated contingent claims defined on a Markovian state variable process.

4.3 Fractional recovery and price-dependent intensities

The contents of this section follow Duffie and Singleton (1995b) and Duffie, Schroder and Skiadas (1995).

Consider a contingent claim which promises to pay X at time T and which if default occurs at time $\tau \leq T$ pays δ_τ at the default date instead. Let

$N_t = 1_{\{\tau \geq t\}}$ be the indicator process of default and assume that $N_t - \int_0^t \lambda_s ds$ is a martingale, i.e. λ is the hazard rate of the default process. Note that λ is zero after the default time.

It then follows directly using the decomposition of N that the price process of the defaultable claim (to be exact, the claim to the remaining dividend process generated by that claim) is given as

$$S_t = E_t \left(\int_t^T \exp\left(-\int_t^u r_v dv\right) \delta_u \lambda_u du \right. \tag{13}$$

$$\left. + \exp\left(-\int_t^T r_v dv\right) X 1_{\{\tau \geq T\}} \right).$$

In the Cox process framework presented above we can combine (7) and (9) and obtain an expression for this. But what if we wish to let the price process of S enter into the expression on the right hand side? It could be the case, as in Duffie and Singleton (1995b), that we wish to model the recovery at the default date as being a fraction of the value of the claim right before default in which case δ_t depends on S_{t-}. Or we might wish more generally to let the settlement payment and the hazard rate at time t depend on S_t. We cannot use (13) in the way we did in the Cox framework – the process we wish to derive an expression for enters into the right hand side as well. Technical conditions are needed to ensure that a solution exists at all. To solve this consistency problem in general, recursive methods are needed. Here we focus particularly on two cases in which the solution becomes very tractable:

1. Recovery at default is given as a fraction of the predefault value of the claim. The hazard rate does not depend on S.

2. Recovery is given as a fraction of the predefault value, the hazard rate depends on S. Payoffs, hazard rates, spot rates are defined through Markovian state variables.

First, it will be convenient to introduce a pre-default process V: As shown in Duffie, Schroder and Skiadas (1995), if default has not occurred at time t, one may rewrite (13) as

$$S_t = V_t - E_t \left(\exp\left(-\int_t^u r_v dv\right) \Delta V_\tau 1_{\{\tau \leq T\}} \right) \tag{14}$$

where

$$V_t = E_t \left(\int_t^T \exp\left(-\int_t^u r_v + \lambda_v dv\right) \delta_u \lambda_u du + \exp\left(-\int_t^T r_v + \lambda_v dv\right) X \right).$$

In the case where V does not jump at τ, i.e. $\Delta V_\tau = 0$ with probability 1 (which will be the case if we use the Cox process setting), or when V is predictable (for which continuity is sufficient), (14) reduces to the equality $S_t = V_t 1_{\{\tau < t\}}$. and then we may consider V to be a value process of a claim which has not yet defaulted. As long as there is no default $S_t = V_t$. Suppose that for some predictable process ϕ we have $\delta_t = \phi_t V_t$, i.e. the recovery at the default date is equal to a fraction ϕ_τ of V_τ. This by the assumption that V does not jump at time τ is then equal to a payment of $\phi_\tau S_{\tau-}$.

Now it is convenient to define the discounted gains process associated with holding S in terms of the process V:

$$G_t = \exp\left(-\int_0^t r_s ds\right) V_t (1 - N_t) + \int_0^t \exp(-\int_0^s r_u du)\phi_s V_{s-} dN_s \text{ for } t \le T$$

Note that the last integral is just the discounted value of a lump sum payment at the default date. Using Ito's lemma and the fact that V does not jump at the default date one may show that the gains process is a martingale precisely when V satisfies

$$V_t = \int_0^t R_s V_s ds + m_t$$

where m is some martingale and

$$R_t = r_t + \lambda_t (1 - \phi_t). \tag{15}$$

Letting $C_t = \exp(\int_0^t R_s ds)$, i.e. $dC_t = R_t C_t dt$ we may rewrite this equation as

$$dV_t = R_t V_t dt + C_t dM_t \tag{16}$$

where $dM_t = (1/C_t)dm_t$. Using partial integration and the fact that C is continuous and of finite variation, we see that $V_t = C_t M_t$ solves (16). Hence, if we are willing to accept that $M = V/C$ is a true martingale, the boundary condition $V_T = X$ tells us that

$$V_t = E_t\left(\exp\left(-\int_t^T R_s ds\right) X\right).$$

This certainly gives a solution to the SDE (16) with the boundary condition $V_T = X$. With this formula we can state the result in Duffie and Singleton (1995b) as follows:

If $S_t = V_t 1_{\{\tau < t\}}$ and $\delta_t = \phi_t V_t$, then S satisfies (13).

In other words, the value of the defaultable claim X at time t given that default has not yet occured is given by

$$S_t = E_t\left(\exp\left(-\int_t^T R_s ds\right) X\right)$$

where R is given by (15). Note that the key trick is to let recovery be formulated in terms of a fraction of predefault value. This allows us to model the price of a credit risky derivative with fractional recovery of the pre-default value using one process for the adjusted short rate process R which incorporates spot rate, default intensity and the fractional recovery rate into one expression. To be precise, any claim X which is measurable with respect to the history of R (and hence does not for example depend on the hazard rate alone) can be priced in this fashion. From a mathematical viewpoint, the mathematics of pricing the defaultable claim X, reduces to that of pricing contingent claims in ordinary term structure models, as long as the spot rate is replaced by the adjusted process R. If we view R as a LIBOR rate, swap contracts can be viewed in this fashion since the payment of the floating payer will be a function of R and the payment of the fixed payer is just a constant set at the initiation date, see Duffie and Singleton (1995a).

A heuristic explanation of the valuation formula

$$S_t = E_t\left(\exp\left(-\int_t^T R_s ds\right) X\right) \quad \text{for } t < \tau \tag{17}$$

in the case where $\delta_t = \phi_t S_{t-}$ (and where we still assume that the hazard is not a function of S) may also be given in terms of thinning of a Poisson process: Consider a fictitious claim which, instead of providing the holder with the fractional recovery ϕ at default, lets the holder of the defaulted claim enter a lottery which with probability $1 - \phi$ allows the holder of the claim to retain the full predefault value and with probability ϕ declares a complete default with zero recovery. By the risk neutrality property of the martingale measure, S and this artificially defined security have the same price. But it is also clear that the fictitious claim corresponds to having a default process whose default intensity has been thinned by the process $1 - \phi$ and whose recovery has been made zero. Hence this may be valued directly using (in the Cox process case) (7) with the thinned intensity $(1 - \phi)\lambda$.

Before considering price dependent hazard rates, two more comments might be helpful: First, one may note that in Jarrow and Turnbull (1995) and Jarrow, Lando and Turnbull (1997) an adjusted spot rate of the form $r + (1 - \delta)\lambda$ also appears. This may be surprising given that the recovery characteristics used there are fractional recovery of a treasury zero coupon bond of equal maturity, not of predefault value. The explanation is as follows: Note that if the value of a corporate bond is given by the expression

$$v(t, T) = \delta p(t, T) + 1_{\{\tau > t\}}(1 - \delta)S(t, T)p(t, T).$$

then as $T \downarrow t$, and given that default has not yet occured, the value of $v(t, T)$ approaches $p(t, T)$ (which is close to 1) and if a default occurs this value is reduced to $\delta p(t, T)$ (which is close to δ). Hence for very short maturities the recovery mechanism becomes a fractional recovery of predefault value. But note that this short rate should not be used in that setup to price contingent claims as in (17), unless the recovery δ is changed from a fractional payout at maturity to a fractional payout of predefault value at the default date.

Second, the formula (17) illustrates clearly the point made earlier that if we alter hazard rates and recovery rates but preserve the distribution under the equivalent martingale measure of R, it will be possible to create models where hazard rates are different but bond prices are the same. It is therefore not generally possible to infer hazards from prices unless additional claims are introduced.

In Duffie, Schroder and Skiadas (1995) and Duffie and Singleton (1995b) the above results are generalized to consider the case where in addition to the recovery rate the hazard rate λ is a function of the price of the defaultable claim itself:

$$\lambda_t = \lambda(\omega, t, S_{t-})$$

where $\lambda(\cdot, \cdot, x)$ is an adapted process for each x. Here, we mention the special case where an underlying Markov process process X is given such that the defaultable claim has the form $g(X_T)$ and the adjusted process R given by (15) has the form $R(\omega, t, S_{t-}) = \rho(X_t, S_{t-})$. In this case the price of the defaultable security may under technical conditions be expressed as a solution to the following recursive equation

$$S_t = E_t \left(\int_t^T -S_u \rho(X_u, S_u) du + g(X_T) \right)$$

which in this Markovian case has a PDE representation: This representation, as given in Duffie and Singleton (1995b), tells us that $S_t = f(X_t, t)$ where f satisfies the so called quasi-linear PDE

$$\mathcal{D}f(x, t) - \rho(x, f(x, t)) f(x, t) = 0 \qquad (18)$$

with the boundary condition

$$f(x, T) = g(x) \qquad x \in \mathbb{R}^n.$$

Duffie and Huang (1995) apply numerical solution of this PDE in the study of swaps with two-party default risk.

4.4   Linear versus non-linear valuation

All pricing in this arbitrage based framework is of course linear, so the distinction here refers to non-linearity of pricing in terms of the promised payments. In other words, should a defaultable claim which promises to pay twice the amount also be twice as valuable? It is easy to see that the PDE (12) is linear in the promised payoff. If we multiply g and h by a factor, this factor is multiplied onto the solution as well. On the other hand in (18) there is non-linearity: If we multiply g by a factor, it is not true in general that the solution gets multiplied by that same factor simply because of the way that

f itself enters into the coefficient R. The fact that the hazard depends on S implies that the true payoff of a claim whose promised payment is $cg(X_T)$, is not c times the true payoff of the promised payment $g(X_T)$.

Whether linearity is a feasible approximation to reality is difficult to say. Certainly, if a derivative is a very small part of the firms' operations it may be reasonable to assume that the value of the contract does not really influence the default probability. On the other hand there is of course a limit: If a bond issue very heavily influences the cash flow of a firm, it may be necessary to include the non-linear effect.

Another question which arises frequently is whether, even in the linear case, it is reasonable to price several coupons promised by the same firm as a sum of prices of the individual (promised) zero coupon bonds. This is no problem. One may easily check that given three firms A,B,C with the same hazard rate but whose defaults are generated by (say) independent jump processes, a portfolio consisting of two zero coupons with maturity T and U issued by firm A, should have the same price as a portfolio consisting of one maturity T zero coupon issued by firm B plus one maturity U issued by firm C. The fact that payments in the first portfolio have a different correlation structure will be relevant for questions of how well-diversified the portfolio is but not for pricing.

5   Applications and Future Directions

We conclude by mentioning some directions that research in the credit risk area is taking or might take.

First, as credit risk models have become more advanced, our understanding of swap rates between counterparties with default risk has been greatly improved. It has been known for a while that the classical 'comparative advantage' scenario often described in introductions to swap markets should be treated with some caution. While it is possible that asymmetric information, differences in tax regimes or other imperfections may cause the 'comparative advantage' of borrowing on different terms (fixed vs. floating, for example) to arise[5] it is also quite possible to have this scenario arise in a competitive, arbitrage free market. In this case the perceived comparative advantage actually represents a compensation for risks taken, and a party entering into a swap should understand this risk. Some works dealing with this problem include Turnbull (1987); Cooper and Mello (1991); Rendleman (1993); Artzner and Delbaen (1990); Duffie and Singleton (1995a); and – in an intensity based framework with two defaultable counterparties – Duffie and Huang (1995).

---

[5]This means in the typical presentation that there exists a firm whose spread between the fixed rate and the floating rate at which it can borrow at a fixed maturity is different than the corresponding spread of another firm

Second, it is important to learn what sort of dependencies exist between defaults of different firms. If defaults are not completely independent, the question is what type of dependence exists. In the Cox process framework we obtain conditional independence given the information in $\mathcal{G}$ as long as we choose the exponential defining default of individual firms to be independent. This approach is considered in Lando (1994b). In this case the environment specified by, say, the stage in the business cycle or the level of interest rates determine the likelihood of default of all firms, but one firm's default does not increase or decrease another firm's likelihood of default. This allows an APT like argument to be applied in the determination of the equivalent martingale measure in incomplete models.

A stronger type of dependence would be similar to the 'weakening by failure' notion of reliability theory, in which the default intensity of one firm increases (or decreases) at the default date of another firm. A manifestation of this in market prices would be jumps in the prices of other defaultable securities as a reaction to one default. Finally, there is the strongest type of dependence in which simultaneous defaults occur, i.e. defaults are induced by a common jump process. This type of dependence is of course the domino effect which – if it exists – would cause some to worry about the stability of the financial system.

Dependence between different issues is analyzed in Duffie, Schroder and Skiadas (1995) who also consider the effects of changes in the underlying filtration on pricing of defaultable securities.

Third, it will be natural to include the techniques of hazard rate estimation into empirical work in the area. This framework allows almost all conceivable hypotheses to be formulated and tested in a unified framework. It will allow identifying relevant covariates for predicting defaults and, combined with models of these covariates' evolution, provide a help in pricing.

References

Artzner, P. and F. Delbaen (1990). 'Finem lauda or the risks in swaps', *Insurance, Mathematics and Economics* **9**, 295–303.

Artzner, P. and F. Delbaen (1995). 'Default risk insurance and incomplete markets', *Mathematical Finance* **5**, 187–195.

Black,F. and J. Cox (1976). 'Valuing corporate securities: Some effects of bond indenture provisions', *Journal of Finance* **31**, 351–367.

Brémaud, P. (1981). *Point Processes and Queues - Martingale Dynamics*. Springer.

Brennan, M. and E. Schwartz (1980). 'Analyzing convertible bonds', *Journal of Financial and Quantitative Analysis* **15**, 907–929.

Buonocore, A., Nobile, A.G. and L.M. Ricciardi (1987). 'A new integral equation for the evaluation of first-passage-time probability densities', *Advances in Applied Probability* **19**, 784–800.

Cooper, I. and A. Mello (1991). 'The default risk of swaps', *Journal of Finance* **46**, 597–620.

Crabbe L. and M. Post (1994). 'The effect of a rating downgrade on outstanding commercial paper', *Journal of Finance* **49**, 39–56.

Das, S. and P. Tufano (1995). 'Pricing credit-sensitive debt when interest rates, credit ratings and credit spreads are stochastic', Working Paper, Harvard Business School.

Duffee, G. (1995). 'The variation of default risk with treasury yields', Working Paper, Federal Reserve Board, Washington DC.

Duffie, D. (1992). *Dynamic Asset Pricing Theory.* Princeton University Press.

Duffie, D. and K. Singleton (1995a). 'An econometric model of the term structure of interest rate swap yields', Working Paper. Stanford University.

Duffie, D. and K. Singleton (1995b). 'Modelling term structures of defaultable bonds', Working Paper. Stanford University.

Duffie, D. and M. Huang (1995). 'Swap rates and credit quality', Working Paper. Stanford University.

Duffie, D., Schroder, M. and C. Skiadas (1995). 'Recursive valuation of defaultable securities and the timing of resolution of uncertainty', Working Paper. Stanford University and Northwestern University.

Geske, R. (1977). 'The valuation of corporate liabilities as compound options', *Journal of Financial and Quantitative Analysis* **12**, 541–552.

Hand, J., Holthausen R. and R. Leftwich (1992). 'The effect of bond rating announcements on bond and stock prices', *Journal of Finance* **47**, 733–750.

Harrison, J.M. (1990). *Brownian Motion and Stochastic Flow Systems.* Krieger Publishing Company.

Ho, T. and R. Singer (1982). 'Bond indenture provisions and the risk of corporate debt', *Journal of Finance* **41**, 375–406.

Ho, T. and R. Singer (1984). 'The value of corporate debt with a sinking fund provision', *Journal of Business* **57**, 315–336.

Hull, J. and A. White (1995). 'The impact of default risk on the prices of options and other derivative securities', *Journal of Banking and Finance* **19**, 299–322.

Jarrow, R. and S. Turnbull (1995). 'Pricing options on financial securities subject to credit risk', *Journal of Finance* **50**, 53–85.

Jarrow, R., Lando D. and S. Turnbull (1997). 'A Markov model for the term structure of credit risk spreads', *Review of Financial Studies* **10**, (2).

Johnson, H. and R. Stulz (1987). 'The pricing of options with default risk', *Journal of Finance* **42**, 267–80.

Kim J., Ramaswamy, K. and S. Sundaresan (1993). 'Does default risk in coupons affect the valuation of corporate bonds', *Financial Management*, Autumn, 117–131.

Lando, D. (1994a). 'On Cox processes and credit risky bonds', Preprint No.9, 1994, Institute of Mathematical Statistics, University of Copenhagen.

Lando, D. (1994b). *Three essays on contingent claims pricing.* PhD Dissertation, Cornell University.

Leland, H. (1994). Corporate debt value, bond covenants, and optimal capital structure. *Journal of Finance* **49**, 1213–52.

Litterman, R. and T. Iben (1991). 'Corporate bond valuation and the term structure of credit spreads', *The Journal of Portfolio Management* **17**, 52–64.

Longstaff, F. and E. Schwartz (1995). 'A simple approach to valuing risky fixed and floating rate debt', *Journal of Finance* **50**, 789–819.

Madan, D. and H. Unal (1995). 'Pricing the risks of default', Working Paper, University of Maryland.

Merton, R.C. (1974). 'On the pricing of corporate debt: The risk structure of interest rates', *Journal of Finance* **2**, 449–470.

Nielsen, L.T., Saá-Requejo, J. and P. Santa-Clara (1993). 'Default risk and interest rate risk: The term structure of default spreads', Working Paper, INSEAD.

Rendleman, R. (1993). 'How risks are shared in interest rate swaps', *The Journal of Financial Services Research* **7**, 5–34.

Shimko,D., Tejima, N. and D. van Deventer (1993). 'The pricing of risky debt when interest rates are stochastic', *The Journal of Fixed Income* **3**, 58–65.

Standard and Poor's (1993). 'Corporate default, rating transition study updates', *Standard and Poor's CREDITREVIEW*, January 25 1993, 1–12.

Titman, S. and W. Torous (1989). 'Valuing commercial mortgages: An empirical investigation of the contingent claims approach to pricing risky debt', *Journal of Finance* **44**, 345–373.

Turnbull, S. (1987). 'Swaps: A zero sum game', *Financial Management* **16**, 15–21.

Term Structure Modelling Under Alternative Official Regimes

Simon Babbs and Nick Webber

Abstract

Monetary authorities exercise control of domestic short term interest rates. We argue that models of the term structure of interest rates must take into account the consequences of this control if they are to capture important empirical features of interest rate dynamics. In particular, previous one or two factor models of the term structure may not adequately describe the short rate process. Interest rate regimes in the United Kingdom, the United States, France and Germany are described. A common characteristic is the presence of 'non-effective' official interest rates used for signalling purposes. We construct a term structure model, consistent with general equilibrium, that captures, in idealised form, some important features of these regimes. The chapter concludes with an illustrative example of a regime in which the short rate is constrained to lie within an officially controlled corridor.

1 Introduction

A recent paper by Babbs and Webber (Babbs and Webber (1994)) developed a term structure model in which the rate setting behaviour of the monetary authorities played an explicit role. Monetary authorities within a currency's domestic market attempt to control the level of short term interest rates in that currency. Although a substantial literature exists on exchange rates and monetary policy where official intervention is explicitly modelled (see, for instance, Krugman and Miller (1992) and Svensson (89)) very few authors have attempted to take official intervention into account in the development of arbitrage-free term structure models[1].

Focusing on the behaviour of the United Kingdom authorities over recent decades, Babbs and Webber were motivated to develop a theoretical framework in which the short term interest rate, r, follows a pure jump process, with fixed jump sizes, and with stochastic intensities influenced by a diffusion

---

[1]A step in this direction was taken by Balduzzi, Bertola and Foresi (1993). See Babbs and Webber (1994) for a fuller discussion. Lindberg and Perraudin (1995) present a model that is a special case of Babbs and Webber. They are able to obtain computationally efficient solutions within their framework.

state variable, x. Simulations of an illustrative model, constructed within this framework, mimicked successfully various empirical regularities of UK interest rates. Diffusion models of the term structure such as Duffie and Kan (1993), Heath, Jarrow and Morton (1992), or extended Vasicek models due to several authors, cannot satisfactorily capture the observed jump-diffusion behaviour of short maturity rates in the UK regime, such as three-month Libor. Jump-diffusion extensions, such as Shirakawa (1991), Ahn and Thompson (1988), or Bjork, Kabanov and Runnaldier (1995) do not attempt to take rate setting behaviour into account.

As Babbs and Webber pointed out, while the existence of rate setting behaviour by the monetary authorities is common to many economies, the precise form of this behaviour differs substantially from country to country, necessitating generalisations of Babbs and Webber's theoretical framework.

In this article we present a generalisation to multiple jump and diffusion state variables, and in which the short rate, r, is a specified function of the state variables, rather than as in Babbs and Webber where r is directly equal to a particular state variable following a pure jump process.

In the chapter we motivate the generalisation by a desire to take account of the existence of additional officially set 'discount' and 'Lombard' rates (however termed in particular countries) which, as discussed below, enable the monetary authorities to signal to the markets their intentions concerning the future levels of the short rate. In practice these rates provide 'soft' floors and ceilings for short term interest rates.

While modelling rate setting behaviour has theoretical interest, it also has significant consequences for pricing and hedging interest rate instruments. The presence, for instance, of an officially controlled ceiling on the short rate may be expected to have a significant effect upon the valuation of short dated instruments, particularly when the short rate is close to the ceiling. There is empirical evidence, for instance, that the German Lombard rate has a significant effect upon the distribution of German short rates, when they are close to the ceiling[2]. Existing models are unable to take such effects into account.

We provide simplified specialisations of this framework to provide stylised representations of the rate setting behaviour of the monetary authorities in the United States, France and Germany. The original Babbs and Webber framework for the UK emerges as a special case. The focus in the present paper is upon signalling aspects of discount and Lombard rates. It does not deal explicitly with other official actions, such as repo auctions.

The results of this article are of relevance to the management of interest rate risk, and the pricing of interest rate derivatives. The results are of interest in the manner in which changes to boundaries may influence the shape of the term structure, in other words, the effect of signalling behaviour upon the

---

[2]Honore (1996).

term structure. Although the exact mechanisms will vary from economy to economy, a change to an official rate may cause the term structure to alter, even if the short rate remains unchanged. This is supported by observations of market practice, where the likelihoods of changes to official rates are studied and taken into account.

The plan of the chapter is as follows. In section two we describe the main features of the short term interest rate regimes in several currencies. Drawing from these examples, in section three we construct a model that captures in mathematical form important characteristics of the regimes. In section four we discuss the form of the state variables that might be present, and in section five we show how idealised versions each of the regimes discussed in section two might be incorporated within the framework. Section six concludes with a numerical illustration.

Acknowledgements

The authors are grateful for discussions with Jorgen Nielsen and for thoughtful comments from an anonymous referee.

Disclaimer

The views expressed in this paper are those of the authors, and not necessarily those of any organisations to which they are affiliated.

2 Monetary Authorities and Interest Rate Regimes

We briefly examine the regimes imposed in several economies. Except where otherwise noted, the descriptions are based upon those given by Schnadt (94). The descriptions are by necessity a summary only, outlining only those features that we believe are most pertinent to term structure modelling. Furthermore, the descriptions in this paper are based upon the regimes in existence in the period covered by Schnadt.

Monetary authorities exist because of the need to service the economic and regulatory requirements of the domestic banking system, and to service the needs of the domestic government. Regulatory structures may explicitly exist to control the levels of bank reserves. For instance, banks may be required to maintain operational deposits of liquid assets with the monetary authorities, above a minimum percentage of their capital base. These reserve levels are frequently calculated on an averaging basis that may have some degree of complexity. Since reserves may be non-interest bearing, and in any case represent an opportunity cost, banks are motivated to keep reserve levels to a minimum, and reserve management may be a significant aspect of day-to-day banking activity.

Because levels of short term liquidity varies exogenously, the aggregate level of reserves is subject to exogenous pressure. There may be significant fluctuations in the demand for short term funds. The sole net supplier, or purchaser, of marginal reserves within an economy is the monetary authorities. The level of interest rates at which marginal liquidity is supplied, or removed, from the banking system is set at the discretion of the monetary authorities. This provides the mechanism by which the authorities are able to set the level of short term interest rates as part of a macro economic management policy. This freedom to set rates is however constrained. If rates are set at levels that result in liquidity flows that are predominately into, or conversely, predominately out of, the monetary authorities, one party or the other to the transaction will eventually be unable to sustain the flows.

The mechanisms by which the roles sketched above are implemented varies from economy to economy. In practice a monetary authority, on a day-to-day basis, must forecast the net liquidity requirements of the domestic banking system. Based on its forecasts it will then make funds available to the market, or absorb funds in the event of a surplus. In the examples described below funds are frequently managed by the authorities' open market operations in the sale or purchase of repos, or 'eligible bills'. Supply of funds is often accomplished by a tendering system. Purchase of funds may be accomplished at a published fixed rate. The frequency and volume of intervention varies widely from economy to economy. In the UK there may be several auctions per day. In Germany auctions are held once a week, although there may be additional daily fine tuning through, for instance, switching public sector deposits between the Bundesbank and the commercial banks.

Any residual imbalance between reserves supply and demand, after the authorities' intervention, is met on the interbank market. Overnight interbank rates may fluctuate considerably in response to short term day-to-day residual liquidity requirements. Overnight and short term interbank rates therefore reflect the degree and direction of the monetary authorities residual forecasting and supply errors. These effects are purely short range and may not be expected to effect market expectations of future levels of rates. We argue that overnight interbank rates do not reflect the underlying levels of the short term interest rate, and that the short rate is better represented by the official interest rates set by the monetary authorities in supplying money to the market. We refer to the effective rate at which the authorities supply or remove funds as the intervention rate. The overnight interbank rate typically fluctuates about the level of the intervention rate.

Even though the intervention rate may, in some regimes, be a two week rate, and may not be continuously available, nevertheless the intervention rate effectively sets the level of short term interest rates. Market expectations of the prospective levels of the intervention rate are widely discussed both in practitioner publications and the popular press, and it may be supposed that these expectations are impounded into the spot rate curve. In terms of term structure formation, the intervention rate behaves as if it were the short rate. It does not seem unreasonable to identify the short rate with the intervention rate.

Signalling

In addition to rate setting behaviour, the authorities may consider that part of their role is to smooth interest rate variability by signalling in advance to the markets their future intentions in interest rate transactions. This allows the authorities to more effectively control rates beyond simple direct rate setting.

Signalling may be facilitated by publishing interest rates at which alternative levels of intervention might take place. Typically, these rates are floors or ceilings to the rate at which the authorities may immediately intervene in the markets. A change to such a rate is a signal of the authorities' future intentions, even though that rate may not be one at which a large volume of transactions is normally made. For instance in Germany banks do not normally wish to borrow at the Lombard rate because it is expensive, and they do not borrow significantly at the cheap discount rate because strict quotas are applied to its use. The Bundesbank may therefore alter the discount rate or the Lombard rate at will without necessarily altering the value of the short rate in the DM money market. It uses changes in these two rates to signal to the market its intentions concerning future levels of the short rate. For instance, if the discount rate is raised, this may be interpreted as a signal that the authorities do not expect, or will not allow, rates to fall below the new level in the near future. Similarly, the Lombard rate acts as a ceiling to the short rate, an with analogous signalling interpretation.

Signalling by changing floor and ceiling rates will affect the term structure, even through the intervention rate itself may not have changed, because the market's expectations of future levels of the short rate will have been altered. We contend that these signalling aspects should be modelled to more fully account for changes in the term structure at short maturities. For instance, in the DM money market the Lombard and discount rates are closely monitored and the likelihood of possible changes is much discussed. These likelihoods are inevitably impounded into interest rate expectations and hence into the term structure itself.

In the examples below, floor and ceiling rates typically remain fixed for extended periods, being changed from time to time by discrete amounts. These rates may thus be modelled as jump processes. The floor and ceiling rates may coincide, forcing the short rate to be a jump process.

We now briefly discuss in turn the regimes operated in the United Kingdom, Germany, France and the United States, as reported by Schnadt, to see how these features are implemented in specific economies.

United Kingdom

Since the early 1980s the banking system has been purposefully kept short of funds[3]. The Bank of England injects money into the system on a daily basis. This is done by the purchase of eligible bills. The Bank accepts tenders above a minimum rate, called the stop rate, in each of a number of maturity bands, up to a level of forecasted cash requirements. The band of shortest maturity is called band one. It covers bills maturing in between one and fourteen days, although the greatest number of transactions are for bills of 3 or 4 days maturity. Since banks must obtain residual funds through the interbank market, overnight and very short term interbank rates are especially sensitive to residual cash requirements resulting from the inevitable mismatch between the Bank of England forecast of cash requirements and actual market requirements.

In Babbs and Webber it was argued that the major determinant on money market interest rates was the rate at which the Bank of England offers assistance to the market, the band one stop rate. This remains fixed from day to day over periods of weeks or months. When it is changed it jumps to a new level. Babbs and Webber argued that UK interest rates could be modelled by requiring the short rate to follow a pure jump process.

A plot of three month sterling libor is shown in Figure 1, taken from Babbs and Webber (1994). The plot also shows the value of UK base rate, which closely follows the band one stop rate. It is clear from the graph that the values of libor and the base rate are closely linked.

Germany

Banks are required to maintain high average reserves, relative to countries such as France or America. Cheap funding may be obtained from the Bundesbank by borrowing at the discount rate. However, there is a quota imposed on how much may be borrowed at this rate, and the level of this quota has been declining. More expensive funding may be obtained at the Lombard rate. This is set at a level one percent or so above the discount rate, but funds are relatively freely available. The main method by which the Bundesbank makes funds available to banks is by operations in the repo market, although other forms for 'fine-tuning' are available. It holds weekly auctions of repos, predominantly of 2 week maturity. It does so in one of two ways, each way leading to qualitatively different behaviour of the short rate. The Bundesbank invites tenders from the banks on either on a fixed rate basis,

---

[3]For an early reference see the Quarterly Bulletin of the Bank of England, 1982.

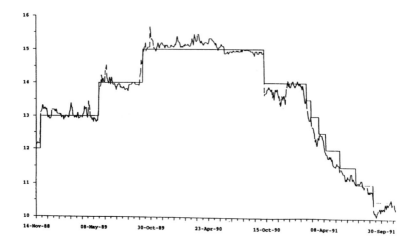

Figure 1: 3–month LIBOR and base rates: actual data

'mengentender', where it announces a rate and allows banks to tender on a volume basis only, or on a floating basis, 'zinstender', where it allows the market to tender both for volumes and for rates. Tenders above a minimum rate, chosen by the Bundesbank, are accepted.

In the first regime the level of the repo rate is determined by the Bundesbank. The repo rate is thus a third official rate, lying between the discount Rate and the Lombard Rate. In the second regime, the level of the repo rate is effectively determined by the demand for funds in the market. The Bundesbank may switch between regimes at its discretion, and speculation about its intentions is an important topic in the DM money markets.

The overnight money market rate, the Tagesgeld rate, is not very volatile. Figure 2, taken from Schnadt, shows the Tagesgeld rate, the Lombard rate and the discount rate for the period 1979 to 1989.

France

Average reserve requirements in France are low to the point that day to day fluctuations in banks' cash positions due to trading requirements may significantly interfere with the banks' average reserves position. To maintain the benefits of low reserve requirements while avoiding negative balances presents special cash management problems. This means that marginal funding requirements, normally met through the interbank market, cause the overnight interbank rate to be very sensitive to reserves pressure.

The Banque de France allows banks to borrow from it by inviting tenders for 7 day repos. Auctions are held twice a week. It accepts bids above a

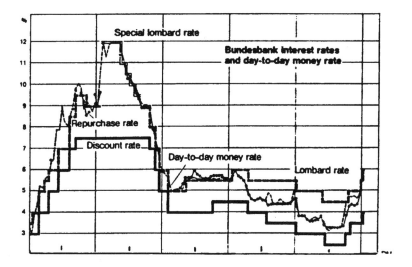

Figure 2: German Day-to-day Interbank Interest Rate 1979–1989
(Schnadt (1994), Source: Deutsche Bundesbank)

minimum rate, uniformly at this rate, the 'taux des pensions appel d'offres'.
Bids are scaled so that the volume matches the Banque de France's forecast
of demand. Additional funding may be obtained on the banks' initiative, at
the 'taux des pensions de 5 a 10 jour'. This rate has usually been around
0.75% greater than the taux des pensions appel d'offres.

Figure 3, taken from Schnadt, shows the taux des pensions appel d'offres,
the taux des pensions de 5 a 10 jour, and the overnight interbank rate in
1991 and 1992. On only one occasion did the two official rates not change
simultaneously.

United States

Banks must maintain reserves over a two week averaging period. There is a
very active interbank market through which most funds are obtained. The
Fed permits banks to borrow at an official discount rate, but this form of
funding is limited and discouraged by non-price mechanisms. The Fed sets
a 'fed funds target rate' and by active open market operations attempts to
keep the fed fund overnight rate close to the target rate.

The main supply of liquidity is via overnight repos. Tenders are invited
and bids are accepted down to a minimum 'stop-out rate', or until the target
volume is achieved. Borrowing at the discount rate is discouraged. The fed
funds rate is normally greater than the discount rate. This is not absolute,
however; the fed funds rate may fall below the discount rate.

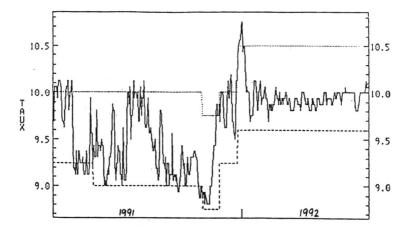

Figure 3: French Interbank Rates 1991–1992 (Schnadt (1994),
Source: Banque de France (1992))

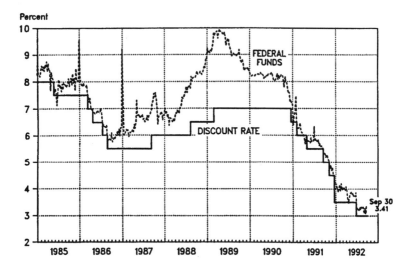

Figure 4: Fed Funds Rate and Discount Rate 1985–1992 (Schnadt
(1994), Source Federal Reserve)

Figure 4, taken from Schnadt, shows the fed funds rate and the discount
rate from 1985 to 1992.

3 Mathematical Formulation

We wish to model key features of the interest rate regimes that have been described in section 2. In a general formulation the term structure will need to be a function of several state variables. The discussion above has mentioned informally potential factors such as 'reserves pressure', the intervention repo rate, the prospective intervention rate, etc. Variables such as 'reserves pressure' may well exhibit a cyclical pattern, becoming particularly pronounced towards the end of a reserves maintenance period. This effect may have practical significance, because of its possible implications for interest rate behaviour, and could be explicitly modelled. We shall ignore it in this article.

We have also seen the need to include jump variables into the model, representing, for instance, potential floor and ceiling rates. In the examples we saw above, neither the intervention rate nor the floor or ceiling rates need be as short as overnight. In Germany the intervention rate is normally a two week rate, while in France the ceiling rate is approximately a one week rate. This leads to the existence of 'soft' floors or ceilings. In this paper we assume that all official rates are instantaneous rates. Thus we cannot explicitly account for the existence of 'soft' floors or ceilings. We can do so implicitly, however, by allowing bounds to be breached because of 'liquidity factors'. Where an official short rate, realised through the use of overnight repos, for instance, is constrained to admit bounds, an overnight interbank rate may not be constrained to lie within these bounds. This is discussed more fully in section 4.

We allow all coefficient functions and intensities to be functions of the state variables, the official rates, and time. We suppose that there are N diffusion state variables in the economy,

$$X_i, \quad i = 1, \ldots, N.$$

In addition we suppose there are M state variables in the economy following pure jump processes,

$$Y_j, \quad j = 1, \ldots, M.$$

In this article we restrict ourselves to jump processes each of whose jump sizes belong to a known finite set. This is not a severe restriction since we shall use the jump variables to model official interest rates such as the Lombard or discount rate. These rates typically jump in a small number of multiples of a minimum jump size, such as 0.25%.

We allow the short rate r to be a function of the state variables. Babbs and Webber took $M = 1$ and identified the short rate with $Y = Y_1$. Other one or two factor models of the term structure may identify r with a diffusion state variable. For instance, both Fong and Vasicek (1991, 1992) and Brennan and Schwartz (1979) identify one of their two state variables with the short rate. The second state variable is taken to be the short rate volatility for Fong and

Vasicek and the yield on a perpetual coupon bond for Brennan and Schwartz[4]. Longstaff and Schwartz (1991) set up a model in terms of two state variables representing asset returns related and unrelated to volatility, respectively. In their model the short rate is expressible as a linear combination of the two state variables. However there is no reason why r should not be a complicated function of the state variables. Longstaff and Schwartz are a special case of the general equilibrium framework of Cox, Ingersoll and Ross (1985a,b). In that framework the short rate may be expressed as a function of a set of underlying state variables. r is given by

$$r = \frac{1' \cdot \Omega^{-1} \cdot \alpha - 1}{1' \cdot \Omega^{-1} \cdot 1}$$

where α is a vector of returns to a set of production processes and Ω is the covariance matrix of returns to the set of production processes. Both α and Ω are functions of the set of underlying state variables. It may be seen that in this framework the short rate r may be an extremely complex, and in general non-linear, function of the state variables. Ahn and Thompson generalise Cox, Ingersoll and Ross by admitting jump processes within their specification. It can be shown that with non-stochastic jump magnitudes the short rate is given by

$$r = \frac{1' \cdot \Omega^{-1} \cdot \alpha - 1}{1' \cdot \Omega^{-1} \cdot 1} + \frac{1' \cdot \Omega^{-1} \cdot \delta}{1' \cdot \Omega^{-1} \cdot 1},$$

where δ is a vector whose components are related to the jump magnitudes and jump intensities.

In our general framework, rather than on the one hand requiring r to be identified as a state variable, or on the other requiring r to have a particular functional dependence upon the state variables determined within an explicit general equilibrium framework, we allow r to be a relatively arbitrary function of the state variables.

In our context the short rate may either be an official rate controlled by the authorities, or it may be a rate partially determined by the market. In the former case the functional form of r represents the authorities' response to market pressure as described by the state variables. In the latter case the market's beliefs about the pattern of the authorities' future behaviour influences the demand for funds of various maturities.

The process for X_i will be written

$$dX_i = a_i(X_1, \ldots, X_N, Y_1, \ldots, Y_M, t) \cdot dt + b_i(X_1, \ldots, X_N, Y_1, \ldots, Y_M, t) \cdot dz_i \,,$$

where a_i and b_i are functions of the X_i and Y_j, z_i is a standard Wiener process, and

$$dz_i \cdot dz_j = \rho_{ij} \cdot dt.$$

---

[4]Hogan subsequently showed the Brennan and Schwartz model to be unstable.

Each Y_j will be represented as a sum of point process

$$Y_j(t) = Y_j(0) + \sum_{k=1}^{N_j} c_{jk} \cdot N_{jk}(t),$$

where c_{jk}, $k = 1, \ldots, N_j$ are constants, and each N_{jk} is a counting process. c_{jk} is the size of the jump to Y_j that occurs when N_{jk} jumps. Each Y_j has a finite number N_j of allowable jump sizes. Each N_{jk} has an associated jump intensity, λ_{jk}. We allow each intensity to depend upon the full set of state variables and jump processes, and current time,

$$\lambda_{jk} = \lambda_{jk}(X_1, \ldots, X_N, Y_1, \ldots, Y_M, t).$$

We write

$$B(t, T) = B(t, T, X_1, \ldots, X_N, Y_1, \ldots, Y_M)$$

for the value at time t of a pure discount bond yielding 1 at time T. Within this framework, under a number of mild restrictions, we are able to price $B(t, T)$ by solving a set of partial differential difference equations.

We wish to allow for the possibility of more than one variable jumping simultaneously. For instance, a monetary authority may wish to change both a ceiling rate and a floor rate at the same time. We incorporate this into the framework as follows. We define an 'event' to be the occurrence of an allowable set of simultaneous jumps. A description of an event must specify which of the variables Y_j jump in the event, and by how much. An event v may thus be identified with a vector $v = (v_1, \ldots, v_M)$, where

$v_j = 0$, if Y_j does not jump,

$v_j = k$, if Y_j jumps to $Y_j + c_{j,k}$ in the event v.

For notational convenience we define $c_{j0} = 0$, so that the event $v = (v_1, \ldots, v_M)$ defines the simultaneous jump

$$Y_j \to Y_j^v = Y_j + c_{j,v_j} , \quad j = 1, \ldots, M.$$

The set of all allowable events, V, is a subset of the set of all possible combinations of jumps

$$V \subseteq \prod_{j=1}^{M} \{0, \ldots, N_j\},$$

where the right hand side contains the 'null event' $(0, \ldots, 0)$.

For each $v \in V$, we define the new value of the bond $B(t, T)$ after the set of jumps v has occurred to be

$$J_v(B) = B(t, T, X_1, \ldots, X_N, Y_1^v, \ldots, Y_M^v).$$

With this notation we then obtain the following theorem:

Theorem

$$B(t, T) = B(t, T, X_1, \ldots, X_N, Y_1, \ldots, Y_M)$$

obeys the following partial differential difference equation:

$$\frac{1}{2} \sum_{i=1}^{N} b_i^2 \frac{\partial^2 B}{\partial X_i^2} + \sum_{i,j=1}^{N} \rho_{ij} \cdot b_i \cdot b_j \cdot \frac{\partial^2 B}{\partial X_i \partial X_j} + \sum_{i=1}^{N} a_i^* \cdot \frac{\partial B}{\partial X_i} + \frac{\partial B}{\partial t} +$$

$$\sum_{v \in V} \lambda_v^* \cdot [J_v(B) - B] = r \cdot B$$

where

$a_i^* = a_i - \theta_i \cdot b_i$

ρ_{ij} is the instantaneous correlation between z_i and z_j,

θ_i, $i = 1, \ldots, N$, are (relatively arbitrary) price of risk functions, and

λ_v^*, $v \in V$, are risk adjusted jump intensities.

This theorem is a simple extension of theorem 6.2 in Babbs and Webber (1994). Its proof is omitted. Note that there is one equation for every possible set of values that the set of jump variables $\{Y_1, \ldots, Y_M\}$ may take.

The mild restrictions mentioned above refer chiefly to issues of limiting behaviour and differentiability. For the purposes of the present paper none of them are crucial.

The importance of the theorem is that we may choose the functions θ_i and λ_v^* essentially freely, while being guaranteed, under the conditions of the theorem, that we have not introduced the possibility of arbitrage. This approach, developed by Babbs and Selby (1993), allows θ_i and λ_v^* to be determined on essentially mathematical grounds, without reference to an explicit underlying economic framework.

If jumps in two or more variables cannot occur simultaneously the differential equation simplifies to

$$\frac{1}{2} \sum_{i=1}^{N} b_i^2 \frac{\partial^2 B}{\partial X_i^2} + \sum_{i,j=1}^{N} \rho_{ij} \cdot b_i \cdot b_j \cdot \frac{\partial^2 B}{\partial X_i \partial X_j} + \sum_{i=1}^{N} a_i^* \cdot \frac{\partial B}{\partial X_i} + \frac{\partial B}{\partial t} +$$

$$\sum_{j=1}^{M} \sum_{k=1}^{N_j} \lambda_{jk}^* \cdot [B(t, \ldots, Y_j + c_{jk}, \ldots) - B(t, \ldots, Y_j, \ldots)] = r \cdot B,$$

where

$\lambda_{jk}^* = (1 - \theta_{jk}) \cdot \lambda_{jk}$, and

θ_{jk}, $j = 1, \ldots, M, k = 1, \ldots, Nj$, are price of risk functions.

When there is one diffusion variable and one jump variable, $N = M = 1$. We reduce to the result of Babbs and Webber (1994). The differential equation in this case is:

$$\frac{1}{2} \cdot b^2 \cdot \frac{\partial^2 B}{\partial X^2} + a^* \cdot \frac{\partial B}{\partial X} + \frac{\partial B}{\partial t} + \sum_{k=1}^{N} \lambda_k^* \cdot [B(t, X, Y + c_k) - B(t, X, Y)] = r \cdot B,$$

where

$$a^* = a - \theta_0 \cdot b,$$
$$\lambda_k^* = (1 - \theta_k) \cdot \lambda_k, \text{ and}$$
$$\theta_0, \text{ and}$$
$$\theta_k, \quad k = 1, \ldots, N,$$

are the price of risk functions. In Babbs and Webber r was identified with Y.

Explicit solutions to these equations are unavailable in general[5], but they may be solved using numerical methods. Note that if only a limited number of jump sizes are permitted by the model, introducing additional jump variables is far less expensive computationally than introducing additional diffusion variables[6]. Thus, from the point of view of solving the differential equations, we are less reluctant to increase the number of jump variables than to add diffusion variables.

4   Identification of the State Variables

In this section we set up a general model for which the illustrative regimes may be considered as special cases. A set of possible state variables is identified, and each variable is discussed in turn.

We shall consider two possible diffusion state variables. As in Babbs and Webber (1994), the most important diffusion state variable will be the 'prospective' level of the intervention rate within the economy: $X = X_1$ will follow a mean reverting process, similar to an interest rate. Jump intensities are set so that the short rate tends to jump to the vicinity of X. Thus X represents a likely level of the short rate in the near future. We also discuss the possibility of introducing a second diffusion state variable, $F = X_2$, to model a money market liquidity factor representing the residual demand for funds in the economy remaining after official intervention. This second diffusion factor will not be employed in the illustrative example of section 6.

We identify $D = Y_1$ with an official floor, and $L = Y_2$ with an official ceiling, where this exists. The notation D and L is adopted because of the use of the term 'discount rate' for the floor rate, and because in the German system the ceiling rate is called the Lombard rate. In an economy where the short rate

---

[5]For an example of a special case that may be efficiently solved, see Lindberg and Perraudin (1995)

[6]For instance, using finite difference techniques an extra diffusion variable may increase the number of grid points by a factor of the order of 10^3. An extra jump variable increases the grid size by a factor equal to the number of possible values the new jump variable might take. For a jump variable taking values at 0.25% increments in the range 4% to 16% this is a factor of about 50, a twentieth of the increase a diffusion variable would cause.

is bounded by official action we allow the short rate and its intensities to be functions of X, D and L. The functional form of r, and the structure of the jump intensities for D and L, will ensure that r remains bounded by D and L. We do not exclude the case where the short rate is an official rate following a third jump process. We then set $R = Y_3$, and

$$r = r(X, F, D, L, R) \equiv R.$$

In this case we may wish to constrain the process for R to ensure that r cannot leave the interval $[D, L]$.

The dynamics of the model are contained in the specification of the processes of X, D, L, and any other state variables. Outside of the scope of our discussion is the effect of changes to the monetary regime. For instance, in periods of intense pressure the Banque de France may temporarily suspend the ceiling rate, allowing it to fix the short rate at arbitrarily high levels using overnight repos. However, by leaving the floor rate unchanged, it signals that its underlying stance remains unaltered.

Similarly, the Bundesbank adopts one of two possible regimes, fixed or floating, and may switch between them. In this paper we consider each regime separately. A full treatment would require the possibility of regime switches to be explicitly considered and incorporated within the model.

We now discuss in turn the prospective intervention rate, the liquidity factor, and the floor and ceiling rates.

The Prospective Intervention Rate

The prospective interest rate, as described above, represents the market's view on anticipated levels of interest rates. The prospective rate reflects the market's beliefs of the level of the intervention rate in the near future. Alternatively, the prospective rate may be viewed as a quantification of the notion frequently expressed by market practitioners that 'interest rates are too low' or 'too high'. It also may be interpreted as the level the short rate would be expected to have, were it not constrained by official action.

We require the prospective rate to be a diffusion process. It is reasonable to write the process for X as a standard mean reverting process. However, we explicitly allow the reversion level of X to depend upon D and L. This permits feedback between the controlled levels of D and L, and the state of the economy as epitomised by X. Feedback may be posited to take two forms. Firstly the long range effects on the economy caused by shifts in levels of rates. Secondly, and perhaps more importantly in the short term, the effect that changes in D and L have upon market expectations of future interest rate levels. This signalling affect is widely recognised as being important by practitioners.

For illustrative purposes we adopt a process for X of the following form:

$$dX_1 = a \cdot (b \cdot \mu + c \cdot D + d \cdot L + e \cdot r - X) \cdot dt + \sigma_1 \cdot X \cdot dz_1,$$

$$a, \ b, \ c, \ d, \ \sigma_1 > 0,$$

$$b + c + d + e = 1.$$

X reverts to a level given by a weighted average of D, L, r, and constant μ. An increase in either D or L causes X to revert to a higher level. A decrease in D or L causes X to revert around a lower level. Adjusting the weightings allows the strength of a signal to be varied to suit a particular regime. r is included to represent the feedback between the level of r in the economy and the process followed by the state variable denoting the state of the economy. e may be positive of negative depending on the nature of the feedback. The form of the volatility term is not central to this article.

The Liquidity Factor

An overnight rate may be modelled by including a second diffusion factor $F = X_2$. F may be taken to represent either the actual liquidity pressure within the economy, or it may simply be taken to be a noise term added on to the official short rate to give the overnight rate. The latter route has the advantages of simplicity. The former may allow a deeper analysis into the formation of the overnight rate.

Overnight rates tend to fluctuate around the level of the intervention rate. If the liquidity factor is represented purely as an error term we could write

$$dF = -a_2 \cdot F \cdot dt + \sigma_2 \cdot dz_2,$$

and set

$$r_{overnight} = r + F.$$

$r_{overnight}$ could then be used in the differential equation instead of r. However, this may not lead to further explanatory power. We anticipate that the reversion rate a_2 is high, and that the volatility σ_2 is comparatively high, compared to estimates of the drift and volatility of three-month money market rates, for instance.

Although an official short rate may be constrained to lie within the interval $[D, L]$, an overnight rate may not be so constrained. On occasion, overnight rates in both Germany and France have been observed at values outside the interval, and for the dollar, the fed funds rate has fluctuated below the floor level defined by the discount rate. For sterling, the overnight rate is not a pure jump process.

We discuss a liquidity factor for generality only. In practice we do not expect this factor to significantly influence the term structure beyond the extremely short end, although it could be relevant to the valuation of derivatives on repos or overnight rates[7].

The Floor and Ceiling rates, D and L

We consider D and L together, although L may be absent in some economies, notably the US. All of the economies considered in section 2 maintain an official interest rate at levels less than the official short rate. The floor rate, D, and the ceiling rate L, where present, are pure jump processes, but their jump intensities will be influenced by the levels of the state variables in the economy. For instance, we prohibit jumps by D that would leave it at a value greater than that of the short rate r. Jumps are not precluded, however, if a functional dependence of r on D takes r to a level in excess of D when a jump occurs. Nor is D allowed to jump so that $D > L$, when a ceiling level L exists. Similar observations apply to L. The form of the intensity functions are highly dependent upon the particular interest regime. See Babbs and Webber (95) for a discussion in the UK case.

Write λ_{Dj} for the jump intensity functions of D, $j = 1, \ldots, N_j$ and write λ_{Lj}, for the jump intensity functions of L. We shall consider the functions λ_{Dj}, but similar arguments apply to λ_{Lj}.

One may suppose that differing policies could result in several alternative forms of intensities λ_{Dj}. It seems reasonable to express λ_{Dj} in terms of the offset between D and each of X_1, r, and L. L would be included to reflect the possibility that D and L tend on average to have a fairly constant spread, that is, $L - D$ is approximately constant. One may set up the jump intensities to reflect this. A model that includes a liquidity factor F might allow λ_{Dj} to depend on F. For instance, if an overnight rate $r + F$ were persistently below D, a monetary authority might be unable to maintain control over the official short rate. If might then be forced to reduce D. The market's expectation of a change in D, reflected in this model by the specification of the intensities for D, would legitimately depend upon F as well as X.

In the appendix we illustrate how several differing policy stances by the monetary authorities could be modelled. These policies are:

1. A 'smoothing' policy. The authorities wish to allow the short rate to generally follow the rate X_1, but wish to reduce the volatility.

2. A 'static' policy. The authorities wish to keep the short rate within a static band, now and again permitting the band to shift when it can no longer be sustained.

---

[7]Balduzzi, Bertola and Foresi (1993) essentially consider a noise term of this sort. However, their analysis is not set in a context that guarantees the absence of arbitrage.

3. A 'nudging' policy. The authorities wish the short rate to tend in the long term to some level μ_2. Moves towards μ_2 are encouraged. Those away from μ_2 are discouraged.

These three possibilities are included to show the range of behaviour that might plausibly be modelled within this context.

5 Applications to Selected Regimes

We now briefly consider each of the economies described in section 2 with a view to incorporating them into the context of the general model described in section 4. Since our interest is on the yield curve beyond very short maturities, in the illustrations below we prefer to exclude a 'liquidity factor' state variable, F. However we shall mention this factor where it may shed light upon the behaviour of rates of interest to us. We assume that all rates are instantaneous, and that they are continuously available. This implies that all floor and ceiling rates are 'hard', since they have the same maturity as the short rate.

United Kingdom

Take $N = M = 1$; one diffusion variable and one pure jump variable. Set $r = Y_1 = Y$. We identify the short rate with the jump variable. In Babbs and Webber the diffusion variable X was taken to be a 'prospective' interest rate. Jump intensities reflect market expectations of jumps occurring as functions of the difference in level between the prospective level of interest rates and the current value of the short rate.

France

Take D to be the taux des pensions appel d'offres. We have seen that the taux des pensions de 5 a 10 jour tends to remain at a relatively fixed premium above D. It is therefore not an unreasonable approximation to suppose that the ceiling rate L is a fixed amount above D. As a consequence only one jump variable, D, is required. The diffusion variable X is taken to be a prospective level of interest rates.

We identify the overnight rate with the short rate, and set

$$r = \max(D, \min(X, L)).$$

If we were to allow the presence of a significant liquidity factor, F, following a process mean reverting to zero, one might set r to be

$$r = \max(D, \min(X + F, L)).$$

Prior to a regime shift in the French market towards the end of 1991 the overnight rate was very volatile. With a liquidity factor this could be modelled by attributing most of the volatility in the overnight rate to the liquidity factor. The regime shift is then consistent with a reduction in the volatility of the liquidity factor.

Germany

D and L provide hard boundaries, so that the short rate is constrained to lie between them. We identify the short rate with the intervention rate. In the floating rate regime the intervention rate may be modelled as the diffusion variable, $X = X_1$. In the fixed rate tender regime, the short rate is effectively a third pure jump process, R. In either case the short rate may be set equal to the middle of D, L, and the intervention rate;

$r = \max(D, \min(X, L))$, in a floating regime,

$r = \max(D, \min(R, L))$, in a fixed regime.

United States

Two alternative models may be considered for the fed funds rate. The first ignores the discount rate and models the fed funds rate as a pure jump process with liquidity factor noise. There is a single jump variable, the fed funds target rate, R, and two diffusion processes, the prospective rate, X, and the liquidity factor, F. The liquidity factor may be dropped if one is not interested in rates of very short maturities. In this case we are reduced to the situation of Babbs and Webber. The second takes the discount rate into account as a signalling factor. Then there is an additional jump variable, D. Depending on the regime operated by the fed, D may sometimes be considered as a hard lower bound on the short rate. At other times the regime in operation may permit the short rate to fall below D.

6 Illustrative Example

As an illustration of the model we solve numerically the partial differential difference equation in a case where the short rate is constrained to lie within an official corridor. In practice, floor and ceiling rates tend to maintain an approximately constant interval between them. For simplicity, in this illustration, we presume that D and L always jump simultaneously and by the same amounts. In effect there is a single jump variable, which we take to be D. We set

$$L = D + 0.01$$

to maintain a corridor with a constant width of 1%. This form of regime is similar to the French and German regimes. Although in neither case is the assumption exact, nevertheless we believe it to be a reasonable approximation, particularly for the French case.

We take a single diffusion state variable, $X = X_1$, representing a prospective interest rate. The process for X is taken to be

$$dX = 0.1 \left(\frac{1}{2} \cdot (0.085 + D) - X \right) \cdot dt + 0.015 \cdot dz.$$

Feedback between the official rate D and the state variable X is incorporated by allowing X to revert to a weighted average of D and a long term level of 8.5%.

The short rate is set equal to X restricted to a range determined by D and L:

$$r = \max(D, \min(X, L)).$$

This corresponds to a situation where the authorities allow the short rate to be determined by market forces while it lies within the official interval. When the short rate reaches the boundary of its permitted band the authorities intervene to prevent it moving outside the band. This is an extreme case. In practise the authorities may decide to intervene before the short rate reaches the edges of the band. In France and Germany the authorities may attempt to maintain the short rate near the centre of the band.

We choose this functional form for r since it is our intention, in this example, to illustrate the affect of signalling upon the term structure. More generally one might allow r to jump when D jumped. For instance we could set

$$r = \max \left(D, \min \left(\frac{1}{2}(X + L), L \right) \right).$$

This describes a situation in which the market puts equal weighting to X and L to determine the short rate. A jump in D causes r to jump in the same direction, by half the amount (if this leaves r still inside the corridor). Other formulations for the functional form of r are given in the appendix.

We allow D to jump by four possible magnitudes; $\pm 1\%$ and $\pm \frac{1}{2}\%$. These sizes are large compared to values typically seen in France and Germany, but they suffice for this illustrative example. The jump intensities are taken to be functions of $X - \frac{1}{2} \cdot (D + L)$, that is, functions of the offset between X and the centre of the corridor. The functional form of the intensities λ_j are shown in figure 5. These are set up so that the jump which is most likely to occur is the one that takes X to being closest to the centre of the corridor. The parameter h determines the expected number of jumps per year. It was given a value corresponding to a small number, 2 or 3, jumps per year on average.

The PDEs were solved for various initial values of X and D. The results are presented in figures 6, 7 and 8, where the spot rates derived from the computed values of pure discount bonds are shown. For these figures the prices of risk were set to zero[8].

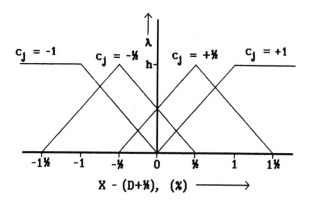

Figure 5: λ a function of $X - D$.

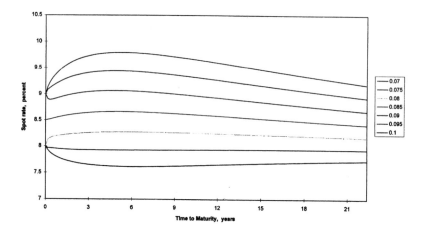

Figure 6: Term structures in a corridor, width 1%, $D = 8\%$, X from 7% to 10%, $\mu = 8.5\%$

The short rate is constrained to lie inside the official corridor, and the asymptotic values of the term structure at the long end are determined by the value of the long run reversion level, 8.5%. Medium term levels of spot rates are determined by the initial value of X. X denotes a prospective short rate, so the numerical results informally bear out a version of an expectations

---

[8]Other term structures, with non-zero values for prices of risk, were computed. The results were qualitatively similar to those in the figures presented here, which may be taken to be representative.

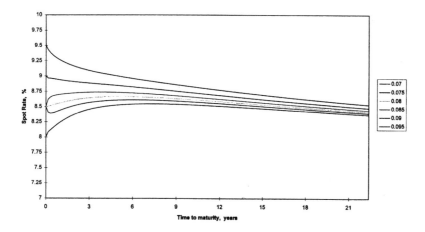

Figure 7: Term structures in a corridor, width 1%, $X = 8.5\%$, μ = 8.5%, D = 7% to 9.5%

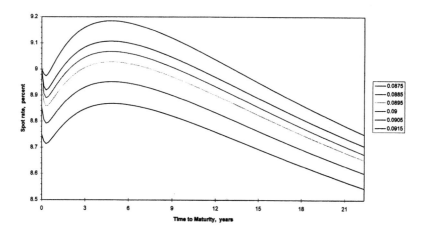

Figure 8: Term structures with a corridor, floor = 8%, ceiling = 9%. X varying from 5.75% to 9.15%, μ = 8.5%

hypothesis that equates spot rates to the expected future levels of the short rate.

Figure 6 shows a number of term structures for an initial corridor of [8%, 9%], and various initial values of X, ranging from 7% to 10%. The long run value of 8.5% lies within the initial range of the corridor. For these values the term structures are relatively flat.

The effect of jumps in the value of the corridor may be seen in figure 7. Each term structure in figure 7 has the same initial value of X, but is constrained by a different initial corridor. The bottom term structure belongs to the interval [7%, 8%]. The top term structure to the interval [9.5%, 10.5%]. The three term structures with the same short rate value of 8.5% lie in the intervals [7.5%, 8.5%], [8%, 8.5%], and [8.5%, 9.5%]. Jumps from one of these intervals to another do not cause the short rate to jump, but nevertheless cause the term structures as a whole to shift. Jumps in D that leave the value of r unchanged will cause the term structure to change shape. This corresponds to observed behaviour in the markets, where changes in D are interpreted as signals about future intentions. We have built the market's interpretation into the formulation of the illustrative model. The effects observed in this illustrative example are not great, but they are real. Taking other choices of parameters leads to more acute behaviour.

A special effect may be seen at the short end when the short rate lies close to D or L. For example, if r has a value just less than L when X is larger than L, then the term structure may dip at very short maturities before rising towards X. This is illustrated in figure 8, which shows a number of term structures for a corridor of [8%, 9%]. X ranges between 8.75% and 9.15%. A dip can be observed even when X lies 0.15% above L. The size of the dip may exceed 10 basis points. This surprising effect may be interpreted in terms of the likely future value of the short rate. r is capped by L. Because only a few jumps per year are anticipated in the short run r is unlikely to be permitted to exceed the current value of L. Hence the expected future value of r, in the short run, may be less than its current value. In the medium term a jump is increasingly likely to have occurred, and the expected future value of r may rise to the vicinity of X.

Intuitively, the level of the long end of the yield curve is determined by the parameter μ, 8.5% in this illustration. At medium maturities the term structure may possess a hump. The size and location of the hump will depend upon the value of X relative to D, and the rates at which X moves towards μ and L moves towards X. At the very short end we have seen that a dip may exist. The short rate itself is, of course, fixed by r.

The existence of dips is a desirable feature of the model. So called 'spoon shaped' yield curves are frequently observed in the market in various currencies. An extreme example was observable in the sterling gilt market in July 1991, for instance. Models may be rejected by market practitioners, for some purposes, if they cannot fit term structures of this sort.

A further feature of this illustrative example, shared with illustrations in Babbs and Webber, is that there is a natural positive correlation between the jump and diffusion components of short maturity interest rates. This is implicit, as a consequence of the structure of the jump intensities, rather than being explicitly imposed. To see this, note that a high jump intensity causes the term structure to bow in the direction of the 'anticipated' jump. Hence high jump intensities are associated both with jumps and with movements of the term structure in the same direction, resulting in a positive correlation.

7 Conclusion

We have seen how within several important interest rate regimes the short rate may be modelled as a process bounded by officially controlled levels. Changes to these levels pass signals to the markets concerning anticipated future interest rate levels. Although the details vary from regime to regime all of the examples given here fit into a common pattern. We have seen how very short term money market rates are sensitive to 'liquidity' or 'forecast error' pressure. In particular it seems appropriate to use as a surrogate for the short rate not an overnight money market rate, but the intervention level of the monetary authorities within the economy. An overnight rate can then be modelled as fluctuations away from the official short rate. We have seen how changes to a boundary may act as a signal to the market and cause a change to the shape of the term structure, even through the monetary authorities may not have altered the level of the official short rate. We obtain qualitatively different term structure behaviour from this model to that seen in other formulations. We expect our formulation to have implications for hedging and valuing derivative securities based on short maturity interest rates.

Appendix

In this appendix we illustrate how three different policy stances might be modelled. The three policies are:

1. A 'smoothing' policy. The authorities allow the short rate to follow X_1 but with reduced volatility.

2. A 'static' policy. The authorities wish to keep the short rate within a static band, shifting the band when it can no longer be sustained.

3. A 'nudging' policy. The authorities wish the short rate to tend in the long term to some level μ_2.

These different policies might be reflected in the model as follows. It is assumed that the markets are aware of the nature of the policy stance adopted by the authorities, and that signals are interpreted correctly. Authorities may attempt to 'defend' boundaries, that is, attempt to prevent the short rate from crossing them, with various degrees of vigour. Although an authority may always defend a boundary, we have previously noted that there is a cost to

doing so. An authority may not be able to sustain a boundary. We assume that
if this is the case then the boundary is shifted, permitting the authorities the
opportunity to allow the short rate to cross the previous boundary level. We
assume that the effect of a defence of a boundary is to dampen the volatility
of r as x approaches the boundary, D say. We interpret this as follows. The
volatility of r as a function of that of X, in the absence of jumps, is given by

$$\sigma_r = \frac{\partial r}{\partial X} \cdot \sigma_X$$

If we assume that $\partial r / \partial X$ is positive, and intervention reduces the volatility
of r as r approaches D, then $\partial r / \partial X$ must decrease as r decreases towards D,
that is,

$$\frac{\partial^2 r}{\partial X^2} > 0.$$

Similarly, as r approaches a ceiling, L, we have $\partial r / \partial X$ decreases as r
increases towards L, that is,

$$\frac{\partial^2 r}{\partial X^2} < 0.$$

We assume that the greater the intervention by the authorities, that is,
the stronger the defence of the boundary, the greater the dampening effect.
With no attempt to defend, r might depend on D, L and X as

$$r = \max(D, \min(X, L)).$$

If both D and L were equally defended, r might take the form

$$r = X + D - L + c(X, D, \sigma) - c(X, L, \sigma),$$

where

$$c(X, D, \sigma) = X \cdot N(d_1) - D \cdot N(d_2)$$

$$d_1 = \frac{1}{\sigma} \cdot \ln \frac{X_1}{D}, \quad d_2 = d_1 - \sigma,$$

and $N(d)$ is the cumulative normal density function.

This functional form has been chosen for illustrative purposes only. The
value of the parameter σ reflects the amount of the dampening caused by the
defence of D and L. When $\sigma \to 0$, r tends to the function

$$r = \max(D, \min(X, L)).$$

As σ increases, σ exhibits greater and greater dampening as it approaches D
or L. The formula may be modified to allow differing degrees of dampening
as r approaches D and L. In this case there will be two parameters, σ_D and
σ_L, determining the dampening at D and L respectively. The more strongly

defended a boundary is, the less likely the authorities are to shift the boundary when approached by r.

(i) A Smoothing Policy

The authorities may defend D and L only lightly, being prepared to shift them if X remains away from defensible levels. Shifts in D and L are not seen as strong signals. Set

σ_D and σ_L small:

Little dampening near D or L.

$$dX = a \cdot (b \cdot \mu + c \cdot D + d \cdot L + e \cdot r - X) \cdot dt + \sigma_X \cdot X \cdot dz,$$

with c and d small:

Little feedback between D and L, and the process for X.

$\lambda_{Dj} = 0$ for jumps up, unless

(i) r is close to L,

or

(ii) r is further away from D than some threshold

λ_{Dj} is small for jumps down, unless r approaches D

When r approaches close to D, jumps down become increasingly certain. r is dampened only slightly by the actions of the authorities to defend the official rates. However, when a jump to D or L occurs r will jump, if only slightly.

(ii) A Static Policy

The authorities may defend D and L strongly, being prepared to shift them only if X approaches close to D or L. Shifts in D and L are seen as strong signals. Set

σ_D and σ_L large:

Significant dampening near D and L.

$$dX = a \cdot (b \cdot \mu + c \cdot D + d \cdot L + e \cdot r - X) \cdot dt + \sigma_X \cdot X \cdot dz_1,$$

with c and d large:

Significant feedback between D and L, and the process for X.

$\lambda_{Dj} = 0$ for jumps up, unless:

(i) r is close to L,

or

(ii) r is further away from D than some threshold

$\lambda_{Dj} =$ is small for jumps down, unless r approaches close to D.

When r approaches sufficiently close to D, jumps down become increasingly certain.

When a jump occurs both L and D jump to place r in roughly mid-band. r is dampened considerably by the actions of the monetary authorities. When a jump occurs the largest reasonable jump is made. The rate structure is re-aligned.

(iii) A Nudging Policy

We suppose that the desired level of r, μ_2, is outside the current band,

$$\mu_2 \notin [D, L].$$

If the short rate is moving towards the desired direction when it reaches a boundary (the 'near' boundary) that boundary will not be defended. If the boundary in the direction away from the desired level (the 'far' boundary) is approached the authorities will defend it to some extent. Shifts in the far boundary are seen as strong signals, shifts in the near boundary as weak signals. (We do not consider the form of action the authorities might take if μ_2 lies within the current band). Set

σ_{far} large, σ_{near} small:

Significant dampening at the far boundary, little at the near.

$$dX = a \cdot (b \cdot \mu + c \cdot D + d \cdot L + e \cdot r - X_1) \cdot dt + \sigma_X \cdot X \cdot dz_1,$$

with c and d functions of r, D and L so that c or d large for the far boundary, small for the near boundary. Signalling is significant when r is close to the far boundary, not so important when r is close to the near boundary.

$\lambda_{far,j} = 0$ for jumps towards, unless:

(i) r is close to L,

or

(ii) r is further away from D than some threshold

$\lambda_{far,j}$ is small for jumps away, unless r approaches close to the far boundary. When a jump to the far boundary occurs, the near boundary may jump by a similar amount.

$\lambda_{near,j} = 0$ for jumps away, unless r is close to the far boundary

$\lambda_{near,j}$ is small for jumps towards, unless r approaches to the near boundary.

When a jump by the near boundary occurs, the far boundary may jump by a similar amount.

References

Ahn, C.M. and Thompson, H.E. (1988). 'Jump-Diffusion Processes and the Term Structure of Interest Rate', *Journal of Finance* **431**, 155–174.

Babbs, S.H. and Selby ,M.J.P. (1993). 'Pricing by Arbitrage in Incomplete Markets', working paper, Midlands Global Markets.

Babbs, S.H. and Webber, N.J. (1994).'A Theory of the Term Structure with an Official Short Rate', FORC working paper, 94/49.

Balduzzi, P., Bertola, G. and Foresi, S. (1993). 'A Model of Target Changes and the Term Structure of Interest Rates', National Bureau of Economic Research, working paper, 43–47.

Bjork, T., Kabanov, Y. and Runnaldier, W. (1995). 'Bond Markets when Prices are Driven by a General Marked Point Process', Stockholm School of Economics, Working paper.

Brennan, M. and Schwartz, E.S. (1979). 'A Continuous Time Approach to the Pricing of Bonds', *Journal of Banking and Finance, 3*, 133–155.

Cox, J.C., Ingersoll, J.E. and Ross, S.A. (1985). 'An intertemporal general equilibrium model of asset prices', *Econometrica* **53**, 363–384.

Cox, J.C., Ingersoll, J.E. and Ross, S.A. (1985). 'A theory of the term structure of interest rates', *Econometrica* **53**, 385–407.

Duffie, D. and Kan, R. (1993). 'A Yield Factor Model of Interest rates', GBS Stanford University working paper.

Fong H.G. and Vasicek, O.A. (1991). 'Fixed Income Volatility Management', *Journal of Portfolio Management*, Summer, 41–46.

Fong H.G. and Vasicek, O.A. (1992). 'Interest Rate Volatility as a Stochastic Factor', Gifford Fong Associates Working Paper.

Heath, D.C., Jarrow, R.A. and Morton, A. (1992). 'Bond Pricing and the Term Structure of Interest Rates: A New Methodology for Contingent Claims Evaluation', *Econometrica* **60**, 77–105.

Hogan, M. (1993). 'Problems in Certain Two-Factor Term Structure Models', *Annals of Applied Probability* **3**, 576-581.

Honore, P. University of Aarhus, private communication.

Krugman, P. and Miller, M. (eds.) (1992). *Exchange Rate Targets and Currency Bands*, Cambridge University Press.

Lindberg H. and Perraudin, W. (1995). 'Yield Curves with Jump Short Rates', working paper, Birkbeck College.

Longstaff, F.A. and Schwartz, E.S. (1991). 'Interest Rate Volatility and the Term Structure: A Two Factor General Equilibrium Model', *Journal of Finance* **47**, 1259–1282;

'The Role of the Bank of England in the Money Market', *Quarterly Bulletin of the Bank of England*, March 1982, 86–94.

Schnadt, N. (1994). 'The Domestic Money Markets of the UK, France, Germany and the US', The City Research Project, Subject report VII, Paper I, Corporation of London.

Shirakawa, H. (1991). 'Interest Rate Option Pricing with Poisson–Gaussian Forward Rate Curve Processes', *Mathematical Finance* **1**, 77–94.

Svensson, L.E.O. (1989). 'Target Zones and Interest rate Variability', CEPR discussion paper No. 372.

Interest Rate Distributions, Yield Curve Modelling and Monetary Policy

Lina El-Jahel, Hans Lindberg and William Perraudin

Abstract

The way in which countries conduct monetary policy significantly affects the distribution of their short interest rates and hence the pricing of bonds further along the yield curve. This chapter examines the impact of monetary policy on the higher moments and persistence of interest rate changes. We present a simple version of an analytic model of interest rate setting by monetary authorities and show that this can mimic properties of interest rates that standard models tend to fit badly.

1   Introduction

1.1   Sources of Interest Rate Changes

In many financial applications, one can afford to be agnostic about the ultimate source of asset price changes. Widely applied finance models such as the CAPM and APT derive implications for relative asset prices from assumptions about the joint distributions of asset returns. Arbitrage models such as the Black–Scholes model of European option valuation proceed from an even tighter distributional assumption, namely, that changes in the prices of the cash security and the option are perfectly locally correlated.

What one might call the 'statistical approach', of concentrating exclusively on the joint distribution of asset returns while ignoring the underlying causes of asset price changes, has its limits, however. In government bond markets, all prices ultimately depend on the distribution of the short-term interest rate[1] which, in turn, is largely determined by the actions of the monetary authorities. Understanding the links between policy and interest rate distributions is important because, when policies shift, valuation must be performed before a usable amount of data has accumulated under the new regime.

---

[1]Note that this distribution may depend on multiple state variables, not just on the level of the short rate. Also, note that strictly speaking, it is, of course, the risk adjusted distribution of the short rate that matters.

1.2   Rate Pegging and Leptokurtosis

In this article, we focus on two aspects of monetary policy and two corresponding features of interest rate distributions. The first is the practice of many monetary authorities of pegging a reference rate at the short end of the yield curve and periodically adjusting it in discrete jumps. For example, in Germany, the reference rate is the repo rate, while in the UK it is the Bank of England's band 1 stop rate. Both rates are held constant for periods of time and the market regards changes as indicating significant policy shifts. By contrast, the US reference rate, the rate on Federal funds, fluctuates around intervention trigger levels set by the authorities.

An important implication of pegging the reference interest rate is that short-term interest rate changes become increasingly leptokurtic, both conditionally and unconditionally, as one considers higher frequency data. Typical term structure models presume that short-term interest rates follow diffusion processes. It is, of course, possible to specify diffusion processes with highly leptokurtic unconditional distributions but this is likely to require fairly extreme parameter values. Conditionally, all diffusion processes are locally Gaussian in their increments, however. So, as the frequency of data increases, returns with jumps exhibit quite different distributional properties from those generated by diffusions processes.

1.3   Persistence

The second feature of monetary policy on which we shall focus is the resolution that central banks exhibit in their reactions to inflationary shocks. We shall argue that such persistence will show up as persistence in shocks to interest rates. Clearly, monetary authorities in different countries differ greatly in their 'reaction functions', and in particular in the speed with which they relax the tightening or loosening of monetary policy that they adopt in the face of a monetary shock. Numerous studies have examined central bank reaction functions. Among many examples, one might mention McCallum (1994) who examines the constraints on interest rate setting imposed by the need to maintain banking sector stability; Dotsey & King (1986) who look at interest rate setting under differential information; and Siegel (1983) who examines operational interest rate rules.

Few studies, however, have stressed the important implications of central bank reaction functions for the distribution of short interest rates. Stochastic models of interest rates generally have parameters that describe (i) the long run mean interest rate level, and (ii) how rapidly interest rates return to this mean after a shock (the third crucial element in such models is, of course, the parameter or parameters that describe instantaneous volatility). Both mean and 'reversion rate' parameters are crucial for many fixed income derivative pricing problems. The rate of reversion to normal interest rate levels directly reflects the speed with which monetary authorities relax their reaction to monetary shocks. Exercise of resolute monetary policy is likely to be associated with slower reversion to 'normal' interest rate levels and hence greater persistence.

1.4 The Contribution of the Article

In this article, we begin by discussing the techniques of monetary control employed in Germany, the UK and the US. We contrast the approaches taken and suggest ways in which they are likely to influence the statistical behaviour of interest rates. In Section 3, we examine distributional properties of short term interest rates from Germany, the UK and the US employing a range of econometric techniques. In Section 3.2, we study the higher moments of interest rate changes using non-parametric kernel estimates of their distributions. Our results suggest that the degree to which distributions are fat-tailed depends on the monetary control regime. This is especially apparent in the US where two distinct periods of interest-rate- and monetary-targeting are compared. In Sections 3.3 and 3.4, we investigate the persistence of interest rate changes by considering unit root tests and simple autoregressions. The specification of the lag structure in interest rates affects the estimated persistence one obtains in autoregressions in that, in almost all cases we examine, allowing for more lags generates more rapid reversion to the long run mean.

Informed by the kernel estimates and the autoregressions, in Section 4.3, we report the results of Maximum Likelihood estimates of two commonly applied single state variable yield curve models, the Vasicek and the Cox, Ingersoll and Ross models. In each case, we perform the estimation for nine data sets: Sterling, Deutschmark and US dollar Euro-deposit rates of one-, six- and twelve-month maturities. Our approach to estimation fully allows for temporal aggregation in that the likelihoods we employ are constructed using the conditional densities of the discretely sampled data. We also take account of the fact that the theoretical interest rates implied by the models are affine functions of the underlying instantaneous interest rate, and that the affine mappings involved depend on the parameters of the processes.

The estimates we obtain illustrate the difficulties of fitting short-term interest rate data using standard diffusion models. The Vasicek model results are unstable in that parameters estimated from interest rate data of different maturities (which should be equal) differ significantly. The Cox-Ingersoll-Ross model estimates are more homogeneous across maturities but suffer from the well-known failing of empirical estimates of this model (see Gibbons & Ramaswamy (1993) and Ball & Torous (1995)) that the estimated reversion rates are implausibly large. Again, we interpret this finding as evidence of specification problems. The two models differ only in their specifications of the instantaneous volatility of the short rate. To the extent that these specifications are inaccurate, this is likely to show up in distorted and implausible estimates of 'persistence parameters' like reversion rates which affect volatility over finite periods of time.[2]

In Section 4 of this chapter, we exposit a new approach to yield curve modelling, described in more detail in El-Jahel, Lindberg & Perraudin (1996). Under this approach, short term interest rates are modelled as pure jump processes of which the rate of jump is a function of a diffusion process. Such models, which have been discussed by Babbs & Webber (1994), have the potential to explain the higher moment behaviour of interest rate changes, since the basic assumptions accurately mimic actual practice in most bonds markets, i.e. that the authorities peg a particular key rate at the short end of the yield curve. Moments of simulated bond prices generated from the jump yield curve model are compared to moments of actual bond prices. We show that the characteristic of actual data that kurtosis increases rapidly as data frequency rises is faithfully captured by the model.

2 Monetary Policy in Three Countries

2.1 A Schematic Representation

We begin by discussing the monetary control systems of the three countries whose interest rates we shall analyse statistically, Germany, the United Kingdom and the United States. The approaches the three countries take to monetary control differ in many specific aspects but possess important common elements. In general, monetary control systems may be represented with the following simple schema:[3]

$$\text{instrument} \quad \rightarrow \quad \begin{matrix} \text{operational} \\ \text{target} \end{matrix} \quad \rightarrow \quad \begin{matrix} \text{intermediate} \\ \text{target} \end{matrix} \quad \rightarrow$$
$$\begin{matrix} \text{ultimate} \\ \text{goal} \end{matrix}$$

Our diagram identifies four distinct elements: (i) the instruments directly controlled by the central bank (its portfolio and the terms of its credit facilities); (ii) the operational target on which the instrument changes operate (usually a very short-term market interest rate); (iii) the intermediate target to which the monetary authorities may commit themselves (money or credit or nominal spending or the exchange rate); and (iv) the ultimate goals that they hope to achieve in the long run. How do our three countries' monetary

---

[2]The extended Vasicek and Cox-Ingersoll-Ross models discussed in Hull & White (1990) allow the unconditional mean of the short rate to depend upon time. The motivation is that such models permit one to fit the entire structure of discount bond prices at some instant of time. In principle, however, allowing the mean to change could improve the model's ability to mimic the time series properties of short interest rate.

[3]The terminology we employ here draws on Freedman (1990).

Table 1: Monetary Policy and Interest Rates

Characteristics	Germany	United Kingdom	United States
Major instrument	Repo rate	Band 1 stop rate	Adjustment of non-borrowed reserves
Maturity of major instrument	Usually 2 to 8 weeks	1 to 14 days	1 to 14 days
Other instruments	Discount rate and lombard rate	2.30 pm lending rate and minimum lending rate	Discount rate
Operational target rate	Call rate	Base rate	Fed funds rate
Reserve requirements	Yes, averaging over the month	Only clearing balances	Yes, instantaneous reserve accounting
Interest elasticity of reserve demand	High	Low	Low
Fine tuning activity	Low	Very high	High

Sources: Batten, Blackwell, Kim, Nocera & Ozeki (1990), Freedman (1990) and Kneeshaw & den Bergh (1989).

arrangements fit into this schematic representation?[4] Three important points should be mentioned.

2.2 Interest Rate Instruments and Targets

First, each country has a key interest rate which serves as its operational target. (Summaries of the key interest rates and the ways in which they are manipulated by the authorities may be found in Table 1.) One could perhaps argue that base and call rates play this role in the UK and Germany, while the

---

[4]Detailed discussions of implementation procedures in a number of countries can be found in Batten, Blackwell, Kim, Nocera & Ozeki (1990), Freedman (1990) and Kneeshaw & den Bergh (1989).

operational target in the US is the rate on Federal funds. In Germany and the UK, the authorities tightly control their operational target rate by adjusting respectively the repo rate (the rate at which the Bundesbank is willing to enter repurchase agreements in government bonds with money market participants) and the band 1 stop rate (the rate at which the Bank of England is willing to discount Treasury bills).

Typically, the underlying rates are constant over quite long periods of time and when adjusted move up or down in a discrete jump. The discrete nature of these adjustments means that changes in interest rates further along the yield curve are likely to exhibit leptokurtic behaviour. In fact, if very short interest rates jump by discrete amounts, the shorter the frequency of data one examines, the more fat-tailed the distributions of interest rate changes will appear.

The US system is somewhat different in that the authorities influence their operational target rate, the rate on Federal funds, not by adjusting some other rate but by altering the supply of non-borrowed reserves through open market operations. In consequence, the Fed funds rate shows somewhat more short-term variability than the German and UK operational target rates. Even so, the US authorities adjust non-borrowed reserves with the aim of keeping the Fed funds rate in a narrow corridor around an implicit target rate and the latter tends to adjust discretely when the authorities alter their policy stance.

2.3   Intermediate Targets and Policy Goals

Second, the central banks of the three countries employ different approaches to the use of intermediate targets. The Bundesbank focuses on money growth whereas the Federal reserve and the Bank of England rely on a wide range of leading indicators of the course of inflation and the general economy. However, it is important to distinguish reality from public utterances and realise that adherence to intermediate targets may be more or less strict. For example, the Bundesbank allows large discrepancies to open up between monetary aggregates and its targets and some commentators have argued that one should regard German policy as reacting to the general state of inflationary pressures in the economy.

Third, the three countries considered differ significantly in the way in which they adjust interest rates in the face of inflationary shocks. Such shocks might involve an increase in velocity associated with increases in private or overseas sector demands for domestic goods and services. All three countries broadly-speaking espouse price stability as the ultimate goal of monetary policy,[5] but

---

[5]The Bundesbank Act of 1957 stipulates that the central bank's fundamental task is to achieve stable prices. Central bank legislation in the United Kingdom and the United States does not explicitly state that monetary policy should aim primarily at one price stability.

differ in the extent to which they allow this ultimate aim to influence more short term policy. Hard money policies such as that of the Bundesbank may be characterised as reacting firmly to inflation shocks and then resolutely adhering to the higher interest rates adopted. Such behaviour generates slow reversion of interest rates to their long run means. Monetary policy in the UK by contrast typically involves sharp movements in interest rates as inflation picks up but quick downward adjustments in interest rates as soon as this is feasible. Such policies are likely to generate rapid reversion to the mean in interest rates.

3   Interest Rate Policies and Distributions

3.1   Key Rates in Three Markets

In Figure 1, We plot daily data on interest rates from three different bond markets we shall study in our empirical work. Those of Germany, the UK and US[6] In each market, the authorities control a key short term interest rate, thereby aiming to influence the market more generally. In the Appendix, we describe the operation of monetary policy in the above three countries and the role played by various different interest rates. We argue that the key short rates are respectively the repo rate in Germany, the band 1 stop rate in the UK market and the Fed funds rate in the US. In Figure 1a, we show the German repo rate and the UK band 1 stop rate. It is apparent from the figure that these rates are pure jump processes, varying only through sharp, discontinuous movements. The sample paths of the German and UK reference rates shown in the Figure possess a number of interesting characteristics.[7]

First, the inter-jump times are highly variable. On occasion, the authorities alter rates through a rapid sequence of moves, while at other times, quite substantial rate changes suddenly occur after a long period of stability. Second, a striking aspect of the sample paths is the very considerable auto-correlation in the signs of interest rate changes, in that there are long periods in which rates move consistently in one direction. Third, step-sizes vary also in a highly dependent way. At times, the authorities have pushed rates grad-

---

[6]In the plots and descriptive statistics we report in earlier sections of the paper, we employ daily data. When we estimate parametric yield curve models (the Vasicek and Cox-Ingersoll-Ross models), it makes more sense to use weekly or lower frequency data since the non-linear, iterative Likelihood Maximization technique we employ performs much better numerically when the data is not too noisy. To facilitate comparisons, we also use weekly data in the unit root tests and autoregressions reported below.

[7]The figure shows data from the period since the break-up of the Exchange Rate Mechanism in 1992. The behaviour of rates prior to this date was somewhat different although, broadly-speaking, the distributional features we describe below were present in the earlier period.

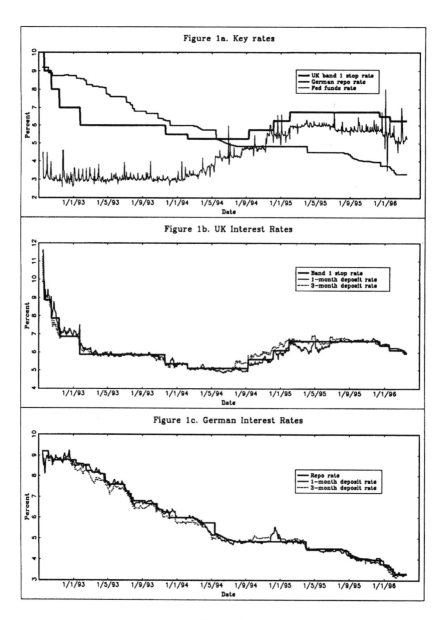

Figure 1

ually up or down through extremely small changes in the level whereas, in other periods, adjustments have been more brusque.

Figure 1a also shows the key US interest rate, the rate on Fed funds. Market-determined to a greater degree than the German or UK key rates, the Fed funds rate nevertheless appears from the plot to move in a reasonably narrow band around a stable, underlying level which changes periodically.[8] Figures 1b and 1c show the behaviour of German and UK 1– and 3–month deposit rates compared to that of the respective key rates. It is interesting to chart the evolution of the market rates around the key rates.

3.2 Leptokurtosis and Official Rates

Now, consider ways in which the distribution of interest rate changes reflects the monetary control arrangements adopted by the authorities. As mentioned in the Introduction, two features of monetary policy are likely to show up in stochastic properties of interest rates (i) the way in which the authorities' control of key rates at the short end of the market will influence higher moments of interest rate changes further down the yield curve, most notably kurtosis and skewness, (ii) the authorities' resolution in adhering to policies will affect the speed of reversion towards the long run mean interest rate.[9]

First, consider the higher moments of interest rate changes. Figure 2a shows estimated densities for daily changes in German and UK 3-month interest rates. The densities depicted are based on non-parametric kernel estimates. The estimates are calculated using a Gaussian kernel and a window size of

$$1.06 \times \text{standard deviation}/(\text{sample size} * (1/5)).$$

For details, see Silverman (1986). For each series, we standardize the data, demeaning and scaling it by the sample standard deviation. This enables us to compare the densities with a standard normal density also shown in Figure 2a.

Evident from the plotted densities is the unconditional leptokurtosis of the interest rate changes. The UK interest rate changes are *less* fat-tailed than those of Germany, in that their respective kurtosis coefficients are 19.2 and 44.7[10]. Recall that the kurtosis coefficient, the ratio of the fourth central moment to the square of the second, is 3 for a normally distributed random

---

[8]Such behaviour, indeed, is consistent with the US authorities' avowed policy of intervening to keep Fed funds in a band around an implicit target level. For example, for most of 1993, the official target level was 3%.

[9]Other important feature of distributions that will be affected by monetary policy are the long run average interest rate and the variance of rates. While important, these aspects have been widely discussed in previous studies and are not dwelt on here.

[10]These are calculated using data on raw daily interest changes from January 1980 to March 1996.

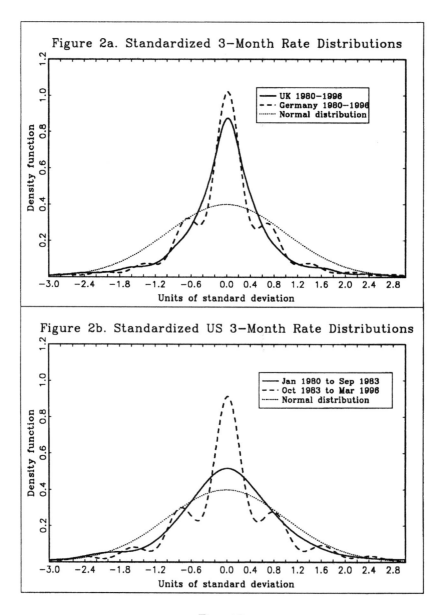

Figure 2

variable. It is also interesting to note that the German rate density is tri-modal, reflecting the important influence on the distribution of large jumps in rates. We should stress that this feature of the estimates is not simply an artifact of our kernel estimation techniques. The frequency histogram of the raw data without any smoothing shows a similar tri-modal pattern.

Figure 2b shows most clearly perhaps the impact on rate distributions of different monetary arrangements in that the two kernel estimates depicted are the densities of US interest rate changes in two periods, January 1980 to September 1983, and October 1983 to March 1996. The earlier period is unusual in that the Federal Reserve was following a policy of targeting Non-Borrowed Reserves, effectively a measure of base money, and allowing the market to determine the Fed funds rate.[11] Since October, 1983, the Federal Reserve has targeted Borrowed Reserves. In the US system Borrowed Reserves are directly proportional to the Fed funds rate, so this policy has effectively meant targeting interest rates.

The density for the earlier period appears somewhat closer to that of the normal distribution also shown in the figure. In fact, the kurtosis for the early period was just 6.3. While this probably represents a statistically significant deviation from the kurtosis of a normally distributed random variable, it is far less than the sample kurtosis of 13.2 that we calculate from the data after October 1983. Finally, the density for US data from the later period exhibits the same kind of tri-modal configuration commented on in the case of Germany above. Once again, sharp and comparatively large interest rates adjustments appear to account for this.

3.3 Interest Rate Persistence

Now consider a second way in which monetary policy affects the distribution of interest rate changes, namely the degree of persistence they exhibit. In this section, we shall examine two different statistical measures of persistence, first, unit root tests, and, second, the rate at which shocks die out as shown by estimated autoregressions.

3.3.1 Unit Root Tests

Unit roots in interest rates imply that there is no tendency for the effects of innovations to die out over time. The presence of a unit root is therefore an extreme form of persistence. Ball & Torous (1995) argue that a problem with recent empirical work on interest rate distributions is that it fails to allow for unit roots in interest rates. To a large extent, the notion that there are

---

[11]Even in this period, it should be noted, the US authorities adjusted their Non-Borrowed Reserve target in part to take account of interest rate developments, so interest rates remained to some extent a target variable.

Table 2: Unit Root Tests

Interest		One Lag Model		
Rates		Germany	UK	US
6 month	DF	-1.149	-1.629	-1.436
rate	PP	-1.144	-1.591	-1.420
		Five Lag Mode		
		Germany	UK	US
6 month	DF	-1.097	-1.494	-1.474
rate	PP	-1.099	-1.489	-1.478
		Monte Carlos		
		Germany	UK	US
Power	p	13.4%	25.1%	24.9%

Notes:
DF: Dickey-Fuller test and PP: Philipps-Perron
test. Critical value at a 5% level is -1.95.
DF and PP with constants give similar results.
Monte Carlos are based on 1000 replications
using the estimated parameters from the
five-lag autoregressions.
p measures the power of the DF test using
the 6 month rate, i.e. one minus the
probability of accepting a unit root when
rates are stationary.

unit roots in short-term interest rates is a matter of faith. Standard tests for
unit roots have so little power against interesting alternative hypotheses (in
particular, that interest rates are stationary but converge somewhat slowly)
that no firm conclusion can be reached.

To illustrate this claim, we performed Augmented Dickey–Fuller and Phil-
lips–Peron tests for unit roots on our nine interest rate series. The values of
the test statistics are given in Table 2. As one may see, unit roots cannot be
rejected for any of our nine interest rates. Rather than immediately concluding that the series is non-stationary, however, one should examine the results
of Monte Carlos reported in the lower half of the table. Using interest rate
parameters in an economically interesting range, we calculate the proportion
of times that one would incorrectly fail to reject a unit root even when the
series is stationary.[12] Repeating the experiment 1,000 times, we found that
the relevant percentages were 86.6, 74.9, 75.1.

---

[12]The data for the Monte Carlos was generated assuming the interest rate levels are order
5 autoregressions with normally distributed innovations. The autoregression parameters
and the variance of the innovations were set equal to the parameter estimates obtained in
Section 3.4 below.

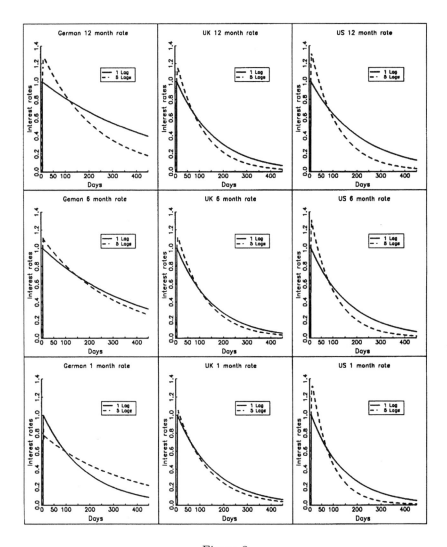

Figure 3

3.3.2 Autoregressions

Under the assumption that interest rates are stationary, we performed a series
of autoregressions for the nine interest rate series. In each case, we ran
regressions of interest rate levels on lagged levels using weekly data from
2/1/80 to 13/03/96. In each case, the regression was performed with one and
five lags. We report the results in the form of the plots shown in Figure 3.
Each plot represents the path followed on average by interest rates according
to the estimated model following a one percent positive shock.

The ordering of the degree of persistence across countries is consistent with one's intuitive understanding of the degree to which the three countries in our sample are resolute in their adherence to monetary policies. Shocks to German rates are more persistent than those to UK and US rates, and UK shocks exhibit the least persistence. It is interesting to note, however, that German rates have greater 'long-run' or unconditional variance because of this since the lower degree of reversion is offset by a lower amount of instantaneous volatility.

One way to measure the contrary effects of reversion and instantaneous volatility on unconditional variance is to take the ratio of the log of the AR(1) coefficient to half the variance of the residuals from the interest rate regression. This measure is suggested by the fact that when interest rates follow a continuous time AR(1) process with normal increments (i.e. an Ornstein-Uhlenbeck Process), the unconditional variance of the interest rate is the inverse of this quantity. Hence, we may regard our persistence measure as calculated 'per unit of volatility'. On this basis, Germany does have more rapid reversion in that the measure is 0.133 for Germany and 0.127 and 0.082, respectively, for the UK and US.

3.4   Parametric Yield Curve Models

As the last stage of our empirical investigation of interest rate distributions, we shall implement two standard single state variable yield curve models, those developed by Vasicek (1977) and Cox, Ingersoll & Ross (1985). These models make simple assumptions about the stochastic process followed by the default-free instantaneous interest rate and the price of interest rate risk and then derive the price of finite maturity bonds consistent with these assumptions.

More specifically, Vasicek (1977) assumes that instantaneous interest rates follow a Gaussian mean-reverting process, the Ornstein-Uhlenbeck process:

$$dr_t = \alpha(\theta - r_t)dt + \sigma dW_t, \tag{1}$$

for constants α, θ, and σ and a standard Brownian motion, W_t. Cox, Ingersoll & Ross (1985), on the other hand, suppose that instantaneous interest rates satisfy a square-root process:

$$dr_t = \alpha(\theta - r_t)dt + \sigma\sqrt{r_t}dW_t \tag{2}$$

again for constant parameters α, θ, and σ. Discrete time increments in square-root processes of this kind are conditionally distributed as independent non-central chi-squared random variables.

In the Appendix, we provide more details of our empirical implementation of these models. Both models have the convenient feature that interest rates over finite periods of time are affine functions of the instantaneous interest rates where the coefficients in the affine mapping are functions of the parameters α, θ, and σ. This means that is is easy to construct a likelihood for discretely sampled observations of the finite maturity interest rates implied by the two models simply by employing the conditional densities of finite maturity changes in the instantaneous interest rate processes (Gaussian for the Ornstein-Uhlenbeck process and non-central chi-squared for the square-root process), and incorporating a simple Jacobian adjustment term to allow for the fact that the observable is an affine function of the latent variable in question. Note that in our econometric implementation of these models, we suppose that the price of risk is zero. Otherwise, the drift parameters as they enter the affine mapping linking instantaneous with finite maturity interest rates would differ from the corresponding parameters in the distribution of the short interest rate.[13]

We estimate the Vasicek and Cox-Ingersoll-Ross models using interest rate data from our three markets, US dollar, Deutschemark and British pound, and for three maturities, one, six and twelve months. The data consists of Wednesday morning observations on Euro-deposit interest rates over the period 2/1/1980 to 15/03/1996, taken from the Bank for International Settlements database.[14] In all, this yielded 846 observations for each time series. We chose to focus on short-term interest rates rather than using series further along the yield curve since no single-state-variable yield curve model is likely to explain both short and long bond returns very satisfactorily (see the discussions in Litterman & Scheinkman (1991), Gibbons & Ramaswamy (1993), and Pearson & Sun (1994)).

3.5 Parameter Estimates

Table 3 contains our Maximum Likelihood parameter estimates for the Vasicek and Cox-Ingersoll-Ross models. Estimates are reported on an annualized basis and t-statistics appear in parentheses. In estimating these models on finite maturity interest rates, we effectively invert the relationship between the observed interest rates and the implicit instantaneous interest rates assumed by the models. Hence, if either the Vasicek or the Cox-Ingersoll-Ross model were correct, the parameter estimates for that particular model based on one-, six- and twelve-month interest rates should be the same.

The results for the Vasicek model are notable for the instability of parameters estimated from interest rate data of different maturities. The reversion parameter for Deutschemark interest rates is 1.2, 0.7 and 0.6 depending on

---

[13]An alternative might be to infer the value of the risk adjustment from the price of some yield-curve-related asset. Attempting to estimate the risk adjustment simply using the time series data we here employ is not feasible since it is scarcely identified.

[14]This data consists of quotes fixed at 11am Swiss time. It is preferable to use Wednesday observations since relatively few public holidays fall on this day.

which data one employs. It is also noticeable that the unconditional mean interest rates according to the model, the θ parameters, differ significantly from the actual unconditional mean of interest rates over the period covered by the sample. Both of these features of the estimates is suggestive of misspecification. This, after all, would not be surprising given that Ornstein-Uhlenbeck processes have normally distributed discrete time increments (implying kurtosis of 3) yet the sample kurtosis coefficients of the data for changes in the Deutschemark, sterling and US dollar one-month interest rates respectively are 15.0, 8.8, and 19.2.

On the other hand, the results for the Cox-Ingersoll-Ross model reported in the lower half of Table 4 exhibit remarkable homogeneity in parameter estimates for different interest rate maturities. Also, the unconditional mean interest rates, θ, are quite close to the corresponding sample means.[15] Both these aspects of the parameter estimates suggest that the square root process may provide a good model of the data. However, the parameter estimates of the square root process we report suffer from the failing noted in past empirical studies of the Cox-Ingersoll-Ross model such as Gibbons & Ramaswamy (1993) and Ball & Torous (1995), namely that the estimated reversion rates, the α parameters, are implausibly large.

In Table 3, the reversion rates range from 34 to 52. Recall that the parameters are reported on an annualized basis so the results imply that the interest rate should revert to its unconditional mean within a couple of weeks of a shock. These rates of reversion may be compared with the far lower reversion rates obtained in the autoregressions reported in Section 3.3.2.

Ball & Torous (1995) argue that estimates of autoregression parameters may be biased up in small samples if the series in question is close to a unit root. However, the magnitude of the biases that Ball and Torous generate in Monte Carlo simulations is too small to account for the very rapid mean reversion rates obtained by Gibbons & Ramaswamy (1993) and others including ourselves. Also, such biases would affect estimates not just of the continuous time models investigated in this section but also the autoregression models of Section 3.3.2, whereas the latter yielded quite low reversion rates.

3.6   Specification Problems

The obvious difference between the continuous time models described in Sections 3.4 and 3.5 and the autoregressions of Section 3.3.2 is that the former impose restrictions across the conditional means and higher moments of interest rate changes while the latter impose no restrictions on the second or

---

[15]One may compare the estimates of θ in the table with the sample means of the one, six and twelve month rates. For Deutschemark rates, these were: 6.7%, 6.8% and 6.8%. For sterling rates, they were 10.7%, 10.7% and 10.6%. For US dollar rates, they were 8.2%, 8.5%, and 8.6%.

Table 3: Yield Curve Model Estimates

		Vasicek Model					
		Germany		UK		US	
One	α	1.248	(1.447)	1.225	(1.721)	1.662	(1.943)
month	θ	0.057	(3.299)	0.086	(3.809)	0.069	(3.395)
rates	σ	0.040	(20.163)	0.046	(20.266)	0.064	(20.163)
Six	α	0.723	(1.052)	1.561	(1.899)	1.336	(1.740)
month	θ	0.050	(1.805)	0.090	(4.871)	0.068	(2.900)
rates	σ	0.032	(20.280)	0.050	(20.188)	0.057	(20.224)
Twelve	α	0.631	(0.888)	1.359	(1.660)	1.084	(1.473)
month	θ	0.050	(1.548)	0.089	(4.473)	0.069	(2.584)
rates	σ	0.030	(20.258)	0.047	(20.195)	0.051	(20.247)
		Cox–Ingersoll–Ross Model					
		Germany		UK		US	
One	α	44.670	(11.019)	46.251	(22.791)	33.990	(17.894)
month	θ	0.067	(27.890)	0.107	(30.355)	0.082	(18.823)
rates	σ	0.839	(38.558)	1.008	(39.495)	1.051	(38.302)
Six	α	46.316	(11.289)	48.950	(11.817)	35.034	(9.230)
month	θ	0.068	(29.316)	0.107	(32.618)	0.084	(19.749)
rates	σ	0.833	(38.626)	0.990	(38.883)	1.047	(39.309)
Twelve	α	48.299	(11.596)	52.150	(12.271)	36.914	(9.672)
month	θ	0.068	(30.970)	0.106	(35.568)	0.086	(21.531)
rates	σ	0.823	(38.340)	0.964	(38.418)	1.022	(39.720)

Notes: All parameters are reported on an annualised basis.
T-statistics appear in parentheses after the parameter.
Maximum Likelihood estimation was performed using weekly
data on discount bond yields from 2/1/80 to 13/3/96.
The CIR model assumes the short rate, r_t, follows:
$dr_t = \alpha(\theta - r_t)dt + \sigma\sqrt{r_t}dW_t$.
The Vasicek model assumes r_t follows:
$dr_t = \alpha(\theta - r_t)dt + \sigma dW_t$.

higher moments. To see the potential significance of this, consider the way in which parameters α and σ interact in diffusion models of the kind we are investigating. Conditional means are relatively ill-determined statistically in data sets of interest rate changes so the estimates of the parameter α will be most strongly affected by its contribution to the variance and higher moments.

A given variance in discretely-sampled data on interest rate changes may be matched by different combinations of α and σ since increasing the two parameters together generates greater instantaneous volatility but more rapid reversion within the finite period in question. One may then fit higher moments such as kurtosis as well as a given variance by selecting some combination of α and σ values that can mimic both statistical properties simultaneously. Leptokurtic behaviour will be captured in such models by high instantaneous volatility combined with high reversion rates. Such a combination of values will generate substantial interest rate shocks without boosting variance.

If the model were correctly specified, then it would not matter that a parameter like α which influences both the conditional means and the higher population moments of interest rate changes be largely determined by the higher sample moments. However, if the model is misspecified, as it undoubtedly is, the estimates of the mean that one obtains will be highly distorted. Unfortunately, the mean is crucial for most pricing applications and a model that cannot fit the evolution of the conditional mean is therefore unlikely to be of little use.

4   A Jump Model of Interest Rates

4.1   Basic Assumptions

In this section, we suggest a rather different approach to modelling interest rates and the term structure than that followed in the classic diffusion models of Vasicek (1977) and Cox, Ingersoll & Ross (1985). A general version of the approach we describe has been discussed by Babbs & Webber (1994) while El-Jahel, Lindberg & Perraudin (1996) obtain analytical solutions under specific distributional assumptions. The model of this section is taken from El-Jahel, Lindberg & Perraudin (1996)[16].

The basic approach consists of assuming that short-term interest rates are a pure jump process, pegged by the monetary authorities and periodically adjusted by discrete amounts. Of course, bond prices further along the yield curve diffuse as well as jump so there is need to introduce a state variable that can be regarded as summing up information about future government interest rate policies. We therefore assume that the rate of jump, γ_t, for the government-controlled short interest rate, r_t, is a function of the current level level of a diffusion process, X_t,

$$\frac{d}{d\Delta}\mathrm{E}_t\left(r_{t+\Delta}\right)\Big|_{\Delta\downarrow 0} = \delta_t\gamma(X_t) \tag{3}$$

where $\gamma(X_t)$ is the rate of jump or jump intensity, and δ_t is the expected jump size of a jump in the instantaneous interest rate at t should one occur (where the expectation is conditional on information at $t-$). We further suppose:

---

[16]One should perhaps note that this way of introducing jumps into interest rates differs significantly from earlier work such as Ahn & Thompson (1988) and Shirakawa (1991) which incorporated independent and identically distributed jump components in jump-diffusion models of short-term interest rates and the yield curve.

1. that the diffusion state variable, X_t, is an Ornstein-Uhlenbeck process:

$$dX_t = \alpha(\theta - X_t)dt + \sigma dW_t, \tag{4}$$

2. that the jump rate function, $\gamma(.)$ is quadratic:

$$\gamma(X_t) = \beta X_t^2, \tag{5}$$

3. that future jump sizes are a sequence of known constants, $\delta_1, \delta_2, \delta_3 \ldots$,

4. and that agents are risk neutral.

The assumption of risk neutrality is made here for expositional convenience. One may introduce risk aversion using standard change of measure arguments. The assumption of fixed and known jump sizes is also made for simplicity. There are various ways to relax this and introduce stochastically varying jump sizes. El-Jahel, Lindberg & Perraudin (1996) develop a model in which the jump size at t (should a jump occur) equals the level of an arithmetic Brownian motion, independent of the Ornstein-Uhlenbeck process driving the jump rate function.

4.2 Applying the Karhunen-Loeve Theorem

These assumptions fully specify the stochastic behaviour of short-term interest rates. To derive the price $P_{t,T}$ of a discount bond maturing at date T, we need to evaluate the expectation:

$$P_{t,T} = E_t \left(\exp \left[-\int_t^T r_s ds \right] \right). \tag{6}$$

At any time, t, the level of interest rates, r_t, may be written as:

$$r_t = r_0 + \sum_{j=1}^{N(t)} \delta_j \tag{7}$$

where $N(t)$ is the number of jumps up to and including time t. The integral in eq. (6), can then be written:

$$\int_0^T r_s ds = r_0 t_1 + (r_0 + \delta_1)(t_2 - t_1) + (r_0 + \delta_1 + \delta_2)(t_3 - t_2) + \ldots$$
$$+ \left(r_0 + \sum_{j=1}^{N(T)} \delta_j \right) (T - t_{N(T)}). \tag{8}$$

Cancelling terms, one may then write Eq. 6 as:

$$P_{t,T} = E_t \left[\exp \left(-r_0 T - \sum_{j=1}^{N(T)} \delta_j (T - t_j) \right) \right]. \tag{9}$$

El-Jahel, Lindberg & Perraudin (1996) show how one may calculate for the simple case in which the δ_i are known constants and for another model in which they are proportional to the level of an independent Brownian motion. To give a flavour of their analysis, we shall describe the case in which the δ_i are known constants.

The first step is to solve for the density of the sample paths of the jump process (equivalent here to deriving the joint density of the jump times conditional on information at t). Let $\rho[N_t : 0 \leq \tau < 1]$ denote the path density of the counting process that records changes in interest rates. Here, we have chosen time units so that the horizon $[t, T]$ is normalized to $[0, 1]$. Conditional on the time path followed by the forcing process, X_t, the counting process, N_t for interest rate changes is a Poisson process. By Snyder & Miller (1991) page 358, it follows that:

$$\rho[N_\tau : 0 \leq \tau < 1] = E_0 \left[\exp \left(- \int_0^1 \gamma(\tau, X_\tau) d\tau + \int_0^1 \ln \gamma(\tau, X_\tau) dN_\tau \right) \right] .$$
(10)

To evaluate the expectation in equation (10), El-Jahel, Lindberg & Perraudin (1996) use techniques employed in the physics literature in models of photon emission (see Macchi (1971)). The main step in this is to express the conditional distribution of the Ornstein-Uhlenbeck forcing process, X_t, as an expansion in terms of eigenfunctions, using the Karhunen-Loeve Theorem. In particular, El-Jahel, Lindberg & Perraudin (1996) show that:

Proposition 1 *By the Karhunen-Loeve Theorem, X_t can be written as:*

$$X_t = \sum_{n=1}^{\infty} x_n \phi_n(t) \qquad \text{for} \quad t \in [0, 1]$$
(11)

where the $\phi_n(t)$ for $t \in [0, 1]$ are functions of the form:

$$\phi_n(\tau) = \frac{2 \sin(\omega_n \tau)}{\sqrt{2 - \frac{\sin(2\omega_n)}{\omega_n}}}$$
(12)

Here, the $\omega_1, \omega_2, \omega_3 \ldots$ are the positive roots of the equation,

$$\frac{2\alpha i \omega_n}{\omega_n^2 + \alpha^2} = \alpha \left(\frac{\exp[-\alpha + i\omega_n] - 1}{-\alpha + i\omega_n} - \frac{\exp[-\alpha - i\omega_n] - 1}{-\alpha - i\omega_n} \right)$$
(13)

The x_n are real-valued, normally distributed independent random variables with variances, $v_n^2 = \sigma^2/(\omega_n^2 + \alpha^2)$ while the means, $m_n \equiv E_0 x_n$, are given by:

$$m_n = \frac{\eta_n(X_0 - \theta)}{\omega_n^2 + \alpha^2} \{ \omega_n - \exp[-\alpha](\alpha \sin(\omega_n) + \omega_n \cos(\omega_n)) \} + \frac{\eta_n \theta}{\omega_n}(1 - \cos(\omega_n))$$
(14)

where $\eta_n \equiv 2/\sqrt{2 - \sin(2\omega_n)/\omega_n}$.

Proof: see El-Jahel, Lindberg & Perraudin (1996). □

As El-Jahel, Lindberg & Perraudin (1996) discuss, in any practical application, the infinite sum in equation (11) must be truncated at some finite number N of terms in the summation. In fact, rather than using the sequence of eigenfunctions, ϕ_1, ϕ_2, ϕ_3, ..., ϕ_N, it turns out that it is better in practice to employ as eigenfunctions: ϕ_0, ϕ_1, ϕ_2, ϕ_3, ..., ϕ_{N-1}, where ϕ_0 is defined as the orthogonal projection of a constant onto the ϕ_i, $i = 1, 2, \ldots, N - 1$.

4.3 The Interest Rate Path Density

The orthogonality properties of the Karhunen-Loeve representation enable one to write the expectation in equation (10) in the following simple form:

$$\rho\left[N_\tau : 0 \leq \tau < 1\right] = E_0\left[\exp\left[-\beta \sum_{s=1}^\infty |x_s|^2 + \int_0^1 \ln(\beta X_\tau^2)dN_\tau\right]\right] \tag{15}$$

where N_τ is the counting process associated with the Poisson process. It is much easier to evaluate this and hence obtain the sample path density of interest rates than it is to attempt a more direct evaluation of the expression in (10). In fact, the path density ρ is:

Proposition 2 *If the jump times are denoted t_1, t_2, ..., t_L,*

$$\rho\left[N_\tau : 0 \leq \tau < 1\right] = \sum_{L=0}^\infty \beta^L \sum_{\substack{j_1,\ldots j_L=1 \\ k_1,\ldots k_L=1}}^\infty a_L(j, k) \prod_{l=1}^L \phi_{j_l}(t_l)\phi_{k_l}(t_l) \tag{16}$$

where:

$$a_L(j, k) = \prod_{n=1}^\infty E_0\left[x_n^{p_n(j)}x_n^{q_n(k)}\exp\left[-\beta x_n^2\right]\right]. \tag{17}$$

The vectors $(j_1, \ldots j_L)$ and $(k_1, \ldots k_L)$ are L-dimensional permutations of the positive integers. The inner summation in equation (16) is thus over all these permutations. $p_n = p_n(j)$ and $q_n = q_n(k)$ are defined as the number of elements in permutations j and k that equal a given integer, n. [17]

Proof: see El-Jahel, Lindberg & Perraudin (1996). □

Once one has the path density of the interest rate process, it is simple to obtain the bond price by evaluating the expectation in equation (6). Further details are spelt out in El-Jahel, Lindberg & Perraudin (1996).

---

[17]The evaluation of the $a_L(j, k)$ is described in detail in El-Jahel, Lindberg & Perraudin (1996).

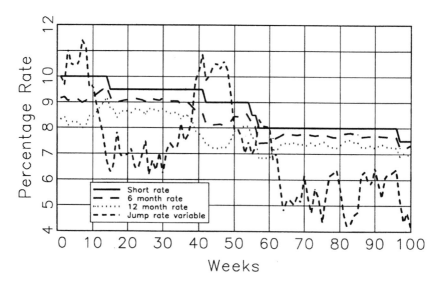

Figure 4

4.4 Bond Yields and Sample Paths

Figure 4 shows a typical set of sample paths for the state variable, X_t, the instantaneous interest rate, r_t and the yield to maturity, $R_{t,T}$ on discount bonds with six and twelve month maturities. The solutions are calculated assuming that the size of interest rates changes is deterministic and given by $\delta = -0.5\%$. Of course, if we were seeking to model anything except short term bonds, this assumption would be very restrictive. Over a horizon of six months or a year, however, it seems reasonable to suppose that market participants belief that rates can only move in one direction and that the jump size is some 'conventional' amount like half a percentage point. The other parameters used in calculating the interest rates in Figure 4 are: $\alpha = 0.5\sqrt{700}$, $\theta = 0.05$, $\sigma = 0.05\sqrt{700}$, $r_0 = 0.1$, and $X_0 = 0.1$. We also assume that the rate of jump equals $\beta \times X_t^2$ where $\beta = 700$.

The sample path shown may be interpreted as follows. The jump rate starts at its unconditional mean, 0.1. This produces a steeply downward sloping yield curve since downward movements in interest rates are anticipated. After 15 weeks, the state variable, X_t, falls markedly, leading to a partial flattening of the yield curve and a long period in which rates do not decline. After a year, the chance of a rate decline rises again. Several interest rate alterations do actually occur and again the yield curve becomes steep. Towards the end of the sample period, the state variable, X_t, remains low and the yield curve becomes less downward-sloping.

Table 4: INTEREST RATE MOMENTS

Frequency (weeks)	£	DM	$	Δr	ΔR
VARIANCE					
1	0.07	0.04	0.06	0.02	0.02
2	0.16	0.07	0.12	0.04	0.04
4	0.35	0.10	0.16	0.10	0.11
8	0.72	0.18	0.33	0.22	0.23
13	0.89	0.24	0.39	0.40	0.40
KURTOSIS					
1	8.82	14.99	19.23	14.80	17.75
2	5.40	11.90	14.87	14.42	12.57
4	5.09	6.39	5.35	7.82	7.74
8	2.95	3.73	4.17	5.68	5.43
13	2.18	2.77	2.19	3.61	3.77
SKEWNESS					
1	1.16	0.69	0.50	-0.94	-0.69
2	0.61	-0.21	0.09	-0.84	-0.82
4	0.63	0.80	0.43	-0.29	-0.32
8	0.21	0.53	0.00	0.67	0.55
13	0.25	-0.53	0.28	0.11	0.15

NOTES: Columns 2, 3, and 4 contain moments
of changes in Sterling 3 month ($£$),
Deutschmark 3 month (DM), and US dollar
3 month ($) interest rates.
Columns 5 and 6 contain moments of changes
in the instantaneous and 3 month interest
rates implied by the model with parameters:
$\alpha = 0.5$, $\theta = 0.1$, $\sigma = 0.05$,
$\delta = -0.005$ or 0.005, $\beta = 250$.
Rates are expressed in percent.

4.5 Higher Moments

In Table 4, we compare the higher moments of discount bond yields implied by
a typical sample path from our model with those of the interest rate data used
in Sections 3.4 and 3.5. Again, for simplicity, we take the known jump size
version of the El-Jahel, Lindberg & Perraudin (1996) model. The moments
are calculated for returns covering increasing periods of time. As one may see
from the table, a characteristic of the data is a steep rise in kurtosis as the

frequency of the data shortens. Over quarterly periods, actual interest rate changes have kurtosis of between 3 and 6, whereas weekly changes in interest rates exhibit kurtosis coefficients of 8.8, 15.0, 19.2, and 14.8.[18] The kurtosis coefficients implied by the interest rate sample path generated by our model show a similar steep rise as the data frequency is reduced.

5   Conclusion

In this article, we have investigated the distributions of short-term interest rates denominated in three different currencies, and stressed their dependence on the monetary control regimes operated by the corresponding national monetary authorities. Countries which peg the short interest rate, periodically adjusting it in discrete steps, tend to possess market rates further along the yield curve which have strikingly high kurtosis. Comparison of US interest rate data for periods of money and interest rate targeting provides further evidence that this is the case.

A second distributional property we examine is the speed with which short-term interest rates in different currencies converge to their long run averages. Again, we relate this to aspects of the approach taken to monetary policy, in this case to the degree of resolution shown by the national monetary authority in the face of a monetary shock. To illustrate the importance of the distributional properties we examine, we estimate the parameters of two standard yield curve models, namely the Vasicek and Cox-Ingersoll-Ross models. We discuss the relationship between their empirical failings and our prior examination of leptokurtosis and interest rate reversion rates.

In particular, we argue that the well-known tendency for estimates of reversion rates in the Cox-Ingersoll-Ross to be implausibly high reflects the fact that it is difficult to mimic the characteristic high kurtosis in high frequency short term interest rate data with standard diffusion models of the instantaneous interest rate. As an alternative approach, in the last section of the paper, we describe a new approach to yield curve modelling developed by El-Jahel, Lindberg & Perraudin (1996) in which the instantaneous interest rate is taken to be a pure jump process of which the rate is fixed and occasionally changed in discrete steps by the monetary authorities.

---

[18]These differ from the kurtosis coefficients reported in Section 3.2 because the latter are based on daily interest rate changes.

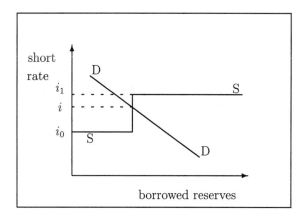

Figure 5: German Monetary Policy

6 Appendix

6.1 Monetary Policy in Three Countries

We here describe the operation of monetary control and its impact on money markets in Germany, the UK and the US. In particular, we focus on the elasticities of money market supply and demand and their dependence on the institutional arrangements.

6.1.1 Germany

Figure 5 illustrates the German system of monetary policy instruments. Banks are able to borrow a limited amount of reserves at the relatively low discount rate (i_0 in the figure) and unlimited amounts at the higher lombard rate (i_1). The two rates provide lower and upper bounds for the range within which the call rate (the overnight rate in the interbank market) fluctuates.

The high elasticity of demand for reserves in the German system is largely a consequence of the fact that bank reserve requirements are based on monthly averages. The averaging implies that unexpected temporary shifts in either the demand or supply of reserves will affect the overnight rate relatively little since any excess or shortage of reserves can be made up in the course of the month.

In the past decade, open market operations involving repurchase agreements or repos have been the most important monetary policy instrument of the Bundesbank both for controlling the supply of reserves and adjustments in interest rates. The Bundesbank uses the repo rate to guide the call rate below the lombard rate. Hence, one may view the call rate as an operational target of the Bundesbank.

Until September 1992, repos were usually of one or two months maturity. But, in an effort to gain more operating flexibility, the Bundesbank now employs two week repos. Two different kinds of repos are employed: fixed-rate repos with volume tenders and variable-rate repos with interest rate tenders. Interest-rate tenders have been more common but the Bundesbank occasionally prefers volume tenders, especially if money market rates or liquidity allocation has deviated significantly from the levels desired by the authorities. Volume tenders allow the Bundesbank to signal preferred money market rates. Repo bidding usually occurs every Tuesday and allotments are announced around 10.00 a.m. the following day.

In addition to securities repurchase agreements, the Bundesbank employs a number of so-called "reversible fine-tuning" measures to adjust the supply of reserves on a day-to-day basis, thereby avoiding excessive fluctuations in short-term interest rates. These include outright transactions in short-term treasury bills, repo operations with maturities down to two days and swapped foreign exchange transactions.

6.1.2 United Kingdom

The British system differs significantly from that of Germany. First, the ultimate authority for interest rate changes lies with the Chancellor of the Exchequer and the role of the Bank of England is limited to advising on policy formation and implementing policy changes. Second, so-called discount houses occupy an important position between commercial banks and the Bank of England. The discount houses are private companies whose role is to intermediate surpluses or shortages of funds by discounting treasury or commercial bills. In normal day-to-day operations, the Bank of England's only counterparties are the discount houses. Lending to discount houses is usually at maturities from overnight to fourteen days. The UK does not operate reserve requirements, but banks hold a small amount of clearing balances (represented in Figure 6 by the vertical DD curve).

In principle, the Bank of England determines rates by selling or buying securities at terms of its choosing. The supply curve of reserves is therefore horizontal. There are two principal ways in which changes in monetary policy are signalled. First, by changes in the stop rate for Band 1 bills which is the lowest rate of discount at which the Bank is prepared to buy bills with maturities of up to 14 days from discount houses. A significant fall or rise in the stop rate is associated with a corresponding movement in the base lending rate of commercial banks.[19] Second, on some occasions the Bank

---

[19]Base rate are administered interest rates for commercial loans, in some respect analogous to prime rates in the US.

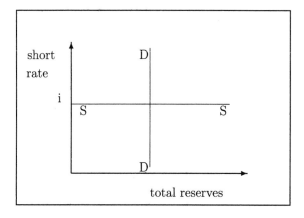

Figure 6: United Kingdom Monetary Policy

does not provide sufficient funds to the market at the rates on bills offered by the discount houses.

The discount houses are then invited to borrow at 2.30 p.m. at the 2.30 p.m. lending rate. If this rate differs significantly from the most recent dealing rates on eligible bills, a clear signal is given to the commercial banks and a change in their base lending rates generally follows. An even stronger signal is given if the Bank announces a fixed minimum lending rate for all its dealing with discount houses. In principle, any rate at which the Bank of England deals could be viewed as a key rate, but de facto the base lending rate of banks, which is largely determined by the Bank, plays this role.

6.1.3   United States

The nominal operational target is the level of borrowings at the Federal Reserve's discount window, which is associated with a desired path of non-borrowed reserves. However, the actual operational target is the level of the Fed funds rate, which is the interest rate in the overnight interbank market on deposits at the Fed that depository institutions with reserve deficiencies borrow from institutions with excess reserves. The demand for reserves is fairly interest inelastic being the sum of required reserves, which are very inelastic, and excess reserves which are slightly elastic.

A key factor in the US procedures is the nature of commercial banks' access to borrowed reserves (BR), that is, discount window borrowing. Such reserves are provided at the discount rate but are rationed by virtue of the administrative guidelines on borrowing imposed by the Federal reserve . From

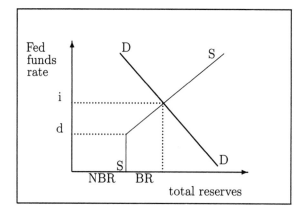

Figure 7: United States Monetary Policy

the point of view of the banks, there is a trade off between the pecuniary gains and the non-pecuniary costs of borrowing. Hence, the higher the federal funds rate relative to the administrative discount rate, the more banks are willing to borrow at the Fed.

The Fed influences the Fed funds rate (i) through the discount rate and (ii) by adjusting the supply of non-borrowed reserves (NBR). The primary monetary instrument is open market operations in which the Fed buys or sells securities, thereby adding or draining non-borrowed reserves from, the banking system. The Fed rarely performs outright purchases or sales. More frequently, it uses repurchase or matched sale agreements. These agreements typically last for only 1 day, rarely for more than 7 days, and never exceed 15 days. The discount rate is of secondary importance, but is the floor for the Fed funds rate and is the only rate that the Fed formally admits to controlling. Thus, when it changes the discount rate, the Fed is making a particularly strong statement about the direction in which it seeks to move the Fed funds rate.

6.2   Estimation of Parametric Yield Curve Models

In this Appendix, we sketch the approach we took in estimating the parameters of the parametric yield curve models of Section 3.4. Since the transition densities of the state variable in both the Vasicek and the Cox-Ingersoll-Ross models is known, it is straightforward to derive a Maximum Likelihood estimator. The conditional densities of the two interest rate processes, $\rho(r_t|r_{t-1})$, may be summarized as:

1. Vasicek Model (normal density)

$$\rho(r_t|r_{t-1}) \quad = \quad \frac{1}{\sqrt{2\pi v}} \exp\left[-\frac{1}{2}\frac{(r_t - r_{t-1} - m_t)^2}{v^2}\right] \tag{18}$$

$$m_t \quad \equiv \quad (\exp[-\alpha] - 1)\,(r_{t-1} - \theta) \tag{19}$$

$$v^2 \quad \equiv \quad \frac{\sigma^2}{2\alpha}\,(1 - \exp[-2\alpha]) \tag{20}$$

2. Cox–Ingersoll–Ross Model (non-central chi-squared density)

$$\rho(r_t|r_{t-1}) \quad = \quad c\exp[-d_t - e_t]\left(\frac{e_t}{d_t}\right)^{\frac{q}{2}} I_q\left(2\sqrt{d_t e_t}\right) \tag{21}$$

$$\text{where} \quad c \quad \equiv \quad \frac{2\alpha}{\sigma^2(1 - \exp[-\alpha])} \tag{22}$$

$$d_t \quad \equiv \quad cr_{t-1}\exp[-\alpha] \tag{23}$$

$$e_t \quad \equiv \quad cr_t \tag{24}$$

$$q \quad \equiv \quad \frac{2\alpha\theta}{\sigma^2} - 1 \tag{25}$$

and where $I_q(.)$ is the modified Bessel function of the first order.

Both models yield affine term structures. In other words, the yield at time t on any $T - t$ maturity discount bond, $R(t, T)$, is a linear function of the instantaneous interest rate:

$$R(t, T) = \frac{1}{T - t}\left\{B(t, T)r_t - \log A(t, T)\right\} \tag{26}$$

for a pair of functions, $A(t, T)$ and $B(t, T)$. In the simple case in which agents are risk neutral so the price of risk is zero, the functions $A(t, T)$ and $B(t, T)$ for the Vasicek and Cox-Ingersoll-Ross models are:

1. Vasicek Model

$$B(t, T) \quad = \quad \frac{1 - \exp[-\alpha(T - t)]}{\alpha} \tag{27}$$

$$A(t, T) \quad = \quad \exp\left[(B(t, T) - (T - t))\left(\theta - \frac{\sigma^2}{2\alpha^2}\right) - \frac{\sigma^2}{4\alpha}B(t, T)^2\right] \tag{28}$$

2. Cox-Ingersoll-Ross Model

$$B(t, T) \quad = \quad \frac{2\exp[\gamma(T - t)] - 1}{(\gamma + \alpha)(\exp[\gamma(T - t)] - 1) + 2\gamma} \tag{29}$$

$$A(t, T) \quad = \quad \left(\frac{2\gamma\exp[(\alpha + \gamma)(T - t)/2]}{(\gamma + \alpha)(\exp[\gamma(T - t)] - 1) + 2\gamma}\right)^{\frac{2\alpha\theta}{\sigma^2}} \tag{30}$$

$$\text{where} \quad \gamma \quad \equiv \quad \sqrt{\alpha^2 + 2\sigma^2} \tag{31}$$

To formulate a likelihood for a sample of finite maturity interest rates sampled at discrete intervals of time, one may infer the underlying interest rate, r_t, by inverting the relationship in equation (26) and then evaluate the densities in equations (18) and (21) multiplied by Jacobian adjustment terms to take account of the change of variable.

Acknowledgements

We thank Mike Orzsag for helpful discussions and an anonymous referee for valuable comments. Opinions expressed in this paper are solely those of the authors and do not necessarily reflect those of the Sveriges Riksbank. Correspondence regarding this paper should be addressed to William Perraudin, Department of Economics, Birkbeck College, 7–15, Gresse Street, London, W1P 2LL, UK. Much of the work was completed while Perraudin was a visiting scholar at the Sveriges Riksbank and he thanks the Economics Department of the Riksbank for its hospitality.

References

Ahn, C. M. & H. E. Thompson (1988): 'Jump-Diffusion Processes and the Term Structure of Interest Rates', *Journal of Finance* **43**, 155–174.

Babbs, S. H., & N. J. Webber (1994): 'A Theory of the Term Structure with an Official Short Rate', FORC working paper, 94/49.

Ball, Clifford, A., & W. N. Torous (1995): 'Units Roots and the Estimation of Interest Rate Dynamics', mimeo.

Batten, D. S., M. P. Blackwell, I.-S. Kim, S. E. Nocera, & Y. Ozeki (1990): 'The Conduct of Monetary Policy in the Major Industrial Countries: Instruments and Operating Procedures', Occasional paper no. 70, International Monetary Fund, Washington, DC.

Cox, J. C., J. E. Ingersoll, & S. A. Ross (1985): 'A Theory of the Term Structure of Interest Rates', *Econometrica* **53**, 385–407.

Dotsey, M., & R. G. King (1986): 'Informational Implications of Interest Rate Rules', *American Economic Review*, **76**, 33–42.

El-Jahel, L., H. Lindberg, & W. R. Perraudin (1996): 'Yield Curves with Jump Short Rates', Institute for Financial Research, Birkbeck College Working Paper.

Freedman, C. (1990): *Implementation of Monetary Policy* Reserve Bank of Australia.

Gibbons, M. R., & K. Ramaswamy (1993): 'A Test of the Cox, Ingersoll and Ross Model of the Term Structure', *Review of Financial Studies* **6**, 619–658.

Hull, J., & A. White (1990): 'Pricing Interest Rate Derivative Securities', *Review of Financial Studies* **3**, 573–592.

Kneeshaw, J. T., & P. V. den Bergh (1989): 'Changes in Central Bank Money-Market Operating Procedures in the 1980s', *Bank for International Settlments Economics Papers* **23**.

Litterman, R., & J. Scheinkman (1991): 'Common Factors Affecting Bond Returns', *Journal of Fixed Income* **1**, 54–61.

Macchi, O. (1971): 'Distribution Statistique des Instants d'Emission des Photoelectrons d'une Lumiere Thermique', *Cahiers de Recherches de l'Academie Scientifique de Paris* **272**A, 437–440.

McCallum, B. T. (1994): 'Monetary Policy Rules and Financial Stability', Working paper no. 4692, National Bureau of Economic Research, Cambridge, MA.

Pearson, N. D., & T.-S. Sun (1994): 'Expoiting the Conditional Density in Estimating the Term Structure: An Application to the Cox, Ingersoll, and Ross Model', *Journal of Finance* **49**, 1279–1304.

Shirakawa, H. (1991): 'Interest Rate Option Pricing with Poisson-Gaussian Forward Rate Curve Processes', *Mathematical Finance*, **1**, 77–94.

Siegel, J. (1983): 'Operational Interest Rate Rules', *American Economic Review* **73**, 1102–1109.

Silverman, B. W. (1986): *Density Estimation for Statistics and Data Analysis*. Chapman Hall, London.

Snyder, D. L., & M. I. Miller (1991): *Random Point Processes in Time and Space*. Springer-Verlag, Berlin, second edn.

Vasicek, O. A. (1977): 'An Equilibrium Characteristic of the Term Structure', *Journal of Financial Economics* **5**, 177–188.

Part V
Numerical Methods

Numerical Option Pricing using Conditioned Diffusions

S.K. Gandhi and P.J. Hunt

Abstract

We demonstrate, by considering a specific option pricing problem, how to use closed form expressions or approximations to improve the speed, stability and accuracy of pricing and hedging algorithms. A tree, lattice or simulation pricing algorithm is effectively, in both time and space, a discrete approximation to some underlying continuous stochastic process. The main idea of this paper is to perform explicit calculations concerning the behaviour of the process between the discrete grid points generated in the pricing algorithm, conditional on the values of the process at these grid points, to improve the algorithm's performance. These ideas extend beyond the example treated here and yield highly efficient algorithms.

1   Introduction

Financial derivatives are continually becoming more complex and sophisticated. Furthermore, the numbers of complex deals in trading books throughout the financial world are increasing so that in many cases they are in their hundreds. This has led to a great deal of attention and effort being focused on the speed and accuracy of algorithms used to price these options.

The problems faced by practitioners are worrying. It is vital for a trader to have a detailed understanding of all the trades in his book and the risks to which he is exposed as a consequence. As such he needs to be able not only to calculate the value of all the trades, but also to calculate the sensitivity of these values to input variables – by evaluating a mathematical derivative numerically – and also to be able to revalue the portfolio under different scenarios. He would like to be able to do this real time. The time when he needs the information most is when the market is changing rapidly – precisely when he has the least time to get the information. The magnitude of the difficulty is clear from the fact that at least one US institution currently takes 2–6 hours just to revalue its exotic products portfolio of a couple of hundred deals.

Given these requirements it is not surprising that such considerable effort has been expended on developing fast and accurate algorithms for pricing and hedging exotic products. The problems have been attacked from many angles and viewpoints. For 'American' style products, which require decisions to be made throughout the life of an option, trees or more general lattice algorithms are used. For 'path dependent' products these techniques can sometimes be applied but simulation is also a possibility, and often simulation is then the only possibility. The major difficultly with exotic products, however, is that the calculations being performed are to estimate some value function $V(X_t, t)$ where X_t represents the state of an underlying process. Because options have non-smooth payoffs–often being discontinuous and allowing decisions to be made at discrete time points–the function V is only piecewise C^∞. This causes considerable problems for any algorithm and can reduce the accuracy to be $O((\Delta T) \vee (\Delta X))$ where ΔT and ΔX are, respectively, the time and space discretisations. This is, for example, the case for a standard multinomial tree pricing interest rate products. A tree with K time steps will have $\Delta X = O(K^{-1/2})$, $\Delta T = O(K^{-1})$, a number of nodes $O(K^{-2})$ and an accuracy $O(K^{-1/2})$.

The complex form of the option pricing problem causes difficulties with accuracy and stability of pricing algorithms. One compensation for option problems, however, is that they relate to stochastic processes and there is a great deal of intuition and knowledge about the processes typically encountered which can be brought to bear on the problems–but usually isn't! Huge improvements in pricing algorithms can be achieved by exploiting the structure of a given problem and it is the aim of this paper to demonstrate by considering a specific example how this can be done. In dealing with this example we provide an algorithm which is of interest in its own right. It is new, easy to implement, flexible, easily extended to higher dimensions and addresses real problems of practical interest. However, the main aim of the paper is to make the point that accuracy and stability of pricing algorithms is achieved much more through a thorough understanding of the problem and its structure than through the application of a general solution technique.

Before outlining in more detail the specific problem we consider we should first discuss what errors in pricing are acceptable. In our example, that of a Bermudan swaption, it is possible to calculate the value to a relative error of 10^{-9} in a negligible amount of time (under a second on a Pentium PC). This corresponds to an error of one penny on an option worth £1m. Clearly such accuracy is not required. Indeed the price is model dependent and is also dependent on the prices of the underlying instruments on which the exotic option is based. As such an accuracy of 10^{-1} is more realistic in practice and often even that cannot be achieved. Nonetheless accuracy of closer to 10^{-9} that 10^{-1} is what we must achieve. This is for reasons of consistency and stability.

Consistency is clearly required so that trading books do not exhibit large unexplained swings in value when the input market prices change slightly–a trader must be able to understand his positions. Equally, small changes in the specification of the deal must lead to small changes in value. More critical however is the need for the algorithm to be accurate and well-behaved so that hedges can be calculated. Exotic products need hedging with other products. The complexities of pricing mean that often the only way to calculate the (mathematical) derivative of the exotic's price with respect to an input price is by revaluing the product with the new price as input and taking the difference. If the error in each of the input prices in $O(\varepsilon)$ and the shift size is h, the error in the simplest estimate of the derivative will be $O(\varepsilon/h)$ which is orders of magnitude higher that ε. For second derivative calculations the situation is even worse. The need to keep h small so that $V(x+h) - V(x)$ approximates $V'(x)$ accurately conflicts with the need to keep ε/h small. So the smaller we can make ε the better. Of course, we can improve the accuracy of this numerical estimate by choosing a higher order approximation for $V'(x)$ and by ensuring the two price errors are correlated. However, this is not always possible and, as we have already noted, V is not always smooth so we cannot make h too large.

The example we will consider is pricing Bermudan swaptions using the Hull–White (extended Vasicek) model [3]. This model is particularly tractable since the spot rate is a Gaussian process

$$dr_t = (\theta_t - a_t r_t)dt + \sigma_t dB_t \tag{1}$$

which has solution

$$r_t = \phi_t^{-1}\phi_0 r_0 + \phi_t^{-1}\int_0^t \phi_u \theta_u du + \phi_t^{-1} W\left(\int_0^t \phi_u^2 \sigma_u^2 du\right) \tag{2}$$

where

$$\phi_t = \exp\left(\int_0^t a_u du\right),$$

and W is a standard Brownian motion. We will repeatedly need properties about the path of $\{r_u : t \le u \le T\}$ conditional on r_t and r_T. If we condition a standard Brownian motion on its values at two times the intermediate process is a Brownian bridge and this has been extensively studied. Equation (2) represents our process as a time-changed Brownian motion with a drift, and so enables us to exploit the results about the Brownian bridge.

The property of this problem which makes our algorithm particularly efficient is the fact that the option can only be exercised at a discrete set of time points. By performing explicit calculations between these times we are able to remove the need for nodes at intermediate times and remove all errors caused by discretisation in time. We concentrate then on reducing to a minimum the error caused by space discretisation.

The ideas which we apply to the Hull–White model can also be applied more generally. We have restricted attention to Hull–White because it is particularly tractable. However the ideas apply to other models. The forthcoming paper by Kennedy and Laws [4] shows how similar ideas can be applied to the less tractable log-normal models of Black and Karasinski [2] and Black, Derman and Toy [1]. These results will be of particular practical importance because log-normal models are used more widely and yet implementation is currently much less efficient than for normal models. Other products where improvements in performance of the algorithm are simple to obtain include equity barrier options, interest rate barrier options, either for a sloping barrier or a non-constant discount rate, and American options. In these latter cases it is not possible to remove the 'time' error completely but it is possible to reduce it by a couple of orders of magnitude from the usual situation.

The layout of the rest of the paper is as follows. Section 2 defines a Bermudan option and briefly outlines the pricing algorithm. Sections 3–6 then describe the algorithm in more detail, in particular how to fix the model parameters from market data (Section 3), how the tree is defined (Section 4), and how to calculate option values within the tree (Sections 5 and 6). In Section 7 we present an error analysis for the algorithm and finally, in the Appendix, we present in considerable detail the steps which must be gone through to implement the algorithm.

2   Bermudan Options via Hull–White

We begin with a statement of the problem. For mathematical precision and generality we will describe the product in terms of cashflows and as much as possible avoid market terminology such as 'payers swaptions', 'forward start swaptions', etc. It is the case however that all interest rate Bermudan products that we have come across fit into this framework.

Under the terms of the Bermudan option contract the option holder has the right, but not the obligation, to exercise the option on at most one of a series of dates $T_1, \ldots, T_K$. If the option is exercised at T_i the holder then receives a single payment, the intrinsic value of the option at T_i, of amount

$$I_i := \left(\sum_{j \in J_i} c_{ij} P_{T_i S_{ij}} - K_i \right)_+ \tag{3}$$

where P_{tT} is the value at time t of a cashflow which is contracted for payment at T (a zero coupon bond), and c_{ij} and K_i are arbitrary fixed constants. The constant J_i is an arbitrary integer, and the S_{ij} are an arbitrary set of times greater than T_i. Where there is no risk of confusion in the sequel we will tend to drop some of the subscripts.

We leave aside any discussion as to why it is appropriate to use a one-factor Hull–White model to price such an option. We take as given the fact that this is the model to be used and our aim is then to derive the correct price.

In the following sections we describe in detail the algorithm employed. The general approach, that of using a tree to carry out a dynamic programming calculation via backward induction is standard. The steps we need to go through to implement this approach are the following.

(i) Use a set of market instruments to calibrate the Hull–White model, i.e. to fix a_t, θ_t and σ_t in equation (1).

(ii) Specify the locations of the nodes in the tree and how the intrinsic value of the Bermudan will be calculated at each of these nodes.

(iii) Calculate at each node the discounted future value of the Bermudan, and thus calculate the Bermudan value at this node by taking the maximum of this with the intrinsic value at the node.

We address each of these in turn.

3 Model Calibration

The model needs to be calibrated so that it correctly prices a set of market instruments. These instruments will typically be (pure discount) bonds and either swaptions or caps and floors, depending on the application.

Calibrating a Hull-White model to these products is, because of the one-factor Gaussian structure, a straightforward task. We will for our current purposes assume that the (critical) mean reversion parameter a_t has been chosen and fixed beforehand, although these techniques also apply to the more general situation when a_t depends on σ_t. The first step is to calculate the function σ_t which follows from the prices of the calibrating options, although we do have to choose a functional form for σ_t between option exercise dates. Once this has been achieved the drift function θ_t can be obtained analytically from the parameters already calculated and pure discount bond prices.

Details of the procedures to calculate these parameters are standard and contained in the Appendix. The hardest task in practice is fitting the volatilities but this can be done with a few applications of a safe Newton-Raphson or Brent algorithm [7] and an appeal to the option decomposition approach first introduced by Jamshidian [5]. As we will see in the next two sections we do not need, and should not calculate, the values of a_t, θ_t and σ_t themselves, but instead the following integrals of these quantities, such as those in equation (2) above,

$$\mu_t = \int_0^t \phi_u \theta_u du,$$

$$\xi_t = \int_0^t \phi_u^2 \sigma_u^2 du.$$

We have here skipped over questions of how we would choose the functional form for both a_t and σ_t which in practice have major effects on the price of Bermudan options. We have also avoided the question of which instruments we should use to calibrate the model.

4   Specifying the Tree

The approach we use to price the option is a tree-based approach, as is common for American-style options. As it turns out however the approach is more closely related to a multiple integration. The first step is to allocate the positions of nodes in our tree.

The most striking property of the tree is that we *only* position nodes at times when decisions must be made, namely $T_1, \ldots, T_K$. At intermediate times no decisions need to be made and the value function V is well behaved. We are thus able to calculate any intermediate properties explicitly.

At each grid-time T_i we position nodes to be equally spaced in the variable r_{T_i}. This 'vertical' spacing between nodes is a user input, depending on the accuracy required. The location of the top and bottom nodes in any column we will describe shortly.

The importance of being able to choose the times of nodes should not be overlooked. We have positioned the nodes at precisely those times at which we need to know the option's value and so have not lost accuracy caused by nodes missing decision points – a problem with trinomial trees. The second advantage is that, should we wish to price options in two or more dimensions using a multifactor model, then we can do so – the structure of our tree has not forced us to make a time change of the underlying stochastic process in order to solve the problem. Such a time-change only holds for one dimension.

The choice of the locations of the extreme nodes in each column depends on the required accuracy. We discuss later the various sources of error in the algorithm. To understand the effect of this choice, note that

$$
\begin{aligned}
V(r_0) &= \mathbb{E}\left[\exp\left(-\int_0^\tau r_u du\right) V(r_\tau)\right] \\
&= \mathbb{E}\left[\exp\left(-\int_0^\tau r_u du\right) V(r_\tau) \mid r_{T_i} < M_i\right] \mathbb{P}(r_{T_i} < M_i) \qquad (4) \\
&\quad + \mathbb{E}\left[\exp\left(-\int_0^\tau r_u du\right) V(r_\tau) \mid r_{T_i} \geq M_i\right] \mathbb{P}(r_{T_i} \geq M_i)
\end{aligned}
$$

where τ is the random time at which the option is exercised. By truncating column i of the tree at M_i we are ignoring the second term in equation (4) and also the probability in the first term. Given that the spot rate process is Gaussian, the first probability differs from one by a term $O(\exp(-M_i^2/2\xi_{T_i}))$. The second term decays to zero at the same rate since the payoff is at most $O(\exp(cM_i))$ for some constant c. This argument extends to include all the exercise times T_i and shows that errors resulting from truncation are below polynomial in order of magnitude.

Should we wish to choose the limits to guarantee a certain level of accuracy we can use the above argument to do so. What is important is the number of standard deviations of the Gaussian distribution for r_{T_i} covered by the nodes. The limits should thus be set at a fixed number of standard deviations above and below μ_{T_i}, the mean for r_{T_i}. In practice, because the error decays so quickly, the exact level is not important and a conservative value can be chosen without adversely affecting the algorithm's speed. We have found ± 7 standard deviations to be more that sufficient.

The precise algorithm for placing nodes is as follows. First select your node spread, m, in our case 7 standard deviations. Locate the top node in each column at the value

$$\mathbb{E}[r_T \mid r_0] + m\sqrt{\mathrm{Var}[r_T \mid r_0]}.$$

Then locate nodes at a regular spacing Δ, chosen depending on the accuracy required (as described in Section 7), down from this node until the first node lower than

$$\mathbb{E}[r_T \mid r_0] - m\sqrt{\mathrm{Var}[r_T \mid r_0]}.$$

We do not specify any further tree structure at this point. Strictly speaking the algorithm is not a standard tree algorithm, and we do not assign arcs and probabilities as in a tree. The algorithm is, however, closely related to and very similar to a multinomial tree algorithm.

5 Valuation through the Tree

The theory behind option valuation is standard and well-known. As usual we work back in time from the final exercise date assigning option values at each node in the tree. On the final exercise date, if we have not already exercised, the option value is intrinsic which can be calculated as a closed form expression from (3). The value of this expression can be calculated directly from the familiar representation

$$P_{tT} = A_{tT} e^{-B_{tT} r_t},$$

a derivation of which is contained in the Appendix.

To calculate the value of the option at any other node in the tree, suppose first that we have calculated the value at all later nodes. Let t be the time of this node and T be the time of the next set of nodes. Let $V(r_t)$ be the value at this node, $E(r_t)$ be discounted future value of the option at this node (if we choose not to exercise), and $I(r_t)$ be the intrinsic value at the node. Then

$$V(r_t) = \max(E(r_t), I(r_t)) \tag{5}$$

$$E(r_t) = \mathbb{E}\left[\exp\left(-\int_t^T r_u du\right) V(r_T) \mid r_t\right]. \tag{6}$$

The efficiencies and accuracy of our approach come from the way we evaluate (6). Conditioning on r_T we can rewrite (6) as

$$E(r_t) = \mathbb{E}\left[V(r_T)\mathbb{E}\left[\exp\left(-\int_t^T r_u du\right) \mid r_t, r_T\right] \mid r_t\right].$$

The inner expectation, if the conditioning on r_T is removed, yields the well-known expression of the form

$$\mathbb{E}\left[\exp\left(-\int_t^T r_u du\right) \mid r_t\right] = A_{tT} e^{-B_{tT} r_t}.$$

What is less well known, but nonetheless a standard result following almost immediately (from properties of the Brownian bridge [7]), is that

$$\mathbb{E}\left[\exp\left(-\int_t^T r_u du\right) \mid r_t, r_T\right] = \alpha_{tT} e^{-\beta_{tT} r_t - \gamma_{tT} r_T}.$$

The form for α, β and γ is readily calculated analytically and is given in the Appendix.

Once we have evaluated these conditional discount factors we can represent the discounted future payoff as

$$
\begin{aligned}
E(r_t) &= \mathbb{E}\left[V(r_T)\alpha_{tT} e^{-\beta_{tT} r_t - \gamma_{tT} r_T} \mid r_t\right] \\
&= \alpha_{tT} e^{-\beta_{tT} r_t} \mathbb{E}\left[V(r_T) e^{-\gamma_{tT} r_T} \mid r_t\right].
\end{aligned}
\tag{7}
$$

All that remains is to evaluate (7), and thus $V(r_t)$, efficiently.

6   Evaluation of Expected Future Value

There are many ways to evaluate an expectation for a univariate random variable. There are two important considerations for us. The first is that the integrand in (7), $V(r_T)e^{-\gamma_{tT} r_T}$, is piecewise C^∞, but not in itself C^∞. The problem is, from equation (5), that V takes two distinct functional forms, one below an exercise boundary r_T^*, the other above it. We must first estimate the level r_T^* from the values of $E(r_T)$ and $I(r_T)$ at our nodes. This can be done by fitting a polynomial to E and I about the exercise point and solving, as illustrated below in Figure 1.

Having calculated $\hat{r}_T$, our estimate of r_T^*, we now perform the integration on each side of $\hat{r}_T$, separately. To do this efficiently, over each interval $[r, r+\Delta]$, approximate $V(r_T)$ by a polynomial P, of order N, using neighbouring points. Then approximate the integral of V over this interval by instead calculating the integral of P analytically (we know the integral of the polynomial against a Gaussian).

One further point on efficiency should be mentioned. This issue does not affect the asymptotic order of convergence, rather it affects convergence only through the constant term. The point to consider is that we do not just wish to integrate for one value of r_t but for a whole set of values. Also one of the most expensive calculations in the whole algorithm is to calculate the exponential function which is needed both as part of the integrand and also in

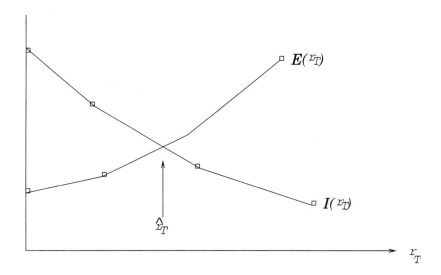

Figure 1. Calculation of crossover point

the evaluation of the normal density function which appears in the expressions for the integrals of a Gaussian against a polynomial. We do not go into the details here but it is a straightforward matter to exploit the fact that the values r_t and r_T are equally spaced to considerably reduce the number of calls to the exponential function. Effectively as much of the calculation that is similar for differing nodes is carried out only once and cached. The speed of this multiple integration also comes from the separation in r_t and r_T that occurs in (7).

7 Error Analysis

There are three (and only three) distinct sources of error within the algorithm. The first, which we have already discussed in Section 4, is that caused by 'clipping' the tree. This is of negligible order compared with the other two.

All the errors come about as part of the integration routine. What we are aiming to calculate for node r_t is

$$
\begin{aligned}
E(r_t) &= \mathbb{E}\left[\exp\left(-\int_t^T r_u du\right) V(r_T) \mid r_t\right] \\
&= \alpha_{tT} e^{-\beta_{tT} r_t} \mathbb{E}\left[F(r_T) \mid r_t\right]
\end{aligned}
\tag{8}
$$

where $F(r_T) = \exp(-\gamma_{tT} r_T) V(r_T)$. The expectation in (8) can be expressed as

$$
\mathbb{E}\left[F(r_T) \mid r_t\right] = \mathbb{E}\left[F_1(r_T) 1_{(r_T > r_T^*)} \mid r_t\right] + \mathbb{E}\left[F_2(r_T) 1_{(r_T \leq r_T^*)} \mid r_t\right]
$$

where r_T^* is the exercise point at which F is non-smooth and F_1 and F_2 are smooth functions (E and I in fact). What we actually calculate (exactly) is

$$\mathbb{E}\Big[G_1(r_T)1_{(r_T > \hat{r}_T)} \mid r_t\Big] + \mathbb{E}\Big[G_2(r_T)1_{(r_T \leq \hat{r}_T)} \mid r_t\Big]$$

where G_1 and G_2 are our polynomial approximations to F_1 and F_2 and $\hat{r}_T$ is our estimate of r_T^*. It follows that the error, the difference between these two, is

$$\mathbb{E}\Big[(F_1 - G_1)(r_T)1_{(r_T > \hat{r}_T)} \mid r_t\Big] + \mathbb{E}\Big[(F_2 - G_2)(r_T)1_{(r_T \leq \hat{r}_T)} \mid r_t\Big]$$
$$+\mathbb{E}\Big[(F_1(r_T) - F_2(r_T))(1_{(r_T > r_T^*)} - 1_{(r_T > \hat{r}_T)}) \mid r_t\Big].$$

Since G is a local polynomial approximation of order N to F it follows that the magnitude of the first two terms is $O(\Delta^{N+1})$, where Δ is, as before, the spacing between neighbouring nodes at time T. The remaining term is the expectation of a function zero outside an interval of size $|\hat{r}_T - r_T^*|$ which is $O(\Delta^{N+1})$ if we calculate $\hat{r}_T$ with a polynomial of order N. Over this interval the integrand $F_1 - F_2$ is also $O(\Delta^{N+1})$, so the net error contribution is $O(\Delta^{2(N+1)})$. Finally, the function F is, of course, not dependent on Δ, thus the net overall error is $O(\Delta^{N+1})$.

In terms of the time to run the algorithm, there are K columns of nodes, each containing $O(m/\Delta)$ nodes where m is the range covered by the columns. Each node performs a calculation involving all the nodes at the next column, so the total number of calculations is $O(m^2/\Delta^2)$, which is thus the order of the time to run the algorithm. Now we must increase m as Δ decreases to ensure that the clipping error is not significant. Suppose we increase m like $\Delta^{-\varepsilon}$ for some $\varepsilon > 0$. The clipping error will then decrease faster than any polynomial order in Δ. Thus the run time is $O(\Delta^{-2(1-\varepsilon)})$, and the error is $o(\Delta^{N+1-\eta})$, any η. This finally yields an error which is $o(T^{-\frac{1}{2}(N+1)+\varepsilon})$ for any $\varepsilon > 0$. This compares with a standard Hull–White tree where the error is $O(T^{-\frac{1}{2}})$.

Appendix

Fixing parameters in the Hull–White model

We now detail how all the parameters of the Hull–White can be calculated. We have been quite explicit to enable the algorithm to be reproduced easily. The starting point is the SDE governing the behaviour of the short rate process, namely:

$$dr_t = (\theta_t - a_t r_t)dt + \sigma_t dB_t. \tag{A1}$$

For the purposes of valuation we need to determine a_t, θ_t and σ_t. However, as we have already seen, it is only certain integrals of these parameters that are needed. In principle it is possible to imply the original defining parameters from these but in practice there is no need and we do not do so.

Before proceeding we note that, even though this model is usually thought of as specifying a continuous process r_t, this is not strictly necessary. There is no reason why r_t cannot exhibit jumps. It is often the case that initial discount curves which are used to calibrate the model are forced to be differentiable just so that r_t will be continuous. However doing this does in itself cause problems when estimating the initial discount curve that is taken as an input and should be avoided.

Throughout this appendix we assume only that the initial discount curve, $\{P_{0T} : T \geq 0\}$, is right-differentiable (the usual case in practice). We also suppose that the prices of the market options that we use to calibrate the model are consistent with the initial assumptions about the form of a_t, i.e. given our choice for a_t there exist θ_t and σ_t consistent with the market prices.

For future reference we make the following definitions, all relating to integrals of the original model parameters.

$$\phi_t = \exp\left(\int_0^t a_u du\right),$$

$$\psi_t = \int_0^t \phi_u^{-1} du,$$

$$\xi_t = \int_0^t \phi_u^2 \sigma_u^2 du,$$

$$\zeta_t = \int_0^t \psi_u \phi_u^2 \sigma_u^2 du,$$

$$\eta_t = \int_0^t \psi_u^2 \phi_u^2 \sigma_u^2 du$$

$$\mu_t = \int_0^t \phi_u \theta_u du,$$

$$\lambda_t = \int_0^t \psi_u \phi_u \theta_u du$$

As we stated in the Introduction the way to solve the SDE (A1) is to consider the process $y_t = \phi_t r_t$. Applying Itô's Lemma to y_t yields

$$dy_t = \phi_t \theta_t dt + \phi_t \sigma_t dB_t$$

and thus

$$y_t = y_0 + \int_0^t \phi_u \theta_u du + \int_0^t \phi_u \sigma_u dB_u.$$

From this it follows that we can write y_t in the form

$$y_t = y_0 + \mu_t + W(\xi_t) \tag{A2}$$

where W is a standard Brownian motion.

The representation for y_t given in equation (A2) allows us to make all the necessary deductions and calculations for the spot process $r_t \; (= \phi_t^{-1} y_t)$. In particular, given r_s, $s \leq t$, r_t and integrals of r_t against deterministic functions will be Gaussian. From this we have immediately the standard result for the value at time t of a pure discount bond paying unity at time T, namely,

$$P_{tT} = \mathbb{E}\Big[\exp\Big(-\int_t^T r_u du\Big) \mid r_t\Big] = A_{tT} e^{-B_{tT} r_t}$$

for suitable A and B to be determined.

The results we need concerning the spot rate process conditioned on its value at two distinct times follow from the following standard result for a Brownian bridge. This follows from, for example, [6].

Lemma 1   *Let*

$$y_t = y_0 + \mu_t + W(\xi_t).$$

The process y_t, $S \leq t \leq T$, conditioned on y_S and y_T, is a continuous Gaussian process with mean and covariance function given by

$$\mathbb{E}\big[y_t \mid y_S, y_T\big] \;=\; \mu_t + \Big(\frac{\xi_T - \xi_t}{\xi_T - \xi_S}\Big)(y_S - \mu_S) + \Big(\frac{\xi_t - \xi_S}{\xi_T - \xi_S}\Big)(y_T - \mu_T)$$

$$\mathrm{Cov}\big[y_s, y_t \mid y_S, y_T\big] \;=\; (\xi_s - \xi_S) \wedge (\xi_t - \xi_S) - \frac{(\xi_s - \xi_S)(\xi_t - \xi_S)}{\xi_T - \xi_S}.$$

Given this result it is straightforward to see that the other quantity used repeatedly throughout this paper, the conditional discount factor, is of the form

$$\mathbb{E}\Big[\exp\Big(-\int_t^T r_u du\Big) \mid r_t, r_T\Big] = \alpha_{tT} e^{-\beta_{tT} r_t - \gamma_{tT} r_T}$$

for suitable α, β and γ.

We are now in a position to calculate all the parameters needed for the tree.

Derivation of ϕ_t, ψ_t and B_{tT}

Given that a_t is a predefined user input (for current purposes) the evaluation of ϕ_t and ψ_t is straightforward. To calculate B_{tT} we note that

$$M_{tT} := \exp\Big(-\int_0^t r_u du\Big) P_{tT}$$

is a martingale in the risk-neutral measure, i.e. the measure in which (A1) holds. Applying Itô's formula to M_{tT} yields

$$
\begin{aligned}
\frac{dM_{tT}}{M_{tT}} \;&=\; -r_t dt + \frac{dP_{tT}}{P_{tT}} \\
&=\; \Big[-r_t + (A'_{tT}/A_{tT}) - r_t B'_{tT} + \tfrac{1}{2}\sigma_t^2 B_{tT}^2 - B_{tT}(\theta_t - a_t r_t)\Big] dt \\
&\quad\; - \sigma_t B_{tT} dB_t
\end{aligned}
$$

and thus, equating terms in dt,

$$-1 - B'_{tT} + a_t B_{tT} = 0 \tag{A3}$$
$$(A'_{tT}/A_{tT}) + \tfrac{1}{2}\sigma_t^2 B_{tT}^2 - B_{tT}\theta_t = 0 \tag{A4}$$

Solving (A3) subject to the condition $B_{TT} = 0$ yields

$$B_{tT} = \phi_t \int_t^T \phi_u^{-1} du = \phi_t(\psi_T - \psi_t).$$

Derivation of ξ_t, ζ_t and η_t

We already know all the terms in each of the integrands except σ_t. It suffices to calculate $\xi_t, t \geq 0$ since this implicitly defines σ_t and thus ζ_t and η_t. In practice we will only have a finite set of market option prices to calibrate from and so will calculate ξ_t for a finite set of times and assume a (simple) functional form for σ_t at intermediate points.

Suppose we have available the price V_0 of an option expiring at T and paying

$$\left(\sum_{j \in J} c_j P_{TS_j} - K\right)_+.$$

Taking P_{tT} as numeraire it is standard that

$$V_0 = P_{0T}\mathbb{E}_\mathbb{N}\left[\left(\sum_{j \in J} c_j P_{TS_j} - K\right)_+\right]$$

where the measure $\mathbb{N}$ is equivalent to the original risk-neutral measure and is such that (P_{tS_j}/P_{tT}) is a martingale. Applying Itô to (P_{tS_j}/P_{tT}) and changing measure yields

$$\left(\frac{P_{tS_j}}{P_{tT}}\right) = \left(\frac{P_{tS_j}}{P_{tT}}\right)(B_{tT} - B_{tS_j})\sigma_t dB_t$$
$$= \left(\frac{P_{tS_j}}{P_{tT}}\right)(\psi_T - \psi_{S_j})\phi_t\sigma_t dB_t$$

and so

$$\left(\frac{P_{tS_j}}{P_{tT}}\right) = \left(\frac{P_{0S_j}}{P_{0T}}\right)\exp\left((\psi_T - \psi_{S_j})\hat{W}(\xi_t) - \tfrac{1}{2}(\psi_T - \psi_{S_j})^2\xi_t\right)$$
$$V_0 = \mathbb{E}_\mathbb{N}\left[\left(\sum_{j \in J} c_j P_{0S_j} e^{(\psi_T - \psi_{S_j})\hat{W}(\xi_T) - \tfrac{1}{2}(\psi_T - \psi_{S_j})^2\xi_T} - K\right)_+\right] \tag{A5}$$

where $\hat{W}$ is a Brownian motion under $\mathbb{N}$.

Given a value for ξ_T evaluation of (A5) is best performed using the decomposition first introduced by Jamshidian [5]. A requirement for this approach to (be guaranteed to) work is that the payoff must be monotone in the realised value of $\hat{W}$. This is always the case for standard options, including those with amortising and forward start features.

The first step in this approach is to find the value $\hat{w}$ for $\hat{W}(\xi_T)$ which sets the term inside the expectation (A5) to be exactly zero. A safe Newton–Raphson algorithm will do this quickly. We can then rewrite (A5) as

$$V_0 = \mathbb{E}_{\mathbb{N}}\left[\left(\sum_{j \in J} c_j P_{0S_j}\left(e^{(\psi_T - \psi_{S_j})\hat{W}(\xi_T) - \frac{1}{2}(\psi_T - \psi_{S_j})^2 \xi_T} - K_i\right)_{\pm}\right]$$

where

$$K_i = \exp\left((\psi_T - \psi_{S_j})\hat{w} - \tfrac{1}{2}(\psi_T - \psi_{S_j})^2 \xi_T\right)$$

and the $\pm$ in the expectation is $+$ if the payoff is increasing in $\hat{W}$ and $-$ if it is decreasing. Solving for ξ_T can now be completed by iterating with either Brent or Newton-Raphson (for those with the stamina!) on $V_0(\xi_T)$.

Before moving on we note that should an alternative model be required in which a_t is not chosen a priori but is instead a function of σ_t then the starting point for the algorithm would be this section, solving for a_t and σ_t simultaneously.

Derivation of μ_t, λ_t and A_{tT}

Starting with Equation (A4) for A_{tT}, substituting for B_{tT} and simplifying yields

$$\log(A_{tT}) = \log(A_{0T}) + \psi_T \mu_t - \lambda_t - \tfrac{1}{2}[\psi_T^2 \xi_t - 2\psi_T \zeta_t + \eta_t]. \qquad \text{(A6)}$$

We know A_{0T} from the values P_{0T}, namely $A_{0T} = P_{0T} e^{B_{0T} r_0}$. Thus we need only calculate μ_t and hence A_{tT} from (A6) (λ_t is defined once μ_t is known). Substituting $t = T$ into (A6) and differentiating with respect to T yields

$$\mu_T = \psi_T \xi_T - \zeta_T - \phi_T \frac{\partial}{\partial T} \log(A_{0T})$$

and thence

$$\lambda_T = \log(A_{0T}) - \psi_T \phi_T \frac{\partial}{\partial T} \log(A_{0T}) + \tfrac{1}{2}(\psi_T^2 \xi_T - \eta_T).$$

Derivation of α_{tT}, β_{tT} and γ_{tT}

We have already derived the functional form

$$\mathbb{E}\left[\exp\left(-\int_t^T r_u du\right) \mid r_t, r_T\right] = \alpha_{tT} e^{-\beta_{tT} r_t - \gamma_{tT} r_T}.$$

To find β and γ note that, conditional on r_t and r_T,

$$\int_t^T r_u du = \int_t^T \phi_u^{-1} y_u du$$

where y_u is a time-changed Brownian bridge as in Lemma A1. It follows that

$$\mathbb{E}\left[\int_t^T r_u du \mid r_t, r_T\right]$$

$$= \int_t^T \phi_u^{-1} \mathbb{E}\left[y_u \mid y_t, y_T\right] du$$

$$= \int_t^T \phi_u^{-1} \left[\mu_u + \left(\frac{\xi_T - \xi_u}{\xi_T - \xi_t}\right)(y_t - \mu_t) + \left(\frac{\xi_u - \xi_t}{\xi_T - \xi_t}\right)(y_T - \mu_T)\right] du$$

$$= C_{tT} + y_t \int_t^T \phi_u^{-1} \left(\frac{\xi_T - \xi_u}{\xi_T - \xi_t}\right) du + y_T \int_t^T \phi_u^{-1} \left(\frac{\xi_u - \xi_t}{\xi_T - \xi_t}\right) du$$

for some C_{tT}. Substituting for r_t, observing that the variance of the same integral is independent of r_t and r_T, and using the standard result

$$\mathbb{E}\left[e^{N(\mu,\sigma^2)}\right] = e^{\mu + \frac{1}{2}\sigma^2}$$

it now follows that

$$\beta_{tT} = \phi_t \left[\frac{\zeta_T - \zeta_t}{\xi_T - \xi_t} - \psi_t\right]$$

$$\gamma_{tT} = \phi_T \left[-\frac{\zeta_T - \zeta_t}{\xi_T - \xi_t} + \psi_T\right].$$

To derive the final term α_{tT} we note that

$$P_{tT} = \mathbb{E}\left[\alpha_{tT} e^{-\beta_{tT} r_t - \gamma_{tT} r_T} \mid r_t\right]$$

which yields

$$\log(\alpha_{tT}) = \log(A_{tT}) + \gamma_{tT} \phi_T^{-1}(\mu_T - \mu_t) - \tfrac{1}{2}\gamma_{tT}^2 \phi_T^{-2}(\xi_T - \xi_t).$$

References

[1] Black F., Derman E. and Toy W. (1990) 'A One-Factor Model of Interest Rates and its Application to Treasury Bond Options', *Financial Analysts Journal*, Jan-Feb, 33–39.

[2] Black F. and Karasinski P. (1991) 'Bond and Option Pricing When Short Rates are Lognormal', *Financial Analysts Journal*, Jul-Aug, 52–59.

[3] Hull J. and White A. (1990) 'Pricing Interest Rate Derivative Securities', *Review of Financial Studies* **3**(4).

[4] Kennedy J.E. and Laws C.N. (1996) 'Efficient Pricing Algorithms for Lognormal and Other Interest Rate Models', *in preparation*

[5] Jamshidian F. (1989) 'An Exact Bond Option Formula', *Journal of Finance* **XLIV**(1).

[6] Karatzas I. and Shreve S.E. (1988) *Brownian Motion and Stochastic Calculus*, Springer Verlag.

[7] Press W.H., Teukolsky S.A., Vetterling W.T. and Flannery B.P. (1988) *Numerical Recipes in C*, (second edition, 1993), Cambridge University Press.

Numerical Valuation of Cross-Currency Swaps and Swaptions

M.A.H. Dempster and J.P. Hutton

Abstract

We investigate numerical valuation of cross-currency interest rate-based derivatives under Babbs' extended Vasicek-style model by numerical solution of the associated partial differential equation (PDE) – in particular, we consider the terminable differential (diff) swap.

Firstly we precisely formulate, in terms of their cash flows, various types of single and cross-currency swaps and swaptions. We describe Babbs' model for the domestic and foreign term structures and the exchange rate, its formulation in terms of three correlated driftless Gaussian processes and the associated three state variable parabolic PDE. We then formulate finite difference approximations to the PDE, and discuss explicit and implicit methods. With this discrete approximation to the valuation problem in a period, we proceed to value the terminable diff swap and other deals numerically by backwards recursion through the payment dates, and investigate the solutions found graphically.

We conclude that it is certainly practical, on a fast workstation, to solve for the value function of a wide range of cross-currency derivative securities by solution of explicit finite difference approximations of the PDE.

1 Introduction

In this chapter we consider the numerical valuation of interest rate-based derivatives, in particular, valuation of cross-currency swap agreements. The motivation behind this choice is that cross-currency interest rate derivatives form a topic of enormous current practical importance, but such derivatives are under-represented in the literature on numerical valuation. These derivatives, assuming single stochastic factors driving the term structures and exchange rate, are dependent on three stochastic state variables, and so the PDE which their value functions must satisfy has three state variables plus time. Swap deals have the added complexity of multiple cash flow dates. The question arises as to whether such deals can be valued to any reasonable accuracy in a reasonable time on a standard workstation, since the size of the matrix representing the discretised PDE is third order in the number of grid points per axis.

In §2, we introduce various swaps, concentrating on the cross-currency case. We consider models where the value function solves a (model-dependent) PDE between cash flow dates, and give precise specifications of boundary and recursive terminal conditions for the value function of these different swap deals. In particular we look at an extension to the diff swap – new to the literature and suggested to us by Simon Babbs – which involves exchange of domestic and foreign LIBOR rates, both paid on the same domestic principle, with the additional feature that the counterparty may terminate the deal at any of the LIBOR payment dates for a fixed cost in his native currency. At the end of a period, the counterparty terminates if the termination cost is less than the continuation cost, which gives a terminal condition for a period that depends on the value at the start of the next period, and so we may solve for the value of the deal by backwards recursion, the solution in the last period giving a terminal condition for the penultimate period, and so on.

In §3 we give a general cross-currency model due to Babbs, with Ito process models of domestic and foreign bond prices, which are consistent with initial term structures, and of the exchange rate. We describe Babbs' specialisation of the general model that produces an 'extended Vasicek' model for the short rate with term structure processes driven by three correlated driftless Gaussian stochastic state variables, and give his PDE with respect to these variables which any European-style derivative must follow.

Then in §4 we discretise a general three state variable backward parabolic PDE, and consider standard finite difference approximations in each period. Preliminary to actually valuing deals numerically using Babbs' model, we describe the data that must be supplied to the model and we derive step function integration formulae for functions in the bond price and exchange rate formulae which facilitate numerical evaluation.

Finally, in §5, we solve numerically for the value function of some of the deals described in §2, including the terminable diff swap, by backwards recursion, and present results on convergence of the solution and timing of the various routines, as well as giving various plots of cross-sections through the (4D) solution surface of a call option on a terminable diff swap.

We conclude that the *explicit method* is the best of the standard methods for this multivariable type of problem, and that with it we may solve for the value function of a wide range of cross-currency derivatives. Note that we immediately obtain, from this value surface, many of the partial derivatives required for hedging (see Carr [8]). There are many possible directions one could take to speed up and increase the accuracy of the solution, and some of these are discussed in §6.

Throughout, we denote random variables by a **bold** typeface.

2 Swaps and Swaptions

The single currency fixed rate-floating rate (*vanilla*) swap is by far the most common among all swaps – Litzenberger [16] claims that, as of 1992, over two-thirds of the current total $3 trillion outstanding interest rate swaps are vanilla fixed-floating swaps. However, cross-currency swaps are becoming increasingly popular.

In the types of swap which we will consider, a floating rate of interest is swapped for another floating or fixed rate, and this floating rate is usually taken to be some margin above the 1, 3 or 6 month LIBOR rate. The swap period, say $[0, T]$, is then divided into periods of the same length as the LIBOR term, with swapped payments made at the end of each period according to the rates prevailing at the start of the period, i.e. *in arrears*. If the zero-coupon bond price at time t for maturity M is $P(t, M)$, then we define the *LIBOR rate $L(t, M)$ for the period* $[t, M]$ by the annualised return on the corresponding zero-coupon bond, specifically

$$L(t, M) := \frac{1}{\delta} \times \frac{1 - P(t, M)}{P(t, M)}, \tag{1}$$

where δ is the *accrual factor* defined by.

$$\delta := \frac{\text{number of days in } [t, M]}{\text{basis}}, \tag{2}$$

where basis is typically 365 for pounds sterling and 360 for U.S. dollars.

2.1 Vanilla floating-fixed swaps and swaptions

We define a *vanilla fixed-floating interest rate swap* as an agreement between two parties, the 'bank' and the 'counterparty', whereby the bank pays the counterparty a floating annualised rate of interest on a cash amount (or *principal*) Z, and the counterparty pays the bank a agreed fixed rate of interest r^* on the same principal amount, all for a fixed period $[0, T]$. Typically, the life of such a swap is anything from 2 to 15 years. Of course, equally the roles of bank and counterparty could be reversed. We adopt the convention throughout of valuation in domestic terms and from the bank's point of view, and denote value to the bank at time t by $V(t)$. It may be the case that the counterparty has an option, typically at no cost, to enter into such a swap contract at some point in the future – this type of deal we refer to as a *swaption*. In addition, the counterparty may have the option at various points to terminate the deal at a cost, in which case we call the swap *terminable*.

To use domestic LIBOR L_d as the floating rate, the swap period $[0, T]$ is divided into the corresponding LIBOR periods, and we denote period j by $[t_{j-1}, t_j)$ for $j = 1, \ldots, N$, where $t_0 = 0$. Swap payments[1] $p_1, \ldots, p_N$ are made

---

[1]Note that periods are defined as closed below and open above and payments are made at t_j- so that the value function is RCLL everywhere.

(to the bank) at $t_1-, \ldots, t_N-$, where $t- := \lim_{s \searrow 0}(t-s)$, and are given by

$$p_j := Z\delta_j \left[r^* - L_d(t_{j-1}, t_j) \right], \tag{3}$$

where δ_j is the swap rate accrual factor for period j, defined as in (2). Note that the LIBOR rate L is determined at the start of the period, but payment is made at the end – this is a path dependence in the payoff, which we may eliminate by reformulating the deal, so that instead present values of p_j are paid at the end of the preceding period $t_{j-1}-$. So in fact, we have a payment of $P_d(t_{j-1}, t_j)p_j$ at each $t_{j-1}-$, for $j = 2, \ldots, N$, which is determined completely by values of appropriate state variables at time $t_{j-1}-$, and now the final period is redundant. We will use this trick repeatedly in the cross-currency swaps of §2.2.

The value at $t < T$ of the swap is simply the sum of the present values $V_j(t)$ of all remaining swap payments after t. In a particular period we have a PDE (depending on a term structure model) in V_j with the terminal condition

$$V_j(t_{j-1}-) = P(t_{j-1}, t_j)p_j \qquad j = 2, \ldots, N. \tag{4}$$

In fact, we may calculate the current plain vanilla swap value, given the current term structure, since receiving $L(t_{j-1}, t_j)Z$ at t_j- is equivalent to receiving Z at $t_{j-1}-$ and paying Z at time t_j-. However, when the cash flows at period dates are more complex, with option structures such as we consider below, we must resort to numerical solution. In general, numerical solution of such deals is rapid and efficient for virtually any single factor model, because of the low dimensionality. Once we extend the idea of a vanilla swap to a swap across currencies, the resulting increase in the number of state variables makes *efficient* numerical solution much more key.

2.2   Cross-currency swaps

We thus turn to swaps where the two interest rates being swapped are in different, *domestic* and *foreign*, currencies. These have the additional complexity of requiring models of the two term structures and the exchange rate between them. Indeed, it might also be appropriate to use a cross-currency model to price single currency swaps, since it incorporates two additional explanatory variables that affect the domestic term structure through correlation.

The most common (*vanilla*) *cross-currency swap* is the exchange of floating or fixed rate interest payments on principals Z_d and Z_f in two currencies, domestic and foreign, which we define as follows. Again, we divide the swap period $[0, T]$ into N periods, and domestic and foreign payments p_{d_j} and p_{f_j} based on LIBOR are made at the end of each period, given by

$$p_{d_j} := \delta_j \left[k_d L_d(t_{j-1}, t_j) + m_d \right] Z_d \tag{5}$$

$$p_{f_j} := \delta_j \left[k_f L_f(t_{j-1}, t_j) + m_f \right] Z_f, \tag{6}$$

where $L_d(t, M)$ and $L_f(t, M)$ are domestic and foreign LIBOR at time t for the period $[t, M]$ respectively, defined as in (1), k_d and k_f are floating rate parameters, m_d and m_f are fixed rate components or margins above domestic and foreign LIBOR respectively and δ_j is the accrual factor (2) – the parameters k_d, k_f, m_d and m_f determine which party receives which rate. Finally, in the case that the swap is with *exchange of principal*, the party paying the domestic rate receives Z_d and pays Z_f at the start of the deal, with the reverse exchange at the end of the deal.

Again the swap value $V(t)$ is the sum of values of individual payments $V_j(t)$, each value function solving a model-determined PDE. We eliminate the path dependence in the payoff by equivalently exchanging the present values, discounted to time t_{j-1} using domestic and foreign term structures accordingly, and so that the terminal condition for the period $[t_{j-2}, t_{j-1}]$ is

$$V_j(t_{j-1}-) = P_d(t_{j-1}, t_j)p_{d_j} - S(t_{j-1})P_f(t_{j-1}, t_j)p_{f_j}, \qquad (7)$$

where $S(t_{j-1})$ denotes the exchange rate in domestic currency prevailing at time t_{j-1}. Since we have no option features, we can again price this deal analytically by equating each LIBOR payment to paying and receiving the principal – we can then see that the vanilla cross-currency swap with exchange of principal has value zero at time zero and we use this as a test of solution accuracy in §5. If in addition we set $Z_f := Z_d/S(0)$, then there is even no initial exchange of principal.

We now consider an increasingly popular variant of the above deal which has the feature that it avoids any explicit exchange rate exposure. Such deals, even without option features, cannot be valued in the simple way above.

Differential swap A *vanilla differential (diff)* or *switch LIBOR* swap is an exchange of domestic and foreign LIBOR, but foreign interest rates are paid on the *same* domestic principal amount Z as the domestic rate, so there is no explicit exchange rate exposure. The payment to the bank at the end of period j is given by

$$p_j := Z\delta_j \left[k_d(L_d(t_{j-1}, t_j) + m) - k_f L_f(t_{j-1}, t_j)\right], \qquad (8)$$

and then the formulation in a particular period as a PDE problem is the same as for the domestic vanilla swap above, with the terminal condition (4).

The diff swap was introduced to the academic literature by Litzenberger [16], who discusses practical estimation and hedging issues, and was taken up by Babbs [5] as an application of his cross-currency model of §3. Under this model, he derives a simple closed form expression for the diff swap using the risk-adjusted valuation formula (27) and calculating the expectation by exploiting the Gaussian state variables. The expression is couched in terms of current bond prices and integrals of the various volatility and correlation functions, and is relatively straightforward to evaluate numerically – we will use this closed-form formula as a check on our numerical procedure.

Terminable diff swap   Consider now the *terminable* diff swap, suggested to the authors by Babbs [4], where the counterparty has the option to terminate the deal at the start of every interest period for a termination cost of X in the counterparty's native currency. This is altogether a more complicated deal than those discussed earlier, and does not have the same simple European-style payoff structure – it is a 'Bermudan' option, which is an American option with only a *finite* number of early exercise dates. We formulate it as follows.

At the end of each period, the counterparty must either terminate the deal, at a cost in foreign currency of X, or continue the deal by making the diff swap payment (8). As usual, we equivalently exchange present values so that the last period $[t_{N-1}, t_N)$ is redundant. In the penultimate period, we have the boundary condition at t_{N-1} for the solution in the penultimate period

$$V(t_{N-1}-) = \min\{XS(t_{N-1}), P_d(t_{N-1}, t_N)p_N\}, \tag{9}$$

since the counterparty terminates if the termination cost is less than the cost of continuing. In an earlier period j, we have the same payment, but we have to take into account the payments still remaining if the counterparty chooses to continue rather than pay to terminate. So we have the same boundary condition as (9), except that the value of the remaining deal periods $V(t_j)$ must be added to the payment p_j as the reward for continuing, thus:

$$V(t_j-) = \min\{XS(t_j), P_d(t_j, t_{j+1})p_{j+1} + V(t_j)\}. \tag{10}$$

We may thus value this terminable diff swap by solving the PDE in the penultimate period $N-1$ with the terminal condition (9), substituting the resulting solution value at t_{N-2} into (10) to give a terminal condition for period $N-2$, and repeating this procedure, stepping backwards in time until we get the solution at $t_0 = 0$. In practice, a terminable diff swap may be sold with the margin m reduced so that the initial value is zero – to find this zero value margin is a root finding problem, albeit simple, on top of numerical valuation, and we do not consider it here.

We may of course allow additional option features. For example, we might consider a call option on a terminable diff swap, with maturity t_1 and exercise price K, so that we have the same terminal condition as the terminable diff swap in each period except in period 1, for which we have the call option payoff

$$V(t_1-) = \min\{K, V(t_1)\}. \tag{11}$$

We value such a deal (with $K := 0$, as is usually the case) in §5.

3   Babbs' Cross-Currency Term Structure Model

To completely specify the valuation problem for any of the deals discussed above, we need to specify a term structure model. The classical term structure models are concerned with contingent claims in only one, so-called *domestic*, economy. Once we include a second economy, which we call *foreign*, we have different term structure processes and risk preferences in each economy, and a rate of exchange between their currencies. Until recently, models for pricing derivatives in this setting either assumed constant interest rates and a stochastic exchange rate, or modelled stochastic interest rates in the same manner as Merton [17]. Neither of these approaches is satisfactory: the first approach is appealing only for its simplicity and cannot be justified empirically; the second suffers all the flaws of the Merton model – it does not model the full term structure, and as a result cannot support American-style payoffs, which require a continuum of bond price maturities. See Amin and Jarrow [1] for a review and references to empirical work.

Amin and Jarrow [1] extend the Heath, Jarrow and Morton [12] Gaussian model, and Babbs [5, 4, 6] applies his similar model of [2], both in an attempt to extend full term-structure models to the cross-currency case. We consider here the Babbs model, in particular his 'extended Vasicek' specialisation. For more details on the model see Babbs [5], or, in the present context, Hutton [14].

3.1 Model structure

We start by specifying term structure dynamics in terms of the zero-coupon bond prices P_d and P_f in both the domestic and foreign economy, and the exchange rate S between their currencies, in terms of the *objective* probability measure. By convention, we value assets and derivative securities in terms of the domestic currency, and our exchange rate is the domestic price per unit of foreign currency.

We specify our bond price and exchange rate Ito processes as satisfying the stochastic differential system

$$\frac{d\mathbf{P}_d(t,T)}{\mathbf{P}_d(t,T)} = [\mathbf{r}_d(t) + \boldsymbol{\theta}_d(t)\boldsymbol{\sigma}_d(t,T)]\,dt + \boldsymbol{\sigma}_d(t,T)d\mathbf{Z}_d(t)$$

$$\frac{d\mathbf{P}_f(t,T)}{\mathbf{P}_f(t,T)} = [\mathbf{r}_f(t) + \boldsymbol{\theta}_f(t)\boldsymbol{\sigma}_f(t,T)]\,dt + \boldsymbol{\sigma}_f(t,T)d\mathbf{Z}_f(t)$$

$$\frac{d\mathbf{S}(t)}{\mathbf{S}(t)} = [\mathbf{r}_d(t) - \mathbf{r}_f(t) + \boldsymbol{\theta}_S(t)\boldsymbol{\sigma}_S(t)]\,dt + \boldsymbol{\sigma}_S(t)d\mathbf{Z}_S(t), \qquad (12)$$

where σ_d, σ_f and σ_S represent bond price and exchange rate *volatilities*, so that $\sigma_d(t,t) = \sigma_f(t,t) = 0$ for all $t \in [0,T]$ and are strictly positive elsewhere; θ_d, θ_f and θ_S are related to the market prices of risk of domestic and foreign bonds and exchange rate[2]; $\mathbf{Z}_d$, $\mathbf{Z}_f$ and $\mathbf{Z}_S$ are imperfectly correlated Wiener

---

[2]θ_d is exactly the market price of risk for domestic bonds, but θ_f is the market price of foreign bond risk in foreign currency.

processes with correlation processes

$$dZ_d(t)dZ_f(t) = \rho_{df}(t)dt$$
$$dZ_d(t)dZ_S(t) = \rho_{dS}(t)dt$$
$$dZ_f(t)fZ_S(t) = \rho_{fS}(t)dt. \tag{13}$$

3.1.1 'Separable Extended Vasicek' restriction

From the above general specification of the term structure dynamics we may derive the resulting process for the short rate in either economy. See Hull [13] for this result, but the process for the short rate may be non-Markovian, because of path-dependent integrals involving bond price volatility in the drift, and thus in general the current short rate is not sufficient to determine the current term structure. However, if we restrict the *deterministic* volatility to be independent of the bond price and of the functional form

$$\sigma_k(t,T) := [G_k(T) - G_k(t)]\,\lambda_k(t) \qquad k = d, f \tag{14}$$

for arbitrary functions G_k and λ_k, we eliminate any path-dependency in the short rate process r_k, which then satisfies

$$dr_k(t) = \left\{\mu'_k(t) - \frac{G''_k(t)}{G'_k(t)}\,[\mu_k(t) - r_k(t)]\right\}dt -$$
$$G'_k(t)\lambda_k(t)\,[\boldsymbol{\theta}_k(t)dt + dZ_k(t)] \quad k = d, f, \tag{15}$$

where

$$\mu_k(t) := F_k(0,t) + G'_k(t)\int_0^t [G_k(t) - G_k(s)]\,\lambda_k^2(s)ds$$

and $F_k(0,t)$ is the instantaneous forward rate for time t at time zero, which is determined by the initial term structure. Babbs [2] shows that this volatility specification is in fact a necessary and sufficient condition for the existence of a single state variable for the term structure. Furthermore, the resulting short rate process (15) is recognisable as an extended Vasicek-type model, i.e. of the (risk-adjusted) form

$$dr(t) = (\alpha(t) - \beta(t)r(t))dt + \sigma(t)dZ(t), \tag{16}$$

where α/β is the *long run mean level*, β is the *mean reversion rate* and σ is the *variability* of the short rate.

Separable models If we also ask for the more easily specifiable property that the bond price volatility be separable into a product of functions of time to maturity $T-t$ and calendar time t, then it follows (see e.g. Babbs [2]) that this is equivalent to requiring

$$G_k(t) = \frac{1 - e^{-\xi_k t}}{\xi_k} \qquad k = d, f, \tag{17}$$

for some constants ξ_d and ξ_f, and

$$\lambda_k(t) = e^{\xi_k t} \kappa_k(t) \qquad k = d, f, \tag{18}$$

where κ_k is the afore-mentioned function of calendar time. These constants may be interpreted in the context of the extended Vasicek short rate process (15) thus: the *mean reversion rate* of the short rate is $-G_k''(t)/G_k'(t)$, which on substituting for G_k from (17) yields ξ_k; the *variability* of the short rate is $G_k'(t)\lambda_k(t)$, which on substituting from (17) and (18) yields $\kappa_k(t)$.

3.2 Risk-adjusted processes in domestic terms

We now express the three price processes in domestic terms. First, we write each of the correlated Wiener increments $d\mathbf{Z}_d$, $d\mathbf{Z}_f$ and $d\mathbf{Z}_S$ of (12) as a linear combination of increments of three independent (*orthogonal*) Wiener processes $\mathbf{W}_1$, $\mathbf{W}_2$ and $\mathbf{W}_3$ thus:

$$d\mathbf{Z}_k(t) = \sum_{j=1}^{3} \alpha_{kj}(t) d\mathbf{W}_j(t) \qquad k = d, \ f, \ S. \tag{19}$$

Substituting this into (13) and using the fact that the $\mathbf{W}_j$ are uncorrelated, we see that the matrix $A := [\alpha_{kj}]_{j=1,2,3}^{k=d,f,S}$ must be the square root of the correlation matrix $[\rho_{kl}]_{k,l=d,f,S}{}^3$.

In everything that follows, we value securities under the *numeraire* $P_d(0, H)$, i.e. normalised by the initial domestic bond price at some suitably distant horizon date H, since then price processes are martingales under the risk-adjusted probability measure. Under this measure there are no arbitrage opportunities between domestic and foreign bonds of any maturities up to H.

To specify the risk-adjusted measure on continuous paths of the independent coordinate Wiener process $\mathbf{W} := (\mathbf{W}_1, \mathbf{W}_2, \mathbf{W}_3)'$ in 3-space, we utilize the Radon-Nikodym derivative process

$$\exp\left\{ -\int_0^t [\boldsymbol{\theta}_d(s) + \frac{1}{2}\boldsymbol{\theta}_d^2(s) + \boldsymbol{\theta}_f(s) + \frac{1}{2}\boldsymbol{\theta}_f^2(s) + \boldsymbol{\theta}_S(s) + \frac{1}{2}\boldsymbol{\theta}_S^2(s)]ds \right\}$$

on $0 \leq t \leq H$ and apply Girsanov's theorem to obtain the independent coordinate Wiener process $\tilde{\mathbf{W}} := (\tilde{\mathbf{W}}_1, \tilde{\mathbf{W}}_2, \tilde{\mathbf{W}}_3)'$ under the risk-adjusted measure , where $d\tilde{\mathbf{W}}_k = d\mathbf{W}_k + \boldsymbol{\theta}_k dt$, $k = d, f, S$, and remove the market price of risk terms in the risk-adjusted analogue of (12).

It turns out that the two bond prices and the exchange rate may then be captured by three driftless Gaussian state variable processes, a property which leads to simpler numerical procedures, either for computation of the value by integration or for solution of the PDE.

---

[3]Since the square root has three free parameters, we follow Babbs [5] in choosing $\alpha_{d1} = \alpha_{d2} = \alpha_{d3} = 0$.

Theorem 1 *The domestic and foreign bond prices and the exchange rate are given in terms of driftless Gaussian state variables X_d, X_f, X_S by*

$$
\begin{aligned}
\mathbf{P}_d(t,T) &= \tfrac{P_d(0,T)}{P_d(0,t)} \exp\left\{ [G_d(T) - G_d(t)] \left[\mathbf{X}_d(t) - \int_0^t h_d(t,T,H,s)ds \right] \right\} \\
\mathbf{P}_f(t,T) &= \tfrac{P_f(0,T)}{P_f(0,t)} \exp\left\{ [G_f(T) - G_f(t)] \left[\mathbf{X}_f(t) - \int_0^t h_f(t,T,H,s)ds \right] \right\} \\
\mathbf{S}(t) &= \tfrac{P_f(0,t)S(0)}{P_d(0,t)} \exp\left\{ -G_d(t)\mathbf{X}_d(t) + G_f(t)\mathbf{X}_f(t) + \mathbf{X}_S(t) \right. \\
&\qquad \left. - \tfrac{1}{2} \int_0^t h_S(t,H,s)ds \right\},
\end{aligned}
$$

$$(20)$$

where

$$
\begin{aligned}
h_d(t,T,H,s) &:= \left(\tfrac{G_d(T)+G_d(t)}{2} - G_d(H) \right) \lambda_d^2(s) \\
h_f(t,T,H,s) &= \lambda_f(s)\left[\tfrac{\sigma_f(s,T)+\sigma_f(s,t)}{2} + \sigma_S(s)\rho_{fS}(s) - \sigma_d(s,H)\rho_{df}(s) \right] \\
h_S(t,H,s) &= \sigma_f^2(s,t) + \sigma_S^2(s) - \sigma_d^2(s,t) + 2\sigma_d(s,t)\sigma_d(s,H) \\
&\quad + 2\sigma_f(s,t)\rho_{fS}(s)\sigma_S(s) - 2\sigma_f(s,t)\rho_{df}(s)\sigma_d(s,H) \\
&\quad - 2\sigma_S(s)\rho_{dS}(s)\sigma_d(s,H)
\end{aligned}
$$

$$(21)$$

and the state variables X_d, X_f and X_S are defined by

$$
\mathbf{X}_d(t) := \sum_{j=1}^{3} \int_0^t \alpha_{dj}(s)\lambda_d(s)d\tilde{\mathbf{W}}_j(s)
$$

$$
\mathbf{X}_f(t) := \sum_{j=1}^{3} \int_0^t \alpha_{fj}(s)\lambda_f(s)d\tilde{\mathbf{W}}_j(s) \qquad (22)
$$

$$
\mathbf{X}_S(t) := \sum_{j=1}^{3} \int_0^t \left[\alpha_{dj}(s)G_d(s)\lambda_d(s) - \alpha_{fj}(s)G_f(s)\lambda_f(s) \right.
$$
$$
\left. + \alpha_{Sj}(s)\sigma_S(s) \right] d\tilde{\mathbf{W}}_j(s),
$$

where $\tilde{\mathbf{W}}_j(t)$ is a Wiener process under the risk-adjusted probability measure.

Proof: See Babbs [5] for the original proof, or Hutton [14]. The proof uses Ito's lemma to derive the log-price processes, which have constant coefficients and so are simple to integrate.   ∎

3.3   Pricing European derivative securities

We now give a PDE which any derivative security must satisfy between cash flow dates in Babbs' model. The following lemma gives the variances and covariances of the random variables $\mathbf{X}_d(t)$, $\mathbf{X}_f(t)$ and $\mathbf{X}_S(t)$, the integrands of which will essentially form the PDE coefficients and will also enable us to place sensible bounds on the underlying variables of this PDE when we come to considering numerical solution in § 4.1.

Lemma 1 *The driftless Gaussian processes* $\mathbf{X}_d$, $\mathbf{X}_f$ *and* $\mathbf{X}_S$ *defined by (22) have the following variances and covariances at time* $t \in [0, H]$:

$$\text{var}\left[\mathbf{X}_d(t)\right] = \int_0^t \lambda_d^2(s)ds$$

$$\text{var}\left[\mathbf{X}_f(t)\right] = \int_0^t \lambda_f^2(s)ds$$

$$\text{var}\left[\mathbf{X}_S(t)\right] = \int_0^t H^{SS}(s)ds$$

$$\text{cov}\left[\mathbf{X}_d(t), \mathbf{X}_f(t)\right] = \int_0^t H^{df}(s)ds$$

$$\text{cov}\left[\mathbf{X}_d(t), \mathbf{X}_S(t)\right] = \int_0^t H^{dS}(s)ds$$

$$\text{cov}\left[\mathbf{X}_f(t), \mathbf{X}_S(t)\right] = \int_0^t H^{fS}(s)ds, \tag{23}$$

where the functions H^{SS}, H^{df}, H^{dS} *and* H^{fS} *are defined by*

$$H^{SS}(s) := G_d^2(s)\lambda_d^2(s) + G_f^2(s)\lambda_f^2(s) + \sigma_S^2(s) - 2\rho_{df}(s)G_d(s)\lambda_d(s)G_f(s)\lambda_f(s)$$
$$+ 2\rho_{dS}(s)G_d(s)\lambda_d(s)\sigma_S(s) - 2\rho_{fS}(s)G_f(s)\lambda_f(s)\sigma_S(s)$$
$$H^{df}(s) := \rho_{df}(s)\lambda_d(s)\lambda_f(s)$$
$$H^{dS}(s) := \lambda_d(s)\left[G_d(s)\lambda_d(s) - \rho_{df}(s)G_f(s)\lambda_f(s) + \rho_{dS}(s)\sigma_S(s)\right]$$
$$H^{fS}(s) := \lambda_f(s)\left[\rho_{df}(s)G_d(s)\lambda_d(s) - G_f(s)\lambda_f(s) + \rho_{fS}(s)\sigma_S(s)\right]. \tag{24}$$

Proof: See Babbs [5] for the original proof, or Hutton [14]. However, it is simply an application of Fubini's theorem to take expectations through the integrals $\tilde{\mathbb{E}}\left[\mathbf{X}_k(t)\mathbf{X}_l(t)\,|\,\mathbf{X}(0)\right]$ for $k, l = d, f, S$, with $\mathbf{X}_k(t)$ defined by (22). ∎

We now give a PDE for any European-style derivative security whose payoff is a function of the domestic and foreign bond prices and exchange rate and hence in turn the state variables X_d, X_f and X_S. The closed form expressions of Theorem 1 for the bond prices and exchange rate enable us to express the terminal payoff and boundary conditions, formulated in terms of bond prices and rates, in terms of the state variables X_d, X_f and X_S.

Theorem 2 *Let* $V := V(X_d, X_f, X_S, t)$ *denote the domestic value function of a security with a terminal payoff measurable with respect to information σ-field at T and no intermediate payments. Assume that $V \in C^{2,1}\left(\mathbb{R}^3 \times [0, T)\right)$. Then the normalised domestic value function, defined by*

$$V^*(t) := \frac{V(t)}{P_d(t, T)}, \tag{25}$$

satisfies the PDE

$$\frac{1}{2}\lambda_d^2\frac{\partial^2 V^*}{\partial X_d^2} + \frac{1}{2}\lambda_f^2\frac{\partial^2 V^*}{\partial X_f^2} + \frac{1}{2}H^{SS}\frac{\partial^2 V^*}{\partial X_S^2} + H^{df}\frac{\partial^2 V^*}{\partial X_d\partial X_f}$$

$$+H^{dS}\frac{\partial^2 V^*}{\partial X_d\partial X_S} + H^{fS}\frac{\partial^2 V^*}{\partial X_f\partial X_S} + \frac{\partial V^*}{\partial t} = 0 \qquad (26)$$

on $\mathbb{R}^3 \times [0,T)$, *where* H^{SS}, H^{df}, H^{dS} *and* H^{fS} *are defined by (24).*

Proof: See Babbs [6] for the original proof or Hutton [14] for more details. The proof is straightforward though: under the risk-adjusted measure, the normalised price process of a traded European security is a martingale, so that, since it is an Ito process, it must have zero drift. Calculating the drift from Ito's lemma and setting it to zero gives us the PDE (26).   ∎

Babbs [5] notes that the value $V(t)$ at time t of a derivative security which pays $\psi(X_d, X_f, X_S)$ at time T is the discounted risk-adjusted expected payoff

$$V(X_d, X_f, X_S, t) =$$
$$P_d(X_d(t), t, T)\tilde{\mathbb{E}}\left[\psi(\mathbf{X}_d(T), \mathbf{X}_f(T), \mathbf{X}_S(T))|X_d(t), X_f(t), X_S(t)\right], \quad (27)$$

which is, after normalisation, the solution to (26) with the boundary condition $V(T) = \psi$. From a numerical point of view, for a standard European-style derivative security, we may either integrate (27) numerically, exploiting the Gaussian state variables, or solve the PDE with the appropriate boundary conditions.

3.4   Modelling issues

There is of course as much choice for the term structure model as in the single currency case. An important consideration for the cross-currency case is that of dimensionality – any more than one state variable for each term structure process and for the exchange rate would make numerical valuation computationally very demanding. As it is, a single factor model gives us a three state variable (plus time) PDE to solve, which is computationally non-trivial.

The choice of the extended Vasicek form gives us lognormal bond prices, which holds out the possibility of analytic solutions to many European-style derivatives. Parametrisation of the model is also important – a short rate model would require a two step solution procedure: firstly, solving for the zero-coupon bond price as a function of the short rate, and secondly using this to express the terminal condition as a function of the short rate. Furthermore, the PDE with the short rates as state variables has first and zero order derivatives, and so is more difficult to solve numerically.

Under Babbs' model we avoid this first step, since boundary conditions are expressed in terms of the abstract Gaussian state variables (22) by the bond price formulae of Theorem 1 – although inspection of these formulae reveals that this is not altogether trivial from a numerical point of view, many integrals and exponentials must be calculated and, if this is not done efficiently, can represent a significant overhead. One disadvantage of this parametrisation is that the state variables are not observable in the market, making interpretation of the resulting solution more difficult away from $t = 0$.

Probably the most serious fault in extended Vasicek-style models is that they allow negative interest rates. However, Babbs [2] shows, by valuing a contingent claim that pays only when short rates are negative and finding for certain realistic parameter values very low values relative to the payoff (of the order of a basis point, i.e. 0.01%), that this model feature has a quite small effect on derivative valuation.

We will not discuss the calibration of Babbs' model to market data in detail here, as it is not our area of expertise. Suffice it to say that interest rate volatilities of the form utilized (as in Figure 3) can be fitted independently from analytic formula for suitably liquid instruments, such as foreign and domestic swaptions, and correlation data must be estimated historically.

4 Discretisation and Solution of the PDE

We next describe the numerical solution procedure – including localisation and discretisation of the PDE in a swap period – used to produce a discrete system on a finite domain, as well as the specification of data and the evaluation of bond prices and exchange rate.

4.1 Localisation of the PDE

We restrict the spatial domain $\mathbb{R}^3$ to a finite region, which we denote[4]

$$[L_x, U_x] \times [L_y, U_y] \times [L_z, U_z]. \tag{28}$$

The lower and upper bounds on the space variables L_x, L_y, L_z, U_x, U_y and U_z should be chosen in each period to be 'large enough' so as not to introduce significant errors at the boundary. To specify this precisely requires lengthy analysis, so we take an intuitive probabilistic approach. At any instant the state variables are correlated Gaussian with mean zero, so that we may find a confidence interval about zero in $\mathbb{R}^3$ for their position at any future time, which we take to be our truncated state variable region. We take as our

---

[4]For notational convenience we associate X_d, X_f and X_S with the canonical space variables x, y and z respectively.

confidence level three standard deviations[5], where the required variances are given by Lemma 1, and the resulting monotonic increasing time-dependent confidence intervals are plotted for specimen data in Figure 4. For simplicity in computing the bounds in a period, we take as our standard deviation that at the end of the last non-trivial period, t_{N-1}. Thus in every period we choose

$$
\begin{aligned}
[L_x, U_x] &:= [-3\mathrm{var}[\mathbf{X}_d(t_{N-1})], 3\mathrm{var}[\mathbf{Y}_d(t_{N-1})]] \\
[L_y, U_y] &:= [-3\mathrm{var}[\mathbf{X}_f(t_{N-1})], 3\mathrm{var}[\mathbf{Y}_f(t_{N-1})]] \\
[L_z, U_z] &:= [-3\mathrm{var}[\mathbf{X}_S(t_{N-1})], 3\mathrm{var}[\mathbf{X}_S(t_{N-1})]] ,
\end{aligned}
\tag{29}
$$

where the variances are given by (23).

A more sophisticated approach to bound setting would be to allow for different bounds in each period, increasing according to the variance $\mathrm{var}(\mathbf{X}(t_j))$. This was attempted in Hutton [14] – the differing grid points between successive periods complicates matters when computing the recursive terminal condition between them, necessitating linear interpolation to compute $V(t_j+)$, and this was found to produce numerical difficulties.

This localisation is justified as long as we impose the growth condition that the payoff is at most exponential, but we do not attempt here to formulate this more precisely. Note that bond price, LIBOR rates and exchange rates are exponential functions of X_d, X_f and X_S, so this is not a problem here.

4.1.1   Boundary conditions

We must also specify values on the boundaries of the spatial variables, i.e. at $X_d(t) = L_x, \ldots, X_S(t) = U_z$ for all t in $[t_{j-1}, t_j)$. The difficulty with choosing these boundary conditions is that, for an arbitrary payoff function, they are not known, and if we are not to perform quite detailed analysis for each different type of deal, we can only posit quite general approximate boundary conditions. Examples which we investigate include simply setting first or second derivatives constant at the boundary and a more complicated 'stopped process' boundary condition, where we stop the processes $\mathbf{X}_d(t)$, $\mathbf{X}_f(t)$ and $\mathbf{X}_S(t)$ when one hits the boundary, hence the value on the boundary is simply the discounted payoff for current values of the state variables. In §5.2 we present results from some different specifications, the variation between which proves to be gratifyingly small.

4.2   Discretisation of a general 3-D quasi-linear parabolic PDE with Dirichlet conditions

We now formulate the finite difference discretisation of a general quasi-linear PDE, of which the PDE (26) is a special case. We allow for specification of the

---

[5]The probability that a zero-mean Gaussian random variable lies outside three standard deviations of zero is, from tables in [7], approximately .0026.

discretisation scheme, be it explicit, implicit or Crank–Nicolson, by means of setting a parameter[6] $\theta \in [0,1]$, which may be a function of the discretised state variables and time. To allow for *Alternating Direction Implicit* (ADI) methods, step length on an axis may vary along that axis and the coefficients are functions of at most time – although this is simply for notational convenience and all steps follow through for space-dependent coefficients.

So we have a general PDE

$$\alpha(t)u_{xx} + \beta(t)u_{yy} + \gamma(t)u_{zz} + \delta(t)u_{xy} + \epsilon(t)u_{xz} + \zeta(t)u_{yz} - u_t = 0 \quad (30)$$

on the domain $[L_x, U_x] \times [L_y, U_y] \times [L_z, U_z] \times [L_t, U_t)$, where t now represents time until the end of the period (t_j calendar time); hence the sign on the time partial derivative. We write the value function solution at variable width mesh points

$$u\left(L_x + \sum_{n=1}^{i}(\Delta x)_n, \ L_y + \sum_{n=1}^{j}(\Delta y)_n, \ L_z + \sum_{n=1}^{k}(\Delta z)_n, \ L_t + \sum_{n=1}^{m}(\Delta t)_n\right) \quad (31)$$

as $u_{i,j,k}^m$, adopting a convention that $\sum_{n=1}^{0} := 0$, where

$$i \in \{0, 1, \dots, I\} := \mathcal{I}, \quad j \in \{0, 1, \dots, J\} := \mathcal{J},$$
$$k \in \{0, 1, \dots, K\} := \mathcal{K}, \quad m \in \{0, 1, \dots, M\} := \mathcal{M}.$$

We write $\alpha(t), \dots, \zeta(t)$ as $\alpha^m, \dots, \zeta^m$. Finite difference approximations to the partial derivatives in (30) at the point indexed by (i, j, k, m), in the interior of the index domain $\mathcal{I} \times \mathcal{J} \times \mathcal{K} \times \mathcal{M}$, are given by

$$u_{xx} \approx \theta_1 \left(\frac{u_{i+1,j,k}^m - 2u_{i,j,k}^m + u_{i-1,j,k}^m}{(\Delta x)_i^2}\right)$$
$$+ (1 - \theta_1)\left(\frac{u_{i+1,j,k}^{m-1} - 2u_{i,j,k}^{m-1} + u_{i-1,j,k}^{m-1}}{(\Delta x)_i^2}\right)$$

$$u_{xy} \approx \theta_4 \left(\frac{u_{i+1,j+1,k}^m - u_{i+1,j-1,k}^m - u_{i-1,j+1,k}^m + u_{i-1,j-1,k}^m}{4(\Delta x)_i(\Delta y)_j}\right)$$
$$+ (1 - \theta_4)\left(\frac{u_{i+1,j+1,k}^{m-1} - u_{i+1,j-1,k}^{m-1} - u_{i-1,j+1,k}^{m-1} + u_{i-1,j-1,k}^{m-1}}{4(\Delta x)_i(\Delta y)_j}\right)$$

$$u_t \approx \frac{u_{i,j,k}^m - u_{i,j,k}^{m-1}}{(\Delta t)_m}, \quad (32)$$

with u_{yy}, u_{zz}, u_{xz}, u_{yz} approximated (with parameters θ_2, θ_3, θ_5 and θ_6) in an analogous manner. All of these approximations are accurate to second order in the step length apart from the time derivative, which is first order accurate. The parameter θ_n determines the discretisation scheme: $\theta_n = 0, \frac{1}{2}, 1$

---

[6]Note that θ no longer denotes the market price of risk.

gives the explicit, Crank–Nicolson and implicit finite difference scheme for the corresponding derivative respectively. Substituting the approximations of (32) into (30) gives[7], suppressing arguments of θ_n and $\Delta x, \ldots, \Delta t$,

$$\delta^m \frac{(1 - \theta_4)}{4\Delta x \Delta y} u^{(m-1)}_{i-1,j-1,k} + \epsilon^m \frac{(1 - \theta_5)}{4\Delta x \Delta z} u^{(m-1)}_{i-1,j,k-1} + \alpha^m \frac{(1 - \theta_1)}{(\Delta x)^2} u^{(m-1)}_{i-1,j,k}$$

$$- \epsilon^m \frac{(1 - \theta_5)}{4\Delta x \Delta z} u^{(m-1)}_{i-1,j,k+1} - \delta^m \frac{(1 - \theta_4)}{4\Delta x \Delta y} u^{(m-1)}_{i-1,j+1,k} + \zeta^m \frac{(1 - \theta_6)}{4\Delta y \Delta z} u^{(m-1)}_{i,j-1,k-1}$$

$$+ \beta^m \frac{(1 - \theta_2)}{(\Delta y)^2} u^{(m-1)}_{i,j-1,k} - \zeta^m \frac{(1 - \theta_6)}{4\Delta y \Delta z} u^{(m-1)}_{i,j-1,k+1} + \gamma^m \frac{(1 - \theta_3)}{(\Delta z)^2} u^{(m-1)}_{i,j,k-1}$$

$$+ \left(-2\alpha^m \frac{(1 - \theta_1)}{(\Delta x)^2} - 2\beta^m \frac{(1 - \theta_2)}{(\Delta y)^2} - 2\gamma^m \frac{(1 - \theta_3)}{(\Delta z)^2} + \frac{1}{\Delta t} \right) u^{(m-1)}_{i,j,k}$$

$$+ \gamma^m \frac{(1 - \theta_3)}{(\Delta z)^2} u^{(m-1)}_{i,j,k+1} - \zeta^m \frac{(1 - \theta_6)}{4\Delta y \Delta z} u^{(m-1)}_{i,j+1,k-1} + \beta^m \frac{(1 - \theta_2)}{(\Delta y)^2} u^{(m-1)}_{i,j+1,k}$$

$$+ \zeta^m \frac{(1 - \theta_6)}{4\Delta y \Delta z} u^{(m-1)}_{i,j+1,k+1} - \delta^m \frac{(1 - \theta_4)}{4\Delta x \Delta y} u^{(m-1)}_{i+1,j-1,k} - \epsilon^m \frac{(1 - \theta_5)}{4\Delta x \Delta z} u^{(m-1)}_{i+1,j,k-1}$$

$$+ \alpha^m \frac{(1 - \theta_1)}{(\Delta x)^2} u^{(m-1)}_{i+1,j,k} + \epsilon^m \frac{(1 - \theta_5)}{4\Delta x \Delta z} u^{(m-1)}_{i+1,j,k+1} + \delta^m \frac{(1 - \theta_4)}{4\Delta x \Delta y} u^{(m-1)}_{i+1,j+1,k}$$

$$+ \text{ the same again with } (m - 1) \to (m) \text{ and } (1 - \theta_n) \to \theta_n,$$
$$\text{except for the } u^{(m-1)}_{i,j,k} \text{ term}$$

$$+ \left(-2\alpha^m \frac{\theta_1}{(\Delta x)^2} - 2\beta^m \frac{\theta_2}{(\Delta y)^2} - 2\gamma^m \frac{\theta_3}{(\Delta z)^2} - \frac{1}{\Delta t} \right) u^m_{i,j,k} = 0,$$

$$i \in \{1, \ldots, I - 1\}, \ldots, k \in \{1, \ldots, K - 1\}, \ m \in \{1, \ldots, M\}. \qquad (33)$$

We now write the unwieldy expression (33) in vector form. This is rather complicated algebraically, since we have four variables, so we omit the details – see Hutton [14] for more information. However, we collapse each index by replacing it with a vector, indexed by the remaining indices, each entry of which corresponds to one value of that index, which gives us the following linear equation system.

Putting $u := (u^1 \ldots u^M)'$ and $u^m := (u_{0,0,0}, u_{0,0,1}, \ldots, u_{I,J,K-1}, u_{I,J,K})'$, equations (33) become

$$Hu + \phi = 0, \qquad (34)$$

---

[7]We note that a computer algebra system, such as Maple or Mathematica, is invaluable for such work, especially when the output can be translated into C code.

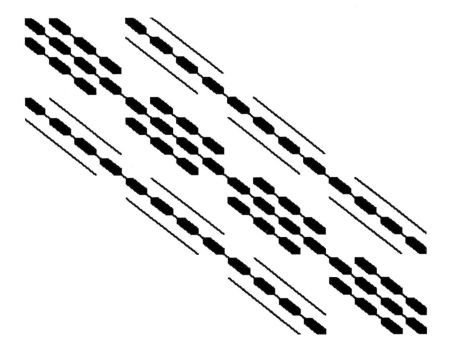

Figure 1: Bitmap of typical F^m or G^m matrix, $I = J = K = 5$

where

$$H := \begin{pmatrix} G^1 & & & \\ F^2 & G^2 & & \\ & \ddots & \ddots & \\ & & F^M & G^M \end{pmatrix} \qquad (35)$$

and ϕ is a vector that contains boundary condition information – for details see Hutton [14]. The matrices F^m and G^m are square, symmetric and of size $(I-1)(J-1)(K-1)$, and their general bitmap (pattern of non-zeroes) is shown in Figure 1. The structure consists of diagonal bands of non-zero elements, arranged in a nested tridiagonal structure.

To solve the linear system (34), we do not solve it directly, but by forward substitution starting from the initial condition[8] u^0 thus:

$$\text{solve } G^{m+1}u^{m+1} = \phi^{m+1} - F^{m+1}u^m \qquad m = 0, \ldots, M-1, \qquad (36)$$

and so existence of a solution is determined by invertibility of the matrix G^m for each $m = 1, \ldots, M$. A well-known sufficient condition for any square

---

[8]Note that we have changed to a backward time variable, so that the initial condition is given by the usual terminal condition in forward time.

matrix to be invertible is that it be strictly *diagonally dominant*[9], and if, for illustrative purposes, we put $\Delta x := \Delta y := \Delta z := \Delta$, then (see Hutton [14]) this condition reduces to the neat form

$$\frac{\Delta^2}{\Delta t} > \theta_4 |\delta^m| + \theta_5 |\epsilon^m| + \theta_6 |\zeta^m|. \tag{37}$$

Diagonal dominance is satisfied in virtually all practical cases, and is always satisfied if the PDE has no cross-derivatives.

4.3   Solution of the Discrete Problem

Precisely how best to solve the discretised problem depends on the discretisation scheme used. In all cases, attention must be paid to exploiting the structure and sparsity of the typically very large square matrices F^m and G^m to achieve reasonable computing time and efficient use of computer memory.

4.3.1   The explicit method

Explicit methods are the simplest to implement and are memory-efficient. If we set $\theta_n := 0$ for all n the matrix G^m is simply the diagonal matrix $\mathrm{diag}(\frac{1}{\Delta t})$ and so is trivial to invert – we see from (36) that u^{m+1} then depends *explicitly* on u^m. Putting $n := (I-1) = (J-1) = (K-1)$, at each time step m we have only to do two matrix multiplications, each of which takes $O(n^3)$ floating point operations since both $\mathrm{diag}(\Delta t)$ and the matrix F^m have $O(n^3)$ non-zero elements. There are M time steps, so the total operations count is $O(Mn^3)$.

The main disadvantage of the explicit method is that it is not necessarily stable. For a version of (26) with no cross-derivatives (i.e. $\delta = \epsilon = \zeta = 0$), the criterion that guarantees stability at each time step m is that

$$\alpha^m \frac{\Delta t}{(\Delta x)^2} + \beta^m \frac{\Delta t}{(\Delta y)^2} + \gamma^m \frac{\Delta t}{(\Delta z)^2} \le \frac{1}{2}. \tag{38}$$

No similar characterisation is known for the case of mixed derivatives, so in §5 we determine the critical time step experimentally – we find that (38) is very nearly sufficient in practice, since the coefficients of the mixed derivatives are relatively small. In any case, we have to take the number of time steps M of the order of the square of the number of space steps, so that the operation count for the explicit method is $O(n^5)$.

The approximation is accurate to second order in space and first order in time, inherited from the finite difference approximations (32). Of course if we take $M = O(n^2)$, as we must for stability, the method is second order accurate in time. Note that computer storage need be allocated only for the current and previous time step solution vectors.

---

[9]A matrix is *diagonally dominant* if the absolute value of the diagonal element is greater than the sum of the absolute values of the off-diagonal elements in each row.

4.3.2 General implicit methods

If $\theta_n(i, j, k, m) > 0$ for some i, j, k, m, n, the matrix G^m is not simply diagonal, and then, from (36), at each time step we have to solve a linear equation system involving the matrix $G^m x = b$. A possible approach is to adapt the general LU decomposition method to take advantage of the band diagonal[10] structure of G^m (see Figure 1) – the resulting L and U factors are both band diagonal lower and upper-triangular, and so computation associated with elements out of the diagonal band may be eliminated and storage requirements reduced. The total operation count is $O(n^7)$, instead of $O(n^9)$ for the standard LU algorithm – note that we have to recompute the LU factors at each time step because the PDE has time-dependent coefficients. The best of the simple implicit schemes (i.e. with θ constant) is the Crank–Nicolson method $(\theta_n = \frac{1}{2})$, which has the advantages of being second order accurate in time and unconditionally stable. However, on comparison with the $O(n^5)$ second order accurate in time explicit scheme, it is clear that implicit methods solved in the manner proposed here are uncompetitive, and storage of dense LU factors is impractical for all but small n, as we discuss in §5.1. We mention some alternatives to LU decomposition in §6.

4.4 Data Functions and Evaluation of Formulae

Before we can proceed with empirical tests of the terminable diff swap problem, we must supply data to the many and various functions involved, and consider how to evaluate the bond price and exchange rate functions.

4.4.1 Data functions

To specify the bond price and exchange rate functions (20), we supply the observed initial exchange rate $S(0)$ and a horizon date $H > t_N$ as positive constants and the observed initial term structures $P_d(0, T)$ and $P_f(0, T)$ as RCLL step functions approximating the observed initial term structures, for which we supply a set of grid points (a *time set*) $0 = \tau_0, \tau_1, \ldots, \tau_n := H$ and corresponding positive bond price values constant for each $t \in [\tau_{j-1}, \tau_j)$. The bond price volatility functions $\sigma_d(t, T)$ and $\sigma_f(t, T)$ are defined by (14), so that we need to supply constant mean reversion rates ξ_d and ξ_f, and the time-dependent variabilities of each short rate, $\kappa_d(t)$ and $\kappa_f(t)$, as step functions, as described above. We also specify the exchange rate volatility $\sigma_S(t)$ as a step function. Finally, we specify the three correlation functions $\rho_{df}(t)$, $\rho_{dS}(t)$ and $\rho_{fS}(t)$ as step functions. Specimen bond prices, exchange rates and short rate variabilities are plotted in Figures 2 and 3.

---

[10]A matrix is *band diagonal* if all non-zeroes lie in a diagonal band containing the diagonal.

4.4.2   Evaluating bond price and exchange rate formulae

All our possible expressions for terminal conditions are in terms of bond prices and the exchange rate, but the PDE (26) has X_d, X_f and X_S as state variables, so we must consider in detail the efficient evaluation of bond prices and exchange rate, given in Theorem 1, as functions of the state variables.

According to Theorem 1, we need to evaluate the following three integrals:

(i) $\int_0^t \lambda_d^2(u)du$,

(ii) $\int_0^t h_f(t, T, H, u)du$,

(iii) $\int_0^t h_S(t, H, u)du$.

The integrands are all products of functions of time with step functions of time, and we may calculate the integrals as a sum of integrals over the intervals in which the step function remains constant. In each integrand, we have a function $f(t)$ and a step function $g(t)$ which is a product of other RCLL step functions and so has a time set $\tau_1 := 0, \tau_2, \dots, \tau_n$ given by the ordered union of the time sets of the step functions which comprise it. Putting $\tau_n := t$, we have $\int_0^t f(u)g(u)du = \sum_{j=1}^n g(\tau_{j-1}) \left[\int_{\tau_{j-1}}^{\tau_j} f(u)du \right]$. In this manner we proceed to calculate integrals (i)–(iii), trying in general to make the resulting expressions amenable to numerical evaluation, for example by multiplying out the product of two exponentials to give one exponential, since exponentials are costly to compute. Thus:-

(i)
$$
\int_0^t \lambda_d^2(u)du = \sum_{j=1}^n \int_{\tau_{j-1}}^{\tau_j} e^{2\xi_d u} \kappa_d^2(u)du
$$
$$
= \sum_{j=1}^n \frac{\kappa_d^2(u)}{2\xi_d} (e^{2\xi_d \tau_j} - e^{2\xi_d \tau_{j-1}}). \tag{39}
$$

(ii) Similarly, although the expressions involved are lengthy and the reader is referred to Hutton [14] for details,

$$
\int_0^t h_f(t, T, H, u)du
$$
$$
= \sum_{j=1}^n \left[\frac{1}{\xi_f^2} (e^{\xi_f \tau_j} - e^{\xi_f \tau_{j-1}}) - \frac{1}{4\xi_f^2}(e^{\xi_f(2\tau_j - t)} - e^{\xi_f(2\tau_{j-1} - t)} + e^{\xi_f(2\tau_j - T)} \right.
$$
$$
\left. - e^{\xi_f(2\tau_{j-1} - T)}) \right] \kappa_f^2(\tau_{j-1})
$$
$$
+ \sum_{j=1}^n \left[\frac{1}{\xi_f}(e^{\xi_f \tau_j} - e^{\xi_f \tau_{j-1}}) \right] \kappa_f(\tau_{j-1}) \sigma_S(\tau_{j-1}) \rho_{fS}(\tau_{j-1})
$$
$$
- \sum_{j=1}^n \left[\frac{1}{\xi_d \xi_f}(e^{\xi_f \tau_j} - e^{\xi_f \tau_{j-1}}) - \frac{1}{\xi_d(\xi_f + \xi_d)}(e^{(\xi_f + \xi_d)\tau_j - \xi_d H} - e^{(\xi_f + \xi_d)\tau_{j-1} - \xi_d H}) \right]
$$
$$
\times \kappa_f(\tau_{j-1}) \kappa_d(\tau_{j-1}) \rho_{fd}(\tau_{j-1}). \tag{40}
$$

(iii) Again, similar manipulations gives us

$$
\int_0^t h_S(t, H, u)du = \sum_{j=1}^n \left[\frac{1}{\xi_f^2} \left(\tau_j - \tau_{j-1} - \frac{2}{\xi_f}(e^{\xi_f(\tau_j - t)} - e^{\xi_f(\tau_{j-1} - t)}) \right. \right.
$$
$$
\left. \left. + \frac{1}{2\xi_f}(e^{2\xi_f(\tau_j - t)} - e^{2\xi_f(\tau_{j-1} - t)}) \right) \right] \kappa_f^2(\tau_{j-1})
$$
$$
+ \sum_{j=1}^n (\tau_j - \tau_{j-1})\sigma_S^2(\tau_{j-1})
$$
$$
+ \sum_{j=1}^n \left[\frac{1}{\xi_d^2} \left(\tau_j - \tau_{j-1} - \frac{2}{\xi_d}(e^{\xi_d(\tau_j - H)} - e^{\xi_d(\tau_{j-1} - H)}) \right. \right.
$$
$$
- \frac{1}{2\xi_d}(e^{2\xi_d(\tau_j - t)} - e^{2\xi_d(\tau_{j-1} - t)})
$$
$$
\left. \left. + \frac{1}{\xi_d}(e^{\xi_d(2\tau_j - t - H)} - e^{\xi_d(2\tau_{j-1} - t - H)}) \right) \right] \kappa_d^2(\tau_{j-1})
$$
$$
+ \sum_{j=1}^n \left[\frac{2}{\xi_f}(\tau_j - \tau_{j-1}) - \frac{2}{\xi_f^2}(e^{\xi_f(\tau_j - t)} - e^{\xi_f(\tau_{j-1} - t)}) \right] \times
$$
$$
\times \kappa_f(\tau_{j-1})\rho_{fS}(\tau_{j-1})\sigma_S(\tau_{j-1})
$$
$$
- \sum_{j=1}^n \left[\frac{2}{\xi_d \xi_f}(\tau_j - \tau_{j-1}) \right.
$$
$$
- \frac{1}{\xi_d}(e^{\xi_d(\tau_j - H)} - e^{\xi_d(\tau_{j-1} - H)})
$$
$$
- \frac{1}{\xi_f}(e^{\xi_f(\tau_j - t)} - e^{\xi_f(\tau_{j-1} - t)})
$$
$$
\left. + \frac{1}{\xi_d + \xi_f} \left(e^{(\xi_d + \xi_f)\tau_j - \xi_f t - \xi_d H} - e^{(\xi_d + \xi_f)\tau_{j-1} - \xi_f t - \xi_d H} \right) \right]
$$
$$
\kappa_f(\tau_{j-1})\rho_{fd}(\tau_{j-1})\kappa_d(\tau_{j-1})
$$
$$
- \sum_{j=1}^n \left[\frac{2}{\xi_f}(\tau_j - \tau_{j-1}) - \frac{2}{\xi_f^2}(e^{\xi_f(\tau_j - H)} - e^{\xi_f(\tau_{j-1} - H)}) \right] \times
$$
$$
\times \sigma_S(\tau_{j-1})\rho_{Sd}(\tau_{j-1})\kappa_d(\tau_{j-1}). \tag{41}
$$

5 Numerical Results and Visualisation

In this section we present the results of the numerical valuation method proposed in §4, using Babbs' model with specimen financial data, applied to specific deals of the type discussed in §2. We also present various cross-sections through the resulting 4-D solution surface in a period.

5.1   Computational details

All results here were computed on an IBM RS/6000 590 serial computer with 128MB of RAM running under AIX 3.2.5. The code was written in C with double precision arithmetic, using the IBM MASS library to speed up computation of exponentials required for bond prices and exchange rates at a grid point according to (20) and in turn (39)–(41), with inlining to speed up calls to these nested functions.

Since the explicit method requires many time steps for stability, it is important to do these efficiently. In the code, the solution vector u^{m+1} is computed from u^m simply by evaluating (33) with $\theta_n = 0$ for each (i, j, k), taking basic precautions to preserve efficiency, such as computing coefficients outside the main loop. Apart from the boundary condition experiment in Table 3, boundary conditions are set by simply extrapolating the new solution vector, i.e. the result of one explicit iteration, linearly to the boundary points. To test implicit methods we used the routine **bandec** and **banbks** for LU decomposition and substitution of band-diagonal matrices in Press *et al* [18]. However, the total memory requirement is prohibitive: the number of elements stored is $3n^5 + 3n^4 + n^3$, so that taking $n = 25$, for example, requires 244MB of RAM for double precision storage. For this reason, we do not pursue implicit methods further here, but for numerical results see Hutton [14]. All results that follow in §5.2 are for the explicit method.

5.2   Numerical Results

All deals valued here are based on 3 month pound sterling and U.S. dollar LIBOR (and hence have quarterly swap payments) and are quoted per unit of sterling (domestic) principal, with initial term structures, volatilities and all other data as specified in Hutton [14][11], except that we take the initial exchange rate to be $S(0) := .64516129$. Some of the data supplied to the model are plotted in Figures 2 and 3 and the resulting 3σ-confidence intervals for the state variables, used to truncate the state variables as described in §4.1, are illustrated in Figure 4.

Vanilla cross-currency swap   In Table 1, we give numerical results for the vanilla cross-currency swap with exchange of principal, defined by the terminal condition (7) with $Z_d := 1$, $Z_f := Z_d/S(0)$, $m_d := m_f := 0$, $\delta_j := .25$, $k_d := k_f := -1$. The state variable bounds for all three deals are fixed to the 10 year value, to aid comparison, and times quoted are for the 10 year deal. To estimate comparative times for the shorter deals, simply scale the times in the ratio of the deal lengths. As discussed in §2.2, this deal has zero initial value, and we see clearly the accuracy of the numerical solution.

---

[11]This data was originally supplied by Simon Babbs, then of Midland Global Markets.

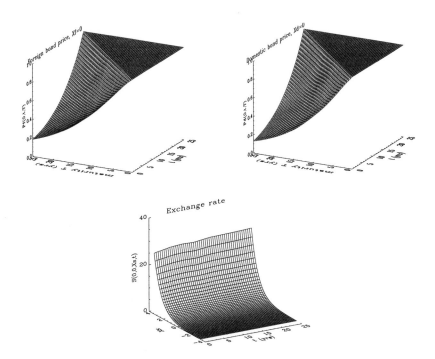

Figure 2: Bond prices $P_d(0,t,T)$ and $P_f(0,t,T)$ and exchange rate $S(0,0,X_S,t)$

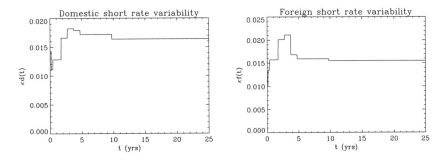

Figure 3: Prospective short rate variabilities $\kappa_d(t)$ and $\kappa_f(t)$

Clearly the accuracy deteriorates as the duration of deal lengthens, although all step widths are constant – this is simply accumulation of standard explicit method discretisation error, which is linear in the total number of time steps, but may also reflect the greater variance of the underlying processes in the later periods. In the case of the 1 year deal, we achieve accuracy of 1 bp (.0001), with $n = 40$ in 25 s, but for the 10 year deal we have to take $n = 160$

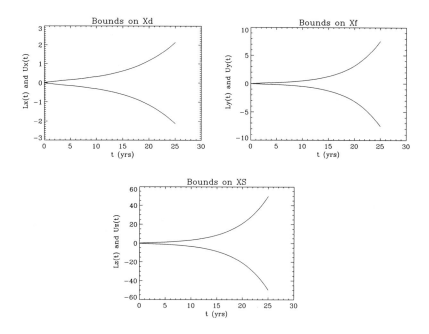

Figure 4: Bounds on Gaussian state variables $\mathbf{X}_d(t)$, $\mathbf{X}_f(t)$ and $\mathbf{X}_S(t)$

discretisation	1 year	5 years	10 years	
$M \times I \times J \times K$	V	V	V	time (s)
20×6^3	.001733	.001733	.154813	0.20
20×10^3	.000564	.009462	.050369	0.63
20×20^3	.000149	.002322	.012407	3.94
20×40^3	.000052	.000629	.003233	31.97
40×80^3	.000014	.000169	.000897	416.64
100×160^3	.000003	.000044	.000294	$\sim$ 7200.00
true value	0	0	0	

Table 1: Vanilla cross-currency swap with exchange of principal, deal value with varying discretisation and deal length.

and hence a solution time of about 2 hours, to approach a similar accuracy. Note that the explicit method stability requirement (38) affects the solution time significantly for higher spatial discretisations, which we need for the 10 year deal.

Diff swap Table 2 gives results for 10 year vanilla and terminable diff swaps, defined by the end-of-period payoff (10), with the known vanilla diff swap

discretisation	vanilla	terminable $X=.01$	
$M \times I \times J \times K$	V	V	time (s)
20×6^3	-.086798	-.124087	0.21
20×10^3	-.086293	-.129086	0.57
20×20^3	-.085919	-.123529	3.90
20×40^3	-.085815	-.123216	31.29
40×80^3	-.085750	-.123057	411.12
100×160^3	-.085721	-.122993	~ 7300.00
true value	-.085712		

Table 2: Vanilla and Terminable Diff swap deal values with varying discretisation.

solution value computed from the formula in Babbs [5]. Solution times are essentially the same as for the vanilla swap of Table 1, and are given for the sake of completeness. We see that we achieve much better convergence than for the vanilla swap with exchange of principal of Table 1, with basis point accuracy in 31s for the vanilla diff swap and apparently in 411s for the terminable version. In both cases this improvement is due to the flatter solution surface than for the vanilla swap – from (10) we see that the vanilla diff swap part of the payoff in each period is flat with respect to the exchange rate, and hence with respect to X_S.

In Table 3 we demonstrate the variation of the numerical solution with the boundary condition type, discussed in §4.1.1, for the 10 year vanilla diff swap. We take six examples, described in the key to the table[12]. Boundary conditions have some effect on solution time – solution time for type 1 (fastest) for the case $n = 80$ is 400s, type 2 (i.e. stopped process, the slowest here) is 420s, with the rest following closely the times in Table 2, so that $n = 80$ takes 410s. We see from the table that, whilst there is some variation in solution value for coarser grids, variation is well within a basis point for grids finer than $n = 20$ apart from type 1. The type 1 case ($V = 0$ on the boundary) corresponds in practical code terms to not specifying a boundary condition, which makes for easy implementation and fast computation but is not reliable, since it is far from convergent as we increase M – in fact $u^M \to 0$ as $M \to \infty$. This remark illustrates the main problem with many approximate boundary conditions, including all those here – the resulting method is convergent with I but not necessarily with M.

---

[12]Note that condition 3 is that used for all other tables, so the column headed '3' in Table 3 is the same as column 2 of Table 2.

discretisation	boundary condition					
	1	2	3	4	5	6
$M \times I \times J \times K$	V	V	V	V	V	V
20×6^3	-.086460	-.087388	-.086798	-.087236	-.087360	-.087362
20×10^3	-.086117	-.086360	-.086293	-.086352	-.086360	-.086357
20×20^3	-.085856	-.085926	-.085919	-.085926	-.085926	-.085925
20×40^3	-.085777	-.085817	-.085815	-.085818	-.085817	-.085817
40×80^3	-.085717	-.085751	-.085750	-.085752	-.085752	-.085751
100×160^3	-.085689	-.085722	-.085721	-.085722	-.085722	unstable
true value	-.085712	-.085712	-.085712	-.085712	-.085712	-.085712

boundary condition key:

1. $V = 0$ on boundary.

2. $V(X_d, X_f, X_S, t) = P_d(X_d, t, U_t)V(X_d, X_f, X_S, U_t)$ on boundary ('stopped process' condition).

3. linear extrapolation to boundary points.

4. quadratic extrapolation to boundary points.

5. $\frac{\partial V(t)}{\partial X_k} = \frac{\partial V(U_t)}{\partial X_k}$ on X_k boundary

6. $\frac{\partial^2 V(t)}{\partial X_k^2} = \frac{\partial^2 V(U_t)}{\partial X_k^2}$ on X_k boundary.

Table 3: Diff swap deal value with varying discretisation and varying boundary conditions.

discretisation	1 year forward	1 year call	
$M \times I \times J \times K$	V	V	time (s)
20×6^3	-.071080	-.071350	0.10
20×10^3	-.067289	-.067973	0.30
20×20^3	-.067963	-.068175	2.03
20×40^3	-.067872	-.068007	16.21
80×80^3	-.067751	-.067871	377.36
200×160^3	-.067732	-.067850	~ 3200.00

Table 4: One year call and one year forward on a five year terminable diffswap with varying discretisation.

Diff swaption   Finally, in Table 4, we give results for a 1 year (zero strike price) call on a 5 year terminable diff swap, i.e. a 1 year into 5 year diff swaption, mainly in order to demonstrate the simplicity of the PDE method when option structure is complicated. We also give, so as to determine the additional value to the counterparty of the call option, a 1 year forward 5 year terminable diff swap. Since the deals are shorter than 10 years, the spatial boundaries are tighter and hence we must take more time steps than for previous deals for stability – although we could have set the spatial boundaries to the 10 year case, as we did for the 1 year and 5 year vanilla swaps of Table 1. Basis point accuracy is apparently achieved in both cases within 380s, and the additional option value is about 1bp to the counterparty. That the difference should be small is unsurprising, since the counterparty has many future termination options and so the deal is already weighted in his favour, even without the additional option – this can be further appreciated by noting the limited range of the effect of the option in period 1 in Figures 5d–5f. We discuss the solution surface of this deal further in §5.3.

5.3   Visualisation of the swaption value surface

In Figure 5 we give various plots of the (4D) value surface[13] as a function of the three state variables X_d, X_f, X_S and time, for the 1 year into 5 year swaption structure of the previous section, results for which are given in Table 4. We choose this deal because it incorporates most features of simpler deals – for example, after period 1 the deal is simply a terminable diff swap. In general, we see good agreement with the theoretical behaviour, which we try to illustrate in the following remarks.

Figures 5a–5c show the value surface for period 20 as a function of X_d, X_f and X_S respectively, with the remaining variables in each case set to their expected value of zero. The termination boundary[14] is clearly visible in Figures 5a and 5b – the payoff, or terminal condition, is 'capped' in the termination region – at $.01S(X_S, t_j)$ at the end of the period. In Figure 5c, variation in X_S cannot take the value into the termination region, but clearly the shape of the surface is influenced by possible termination through variation in X_d and X_f from zero.

Figures 5d–5f show the same plots but for period 1. In Figures 5d and 5e the effect of the option to buy the swap is apparent – the value surface is 'capped' at the end of the period at zero, but in Figure 5f the variables X_d and X_f are set so that the value lies strictly in the buy region[15]. The buy region is simply a section of the terminable diff swap surface of period 2 – it is increasing with X_d, decreasing with X_f and increasing with X_S, since domestic and foreign LIBOR are negative exponential in X_d and X_f respectively, and exchange rate, and hence termination cost, is exponential in X_S.

---

[13]We note here that the use of modern data visualisation computer packages, such as PV-Wave used here, is invaluable for debugging the relatively complex computer code and for understanding the solution produced.

[14]The *termination boundary* is the set of points at the end of a period j at which the counterparty is indifferent between terminating and continuing.

[15]The *buy region* is the set of points at the end of a period j at which the counterparty exercises his option to buy the swap.

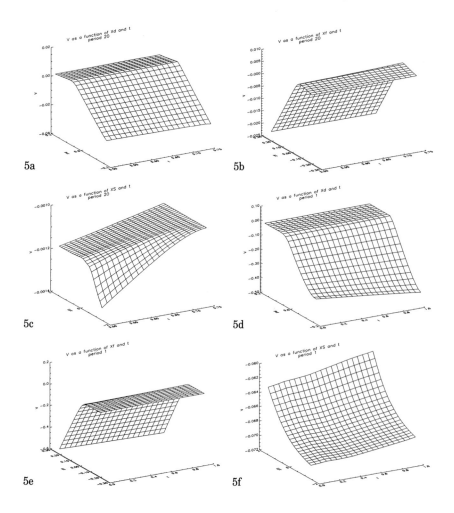

Figure 5: 1 year call on 5 year terminable diff swap: solution surfaces for period 20 (plots a, b, c), $t \in [5.5, 5.75]$ and period 1 (plots d, e, f), $t \in [0, 1]$)

This is clearly not an exhaustive study of the solution surface and there are many other possible cross-sections we could take, but those presented here are fairly representative.

6   Conclusions and Future Directions

Valuation of cross-currency terminable swaps represents a computational task that would usually only be attempted on parallel supercomputers, and as a result we have been restricted to quite coarse grids by the standards of numerical PDE literature – that we get reasonable convergence is due to the fact that the solutions to practical valuation problems do not in general have high curvature. We have in most cases obtained convergence to within a basis point in reasonable computing time.

We conclude that applying band diagonal LU decomposition routines to solving implicit schemes is infeasible for this problem, and it is not clear that any other numerical solution method for implicit schemes could out-perform the ordinary explicit method used here, except in cases where stability is restrictive, such as for the vanilla swap. Obvious methods to try include Successive Over-Relaxation (SOR) and recursive tridiagonal or sparse matrix factorisation techniques (such as that advocated in Keast and Muir [15]) applied to the Crank–Nicolson scheme, offering stable second order accuracy in time and efficient use of memory. For PDEs with fewer state variables such methods may well be more efficient, especially for fine spatial discretisations and single state variable problems.

The explicit method approach advocated here could be further improved. It is possible to linearly transform our state variables (essentially to diagonalise the state variables covariance matrix) so as to eliminate the cross-derivative terms and hence transform the PDE into a time-dependent version of the heat equation, which reduces the number of non-zero bands in the matrix F^m from 19 to 7, with a corresponding reduction in time for matrix multiplication. Preliminary experiments with this approach are underway, but so far we have had numerical difficulties when X_S appears in the terminal condition. Since only matrix multiplications are required, it should also be a simple matter to implement the explicit method on a parallel computer, particularly a *fine grain* parallel or *vector* machine. For example, Ekvall [11] investigated parallelised explicit and ADI methods on a 3-D Black-Scholes-type PDE on a Connection Machine CM200 with 4096 processors. The drawbacks of the explicit method are of course its poor stability characteristics and first order time accuracy, and since we do not usually know the critical mesh ratio in advance, some solution time has to be spent determining it – time which we have not added to our results.

One approach which is worth further investigation for this particular problem is the Fourier method, which uses the Fast Fourier Transform to solve the heat equation – its $O(n^3 \log n)$ solution time for a single time point is very appealing, and further work should investigate whether this is realisable. It cannot be used for state variable-dependent coefficients, so whilst it applies here, it is not immediately applicable to many other models. However, a more general and hence more attractive fast method is that of multi-grid, which is the method choice for many physical applications, and could probably be used to good effect in financial problems.

Of particular interest, given the work in Dempster and Hutton [9, 10] (see also Hutton [14]), would be American-style interest rate derivatives, with numerical valuation via linear programming solution of the finite difference approximation. However, it is clear that the difficulties with implicit methods here would carry over to our LP method for an American derivative, and work should be directed towards producing an ADI or multi-grid linear programming solver for American-style derivatives contingent on up to three stochastic variables.

7   Acknowledgements

Much of the material in this article appears in the PhD thesis [14] of the second author, written under the supervision of the first. We are grateful to the EPSRC (UK), the University of Essex and HSBC Markets for partial support of the research. In particular, we would like to thank Robert Benson, Daniel George and Steven Blyth of the Specialised Derivatives Group at HSBC Markets, London for their support, criticism, encouragement and permission to publish this work. Without Simon Babbs, formerly of HSBC Markets and now at First National Bank of Chicago, London, we would not have undertaken – nor completed – this project; we owe to him a special debt of gratitude. Finally, we would like to acknowledge helpful discussions with and comments from Michael Selby, Stewart Hodges, Mark Davis and participants at seminars at the Isaac Newton Institute for Mathematical Sciences, University of Cambridge, the Futures and Options Research Centre, University of Warwick and NatWest Capital Markets, London.

References

[1]   Amin, K.I. & Jarrow, R.A. (1991). Pricing foreign currency options under stochastic interest rates. *Journal of Banking and Finance* **10** 310–29.

[2]   Babbs, S.H. (1990). The Term Structure of Interest Rates: Stochastic Processes and Contingent Claims. PhD Thesis, Imperial College, London University.

[3]   Babbs S.H.(1993). "Generalized Vasicek" models of the term structure. *Applied Stochastic Models and Data Analysis: Proceedings of the Sixth International Symposium*, World Scientific. To appear.

[4]   Babbs S.H. (1993). Project Proposal for University of Essex Finance Research Group. Midland Global Markets, London.

[5]   Babbs, S.H. (1994a). The valuation of cross-currency interest-sensitive claims with application to "diff" swaps. Working Paper, Midland Global Markets, London, February 1994.

[6] Babbs, S.H. (1994b). Valuation of cross-currency interest-sensitive claims under the cross-currency extended Vasicek model: A PDE approach. Research Note, First National Bank of Chicago, London, December 1994.

[7] Barnett, S. & Cronin, T.M. (1986). *Mathematical Formulae* 4th edition. Longman, Harlow, Essex.

[8] Carr, P. (1993). Deriving derivatives of derivative securities. Working paper, Johnson Graduate School of Management, Cornell University.

[9] Dempster, M.A.H. & Hutton, J.P. (1995). Fast numerical valuation of American options by linear programming. *Mathematical Finance*, to appear

[10] Dempster, M.A.H. and Hutton, J.P. (1997). Fast numerical valuation of American, exotic and complex options. *Applied Mathematical Finance* **4** 1–20.

[11] Ekvall, N. (1993). Two finite difference schemes for evaluation of contingent claims with three underlying state variables. EFI Research Paper 6520, Ekonomiska Forskningsinstitutet, Stockholm.

[12] Heath, D.C., Jarrow, R.A. & Morton, A. (1992). Bond pricing and the term structure of interest rates: a new methodology for contingent claims valuation. *Econometrica* **60** 77–105

[13] Hull, J. (1993). *Options, Futures and Other Derivative Securities,* 2nd edition. Prentice Hall.

[14] Hutton, J.P. (1995). Fast Pricing of Derivative Securities. PhD Thesis, Department of Mathematics, University of Essex.

[15] Keast, P. & Muir, P.H. (1991). EPDCOL: A more efficient PDECOL code. *ACM Transactions on Mathematical Software.* **17** 153–166.

[16] Litzenberger, R.H. (1992). Swaps: plain and fanciful. *Journal of Finance* **47**, Presidential Address, 831–850.

[17] Merton, R.C. (1974). On the pricing of corporate debt: the risk structure of interest rates. *Journal of Finance* **29** 449–470.

[18] Press, W.H., Teukolsky, S.A., Vetterling, W.T. & Flannery, B.P. (1992). *Numerical Recipes in C,* 2nd edition. Cambridge University Press.

Numerical Methods for Stochastic Control Problems in Finance

Harold J. Kushner

Abstract

Control problems for diffusion or jump-diffusion models are occurring with ever greater frequency in modern finance. For example, the problem of pricing American options with a diffusion model for the underlying price and interest rate processes is one of optimal stopping. The forms of the models are often such that analytical methods cannot be used to solve the control problem. The chapter discusses a powerful method for the solution of quite general stochastic control problems. The approach, called the Markov chain approximation method, is the current method of choice for a wide variety of stochastic control problems in continuous time. It is intuitive in that it uses approximations which are 'physically' close to the original model. Convergence can be shown under very weak conditions. The basic approximation method is discussed as well as the ideas used in the proof of convergence.

1 Introduction

The use of controlled or uncontrolled diffusion or jump diffusion models in mathematical finance is now well established. For the derivative pricing problem, the main current control problem is the optimal stopping problem. For American options, where the option can be exercised before the termination time, the problem is to determine the optimal exercise policy, the proper price being the value of the optimal stopping problem. But other types of control problems will likely appear in the near future.

The stochastic models used today or currently considered for future use for the price and interest rate processes go considerably beyond the original simple geometric Brownian motion model of Black and Scholes, and analytical methods can rarely be used to obtain either explicit solutions or adequate qualitative information. Thus, numerical methods are called for. The basic problems of computing values of functionals of jump-diffusion type processes or their optimal values are those of the numerical methods of stochastic control theory. The problems which arise can be quite complicated, involving discontinuities, reflections, corners with discontinuous boundary conditions, degeneracies in the operators, functionals of maxima or minima of paths, singular controls (such as arise when there is continuous trading and transactions costs), etc.

Cost or optimal cost functionals are often representable as the (at least formal) solutions to certain nonlinear partial differential equations or variational inequalities, the so called Bellman equation. For many important cases, the PDE's have only a formal meaning, and either standard methods of numerical analysis are not usable to prove convergence or the classical algorithms will not converge. Additionally, the Bellman equation might not even be formally given by a PDE or variational inequality. In any case, getting the optimal cost and control is the primary interest, and not solving a PDE. The PDE is at best an intermediary in this process. One would like numerical methods which can exploit the intuition inherent in the physical model.

The most powerful current numerical method for stochastic control problems is called the Markov chain approximation method [7, 10, 9]. The basic idea is the approximation of the original controlled (or uncontrolled) problem by a simpler one for which the computation is feasible, and then proving that the computed values converge to that for the original problem as the approximation parameter goes to its limit. The approximating process is a Markov chain indexed by an approximation parameter h, and the only requirement on the chain is 'local consistency,' where the 'local' properties of the chain are close to those of the diffusion for small h, a rather minimal condition. The local properties are just the mean and the mean square change per step, under any control. If the problem has a reflecting boundary, then the mean reflection directions for the approximating chain must approximate the mean reflection directions for the original process. One then chooses an appropriate cost function for the chain and solves the associated control problem. The approximating chain is often easily obtained via standard formulas. The method works with virtually all forms of the stochastic control problem which have been of interest to date for jump-diffusion type models.

Under very broad conditions, one can prove that the sequence of cost or optimal cost functions for the sequence of approximating chains converges to that for the underlying original process as the approximation parameter goes to zero. The proofs are purely probabilistic. One never needs to appeal to regularity properties of or even explicitly use the Bellman equation, whether it is formal or not, although certainly the form of the Bellman equation often plays an important intuitive role in suggesting useful approximations. Additionally, one can take advantage of knowledge of or intuition concerning the physical process in the construction of the approximations and algorithms. The optimal value function for the approximating chain is an optimal value function for a controlled process and cost criterion which are very close to the originally given ones, which makes the entire procedure very natural.

We note that, even when the Bellman equation can be treated by (weak sense) PDE methods, the Markov chain approximation is still the usual numerical approach at this time, and then the convergence theorems provided by the probabilistic approach seem to be much simpler and require weaker conditions than the alternatives; see the last chapter in [10] for a discussion of this point. Some forms of the methods do reduce to standard finite element or finite difference methods. But then, owing to the degeneracies of the operators or to non-standard boundary conditions or controls, the standard methods of proof of numerical analysis often cannot be used.

This article surveys the basic ideas of the Markov chain approximation method. Section 2 describes some examples of current interest in finance. To simplify the presentation, and since the main current control problem in derivative pricing is the optimal stopping problem for diffusion models, the rest of the paper concentrates on that problem. But keep in mind that the general approach works for nearly all current control problems, and with the addition of jumps as well. The simplest stopping problem is where there is no a priori given finite terminal time. Since the procedure is easiest to describe for this case, we start with this case in Section 3. Typical current problems in pricing derivatives do use a fixed terminal time T. But, it is still of considerable theoretical interest to see what happens when T is large. For example, for criterion (2.3), where $T = \infty$, what is the set of (x_1, x_2) at which the option is exercised, and how might it depend on the parameters of the problem? The method, when there is a given upper bound to the terminal time, is easily obtained from the infinite upper limit result, and is developed in Section 6

The basic numerical method is developed in Section 3 together with the usual modifications that allow a solution in a bounded set in the state space. Section 4 outlines the basic ideas of the proof of convergence via the theory of weak convergence of probability measures. To show that the construction of the approximating Markov chain is often easy, an illustration for a one dimensional example is given in Section 5. Section 6 extends the discussions of Sections 3 and 5 to the case where there is a finite upper limit to the time.

There is great deal of art to successful use of any numerical procedure. It is often best to start slowly on simple problems and build an intuition for both the results to be obtained and the appropriate forms of the approximations and algorithms.

2 Some Simple Examples

This section contains some examples of the role of controlled diffusions in finance. The examples are for discussion purposes only and no conditions will be given.

Example 1. Perhaps the simplest example is the classical diffusion model for the pricing of call options. The process is

$$
\begin{aligned}
dx_1 &= b_1(x_1)dt + x_1\sigma_1 dw_1, & x_1 \geq 0, & \quad \text{stock price}, \\
dx_2 &= b_2(x_2)dt + x_2\sigma_2 dw_2, & x_2 \geq 0, & \quad \text{interest rate},
\end{aligned}
\tag{2.1}
$$

where the standard Wiener processes $w_i(\cdot)$ might be correlated. Define $x = (x_1, x_2)$. If the option is to be exercised at time T and the current time is zero, then the present discounted value is

$$W(x, 0) = E_x \exp\left[-\int_0^T x_2(v)dv\right] g(x_T), \qquad (2.2)$$

where $g(x) = [x_1 - K]^+$ and K denotes the strike price. To solve for $W(x, 0)$, one generally needs to solve for the family of functions

$$W(x, t) = E_x \exp\left[-\int_t^T x_2(v)dv\right] g(x_T), \text{ all } t < T. \qquad (2.3)$$

Let $\mathcal{L} =$ denote the differential generator of $x(\cdot)$. Then, at least formally, $W(\cdot)$ solves the parabolic PDE

$$0 = W_t + \mathcal{L}W - x_2 W, \quad t < T, \quad W(x, T) = g(x).$$

Even for this simple case, the numerical problem is not necessarily straightforward. The problem cannot usually be solved numerically in an infinite state space, and some sort of space bounding is needed. This then entails the selection of appropriate boundary conditions for the actual form of the model that is to be solved numerically. The bounding of the space and the boundary conditions (reflecting or absorbing or a combination) are usually a *numerical* requirement, and they need to be done such that the essential features of the solution to the original problem of interest are those of the numerical alteration in the region of state space of greatest concern. This point, which is crucial in all numerical work, will be returned to later.

For American options, the buyer can choose to exercise before the expiration time T, if desired. This leads to the following classical optimal stopping problem. Let $\tau \leq T$ denote an arbitrary stopping time. Since the buyer would exercise the option in a profit maximizing way, the correct price at time zero is then

$$V(x, 0) = \sup_{0 \leq \tau \leq T} E_x \exp\left[-\int_0^\tau x_2(v)dv\right] g(x_\tau).$$

To solve for $V(x, 0)$, one needs to solve for the family of functions

$$V(x, t) = \sup_{t \leq \tau \leq T} E_x \exp\left[-\int_t^\tau x_2(v)dv\right] g(x_\tau), \quad t < T, \qquad (2.4)$$

with $V(x, T) = g(x)$. The optimal policy for the buyer is to exercise the option at the first time (if any) at which $V(x, t) = g(x)$.

At least formally, the function $V(\cdot)$ solves the following variational inequality:

$$0 = V_t + \mathcal{L}V - x_2 V, \text{ on the set where } V(x, t) > g(x), t < T,$$

and we require that $V(x,t) \geq g(x)$ for all x, t.

The strike price can be allowed to depend on the exercise time. Then $K(t)$ replaces K and we need to compute

$$V(x,t) = \sup_{t \leq \tau \leq T} E_x \exp\left[-\int_t^\tau x_2(v)dv\right][x_1(\tau) - K(\tau)]^+. \qquad (2.5)$$

The time varying strike price model is an interesting possibility for the future, and is not generally solvable analytically, even for simple $b_i(\cdot)$ functions.

To account for sudden price or interest rate changes, one can add a 'jump' term to the dynamical equation (2.1). Let $N(\cdot)$ be Poisson measure, and let the jump part of the process be defined by

$$\int_0^t \int q(x(s^-), \rho)N(ds, d\rho), \qquad (2.6)$$

where $q(\cdot)$ is an appropriate function. With the jump part added, the value is not generally computable analytically, and a numerical procedure is required.

Example 2. While the two-dimensional form and the noise terms of the process (2.1) are 'classical', it is not currently clear whether it adequately predicts the actual observed market price fluctuations of the options. Hobson and Rogers [5] introduced an alternative form similar to the following, which 'stochasticizes' the volatility, and argued that it is a better price predictor than the form (2.1):

$$\begin{aligned} dS &= dZ - \lambda S dt, \\ dZ &= \mu(S)dt + \sigma(S)dw. \end{aligned} \qquad (2.7)$$

The criteria to be maximized is the same as that of Example 1. The use of (Z, S) in lieu of x_1 makes an analytical solution even less likely, and the numerical solution more important.

Example 3. A great variety of more exotic options are appearing, each with a use in a particular specialized market. One recent variation is the following, where the model is (2.1), but the payoff depends on the maximum of the price process over a time interval. Define

$$x_3(t) = \max_{s \leq t} x_1(s) \qquad (2.8)$$

Let $\tau \leq T$ be the stopping time at which the option is exercised. Then, with $x = (x_1, x_2, x_3)$, suppose that the payoff at the exercise time τ is

$$G(x(\tau)) = [x_1(\tau) - K]^+ I_{\{x_3(\tau) < v\}} + cI_{\{x_3(\tau) \geq v\}}. \qquad (2.9)$$

In other words, if the maximum price up to the exercise time exceeds v, then there is a payoff of c. Otherwise, the payoff is as in Example 1. Again, since the buyer will act to maximize profit, if the option has not been exercised by time $t < T$, then the value at t is

$$V(x,t) = \max_{t \le \tau \le T} E_x G(x(\tau)).$$

The process (2.1) is degenerate on the zero axes, but here the $x_3(\cdot)$ process is completely degenerate.

Example 4. Transactions costs. At this time, the problem of transactions costs for continuous trading does not apply to the pricing problem for derivatives, which seems to concentrate on the stopping problem alone among the many possible control problems. But it does indicate the potential of the method in finance. The numerical methods to be discussed come into full flower when used on such complex problems, and are generally the only viable approach at this time. This example, which concerns the choice of investment to maximize a discounted utility of consumption over an infinite time horizon comes from [1, 3, 11]. Let $x_1 \ge 0$ denote the assets in a money market, $x_2 \ge 0$ the assets in a stock and $c(\cdot)$ the (to be chosen) consumption rate function. The processes $L(\cdot)$ ($M(\cdot)$, resp.) represent the transfers from the money market to the stock (stock to the money market, resp.). The transactions costs are assumed to be proportional to the amounts transferred, with proportionality constants λ, μ. The model is then

$$\begin{aligned} dx_1 &= (rx_1 - c)dt - dL + (1-\mu)dM, \\ dx_2 &= ax_2 dt + \sigma x_2 dw + (1-\lambda)dL - dM, \end{aligned} \qquad (2.10)$$

with perhaps constraints (e.g., non-negativity) added on the x_i.

The control functions to be chosen are $c(\cdot), L(\cdot)$ and $M(\cdot)$. For a given utility function $U(\cdot)$, the optimal value is

$$V(x) = \sup_{L,M,c} E_x \int_0^\infty e^{-\beta t} U(c(t))dt. \qquad (2.11)$$

For such problems, the optimal $M(\cdot)$ and $L(\cdot)$ are generally singular controls: There is an $x-$set S such that $M(\cdot)$ and $L(\cdot)$ act with 'infinite force' on the boundary of S to keep $x(t)$ in S, and they are non decreasing but not generally absolutely continuous with respect to Lebesgue measure.

Current analytical treatments require strong conditions on $U(\cdot)$, and fail even for relatively simple modifications. However, let us complicate the model by using a more general fee for transactions. If there is a non zero minimum charge (for non zero transactions) added to the proportional charge, then the model is

$$\begin{aligned} dx_1 &= (rx_1 - c)dt - dL + (1-\mu)dM - a_1 I_{\{dM \ne 0\}}, \\ dx_2 &= ax_2 dt + \sigma x_2 dw + (1-\lambda)dL - dM - a_2 I_{\{dL \ne 0\}}. \end{aligned}$$

For a general transactions cost form (perhaps a nonzero minimum charge plus volume discounts) a reasonable model is

$$
\begin{aligned}
dx_1 &= (rx_1 - c)dt - dL + [dM - f_1(dM)],\\
dx_2 &= ax_2dt + \sigma x_2 dw - dM + [dL - f_2(dL)],
\end{aligned}
$$

where the $f_i(\cdot)$ are monotone increasing functions, and concave if there are volume discounts.

For another variation, add a stopping time τ and stopping value (i.e., we can sell and 'leave' the system), with total maximum utility

$$
V(x) = \sup_{\tau,c,L,M} E_x \left\{ \int_0^\tau e^{-\beta t} U(c(t))dt + e^{-\beta \tau} U_0(x(\tau)) \right\}.
$$

For all of the problems described and under quite general conditions, the numerical solutions given by the Markov chain approximation method will converge to the correct optimal solution as the discretization level decreases to zero, irrespective of what we know about the Bellman equation. The important point is that the possibility of analytical solution drops rapidly once we leave certain simple canonical models, but the numerical problem can still be well defined, in that the numerical procedures will yield a (provably) good approximation to the actual solution of the optimization problem.

3   The Infinite Time Optimal Stopping Problem

In this section, we formulate the basic optimal stopping problem on the infinite time interval, and discuss the Markov chain approximation numerical method for its solution. The method will be adapted to the finite time problem in Section 6.

The basic model. The process will be

$$
dx = b(x)dt + \sigma(x)dw, \; x \in \mathbb{R}^k, \tag{3.1}
$$

where $\mathbb{R}^k$ is Euclidean $k-$space. For $\beta > 0$, $g(\cdot), k(\cdot)$ continuous real valued functions, and τ being a stopping time (with possibly infinite values), define the discounted cost

$$
W(x,\tau) = E_x \int_0^\tau e^{-\beta t} k(x(t))dt + E_x e^{-\beta \tau} g(x(\tau)). \tag{3.2}
$$

Define the optimal cost

$$
V(x) = \sup_\tau W(x,\tau). \tag{3.3}
$$

In [10], inf is used in lieu of sup.

Weak sense uniqueness of the solution to (3.1) and the continuity of $b(\cdot)$, $\sigma(\cdot)$ are assumed. Since the problem will be solved numerically in a bounded set, the functions can be supposed to be bounded without loss of generality. By τ being a stopping time, we mean that there is a sequence of nondecreasing $\sigma-$algebras $\mathcal{F}_t$ such that for each t, $\mathcal{F}_t$ measures at least $\{x(s), w(s), s \leq t\}$, where $w(\cdot)$ is an $\mathcal{F}_t-$martingale and standard Wiener process, and that $\{\tau \leq t\} \in \mathcal{F}_t$. The development will be formal, the general aim being the illustration of the basic concepts of the approximation method. The reader is referred to [10] for full mathematical and algorithmic details.

The numerical boundary. For numerical purposes, the state space must be bounded in some way, as is usual with numerical methods for PDE-type problems; i.e., the state must be confined to some compact set G. This alteration is purely for numerical purposes, so that finite algorithms can be used. But it requires a modification of the basic model (3.1). The modification requires that we specify G and the behavior on the boundary ∂G. Sometimes the 'physical' state space has a natural boundary. For example, in many financial models, the state variables are non negative, which yields a 'lower' boundary. In Example 1, the state components are the current stock price and interest rate. If there are (as is common) upper values $\overline{x_1}, \overline{x_2}$ beyond which the components will pass with only negligible probability when starting in the part of the state space of greatest interest (before either the termination time T, or the time after which the discounting renders the future discounted costs insignificant), then these values can be used as an 'upper' boundary. Of course, in any practical algorithm, care must be taken to assure that the numerical problem is not so large that it cannot be solved in a reasonable time. Generally, one has to experiment with the values of the 'nonnatural' or 'numerical' boundaries, which are selected just for numerical purposes. But good values of these quantities tend to be close for closely related problems.

To complete the modified process model, the behavior on the boundary needs to be specified. The simplest boundary behavior is 'absorbtion' (a Dirichlet boundary condition): The process remains at the point of first contact with the boundary ∂G after the time of first contact. If that time is large with a high probability for the states of interest, then the 'numerical' modification will not have a significant effect on the solution for those states. In this case, τ is generally supposed to be no greater than the first time of hitting the boundary of G. An alternative is to use a reflection (a Neumann boundary condition). In this case, if the state is on the boundary of G and attempts to leave G, then it is instantaneously reflected back in the given (and perhaps state dependent) direction. Thus, the process freely 'diffuses' as long as it is in G, but any attempt to leave G is stopped. The appropriate definition of the reflecting process is given via the so-called *Skorohod problem*, and is discussed in detail in [10]. Combinations of reflection and absorbtion can also be used.

In any case, for numerical purposes, the state space must be bounded. If the bounding set G is imposed for purely numerical purposes, then it must be large enough so that the basic features of the solution in the important region of the state space are not much affected. For convenience in exposition only, we will always suppose that G is a rectangle $G = \{x : a_i \le x_i \le b_i, i = 1, \dots, k\}$, where $a_i < b_i$, although the restriction to a rectangle is certainly not a requirement for the method to work. Also, for specificity in this paper, we will suppose that the process is reflected inward orthogonally to the boundary if it tries to escape. This orthogonality is not needed at all for the convergence proofs to be valid, but it does simplify the discussion. We are not suggesting that these particular modifications be used, only that we will use them to simplify the discussion in this paper. Indeed, the natural lower boundary for Example 1 is absorbing, and a reflecting upper (orthogonally reflecting) boundary with levels b_1, b_2 imposed on that problem simply means that the price (interest rate, resp.) is not allowed to go above b_1 (resp., b_2). If the b_i are large enough this is not a problem. In Example 1, if the lower boundary is formed by the absorbing zero axes, then even if we allow a reflection there, there will be no actual reflection since the unreflected process can never be negative.

Non orthogonal reflecting directions (where the direction depends on the face) arise in queueing network problems. In these applications, each face of ∂G corresponds to a different set of full and empty queues, and the reflection direction depends on the routing within the network of the outputs of the affected queues. They also arise in multistock optimal portfolio problems.

The reflected form of (3.1) is

$$dx = b(x)dt + \sigma(x)dw + dL - dU, \tag{3.4}$$

where $L(\cdot) = (L_1(\cdot), \dots, L_k(\cdot))$ and $U(\cdot) = (U_1(\cdot), \dots, U_k(\cdot))$

are the reflection terms. Each process takes the value zero at time zero, is nondecreasing and can increase only when the $x(\cdot)$ process is on the associated boundary: $L_i(\cdot)$ cannot increase unless $x_i(t) = a_i$, and $U_i(\cdot)$ cannot increase unless $x_i(t) = b_i$. If (3.1) has a unique weak sense solution, then so does (3.4) [10].

The Markov chain approximation. For the discretization level $h > 0$, suppose (without loss of generality) that the sides of G are integral multiples of h, and let G_h be the h-grid on G. If the boundary conditions were absorbing, then G_h would be a sufficient state space for the approximating chain. Since we are assuming that the boundary conditions are 'reflecting' on all sides, it is convenient to add a set of states to which the approximating chain goes when it leaves G_h, and from which it is reflected back instantaneously to G_h. These new states are not essential, and with a proper definition of the transition probabilities on the boundary $\partial G_h = G_h \cap \partial G$, they can be eliminated, but they are convenient for both descriptive and algorithmic purposes and are natural from an intuitive point of view.

Let ∂G_h^+ denote the additional set of points which are not in G_h, but are in the h-grid of the set G enlarged by h in all directions. Define $S_h = G_h + \partial G_h^+$. This will be the state space of the approximating Markov chain. The ∂G_h^+, called the *numerical reflecting boundary*, is used solely to approximate the reflecting process. More generally, the discretization level can depend on the direction, in which case h is a vector valued parameter.

Let $\{\xi_n^h\}$ be a finite state Markov chain on S_h, with the following properties for $x \in G_h$. Let $E_{x,n}^h$ denote the expectation given the data at time n and that $\xi_n^h = x$. Suppose that there are intervals $\Delta t^h(x) > 0$ such that

$$E_{x,n}^h[\xi_{n+1}^h - x] = b(x)\Delta t^h(x) + o(\Delta t^h(x)), \qquad (3.5a)$$

$$E_{x,n}^h[\xi_{n+1}^h - x][\xi_{n+1}^h - x]' = \sigma(x)\sigma'(x)\Delta t^h(x) + o(\Delta t^h(x)), \qquad (3.5b)$$

$$|\xi_{n+1}^h - \xi_n^h| \to 0 \qquad (3.5c)$$

as $h \to 0$. (3.5) is all that we require of the approximating chain. The requirement that the intervals go to zero can always be achieved by simply increasing the probability that each state transit to itself. Also, (3.5c) is modified if there are jumps. The interpolation intervals $\Delta t^h(x)$ can always be chosen to be constant if desired. But the allowed state dependence is an advantage. For example, if the mean velocity $b(\cdot)$ is large at some value of x, then considerations of numerical accuracy suggest that the interval be smaller there. The numerical procedures also converge faster when we take advantage of the flexibility given by the variable interpolation intervals.

Note that the chain has the 'local properties' of the diffusion process (1.1) in the sense that

$$E_x(x(\delta) - x) = b(x, \alpha)\delta + o(\delta),$$

$$E_x[x(\delta) - x][x(\delta) - x]' = \sigma(x)\sigma'(x)\delta + o(\delta).$$

This is what 'local consistency' (of the chain with the diffusion) means. It is a rather minimal property for an approximation. The special form of the grid G_h was chosen for ease of visualization. It need not be a grid at all, and any approximating state space for which (3.5) holds will do. Local consistency is not required at all points. For example, if the state space is divided into two parts each with a different grid sizes, then it is hard to assure local consistency on the boundary of the two regions. But, an 'almost everywhere' local consistency might still hold and be enough to guarantee convergence if $x(\cdot)$ itself is 'well behaved' on that boundary.

The intervals $\Delta t^h(x)$ and the transition probabilities are easily produced by many automatic procedures. A full account is in [10, Chapters 5, 12], and an example is given in Section 5 of this chapter.

Local consistency is also needed on the reflecting states. All that is needed in our case is that the chain is sent back to the nearest state in G_h if it ever leaves G_h. For non orthogonal reflection directions, all that is needed is that the mean reflection directions asymptotically be the true directions on the adjacent faces, and that in the corners and edges of G it be in the convex hull of the directions on the faces, again a rather minimal 'consistency' condition. The interpolation intervals for the reflecting boundary states are zero, since the reflection is instantaneous.

Continuous time interpolations. The $\{\xi_n^h, n < \infty\}$ is a discrete time parameter process. To approximate the continuous time parameter process $x(\cdot)$ itself, we need an appropriate continuous time interpolation of the chain. Owing to the properties of the $\Delta t^h(x)$, this provides a natural interpolation interval. Define the *interpolated time* $t_n^h = \sum_{i=0}^{n-1} \Delta t^h(\xi_i^h)$. Define the *continuous parameter interpolation* by $\xi^h(\cdot)$

$$\xi^h(t) = \xi_n^h, , \quad t \in [t_n^h, t_{n+1}^h). \tag{3.6}$$

One can also let the interpolation be a continuous time Markov chain with the mean sojourn time at state x being $\Delta t^h(x)$. This alternative has some mathematical advantages and is heavily used in [10]. In any case, the interpolation is used only for the mathematical proofs, and does not influence the numerical algorithm.

The interpolated process defined by (3.6) is an approximation in G_h to the diffusion (3.1) in G in the sense that the 'local consistency properties' (3.5) hold, along with the consistency on the reflecting boundary states. It is also an approximation in the broader sense: The expectations of continuous and bounded functionals of $\xi^h(\cdot)$ converge to the expectation of that functional of $x(\cdot)$. Indeed, the convergence holds even if the functional is only almost everywhere continuous with respect to the measure of $x(\cdot)$.

The interpolated process $\xi^h(\cdot)$ is piecewise constant. Given the value of the current state and control action, the current interval is known. The interpolation intervals are obtained automatically when the transition functions $p^h(x, y)$ are constructed. See the example in Section 5.

The optimal stopping problem for the approximating chain. Let N_h be a stopping time for the approximating chain $\{\xi_n^h, n < \infty\}$, and let $\tau_h = t_{N_h}$ denote the stopping time for the interpolated process $\xi^h(\cdot)$ which corresponds to N_h. There are several natural and asymptotically equivalent cost functionals for the chain which approximate (3.2), depending on how time is discounted on the intervals of constancy $[t_n^h, t_{n+1}^h)$, one being

$$W^h(x, \tau_h) = E_x \sum_{n=0}^{N_h-1} e^{-\beta t_n^h} k(\xi_n^h) \Delta t^h(\xi_n^h) + E_x e^{-\beta \tau_h} g(\xi_{N_h}^h). \tag{3.7}$$

Then (3.7) can be written as

$$W^h(x, \tau_h) = E_x \int_0^{\tau_h} e^{-\beta t} k(\xi^h(s)) ds + E_x e^{-\beta \tau_h} g(\xi^h(\tau_h)) + \rho^h, \qquad (3.8)$$

where ρ^h is a error due the interpolation of the sum by the integral and goes to zero as $h \to 0$.

The dynamic programming equation for this optimal stopping problem (orthogonal reflecting boundary) is, for $x \in G_h$,

$$V^h(x) = \max \left[\sum_{y \in S_h} e^{-\beta \Delta t^h(x)} p^h(x, y) V^h(y) + k(x) \Delta t^h(x), g(x) \right], \qquad (3.9)$$

and generally one would approximate the exponential by a computationally simpler function. For a state x on a face of ∂G_h^+, $V^h(x)$ takes the value at the point y on the boundary of G_h to which x is reflected.

For the absorbing boundary case, define G_h^0 to be the subset of G_h which does not include the boundary of G. Then, for $x \in G_h^0$, the dynamic programming equation is

$$V^h(x) = \max \left[\sum_{y \in G_h} e^{-\beta \Delta t^h(x)} p^h(x, y) V^h(y) + k(x) \Delta t^h(x), g(x) \right], \qquad (3.10a)$$

and

$$V^h(x) = g(x), \ x \notin G_h^0. \qquad (3.10b)$$

Henceforth, the remarks are confined to the reflecting boundary case.

The reference [10] discusses various methods for the solution of (3.9). A particularly useful approach uses the approximation in policy space method, with multigrid methods with overrelaxed Gauss-Seidel smoothings used to solve the problems for each of the policies. For many problems the simplest forms of approximation in policy space will also work well.

Note the intuitive nature of the entire procedure. The actual computational solution is the optimal solution for a process which is closely related to the original one, hence it has a physical meaning with respect to the original problem. It is not simply an ad-hoc numerical approximation.

4   Comments on Convergence

4.1   Weak Convergence: Definitions and Criteria

Our method for proving convergence of numerical schemes is based on the theory of weak convergence of probability measures [2, 4]. This theory is a powerful extension of the notion of convergence in distribution for finite dimensional random variables and is widely used for the study of process approximation and limits. For the particular problems dealt with in this paper, the probability measures are measures on the path spaces of the $(x(\cdot), \tau)$ or $(x^h(\cdot), \tau_h)$.

Let S denote a metric space with metric d, let $C_b(S)$ denote the set of bounded continuous real valued functions on S, and let $\mathcal{P}(S)$ denote the set of measures on the Borel sets of S. Suppose that we are given S-valued random variables $X_n, n < \infty$, and X, and suppose without loss of generality that they are defined on the same probability space. Let P_n be the measure which induces X_n. If

$$Ef(X_n) \to Ef(X) \tag{4.1}$$

for all $f \in C_b(S)$, we say that the sequence $\{X_n, n < \infty\}$ *converges in distribution* to X and that P_n converges weakly to P, and this defines the weak topology on $\mathcal{P}(S)$. Abusing terminology, we also say that X_n converges weakly to X. This notion of convergence serves our needs, since our approximations $W^h(x, \tau)$ (resp., $V(x)$) to value functions $W(x, \tau)$ ($V(x)$, resp.) are expectations of functionals of the approximating processes. The theory was used in [7, 9, 10] as the basis of the proofs of the convergence of the numerical methods, and gives the convergence and approximation results under the weakest current conditions. Good general references to the theory of weak convergence are [2, 4, 6]. Many applications are in [6, 8].

Let $\{P_n\} \subset \mathcal{P}(S)$. The collection of probability measures $\{P_n\}$ is called *tight* if for each $\varepsilon > 0$ there exists a compact set $K_\varepsilon \subset S$ such that

$$\inf_n P_n(K_\varepsilon) \geq 1 - \varepsilon. \tag{4.2}$$

If the P_n are the measures which induce random variables X_n, then we will also refer to the collection $\{X_n\}$ as tight. Then (4.2) can be written as

$$\inf_n P\{X_n \in K_\varepsilon\} \geq 1 - \varepsilon. \tag{4.3}$$

The basic theorem is the following.

Prohorov's Theorem [2, 4]. *If S is complete and separable, then a set $\{P_n\} \subset \mathcal{P}(S)$ has compact closure in the metric of weak convergence if and only if $\{P_n\}$ is tight.*

Tightness with the spaces $D^k[0, \infty)$ and $\bar{R}$. The criteria (4.2) or (4.3) guarantee that any subsequence of $\{X_n\}$ will have a weakly convergent subsequence, i.e., that there is a subsequence indexed by n_i and a $P \in \mathcal{P}(S)$ such that $\{P_{n_i}\}$ converges weakly to P. If X is the process induced by P, then $\{X_{n_i}\}$ converges weakly to X and (4.1) holds with n replaced by n_i. But the criteria (4.2) and (4.3) are rather abstract. Generally one looks for simpler criteria in terms of the known statistics of the processes X_n. In our application, things simplify considerably and tightness is usually very easy to prove. Let $D^k[0,\infty)$ denote the space of $\mathbb{R}^k$–valued functions which are right continuous and have left hand limits. The paths of our approximating processes $\xi^h(\cdot)$ are in $D^k[0,\infty)$. The particular topology to be used on $D^k[0,\infty)$ is called the *Skorohod topology*. The definition is somewhat technical and the reader is referred to [2, 4] for the precise definition. Loosely speaking, under that topology, the space is a complete and separable metric space. The topology is weaker than that induced by the metric of uniform convergence on bounded time intervals since it allows for function discontinuities to be 'shifted slightly' when measuring the distance between functions. Also, $f_n(\cdot)$ converging to continuous $f(\cdot)$ in this topology actually converges uniformly on each bounded time interval.

The criteria for tightness described below will be quite simple to apply for our problems, and is due to Aldous and Kurtz [6, Theorem 2.7b][4]. Recall that for a sequence of nondecreasing σ–algebras $\mathcal{F}_t$, the random time τ is an $\mathcal{F}_t$–stopping time if $\{\tau \le t\} \in \mathcal{F}_t$ for all $t \in [0,\infty)$.

Theorem. *Let the processes $\{x^n(\cdot)\}$ defined on the probability space $(\Omega, \mathcal{F}, P)$ take values in $D^k[0,\infty)$. Assume that for each t in a dense set in $[0,\infty)$ and each $\delta > 0$ there exists compact $K_{t,\delta} \subset \mathbb{R}^k$ such that $\sup_n P\{x^n(t) \notin K_{t,\delta}\} \le \delta$. Define $\mathcal{F}_t^n$ to be the σ–algebra generated by $\{x^n(s), s \le t\}$. Let T_T^n be the set of $\mathcal{F}_t^n$–stopping times which are less than or equal to T w.p.1. Assume for each $T \in [0,\infty)$ that*

$$\lim_{\delta \to 0} \limsup_n \sup_{\tau \in T_T^n} E\left(1 \wedge |x^n(\tau + \delta) - x^n(\tau)|\right) = 0. \tag{4.4}$$

Then $\{x^n(\cdot)\}$ is tight.

Let $\bar{R}$ denote the closed infinite interval $[0,\infty]$ with the 'compact' topology. Then any sequence of $\bar{R}$-valued random variables is tight. The space S used for our application to the optimal stopping problem is $S = D^k[0,\infty) \times \bar{R}$, the path space of $(x(\cdot), \tau)$ and of $(\xi^h(\cdot), \tau_h)$, and is a complete and separable metric space. To check tightness, one need only check it for each component separately. Tightness of $\{\tau_h\}$ in $\bar{R}$ is trivial since that space is compact. The tightness of $\{\xi^h(\cdot)\}$ is easy to show using (4.4). Given the tightness, we know that each subsequence has a further subsequence which converges weakly. One need only show that the limit of $V^h(x)$ along the subsequence is $V(x)$. But this will follow form the weak sense uniqueness of the solution to (3.1) and the nature of the optimization problem. A few further remarks appear in the next subsection. The entire procedure is quite straightforward. See [9, 10] for full details.

4.2   Convergence for the Optimal Stopping Problem

Only a few comments concerning the main steps to be taken will be made.
Full details are in [9, 10]. Let $\bar{N}_h$ denote the optimal stopping time for the
chain $\{\xi_n^h\}$ with cost (3.2), and let $\bar{\tau}_h$ be the associated interpolated time $t_{\bar{N}_h}$.
The sequence $\{\bar{\tau}_h\}$ is tight since $\bar{R}$ is compact. The sequence $\{\xi^h(\cdot)\}$ is easily
shown to be tight by a direct use of criterion (4.4) and the local consistency
property (both in G and on the reflecting boundary) and the boundedness of
$b(\cdot)$ and $\sigma(\cdot)$ in G: see [10].

Given tightness, one chooses a weakly convergent subsequence of $\{\xi^h(\cdot), \bar{\tau}_h\}$
indexed by h_n. Let $(x(\cdot), \tau)$ denote the limit of this subsequence (in the sense
of weak convergence). Using the martingale method, one shows that there is
a standard Wiener process $w(\cdot)$ and reflection terms $L(\cdot)$ and $U(\cdot)$ such that
(3.4) holds. Furthermore, to assure the correct nonanticipativity, we need
also show the following. Let $\mathcal{F}_t$ be the minimal $\sigma-$algebra which measures

$$\{x(s), w(s), L(s), U(s), s \leq t; \tau I_{\{\tau \leq t\}}\}.$$

Then $w(\cdot)$ is an $\mathcal{F}_t-$martingale and Wiener process. This assures that τ
is a well defined stopping time for the limit. Note that, by the assumed
weak sense uniqueness of the solution to (3.4), the distribution of the limit
$(x(\cdot), L(\cdot), U(\cdot), w(\cdot))$ does not depend on the subsequence which is chosen.
The distribution of $(x(\cdot), L(\cdot), U(\cdot), w(\cdot), \tau)$ might depend on the subsequence,
but the optimality properties guarantee the convergence of the values to the
correct limit.

By the weak convergence and the continuity and boundedness of $(k(\cdot), g(\cdot))$
(recall that we are maximizing, in [10] there is
a minimization),

$$\begin{aligned} V^h(x) &= W^{h_n}(x, \bar{\tau}_{h_n}) \to E_x \int_0^\tau e^{-\beta t} k(x(s))ds + E_x e^{-\beta\tau} g(x(\tau)) \\ &= W(x, \tau) \leq V(x). \end{aligned} \tag{4.5}$$

The convergence (4.5) and the fact that the last inequality holds for all sub-
sequences yields

$$\limsup_h V^h(x) \leq V(x). \tag{4.6}$$

To get the other inequality

$$\liminf_h V^h(x) \geq V(x), \tag{4.7}$$

a slightly indirect approach is used. Given small $\epsilon > 0$, one chooses an
$\epsilon-$optimal stopping time τ^ϵ for (3.4) with cost (3.2), which can be adapted
for use on $\xi^h(\cdot)$ in the sense that there is a τ_h^ϵ such that $\{\xi^h(\cdot), \tau_h^\epsilon\}$ converges
weakly to $(x(\cdot), \tau^\epsilon)$. This procedure, which can be carried out easily and de-
pends only on the weak sense uniqueness of the solution to (3.4) [10, Chapter
10]) leads to

$$V(x) \leq W(x, \tau^\epsilon) + \epsilon = \lim_h W^h(x, \tau_h^\epsilon) + \epsilon. \tag{4.8}$$

The arbitrariness of ϵ in (4.8) and the inequality $W^h(x, \tau_h^\epsilon) \leq V^h(x)$ give (4.7). Finally, (4.6) and (4.7) yield $V^h(x) \to V(x)$.

We have illustrated the procedure for the special case of a simple optimal stopping problem, but the general procedure is very effective on virtually all of the stochastic control problems which have been of interest to date, including singular control, general reflected processes (possibly with 'corners'), jumps, degeneracies, controlled reflection directions, robust nonlinear filters, controlled variance, impulse control; discounted, finite time, stopping time, target set, ergodic costs. Note that the entire procedure is probabilistic, and no analytic properties of the solution to the Bellman equation are used. The procedure has much physical insight and intuition – facilitating proofs as well as algorithm construction.

5   The Markov Chain Approximation: A Simple Example

The book [10, Chapter 5] discusses a variety of straightforward ways of getting appropriate approximating processes. For illustrative purposes, one which is based on a finite difference approximation will be briefly reviewed here in its simplest one dimensional form. When a carefully chosen finite difference approximation is applied to the differential operator of the (unreflected) system process, the coefficients of the resulting discrete equation satisfy the local consistency requirements. With appropriate choices of the finite difference approximations, the method works in any dimension. The validity of this approach does not depend on the validity of the finite difference approach to the solution of the PDE. The finite difference approximations are used as guides to the construction of locally consistent approximating Markov chains only. Once these are available, we use the purely probabilistic methods of weak convergence theory for dealing with them. The transition probabilities obtained with this finite difference method can be altered in many ways, according to numerical convenience, while keeping the local consistency.

Let the system be the one dimensional model

$$dx = b(x)dt + \sigma(x)dw, \qquad (5.1)$$

and for an arbitrary function $k(\cdot)$ (which serves only as a 'place holder') consider the formal expression

$$\mathcal{L}f(x) + k(x) = \frac{1}{2}\sigma^2(x)f_{xx}(x) + b(x)f_x(x) + k(x) = 0, \qquad (5.2)$$

where $\mathcal{L}$ is the differential generator of (5.1). For simplicity in the illustration suppose that $\inf_x[\sigma^2(x) + |b(x)|] > 0$, although this is not necessary. If the expression is zero at some point x_0, then x_0 will be absorbing and the interpolation intervals $\Delta t^h(x)$ will go to infinity as $x \to x_0$. Suppose that for small $h > 0$, $\sigma^2(x) > 2h|b(x)|$. If this condition fails at some point, then a one sided (or 'upwind') difference approximation can be successfully used in lieu of (5.4) below. Use the standard finite difference approximation

$$f_{xx}(x) \to \frac{f(x+h) + f(x-h) - 2f(x)}{h^2}, \tag{5.3}$$

and use the symmetric difference approximation for the first derivative

$$f_x(x) \to \frac{f(x+h) - f(x-h)}{2h}. \tag{5.4}$$

In this one dimensional problem, let G_h be the h-grid on the line.

Now, substituting (5.3) and (5.4) into (5.2), denoting the approximation by $f^h(x)$, collecting terms and dividing by the coefficient of $f^h(x)$ yields the equation

$$f^h(x)$$
$$= \frac{\sigma^2(x) + hb(x)}{2\sigma^2(x)} f^h(x+h) + \frac{\sigma^2(x) - hb(x)}{2\sigma^2(x)} f^h(x-h) + k(x)\frac{h^2}{\sigma^2(x)} \tag{5.5}$$
$$\equiv p^h(x, x+h)f^h(x+h) + p^h(x, x-h)f^h(x-h) + k(x)\Delta t^h(x).$$

The $p^h(x, x\pm h)$ are non negative, and if we define $p^h(x, y) = 0$ for $y \neq x \pm h$, then $\sum_y p^h(x, y) = 1$. Thus the $p^h(x, y)$ are transition probabilities for a Markov chain. The $p^h(x, y)$ and the interpolation interval $\Delta t^h(x)$ can easily be shown to be locally consistent in the sense of (3.5). Note that the interpolation interval $\Delta t^h(x)$ is given automatically by the construction of the transition probabilities.

6   Time Dependent Problems

6.1   Introduction

In the optimal stopping problem of Section 3, the current time did not enter into either the formulation or the analysis. The functions $b(\cdot), \sigma(\cdot), k(\cdot), g(\cdot)$ did not depend on time, and there was no finite upper time limit T. With either (3.1) or (3.4), if $\tau > t_1$ for any $t_1 < \infty$, then the entire problem begins anew with new initial condition $x(t_1)$, but t_1 plays no other role. The interpolation intervals $\Delta t^h(x)$, which were allowed to depend on x, were not given a priori as a discretization of time, but appeared as a natural consequence of the local consistency condition. When the interpolation intervals depend on the state, it is hard to account for the passage of real time in the numerical algorithms. Given any chain which is locally consistent and with an x-dependent interpolation interval, it is always possible to modify it to get a constant interval. To do this, define $\Delta t^h = \min_x \Delta t^h(x)$, increase the probability of a transition from each x to itself (with appropriate renormalization so that the sum is unity) until the interval is Δt^h. This procedure, essentially equivalent to the 'explicit' method discussed below, leads to a less efficient numerical algorithm than the 'implicit' method also discussed below. In the example of the sample construction of Section 5, if $\sigma(\cdot)$ were a constant, then the interpolation interval would not depend on x, and the chain could be used even when there is a finite upper time limit. The transition probabilities for the approximating chain are easiest to construct in the time independent case, and it will be seen in Subsection 6.3 that the probabilities for the implicit case are trivially obtained from them.

We now return to the stopping problem of Example 1 but where the process model is the general form (3.1), with the 'numerical' version (3.4), but where $\tau \leq T$, an a priori given finite upper time limit, and make the necessary modifications. We continue to suppose that $b(\cdot), \sigma(\cdot), k(\cdot), g(\cdot)$ do not depend explicitly on time. (The minor modifications required for the time dependent function case should be obvious.)

Suppose that for $t_1 < T$ we have a stopping time $T \geq \tau > t_1$. Then the problem begins anew at t_1 with initial condition $x(t_1)$, but it is over the new time interval $[0, T - t_1]$; thus, the residual time needs to be accounted for. In order to get an algorithmically convenient approximating Markov chain for this problem and to account for the flow of time in an easy way, both time and space need to be discretized explicitly. There are two ways of doing this, called the *explicit* and the *implicit* corresponding (loosely speaking) to the explicit and implicit finite difference methods of discretizing parabolic equations. As in Sections 3 and 4, $x(t)$ is confined to the box G and the process is reflected orthogonally inward when it tries to leave G. As previously, these assumptions are used only to simplify the discussion. The implicit method is generally preferable in that for similar computational times it gives superior results.

6.2 The Explicit Method of Getting an Approximating Chain

Discretize time and space separately, with spatial interval h and temporal interval δ, and let T be an integral multiple of δ. Again, the spatial discretization level can be a vector. Denote the approximating chain by $\{\xi_n^{h,\delta}\}$ and the transition probabilities by $p^{h,\delta}(x, y)$. Let $E_{x,n}^{h,\delta}$ denote the expectation conditioned on the event $\xi_n^{h,\delta} = x$. The local consistency property (3.5) for $x \in G$ becomes

$$E_{x,n}^{h,\delta}[\xi_{n+1}^{h,\delta} - x] = b(x)\delta + o(\delta), \tag{6.1a}$$

$$E_{x,n}^{h,\delta}[\xi_{n+1}^{h,\delta} - x][\xi_{n+1}^{h,\delta} - x]' = \sigma(x)\sigma'(x)\delta + o(\delta), \tag{6.1b}$$

a minimal condition. Note that the interpolation interval is just δ. Define the continuous time piecewise constant interpolation $\xi^{h,\delta}(\cdot)$ by $\xi^{h,\delta}(t) = \xi_n^{h,\delta}$ on the interval $[n\delta, n\delta + \delta)$. Thus, the time for the interpolated process increases by the constant δ at each step of the discrete time process. Given $\delta \leq \min \Delta t^h(x)$, transition probabilities for a chain which is locally consistent in the sense of (6.1) can easily be obtained from the transition probabilities for any chain which is locally consistent in the sense of (3.5), as noted in Section 6.1. For non degenerate problems, we would have $\delta = O(h^2)$, which can be small. Indeed, when we use a finite difference method on the differential generator to get the transition probabilities, the time interval δ would be same as that required for stability of the discretized parabolic PDE. The interval for the implicit method is larger.

Let N denote a stopping time for the chain where $n\delta \leq \delta N \leq T$. Define the costs, starting at iterate n with $\xi_n^{h,\delta} = x$, to be

$$W^{h,\delta}(x, n\delta, N\delta) = E_{x,n}^{h,\delta} \sum_{i=n}^{N-1} e^{-\beta(i-n)\delta} k(\xi_i^{h,\delta})\delta + E_{x,n}^{h,\delta} e^{-\beta(N-n)\delta} k(\xi^{h,\delta}N), \quad (6.2)$$

$$V^{h,\delta}(x, n\delta) = \inf_N W^{h,\delta}(x, n\delta, N\delta).$$

For x not on the reflecting boundary and $n\delta < T$, the dynamic programming equation is

$$V^{h,\delta}(x, n\delta) = \max \left\{ e^{-\beta\delta} \sum_y p^{h,\delta}(x,y) V^{h,\delta}(y, n\delta + \delta) + k(x)\delta, \ g(x) \right\}, \quad (6.3)$$

with $V^{h,\delta}(x, T) = g(x)$.

6.3   The Implicit Method of Getting an Approximating Chain

The fundamental difference between the so-called explicit and implicit approaches to the Markov chain approximation lies in the fact that in the former the time variable is treated differently than the state variables: It is a true 'time' variable, and its value increases by a constant δ at each step. In the implicit approach, the time variable is treated as just another state variable. It is discretized in the same manner as are the other state variables: the approximating Markov chain has a state space which is a discretization of the (x, t)−space, and the component of the state of the chain which comes from the original time variable does not necessarily increase its value at each step. The implicit method can perhaps be best motivated by use of a finite difference method for the construction of the transition probabilities. Use the one dimensional model (5.1) and the assumptions below (5.2). The method works without these assumptions, with suitable choices of the finite difference approximations. Also, keep in mind that we are illustrating only one way of getting the transition probabilities, and that the proofs are still purely probabilistic and do not depend on the validity of the finite difference approximation to a PDE.

Now, the equation to be discretized is the formal parabolic PDE

$$f_t(x,t) + \frac{1}{2}\sigma^2(x)f_{xx}(x,t) + b(x)f_x(x,t) + k(x) = 0. \tag{6.4}$$

We will use the standard 'implicit' approximations

$$f_{xx}(x,t) \to \frac{f(x+h,t) + f(x-h,t) - 2f(x,t)}{h^2} \tag{6.5a}$$

$$f_x(x,t) \to \frac{f(x+h,t) - f(x-h,t)}{2h}, \tag{6.5b}$$

$$f_t(x,t) \to \frac{f(x,t+\delta) - f(x,t)}{\delta}. \tag{6.5c}$$

Note that $t + \delta$ appears only in (6.5c), and x and t are never incremented simultaneously in any of the expressions in (6.5). This is crucial to the construction. Repeating the procedure which led to (5.5) and letting $f^{h,\delta}(x)$ denote the finite difference approximation yields

$$\left[1 + \sigma^2(x)\frac{\delta}{h^2}\right]f^{h,\delta}(x,n\delta) = \frac{1}{2}\left[\sigma^2(x)\frac{\delta}{h^2} + b(x)\frac{\delta}{h}\right]f^{h,\delta}(x+h,n\delta)$$

$$+ \frac{1}{2}\left[\sigma^2(x)\frac{\delta}{h^2} - b(x)\frac{\delta}{h}\right]f^{h,\delta}(x-h,n\delta)$$

$$+ f^{h,\delta}(x,n\delta+\delta) + k(x)\delta.$$

With the obvious definitions of $\hat{p}^{h,\delta}(\cdot)$ and $\Delta\hat{t}^{h,\delta}(\cdot)$, rewrite the above expression as

$$f^{h,\delta}(x,n\delta) = \sum_{y\neq x}\hat{p}^{h,\delta}(x,n\delta;y,n\delta)f^{h,\delta}(y,n\delta)$$
$$+ \hat{p}^{h,\delta}(x,n\delta;x,n\delta+\delta)f^{h,\delta}(x,n\delta+\delta) + k(x)\Delta\hat{t}^{h,\delta}(x).$$

Let the $\hat{p}^{h,\delta}(x,n\delta;y,n\delta)$ which are not defined above be zero, and let $\hat{p}^{h,\delta}(x,n\delta;y,n\delta+\delta) = 0$ for $y \neq x$. Then the $\hat{p}^{h,\delta}(\alpha;\beta)$ are non-negative and

$$\sum_{y\neq x}\hat{p}^{h,\delta}(x,n\delta;y,n\delta) + \hat{p}^{h,\delta}(x,n\delta;x,n\delta+\delta) = 1.$$

Thus, we can consider the $\hat{p}^{h,\delta}(\alpha;\beta)$ as the one step transition probabilities of a Markov chain $\{\hat{\zeta}_n^{h,\delta}, n < \infty\}$ on the discrete '(x,t)-state space'

$$\{0, \pm h, \pm 2h, \ldots\} \times \{0, \delta, 2\delta, \ldots\}.$$

It is evident that time is being considered as just another state variable, and that the time component does not increase at all steps.

For additional insight into the process, write $\hat{\zeta}_n^{h,\delta} = (\zeta_n^{h,\delta}, \zeta_{n,0}^{h,\delta})$, where the 0th component $\zeta_{n,0}^{h,\delta}$ represents the time variable, and $\zeta_n^{h,\delta}$ represents the original 'spatial' state. Then, for the above construction we have

$$E_{x,n}^{h,\delta} \Delta \zeta_n^{h,\delta} = b(x) \Delta \hat{t}^{h,\delta}(x)$$

$$\text{cov}_{x,n}^{h,\delta} \Delta \zeta_n^{h,\delta} = \sigma^2(x) \Delta \hat{t}^{h,\delta}(x) + \Delta \hat{t}^{h,\delta}(x) O(h), \qquad (6.6)$$

$$E_{x,n}^{h,\delta} \Delta \zeta_{n,0}^{h,\delta} = \Delta \hat{t}^{h,\delta}(x) = \delta \hat{p}^{h,\delta}(x, n\delta; x, n\delta + \delta).$$

Thus, with interpolation intervals $\Delta \hat{t}^{h,\delta}(x)$ the 'spatial' component of the chain is locally consistent in the sense of (3.5). The conditional mean increment of the 'time' component of the state is just the interpolation interval $\Delta \hat{t}^{h,\delta}(x)$. Defining $x_0(t) = t$, $\{\hat{\zeta}_n^{h,\delta}\}$ is also locally consistent with the process $(x(\cdot), x_0(\cdot))$ in the sense of (3.5). Local consistency for the implicit chain is defined by (6.6), modulo $o(\Delta \hat{t}^{h,\delta}(x))$ on the right side.

We have constructed an approximating Markov chain via an 'implicit' method. It is called an implicit method because (6.12) or (6.13) below cannot be solved by a simple backward iteration, since time does not advance for the continuous time interpolation at each step of the discrete time process. At each n, (6.12) determines the values of the $\{V^{h,\delta}(x, n\delta)\}$ implicitly.

A General Construction for the Implicit Case. Locally consistent $\hat{p}^{h,\delta}(\cdot)$ can be easily obtained from locally consistent $p^h(\cdot)$, as will now be seen. First note that for our simple illustrative example we have

$$\hat{p}^{h,\delta}(x, n\delta; y, n\delta) = p^h(x, y) \times \text{normalization}(x), \ x \neq y, \qquad (6.7)$$

where the $p^h(x, y)$ are the transition probabilities constructed in Section 5. The relationship (6.7) can be used to get an implicit Markov chain approximation for any model, starting with any locally consistent approximation in the sense of (3.5) for that model. We now describe the general method.

Let $p^h(x, y)$ and $\Delta t^h(x)$ be a transition function and interpolation interval which are locally consistent in the sense of (3.5) for (3.1). Let the time discretization level $\delta > 0$ be given. Let $\hat{p}^{h,\delta}(\cdot)$ and $\Delta \hat{t}^{h,\delta}(\cdot)$ denote the (to be defined) transition probability and interpolation interval for the implicit approximation. Equation (6.7) implies that for $x \neq y$

$$p^h(x, y) = \frac{\hat{p}^{h,\delta}(x, n\delta; y, n\delta)}{1 - \hat{p}^{h,\delta}(x, n\delta; x, n\delta + \delta)}. \qquad (6.8)$$

Thus, to get the transition probabilities $\hat{p}^{h,\delta}(\cdot)$, we need only get

$$\hat{p}^{h,\delta}(x, n\delta; x, n\delta + n\delta).$$

This will be done via the local consistency requirements on $(p^h(x,y), \Delta t^h(x))$ and the corresponding quantities for the implicit approximation.

The conditional mean one step increment of the 'time' component $\zeta_{n,0}^{h,\delta}$ of $\hat{\zeta}_n^{h,\delta}$ is

$$E_{x,n}^{h,\delta} \Delta \zeta_{n,0}^{h,\delta} = \hat{p}^{h,\delta}(x, n\delta; x, n\delta + \delta)\delta, \tag{6.9}$$

and we define the interpolation interval $\Delta \hat{t}^{h,\delta}(x)$ by the right side of (6.9). Of course, we can always add a term of smaller order. Equating the consistency equation for the spatial component $(\zeta_n^{h,\delta})$ of the chain $\hat{\zeta}_n^{h,\delta}$ with that for the chain ξ_n^h and using (6.8) yields (modulo a negligible error term)

$$
\begin{aligned}
b(x)\Delta t^h(x) &= \sum_y (y - x)p^h(x,y) \\
&= \sum_y (y - x)\frac{\hat{p}^{h,\delta}(x, n\delta; y, n\delta)}{1 - \hat{p}^{h,\delta}(x, n\delta; x, n\delta + \delta)} \\
&= b(x)\frac{\Delta \hat{t}^{h,\delta}(x)}{1 - \hat{p}^{h,\delta}(x, n\delta; x, n\delta + \delta)}.
\end{aligned}
\tag{6.10}
$$

By equating the two expressions for $\Delta \hat{t}^{h,\delta}(x)$ given by (6.9) and (6.10), we have

$$\Delta \hat{t}^{h,\delta}(x) = (1 - \hat{p}^{h,\delta}(x, n\delta; x, n\delta + \delta))\Delta t^h(x) = \hat{p}^{h,\delta}(x, n\delta; x, n\delta + \delta)\delta.$$

Thus,

$$
\begin{aligned}
\hat{p}^{h,\delta}(x, n\delta; x, n\delta + \delta) &= \frac{\Delta t^h(x)}{\Delta t^h(x) + \delta}, \\
\Delta \hat{t}^{h,\delta}(x) &= \frac{\delta \Delta t^h(x)}{\Delta t^h(x) + \delta}.
\end{aligned}
\tag{6.11}
$$

The dynamic programming equation for x not a reflecting boundary state and $n\delta < T$ is

$$
\begin{aligned}
V^{h,\delta}(x, n\delta) = \max \Big\{ &\sum_{y \neq x} \hat{p}^{h,\delta}(x, n\delta; y, n\delta)V^{h,\delta}(y, n\delta) \\
&+ e^{-\beta\delta}\hat{p}^{h,\delta}(x, n\delta; x, n\delta + \delta)V^{h,\delta}(x, n\delta + \delta) + k(x)\Delta \hat{t}^{h,\delta}(x), g(x) \Big\},
\end{aligned}
\tag{6.12}
$$

with $V^{h,\delta}(x, T) = g(x)$.

Approximating the process $\{x(n\delta), n = 0, \ldots\}$: An Alternative chain.
Approximating the transition probabilities of the sampled process $\{x(n\delta), n = 0, \ldots\}$ is of interest. This can be obtained simply by looking at the chain $\zeta_n^{h,\delta}$ at the times that $\zeta_{n,0}^{h,\delta}$ changes. Define $v_0 = 0$, and, for $n > 0$ define

$$v_n = \min\{i > \nu_{n-1} : \zeta_{i,0}^{h,\delta} - \zeta_{i-1,0}^{h,\delta} = \delta\}.$$

Then define the new chain $\tilde{\zeta}_n^{h,\delta} = \zeta_{v_n}^{h,\delta}$. The new chain approximates $\{x(n\delta), n = 0,\ldots\}$. Let $\tilde{p}^{h,\delta}(x,y)$ denote the one step transition probabilities for $\{\tilde{\zeta}_n^{h,\delta}\}$; that is, $\tilde{p}^{h,\delta}(x,y)$ is the probability that the spatial state will be y at the next time that the time component of $\hat{\zeta}_n^{h,\delta}$ changes, given that the spatial state is now x. These probabilities can be calculated as follows. For any function $q(\cdot)$, define $U(\cdot)$ by

$$U(x) = \sum_y \hat{p}^{h,\delta}(x, n\delta; y, n\delta)U(y) + \hat{p}^{h,\delta}(x, n\delta; x, n\delta + \delta)q(x). \qquad (6.13)$$

Now, the values of the $U(x)$ are linear functions of the $\{q(y), y \in S_h\}$. The coefficients in these linear relationships are the desired transition probabilities:

$$U(x) = \sum_y \tilde{p}^{h,\delta}(x, y)q(y). \qquad (6.14)$$

The methods of [10, Chapter 6] can be used to get the $\tilde{p}^{h,\delta}(x,y)$, and solutions of equations such as (6.13). Equation (6.13) gives the value for a stopping problem, where one stops with stopping cost $q(x)$ whenever any state x communicates with itself (or, equivalently, the time component of the state variable advances.)

References

[1] M. Akian, J.-L. Menaldi, and A. Sulem 'On an investment-consumption model with transactions costs'. Research report No. 1926, May 1993, INRIA, France, 1993.

[2] P. Billingsley. *Convergence of Probability Measures*. Wiley, 1968.

[3] M.H.A. Davis and A.R. Norman. 'Portfolio selection with transactions costs'. *Math. of Operations Research* **15**, 676–713, 1990.

[4] S.N. Ethier and T.G. Kurtz. *Markov Processes: Characterization and Convergence*. Wiley, 1986.

[5] D.G.Hobson and L.G.C. Rogers. Complete models with stochastic volatility. Preprint, School of Mathematical Sciences, University of Bath, UK, 1995.

[6] T.G. Kurtz. *Approximation of Population Processes*, volume 36 of *CBMS-NSF Regional Conf. Series in Appl. Math.* SIAM, 1981.

[7] H.J. Kushner. *Probability Methods for Approximations in Stochastic Control and for Elliptic Equations*. Academic Press, 1977.

[8] H.J. Kushner. *Approximation and Weak Convergence Methods for Random Processes with Applications to Stochastic System Theory.* MIT Press, 1984.

[9] H.J. Kushner. 'Numerical methods for stochastic control problems in continuous time'. *SIAM J. Control and Optimization* **28**, 999–1048, 1990.

[10] H.J. Kushner and P. Dupuis. *Numerical Methods for Stochastic Control Problems in Continuous Time.* Springer, 1992.

[11] S.E. Shreve and H.M. Soner. Optimal investment and consumption with transactions costs. *Ann. Appl. Probability* **4**, 609–692, 1994.

Simulation Methods for Option Pricing

John P. Lehoczky

Abstract

This chapter is a summary of a one-hour talk designed to provide a brief survey of some of the major ideas and methods associated with Monte Carlo simulation as a method for computing option prices. The methods described here are very simple and accessible to the beginning student, consequently it is a bit embarrassing to present such mathematically unsophisticated work, in the context of so many mathematically sophisticated papers. Furthermore, this paper does not present any new ideas, rather it merely summarizes some existing work. Still, the practicality of the methods is undeniable, and perhaps some reader will find a new idea or point of view which will make reading this article worthwhile.

1 Introduction

There has been an explosion of research on the use of simulation methods for pricing options. The references are now voluminous dating from the initial paper by Boyle [1], but no attempt is made to provide a comprehensive set of references. The topic is not yet treated in standard simulation texts used in operations research, statistics or computer science, although the book by Kloeden, Platen and Schurz [12] on the numerical solution of stochastic differential equations does present some of this material.

Much of option pricing theory seeks to compute an expected value of a random variable under a risk neutral probability measure. For example, in the case of a European option with strike price K, expiring at time T, on an asset whose price follows the stochastic process $\{S_t, 0 \leq t\}$, one wishes to compute $e^{-rT}E((S_T - K)^+)$ where the expectation is computed under the risk neutral measure. This expectation is merely the integral of a certain function with respect to a well-defined probability distribution; however, it is possible that the dynamics of $\{S_t, 0 \leq t\}$ are sufficiently complex (for example with a stochastic volatility model) to preclude the calculation of a closed-form expression of the p.d.f. of S_T. It is also possible that the integral itself may be difficult to compute analytically, although in such a one-dimensional case, numerical integration methods would likely be the most efficient choice.

When direct calculation of an expectation is not feasible, it may still be possible to generate an arbitrary number of observations of the random quantity. For example, when the asset price is generated by a stochastic volatility model, one can generate a sequence of paths having appropriate finite dimensional distributions. Observing these paths at time T, one has an arbitrarily large sample $\{S_{T_i}, i = 1, 2, \ldots\}$, and if these observations can be considered to be i.i.d., then $\frac{1}{N} \sum_{i=1}^{N} g(S_{T_i}) \to E(g(S_T))$ a.s., as $N \to \infty$. Not only does the Strong Law of Large Numbers guarantee the convergence of these averages to the desired value, the Central Limit Theorem also provides error estimates or asymptotic confidence intervals. That is, suppose that we define $Y_i = g(S_{T_i})$ and estimate $E(g(S_T))$ by $\frac{1}{N} \sum_{i=1}^{N} Y_i$. Defining $s_{\bar{Y}} = \sqrt{\left(\sum_{i=1}^{i=n} (Y_i - \bar{Y})^2 \right) / (n-1)}$, then for large n, $\left(\bar{Y} - z_{\alpha/2} \left(s_{\bar{Y}} / \sqrt{N} \right), \bar{Y} + z_{\alpha/2} \left(s_{\bar{Y}} / \sqrt{N} \right) \right)$ is a $100(1 - \alpha)\%$ confidence interval for $E(g(S_T))$, where $\bar{Y}$ is the sample mean and $z_{\alpha/2}$ is the $1 - \alpha/2$ quantile of the standard normal distribution. The sample size N can be increased until the desired accuracy, as represented by the length of a selected confidence interval, is sufficiently small.

The rate of convergence of $N^{-\frac{1}{2}}$ is based upon the assumption that the $\{Y_i\}$ sequence can be treated as if it consisted of i.i.d. random variables. This assumption is approximately correct if one generates the random variables using pseudo-random numbers. There is an alternative which can create a faster rate of convergence, nearly N^{-1}, if one uses quasi-random numbers. This topic is briefly discussed later in the chapter.

The convergence rate is somewhat more complicated when one considers path dependent options, for example, Asian options. Here, the value of a European-style path dependent option requires knowledge of the entire asset price path, $\{S_t, 0 \le t \le T\}$. For example, the value might depend upon an asset average such as $\frac{1}{T} \int_0^T S_t dt$. If one uses a discrete time model for the asset price process, then rather than generating a single random variable, S_T, and evaluating the option value, one generates a trajectory, $S_0, S_{t_1}, \ldots, S_T$ and then generates the value of the option. If, instead, time is treated as continuous, then one must generate a discrete time approximation to the continuous path. The accuracy of the discrete time approximation can be controlled by altering the order of the approximation used and the mesh size. Thus, there are two distinct aspects to the accuracy and the rate of convergence of the simulation estimator. First, accuracy will increase with the number of replications of each path. Second, accuracy will vary with the order of the discrete approximation and the mesh size. Clearly, both can help; however, it is important to understand the contribution of each to the overall accuracy. In a paper by Duffie and Glynn [6] this problem is studied in detail, and the authors determine the optimal choice of grid size, the order of the discrete approximation and the number of replications for a given fixed computation budget. While we do not discuss this topic any further, the evaluation of path dependent options will invariably result in these computational tradeoffs, and the work of Duffie and Glynn is an important contribution to understanding and optimizing these tradeoffs.

2   Variance Reduction Methods

As noted earlier, Monte Carlo methods have a rate of convergence of $N^{-\frac{1}{2}}$; however, it is possible to improve the accuracy through a number of variance reduction methods. Some basic methods are discussed next.

2.1   Antithetic Variables

One standard variance reduction method is the use of *antithetic variables*, pairs of random variables which have the same marginal probability distribution but are negatively correlated. The average of N pairs of antithetic variables has a variance which is smaller than the average of $2N$ independent variables. This is easy to show and is given below. Let us first cite three simple examples of antithetic variables:

1. Suppose X is a random variable with Uniform$[a, b]$ distribution. Define $X^A = a + b - X$. Now, X^A also has a Uniform$[a, b]$ distribution, and the correlation of X and X^A, $\rho(X, X^A)$, is -1, since these two variables are linearly related. Usually one focuses on the Uniform$[0,1]$ case.

2. Suppose X is a random variable with a normal distribution with mean μ and standard deviation σ. Define $X^A = 2\mu - X$. Now, X^A and X have the same distribution and their correlation is also 1, again because they are linearly related.

3. Suppose X is generated by the probability integral transform, that is, $X = F^{-1}(U)$ where F is the c.d.f. of X and U has a Uniform$[0,1]$ distribution. Define $X^A = F^{-1}(1 - U)$. Now, X^A and X have the same distribution and their correlation is negative although it is not -1 unless F is itself a uniform distribution.

Now, if we wish to estimate $E(h(X))$ where X has distribution F, then by generating $\{X_i, 1 \le i \le N\}$ and $\{X_i^A, 1 \le i \le N\}$, we can construct $\bar{h} = \frac{1}{N}\sum_{i=1}^{N} h(X_i)$ and $\bar{h}^A = \frac{1}{N}\sum_{i=1}^{N} h(X_i^A)$, both method of moments estimators of $E(h(X))$. The antithetic version of this estimator is $\frac{\bar{h} + \bar{h}^A}{2}$. The variance of this estimator is easily computed to be $(\mathrm{Var}(h(X))/2N)\,(1 + \rho)$, where $\rho = \rho(h(X_1), h(X_1^A))$.

We can make the following observations about the use of antithetic variables:

1. If we ignore the extra steps required to compute X_i^A, then no matter what ρ may be, we always end up with a result having variance no greater than the variance of the method of moments estimator with sample size N. Of course, the computation required to evaluate X_i^A from X_i is at least comparable to generating another pseudo-independent X_i.

2. The effective sample size is $2N/(1+\rho)$. Since ρ is negative, this can lead to a substantial increase in the sample size over N.

3. While the use of antithetic variables guarantees a reduction in the variance, these reductions are rather modest, partly because the result is general and is not tailored to the problem itself.

4. Antithetic variables are of limited usefulness in multivariate problems such as path dependent options. This is because the option value is a function of multivariate asset price sequence, and it is difficult to construct "antithetic paths."

Thus, antithetic variables are useful, but often do not apply. Even when they do apply, the overall variance reduction may be small.

2.2 Control Variables

Again, let us suppose that we want to estimate a certain expected value, either $E(h(X))$ or $E(h(X_1, \ldots, X_T))$. This can be done in a straightforward fashion by using the method of moments estimators described earlier. It can also be the case, however, that the same data could be used to estimate a different, but related expectation, one which is *known*. For example, if one were interested in computing the price of an Asian option based on the arithmetic average of the asset price when the underlying asset price was modelled by a standard geometric Brownian motion, then the resulting price could not be computed in closed form. However, the price for an Asian option based on a geometric average could be computed analytically. Since the geometric average Asian option price can be considered to be known, it can be used as a control variable, an idea exploited by Kemna and Vorst [11] and by Fu, Madan and Wang [7]. Generally, if one can compute the price of a related option, then the related option can be used as a control variable.

Suppose that we want to estimate a certain expected value, $E(h(X))$ or $E(h(X_1, \ldots, X_T))$. As mentioned earlier, the standard method of moments approach is to generate an approximate i.i.d. sequence of observations, $\{X_i, 1 \le i \le N\}$ or $\{X_{ij}, 1 \le i \le T, 1 \le j \le N\}$. These observations can be used to compute the desired expected value; however, suppose that these observations could also be used to compute another expectation whose value is already known. For example, suppose in the univariate case that we knew $E(g(X))$ and that $\rho(g(X), h(X)) \ne 0$. Then the errors in the method of moments estimators of $E(g(X))$ and $E(h(X))$ are correlated, and the error in estimating $E(g(X))$ is *known*, since $E(g(X))$ is known. Standard linear estimation or regression methods can be used to reduce the variance in the error from estimating $E(h(X))$. More formally, let $\bar{h} = \frac{1}{N}\sum_{i=1}^{N} h(X_i)$ and $\bar{g} = \frac{1}{N}\sum_{i=1}^{N} g(X_i)$. Define $\sigma_h^2 = \text{Var}(h(X))$ and $\sigma_g^2 = \text{Var}(g(X))$. Consider the modified estimator $\bar{h}_\alpha = \bar{h} + \alpha(\bar{g} - E(g(X)))$. Note that $E(\bar{h}_\alpha) = E(h(X))$, so the modified estimator is still unbiased for $E(h(X))$ for any value of α, because $\bar{g}$ is unbiased for $E(g(X))$. However, $\text{Var}(\bar{h}_\alpha) = \frac{1}{N}(\sigma_h^2 + \alpha^2\sigma_g^2 + 2\alpha\sigma_h\sigma_g\rho(h(X), g(X)))$. If the variances and correlation were known, then the variance minimizing choice of α would be $\hat{\alpha} = -\frac{\sigma_h}{\sigma_g}\rho(h(X), g(X))$. Substituting this choice into the variance expression for $\bar{h}_\alpha$, we find $\min_\alpha \sigma_{\bar{h}_\alpha} = \sigma_h^2(1-\rho^2)/N$. If the squared correlation coefficient is near 1, then the variance of the control variable adjusted estimator can be dramatically reduced.

It is important to realize that in any given problem there can be more than one control variable. An illustrative example of this is given by Clewlow and Carverhill [4]. In their paper the authors illustrate the use hedging portfolios to create a set of control variables. They refer to these as *martingale control variables*. Since this technique involves the hedging portfolios which, in turn, involve the derivatives of the prices of derivative securities, we defer this discussion until the last section of this article. However, it is interesting to note that Clewlow and Carverhill do construct a set of control variables that can be used for variance reduction, and the use of one or more control variables does raise the issue of how to properly weight those variables in adjusting the standard method of moments estimator for the option price.

It must be pointed out that optimal adjustment described above involves parameters σ_h, σ_h and $\rho(\bar{h}, \bar{g})$, which are generally not known. Thus, the optimal value for α is itself unknown. Nevertheless, one can estimate α, although some care is required. The obvious solution is to estimate α using the random variables that are generated to estimate $E(h(X))$. To better understand the issues, let us revisit the formulation of the control variable problem and understand it in the context of regression analysis. Suppose we let $V = h(X), v = E(V), U = g(X)$ and $u = E(U)$. The goal is to estimate v, and u is known. Here U is the control variable, and for each $X_i, U_i - u = g(X_i) - E(g(X))$ is observable. We write

$$V_i = v + \alpha(U_i - u) + \epsilon_i, \quad 1 \le i \le N.$$

Now the $\{\epsilon_i\}$ sequence can be considered to be i.i.d., with $E(\epsilon_i) = 0$. Thus it would appear that standard methods from regression analysis could be used to estimate v. However, there are two important complications: (1) the regressor variable (U) is a random variable, not the constant required in standard regression analysis and (2) the pairs (U_i, V_i) are dependent, equivalently, the ϵ_i have a dependency relationship with both U_i and V_i. Thus, in the context of regression analysis, the problem is known as the "errors in variables" problem, where the independent variables are correlated with the error term. Standard least squares estimators are no longer appropriate in this context. This problem is made more difficult when one considers a set of control variables. When there are K independent variables, the model is expanded to

$$V_i = v + \sum_{k=1}^{K} \alpha_k (U_{ik} - u_k) + \epsilon_i, \quad 1 \leq i \leq N,$$

where u_k is the known expected value of the kth control variable. Here again, the independent variables (the K control variables) are random rather than fixed and the error term is correlated with those variables. We do not describe the potential solutions to this problem, but the interested reader is encouraged to consult standard texts in econometrics where the errors-in-variables problem is widely studied. In the context of simulation, the difficulty is reduced, because it is possible to have large sample sizes, thus asymptotic normal distributions can be assumed (i.e. generally for large N it is reasonable to assume that the error term has a normal distribution, although it will be correlated with the independent variables). Consequently the problem can be treated as a special sort of multivariate analysis problem assuming normally distributed vectors where the goal is to estimate one component of the mean vector, while the other components can be assumed known, but the covariance matrix is not known.

2.3   Importance Sampling and Stratified Sampling

In option pricing a frequent goal is to compute $\theta = E(h(X))$ where X may be either univariate or multivariate (in the case of a path dependent option). In this section, we discuss the simplest case in which X is a random variable with corresponding probability measure μ having p.d.f. $f(x)$. The standard Monte Carlo approach to computing θ is to generate a sequence of i.i.d. random variables $X_1, \ldots X_N$ having p.d.f. $f(x)$ and then estimate θ by $\hat{\theta} = \frac{1}{N} \sum_{i=1}^{N} h(X_i)$. This method of moments estimator is unbiased for θ. Its rate of convergence is $O(N^{-\frac{1}{2}})$, but the constant depends upon $\text{Var}(h(X))$. Now there are important situations in which $\text{Var}(h(X))$ is relatively large, but its size is caused by a special structure of μ. This special case occurs when $h(X)$ has little variability on a set of large μ measure (denoted by A) and very large variability on a set of small μ measure (A^c). Since $h(X)$ has little variability on A, we can approximate $\int_A h(x) f(x) dx$ accurately with a very small sample size. On the other hand, accurate estimation of $\int_{A^c} h(x) f(x) dx$ requires a much larger sample size. Unfortunately, if we sample X randomly from the μ measure, then most of these observations will lie in A (where they are needed least), while relatively few will lie in A^c (where they are needed most). This imbalance can be remedied by changing the measure from which the X_i are selected and adjusting the function we are computing. This approach is known as *importance sampling*.

Suppose we introduce a new probability measure ν which is absolutely continuous with respect to μ and has p.d.f. $g(x)$. Since $\nu << \mu$, we may write $\theta = E(h(X)) = \int h(x)f(x)dx = \int (h(x)f(x)/g(x))\, g(x)dx = E\,(h(Y)f(Y)/g(Y))$ where Y has p.d.f. $g(y)$. The choice of g is arbitrary except that the corresponding measure must be absolutely continuous with respect to μ. Indeed, there are three requirements for ν or the corresponding $g(y)$:

1. $\nu << \mu$,

2. The p.d.f. $g(y)$ should be easy to simulate,

3. The variance of $h(Y)f(Y)/g(Y)$ should be small when Y has p.d.f. g.

Of course, the variance of $h(Y)f(Y)/g(Y)$ is minimized when the ratio is constant, that is when $g(y) = ch(y)f(y)$ a.e. $[\nu]$, where c is chosen to ensure that $g(y)$ is a p.d.f. However, this implies that $c = \theta^{-1}$. Consequently, finding the minimizing $g(y)$ is equivalent to finding θ. Thus it is only reasonable to find a $g(y)$ which reduces the variance, not eliminates it.

One related method arises when the random variable $h(X)$ is a.s. nonnegative with a mixed distribution. In particular, suppose that $h(X)$ takes on the value 0 with probability p and takes on strictly positive values with p.d.f. $f(x)$. In this situation,

$$\theta = E(h(X)) = 0(p) + \int_{x>0} f(x)dx = (1-p)E(h(X) \mid h(X) > 0).$$

To see the variance reduction that is possible if we can restrict attention only to the positive values of $h(X)$, consider the following situation. Let $Y_1,\ldots,Y_n$ have p.d.f. $f(x)$, the p.d.f. of h(X). Moreover, let $Z_1,\ldots,Z_n$ have p.d.f. $f(x)/(1-p)$ restricted to $\{x \mid h(x) > 0\}$. Consider two method of moments estimators, $\hat{\theta} = \frac{1}{n}\sum_{i=1}^n Y_i$ and $\hat{\hat{\theta}} = (1-p)\frac{1}{n}\sum_{i=1}^n Z_i$. It follows that $\mathrm{Var}(\hat{\theta}) = \mathrm{Var}(Y_1)/n$ and $\mathrm{Var}(\hat{\hat{\theta}}) = (1-p)^2\mathrm{Var}(Z_1)/n$. However $(1-p)^2\mathrm{Var}(Z_1) = (1-p)\mathrm{Var}(Y_1) - p(E(Y_1))^2$. Thus, changing the measure to avoid the point mass of p at 0 reduces the variance of the method of moments estimator by a factor of more than $1-p$. Of course, there are some important issues:

1. $1-p$ must be known or readily computable.

2. It is not clear how to apply these ideas in complex cases such as path dependent options.

One of the most important situations in option pricing where importance sampling offers large benefits is the pricing of out-of-the-money options. If we consider a simple European option whose payoff is determined by $(S_T - K)^+$ where S_T is the asset price at time T and K is the strike price, then the payoff will be 0 on the set $A = \{S_T \le K\}$. There is no need to evaluate the option on paths in A, since their value is known to be 0. It is, therefore, useful to change the distribution of S_T to greatly reduce the probability of A. If the option is currently far "out-of-the-money," then the measure should be changed to increase the probability that the option will be "in-the-money" when it expires. The problem was studied by Robert Reider [19] in his 1994 Ph.D. dissertation at the Wharton School of the University of Pennsylvania. He focused especially on the technique of importance sampling for simulating out-of-the-money options. He considers two distinct approaches to creating random variables which down-weight the probability that the option finishes out-of-the money. He changes measure by increasing the drift of the geometric Brownian motion, the scale of the geometric Brownian motion and a combination of both. In a representative selection of out-of-the-money cases for simple European options, Reider shows dramatic improvements in the variance of the estimators, for a factor of 10 to factors up to 10,000. Factors of 100 to 500 were common in Reider's study. Thus, the technique of importance sampling is especially useful for simulating out-of-the-money options.

The case of path dependent options such as Asian options or barrier options is still problematic. The phenomenon of having the option payoff having an atom at 0 still applies, but it has not been well documented how to change the measure to downweight this possibility. Of course, for options which are monotonic in the path, such as Asian options, one would generally increase the drift to reduce the probability of a zero payoff. Not only should this reduce the variance, it is relatively easy to compute the appropriate Radon-Nikodym derivative and to simulate from the new distribution. Nevertheless, more research is needed.

2.4 Stratified Sampling

The situation discussed in the previous section with an option payoff being constant on a set A is a special case of a more general situation in which the random outcome is heterogeneous. That is, there are L distinct sets of values $A_1, \ldots, A_L$, and the payoff exhibits very little variability on these L distinct sets. Thus, the outcomes can be grouped into *strata*. The total variability has two components: the variability within each stratum and the variability of the mean outcomes in the different strata. The solution to the estimation problem is known as stratified sampling wherein a small sample is taken from each of the strata, an estimator is defined for each strata, and those distinct estimators are combined into a single overall estimator by combining the individual estimators using a weighted combination with weights depending upon the probability of the A_i. By estimating separately for each stratum, the overall variability can be greatly reduced.

To formalize this, suppose that a population is composed of L subgroups or strata and that a random variable X sampled from this population has a mixed distribution with L mixture components corresponding to the L strata. The mixing parameters across the L strata are given by $p_1, \ldots, p_L$. If I is an indicator random variable taking values $1, \ldots, L$ which denotes the stratum that is selected, then we denote $E(X \mid I = i) = m_i$ and $\mathrm{Var}(X \mid I = i) = \sigma_i^2$. The moments of X are easily computed to be $m = E(X) = \sum_{i=1}^{L} p_i m_i$ and $\sigma^2 = \mathrm{Var}(X) = \sum_{i=1}^{L} p_i m_i^2 - m^2 + \sum_{i=1}^{L} p_i \sigma_i^2$.

Now, for a given total sample size, n, one can select a sample from the overall population without regard to the strata. In this case, the number in the sample selected from the L strata will be a random vector with a multinomial $(n, \mathbf{p})$ distribution where $\mathbf{p} = (p_1, \ldots, p_L)$. An alternative is to define a deterministic sampling plan in which a sample of size n_i, $1 \leq i \leq L$, is selected from the ith stratum, where $n_1 + \cdots + n_L = n$. In each stratum, we compute the sample mean and sample standard deviation, $\{(\bar{X}_i, s_i), 1 \leq i \leq L\}$. The sample means in each stratum are combined into an overall population estimator $\bar{X} = \sum_{i=1}^{L} p_i \bar{X}_i$. If the $\{(p_i, \sigma_i), 1 \leq i \leq L\}$ are known, then the usual estimate of m is given by $\hat{m} = \sum_{i=1}^{L} p_i \bar{X}_i$. The optimal sampling plan (in the sense of minimizing the variance of the estimate of $\hat{m}$) is found by minimizing $\sum_{i=1}^{L} \left((p_i \sigma_i)^2\right) / n_i$ subject to $\sum_{i=1}^{L} n_i = n$. The solution is $\hat{n}_i = n \left(p_i \sigma_i\right) / \sum_{j=1}^{L} p_j \sigma_j$. The resulting variance of $\hat{m}$ is given by $\left(\sum_{i=1}^{L} p_i \sigma_i\right)^2 / n$ which is smaller than σ^2 by Jensen's inequality. Of course, the above analysis needs to be adjusted when $\sigma_i = 0$ for some i, since the optimal choice of n_i would give $n_i = 0$ if $\sigma_i = 0$. This would not be correct, since at least one observation would be needed to estimate m_i and the variance formula would involve the expression $\frac{0}{0}$. The formal adjustment in this case would be to set $n_i = 1$ for each i for which $\sigma_i = 0$, then use the optimal distribution for the remaining sample size over the remaining strata. Of course, in the case of option pricing, we saw that for an option which finishes out-of-the-money, the value of $m_i = 0$, so *no* observations need to be allocated to this case, because the expected value on the out-of-the-money set is known.

2.5   A note on quasi-random number generators

Another variation on stratified sampling occurs in calculating the value of an integral or expectation by Monte Carlo simulation. Rather than selecting evaluation points at random from an interval (or a rectangle in higher dimensions), it is advantageous to subdivide the region into smaller regions and take a sample of evaluation points within. These become the strata referred to earlier. This can be shown on average to improve the rate of convergence, because the variation in the function being evaluated is less when the function is restricted to a smaller domain. If more information about the function is available, for example regions within which it is nearly constant (or in which it is highly variable), then the appropriate sample size for each region can be calculated using the analysis in the previous section. A recent topic of great interest in using simulation methods to evaluate option prices is the use of *quasi-random numbers* (also referred to as *low discrepancy sequences* by Paskov [16]). Quasi-random numbers give a sequence of numbers which fill out or refine a region in a regular, periodic fashion. The sequence will pass some of the tests for randomness, namely the empirical joint distribution will correspond to those of Lebesgue measure; however, the sequence of quasi-random numbers will not pass any of the short and long run dependency tests. Nevertheless, integration using quasi-random sequences can offer a rate of convergence of $O(n^{-1}(\log n)^d)$ where n is the sample size and d is the dimension of the integration. Thus, quasi-random sequences can greatly improve on the rate of convergence. In addition, quasi-random sequences have an "incremental" property, meaning that for a given sample size, n, if the observed accuracy is insufficient, the sampling can be continued from the point left off until the required accuracy is achieved. In the case of typical grid schemes, the entire grid must be refined, thus requiring a potentially large computational increment. While there are many issues still unresolved (for example ensuring proper convergence behavior for high-dimensional calculations, such as with Asian options), the quasi-random number approach to simulation is clearly of great practical importance. The interested reader should consult the papers by Joy, Boyle and Tan [10], Boyle, Broadie and Glasserman [2], the Ph.D. dissertation research of Paskov [15] and subsequent research [17] and the chapter in this volume by Paskov [16] for excellent treatments of the use of quasi-random sequences in finance. In addition, the reader should consult the books by Niederreiter [14] and Press *et al.* [18] for general treatments of quasi-random numbers.

3   Simulating Derivatives

The final part of this article is an introduction to simulating the derivatives of a derivative security price to determine the hedging parameters ("the Greeks") and sensitivities. This summarizes some of the research by Broadie and Glasserman [3]. In particular, suppose we represent the price of a derivative security by $p = p(S_0, T, r, \sigma) = e^{-rT} E((S_T - K)^+)$ where S_0 is the initial asset price, T is the exercise time, r is the riskless rate of return, σ is the volatility of the asset price, K is the strike price and the expectation is taken with respect to the risk neutral measure. Of course, the model could be more complex, for example σ could itself be a stochastic process (hence a stochastic volatility model) in which case we would be interested in the derivatives of the option price with respect to the parameters of that process as well. Broadie and Glasserman identify three distinct approaches to computing derivatives of option prices: (1) resimulation, (2) calculating pathwise derivatives and (3) likelihood ratio methods. To see how each of these methods works, let us consider a simple European call on an asset whose price process is governed by a geometric Brownian motion process as represented above. Now there is a collection of derivatives of p which are important in hedging including $\frac{\partial p}{\partial S_0}$ (Delta), $\frac{\partial p}{\partial \sigma}$ (Vega), $\frac{\partial^2 p}{\partial S_0^2}$ (Gamma), $\frac{\partial p}{\partial r}$ (Rho) and $-\frac{\partial p}{\partial T}$ (Theta). To consider the general problem, we have a function $p = p(\theta)$ giving a price as a function of a set of parameters $\theta = (\theta_1, \ldots \theta_M)$, and we wish to compute $\frac{\partial p}{\partial \theta_i}$ and possibly a higher derivative as well. For any value of θ one can compute $p(\theta)$ using simulation methods. The question is how to compute the derivative of the price also using simulation.

Resimulation is the most obvious approach. If one begins with the definition of a partial derivative as a limit of difference quotients, then $\frac{\partial p}{\partial \theta_i} = \lim_{h \to 0} \left(p(\theta_{\mathbf{ih}}) - p(\theta) \right) / h$, where $\theta_{\mathbf{ih}}$ denotes θ with h added to the ith component. For suitably small values of h, the ratio should stabilize near the limiting value, and both terms in the numerator can be calculated using simulation, since both represent option prices for a fixed set of parameter values. There are two obvious difficulties. First, it is not clear how to choose h. It should be chosen as small as possible, but numerical instabilities must be avoided. Second, each of the partial derivatives must be estimated separately, since we are computing prices by varying different parameters. Consequently, resimulation can be very inefficient. On the other hand, common random number streams should be used in estimating each of the components of the difference in the numerator. That is, one should generate $S_i(\theta)$ and $S_i(\theta_{\mathbf{ih}})$ and then evaluate $(S_i(\theta) - K)^+$ and $(S_i(\theta_{\mathbf{ih}}) - K)^+$, the payoffs under the two different parameter values. If the same random variates are used, then the two evaluations should be positively correlated and their difference should have less variance. This should serve to substantially reduce the variance of the estimator compared with estimating the two prices separately using different random number streams. The reader should consult the Appendix of the paper by Broadie and Glasserman for more information concerning the resimulation method.

The two other methods involve representing the price, which is an expected value, as an integral. For example, in the standard Black-Scholes situation, the price can be represented as $p = e^{-rT} E((S_T - K)^+)$. Now, one could treat the price as an integral with respect to the density of S_T, namely $p = e^{-rT} \int (s - K)^+ f_{S_T}(s; S_0, T, r, \sigma) ds$. Then the parameters S_0, T, r, σ will reside in the p.d.f. Alternatively, one could represent the random variable S_T as a function of the parameters and other standard normal random variables, not involving the parameters. That is, for the standard geometric Brownian motion model, under the risk neutral measure, we can represent $S_T = S_0 \exp \left((r - \frac{1}{2}\sigma^2)T + \sigma\sqrt{T}Z \right)$ where Z is a standard normal random variable. Consequently, the parameters reside in the function, not the p.d.f. of the random quantities. In both cases, the option price derivative can be computed by executing a series of steps. First, we interchange the order of partial differentiation and integration, passing the partial derivative with respect to the parameter under the integral sign. The justification for this step is given by Broadie and Glasserman. Next, we compute the partial derivative. Either we will differentiate the function or the p.d.f. depending upon where we chose to put the parameters. The next step is to write the resulting integral as an expectation. Finally, this expectation can be evaluated by standard Monte Carlo simulation methods. When the parameters are put in the function and the expectation is taken with respect to a standard normal distribution, the method is called the *pathwise derivative method* by Broadie and Glasserman. When the parameters are put in the p.d.f. of the final stock price, then the method is called the *likelihood ratio method* by Broadie and Glasserman. Let us illustrate the two methods for the case of computing delta, that is, $\partial p / \partial S_0$.

First, consider the pathwise derivative approach. Here, we represent the price of a European call option under a geometric Brownian motion asset price process as $p(S_0, T, r, \sigma) = e^{-rT} \int (S_0 e^{(r - \frac{1}{2}\sigma^2)T + \sigma\sqrt{T}z} - K)^+ \phi(z)dz$ where $\phi(z)$ is the p.d.f. of a standard normal distribution. Interchanging the order of differentiation and integration we find

$$
\begin{aligned}
\frac{\partial p(S_0, T, r, \sigma)}{\partial S_0} &= e^{-rT} \frac{\partial}{\partial S_0} \int (S_0 e^{(r - \frac{1}{2}\sigma^2)T + \sigma\sqrt{T}z} - K)^+ \phi(z)dz \\
&= e^{-rT} \int \frac{\partial}{\partial S_0} (S_0 e^{(r - \frac{1}{2}\sigma^2)T + \sigma\sqrt{T}z} - K)^+ \phi(z)dz \\
&= e^{-rT} \int e^{(r - \frac{1}{2}\sigma^2)T + \sigma\sqrt{T}z} I_{\{S_T \geq K\}} \phi(z)dz \\
&= e^{-rT} E\left(\frac{S_T}{S_0} I_{\{S_T \geq K\}} \right).
\end{aligned}
$$

The final expectation can be calculated using Monte Carlo simulation utilizing all the relevant variance reduction methods described earlier. Similar expressions can be derived in a straightforward fashion for the other option price derivatives.

Second, consider the likelihood ratio approach. In this case, we need to compute f_{S_T}, the p.d.f. of S_T. It is well known that

$$
f_{S_T}(s) = \frac{1}{s\sigma\sqrt{T}} \phi(g(s))
$$

where

$$
g(s) = g(s; S_0, T, r, \sigma) = \frac{\log(s/S_0) - (r - \frac{1}{2}\sigma^2)T}{\sigma\sqrt{T}}
$$

and ϕ is the p.d.f. of a standard normal distribution.

Now,

$$
\frac{\partial p}{\partial S_0} = \frac{\partial}{\partial S_0} e^{-rT} \int (s - K)^+ f_{S_T}(s; \theta)ds
$$

$$\begin{aligned}
&= e^{-rT} \int (s-K)^+ \frac{\partial}{\partial S_0} f_{S_T}(s; S_0) ds \\
&= e^{-rT} \int (s-K)^+ \frac{\frac{\partial}{\partial S_0} f_{S_T}(s; S_0)}{f_{S_T}(s; S_0)} f_{S_T}(s; S_0) ds \\
&= e^{-rT} E\left((S_T - K)^+ \frac{\partial}{\partial S_0} \log f_{S_T}(S_T; S_0) \right).
\end{aligned}$$

The last partial derivative is easily calculated. We write

$$\log f_{S_T}(S_T; S_0) = -S_T \sigma \sqrt{T} - \frac{1}{2} \log 2\pi - \frac{1}{2} g(S_T; S_0)^2,$$

so

$$\begin{aligned}
\frac{\partial}{\partial S_0} \log f_{S_T}(S_T; S_0) &= -g(S_T) \frac{\partial}{\partial S_0} g(S_T; S_0) \\
&= -\frac{\log(S_T/S_0) - (r - \frac{1}{2}\sigma^2)T}{S_0 \sigma \sqrt{T}}.
\end{aligned}$$

Consequently, the likelihood ratio method for computing $\frac{\partial p}{\partial S_0}$ (delta) would be to compute the expectation

$$e^{-rT} E\left((S_T - K)^+ \left(-\frac{\log(S_T/S_0) - (r - \frac{1}{2}\sigma^2)T}{S_0 \sigma \sqrt{T}} \right) \right).$$

Again, this expectation can be calculated by simulation methods including the use of all the variance reduction methods described earlier. Similar expectations can be derived for the other option price derivatives.

It is important to note that all three methods for simulating option price derivatives can be extended to more general situations either involving path dependent options, more complicated price processes or both. Resimulation methods will obviously work, because they simply involve calculating option prices at two nearby parameter points, and this can be done in reasonably general situations. Pathwise derivative methods can also be applied in more general situations. For example, suppose that the asset price process involves a stochastic volatility model[1]

$$\begin{aligned}
dS_t &= \mu S_t dt + \sqrt{V_t} S_t dW_t \\
dV_t &= \phi V_t dt + \zeta V_t dZ_t
\end{aligned}$$

where W and Z are independent Brownian motion processes.

---

[1] The Isaac Newton Programme in Financial Mathematics featured several major presentations on stochastic volatility modelling including presentations by Hodges [9], by Crouhy [5] and by Hobson and Rogers [8]. The interested reader should consult these papers for a full discussion of the appropriateness of the various potential models.

One can develop a discretization scheme to simulate a price process by defining a grid size, $\delta = T/N$. The grid points become $\{t_i, 0 \le i \le N\}$ with $t_i = i\delta$. The discretized version of the asset price process becomes:

$$S_{i+1} = S_i + \mu S_i \delta + \sqrt{V_i} S_i W_i \sqrt{\delta}$$
$$V_{i+1} = \phi V_i \delta + \zeta V_i Z_i \sqrt{\delta},$$

where $\{(W_i, Z_i), 0 \le i \le N - 1\}$ are independent standard normal random variables. Consequently, it is possible to represent S_i in terms of the parameters and a set of standard normal random variables not involving the parameters. If the option payoff is reasonably linear in the $\{S_i, 1 \le i \le N\}$ sequence (for example it depends on S_N or $\frac{1}{N} \sum_{i=1}^{N} S_i$ for an Asian option), then the pathwise derivative can be computed for all the parameters of interest. Broadie and Glasserman give a few examples of computing these more complicated pathwise derivatives. On the other hand, in a stochastic volatility model, the p.d.f. of the random variable used to evaluate the option (e.g. S_T or $\frac{1}{T} \int S_t dt$) will be extremely complicated and the derivatives of the log-likelihood function will often be intractable. Thus, resimulation and the pathwise derivative method appear to offer the most generality in computing the derivatives of option prices. Of course, in the stochastic volatility case with two driving Brownian motion processes, the market is not complete, so perfect hedging is not possible using only the underlying asset. This can be overcome by introducing a second derivative security, for example a standard European option on the final price. Then an Asian option on the average asset price can be perfectly hedged by the underlying asset and a European option on the final price. Again, analytic expressions for the pathwise derivatives can be computed and evaluated using simulation. This has been carried out for the above situation by Koenig [13].

As mentioned earlier, Clewlow and Carverhill [4] use the derivatives of derivative security prices as a method of generating control variables to reduce the variance of the standard method of moments estimator. Furthermore, they show how one might use this idea to reduce the variance of these estimators when the asset price process includes stochastic volatility. We first summarize Clewlow and Carverhill's basic idea for the simple Black Scholes case and then indicate how it can be implemented for more complex problems.

Consider a standard European call option on an asset whose price dynamics are given by a log Brownian motion process. Let the option price be denoted by $p(t) = p(S_t, T, r, \sigma)$, given by the standard Black Scholes formula. Now, one of the derivations of the Black Scholes formula is to create a portfolio, selling one call option and constantly holding $\frac{\partial p}{\partial S_t}$ units of the security. The resulting portfolio has no risk and therefore must earn the rate of return r. Consequently, there is a linear relationship between the value of the call option and a certain option price derivative. Now in simulating this process, one will necessarily discretize the asset price process, so the hedge will not be perfect, but for small time increments, the error will be small and the control variable will be nearly perfect. In this discrete case, one would consider the $\{S_i, 1 \leq i \leq N\}$ price process described above and would construct the control variable

$$X = \sum_{i=1}^{N} \frac{\partial p}{\partial S_{t_i}} \left((S_{t_{i+1}} - S_{t_i}) - E(S_{t_{i+1}} - S_{t_i}) \right).$$

Clearly X will have mean 0 and should serve as an excellent control variable provided the time steps are not too large. Also, one can compute this control variable, since there is a closed form expression available for p and $\frac{\partial p}{\partial S_t}$. Of course, there is no need to simulate to compute p since a closed form expression is already available. However, in more complex situations, no such closed form expression will be known. In such cases, one chooses a related option for which such an expression is available. For example, Clewlow and Carverhill consider a lookback option with a stochastic volatility asset price process. In this case, a formula for the price of a lookback option is known in the standard geometric Brownian motion case with constant volatility. Consequently, if we let L represent that closed form expression, then $\frac{\partial L}{\partial S_{t_i}}$ can be computed. A control variable of the form

$$X = \sum_{i=1}^{n} \frac{\partial L}{\partial S_{t_i}} \left((S_{t_{i+1}} - S_{t_i}) - E(S_{t_{i+1}} - S_{t_i}) \right)$$

can be constructed, with the evolution of the σ process taken into account. In addition, other control variables can be constructed by replacing the $\frac{\partial L}{\partial S_{t_i}}$ factor by other derivatives, such as $\frac{\partial L}{\partial \sigma_{t_i}}$ to hedge against the stochastic volatility.

Clewlow and Carverhill report variance reductions for a simulation experiment involving a lookback call option. Using several control variates, they report a variance reduction factor of 90. Thus, the martingale control variable method can offer a very substantial reduction in variance. It can also be combined with other variance reduction methods, such as importance sampling to offer very rapid ways to compute option prices by Monte Carlo simulation.

References

[1] Boyle, P., 'Options: A Monte Carlo approach', *Journal of Financial Economics* **4**, 323–338, 1977.

[2] Boyle, P., Broadie, M. and Glasserman, P., 'Monte Carlo methods for security pricing', *Working paper, University of Waterloo*, Waterloo, Ontario, Canada, 1995.

[3] Broadie, M. and Glasserman, P., 'Estimating security price derivatives using simulation', preprint, Graduate School of Business, Columbia University, New York, NY, 1993.

[4] Clewlow, L. and Carverhill, A.P., 'On the simulation of contingent claims', *Journal of Derivatives* **2**, 66–74, 1994.

[5] Crouhy, M., 'Extensions of the Black-Scholes model to the case of stochastic volatility', presentation, Issac Newton Institute Programme in Financial Mathematics, 1995.

[6] Duffie, D. and Glynn, P., 'Efficient Monte Carlo simulation of security prices', *Annals of Applied Probability* **5**, 897–905, 1996

[7] Fu, M., Madan, D. and Wang, T., 'Pricing continuous time Asian options: A comparison of analytical and Monte Carlo methods', Working Paper, College of Business and Management, University of Maryland, 1995.

[8] Hobson, D.G. and Rogers, L.C.G., 'Complete models with stochastic volatility', preprint, Isaac Newton Institute Programme in Financial Mathematics, 1995.

[9] Hodges, S., 'New concepts and theories on the treatment of stochastic volatility', presentation, Isaac Newton Institute Programme in Financial Mathematics, 1995.

[10] Joy, C., Boyle, P.P. and Tan, K.S., 'Quasi-Monte Carlo methods in numerical finance', Working paper, University of Waterloo, Waterloo, Ontario, Canada, 1995.

[11] Kemna, A.G.Z. and Vorst, A.C.F., 'A pricing method for options on average asset values', *Journal of Banking and Finance* **14**, 113–129, 1990.

[12] Kloeden, P., Platen, E. and Schurz, H., *Numerical solution of SDE Through Computer Experiments*, Springer Verlag, 1994.

[13] Koenig, M., 'Pricing and hedging options in the presence of stochastic volatility', Working Paper, Carnegie Mellon University, 1995.

[14] Niederreiter, H., *Random Number Generation and Quasi-Monte Carlo Methods*, CBMS-NSF Volume 63, SIAM.

[15] Paskov, S.P., 'Computing high dimensional integrals with applications in finance', Technical report CUCS-023-94, Department of Computer Science, Columbia University, New York, NY, 1994.

[16] Paskov, S.P., 'New methodologies for valuing derivatives', This volume.

[17] Paskov, S, and Traub, J., 'Faster valuation of financial derivatives', preprint, Isaac Newton Institute programme in Financial Mathematics, 1995.

[18] Press, W.H., Teukolsky, S.A., Vetterling, W.T. and Flannery, B.P., *Numerical Recipes in C: The Art of Scientific Computing, Second Edition*, Cambridge University Press, 1992.

[19] Reider, R., 'Two applications of Monte Carlo techniques to finance', PhD dissertation, Wharton School, University of Pennsylvania, 1994.

New Methodologies for Valuing Derivatives

Spassimir H. Paskov

Abstract

High-dimensional integrals are usually solved with Monte Carlo algorithms although theory suggests that low discrepancy algorithms are sometimes superior. We report on numerical testing which compares low discrepancy and Monte Carlo algorithms on the evaluation of financial derivatives. The testing is performed on a Collateralized Mortgage Obligation (CMO) which is formulated as the computation of ten integrals of dimension up to 360.

We tested two low discrepancy algorithms (Sobol and Halton) and two randomized algorithms (classical Monte Carlo and Monte Carlo combined with antithetic variables). We conclude that for this CMO the Sobol algorithm is always superior to the other algorithms. We believe that it will be advantageous to use the Sobol algorithm for many other types of financial derivatives.

Our conclusion regarding the superiority of the Sobol algorithm also holds when a rather small number of sample points are used, an important case in practice.

We have built a software system called FINDER for computing high dimensional integrals. Routines for computing Sobol points have been published. However, we incorporated major improvements in FINDER and we stress that the results reported here were obtained using this software.

The software system FINDER runs on a network of heterogeneous workstations under PVM 3.2 (Parallel Virtual Machine). Since workstations are ubiquitous, this is a cost-effective way to do very large computations fast. The measured speedup is at least $.9N$ for N workstations, $N \le 25$. The software can also be used to compute high dimensional integrals on a single workstation.

1 Introduction

High-dimensional integrals are usually solved with Monte Carlo algorithms. Vast sums are spent annually on these algorithms.

Theory [21], [43] suggests that low discrepancy algorithms are sometimes superior to Monte Carlo algorithms. However, a number of researchers, [14],

[19], [20], report that their numerical tests show that this theoretical advantage decreases with increasing dimension. Furthermore, they report that the theoretical advantage of low discrepancy algorithms disappears for rather modest values of the dimension, say, $d \leq 30$.

We decided to compare the efficacy of low discrepancy and Monte Carlo algorithms on the valuation of financial derivatives. We use a Collateralized Mortgage Obligation (CMO), provided to us by Goldman Sachs, with ten bond classes (tranches) which is formulated as the computation of ten integrals of dimension up to 360. The reasons for choosing this CMO is that it has fairly high dimension and that each integrand evaluation is very expensive making it crucial to sample the integrand as few times as possible. We believe that our conclusions regarding this CMO will hold for many other financial derivatives.

The low discrepancy sample points chosen for our tests are Sobol and Halton points. We compared the algorithms based on these points with the classical Monte Carlo algorithm and also with the classical Monte Carlo algorithm combined with antithetic variables. We refer to the latter as the antithetic variables algorithm. See Section 7 for a discussion of why this algorithm was tested.

An explanation of our terminology is in order here. *Low discrepancy points* are sometimes referred to as *quasi-random* points. Although in widespread use, we believe the latter term to be misleading since there is nothing random about these deterministic points. We prefer to use the terminology *low discrepancy* or *deterministic*.

We assume the finance problem has been formulated as an integral over the unit cube in d dimensions. We have built a software system called FINDER for computing high dimensional integrals. FINDER runs on a heterogeneous network of workstations under PVM 3.2 (Parallel Virtual Machine). Since workstations are ubiquitous, this is a cost-effective way to do large computations fast. Of course, FINDER can also be used to compute high dimensional integrals on a single workstation. The software system FINDER is available and interested readers should contact the author.

A routine for generating Sobol points is given in [28]. *However, major improvements have been incorporated in FINDER. We emphasize that the results reported in this paper were obtained using FINDER.* One of the improvements was developing the table of primitive polynomials and initial direction numbers for dimensions up to 360.

This article is based on software construction and testing which began in the Fall of 1992. Preliminary results were presented to a number of New York City financial houses in the Fall of 1993 and the Spring of 1994. A January, 1994 article in Scientific American discussed the theoretical issues and reported that 'Preliminary results obtained by testing certain finance problems suggest the superiority of the deterministic methods in practice'.

Further results were reported at a number of conferences including [24], [38], [39]. An 'extended abstract' of this article was published in Fall, 1995 [26]. A slightly different version of the chapter appeared as a Columbia University Technical Report in October, 1994.

We summarize our main conclusions regarding the evaluation of this CMO. The conclusions may be divided into three groups.

I. Deterministic and Monte Carlo Algorithms

The Sobol algorithm consistently outperforms the Monte Carlo algorithm. The Sobol algorithm consistently outperforms the Halton algorithm. In particular,

- The Sobol algorithm converges significantly faster than the Monte Carlo algorithm;

- The convergence of the Sobol algorithm is smoother than the convergence of the Monte Carlo algorithm. This makes automatic termination easier for the Sobol algorithm;

- Using our standard termination criterion, see Section 6, the Sobol algorithm terminates 2 to 5 times faster than the Monte Carlo algorithm often with smaller error;

- The Monte Carlo algorithm is sensitive to the initial seed.

II. Sobol, Monte Carlo, and Antithetic Variables Algorithms

The Sobol algorithm consistently outperforms the antithetic variables algorithm, which in turn, consistently outperforms the Monte Carlo algorithm. In particular,

- These conclusions also hold when a rather small number of sample points are used, an important case in practice. For example, for 4000 sample points, the Sobol algorithm running on a single Sun-4 workstation achieves accuracies within range from one part in a thousand to one part in a million, depending on the tranche, within a couple of minutes;

- Statistical analysis on the small sample case further strengthens the case for the Sobol algorithm over the antithetic variables algorithm. For example, to achieve similar performances with confidence level 95%, the antithetic variables algorithm needs from 7 to 79 times more sample points than the Sobol algorithm, depending on the tranche. In a similar comparison, the Monte Carlo algorithm needs from 27 to 607 times more sample points than the Sobol algorithm depending on the tranche. These speedups are measured conservatively, see Section 9;

- Statistical analysis for a large number (1,000,000) of sample points shows that the superiority of the Sobol algorithm is greater than for a small number of points. For example, to achieve similar performances with confidence level 95%, the antithetic variables algorithm needs from 16 to 230 times more sample points than the Sobol algorithm, depending on the tranche. In a similar comparison, the Monte Carlo algorithm needs from 110 to 1769 times more sample points than the Sobol algorithm depending on the tranche. These speedups are measured conservatively, see Section 9;

- The antithetic variables algorithm is sensitive to the initial seed. However, convergence of the antithetic variables algorithm is less jagged than convergence of the Monte Carlo algorithm.

III. Network of Workstations *All the algorithms benefit by being run on a network of workstations.* In particular,

- For N workstations, the measured speedup is at least $0.9N$, where $N \leq 25$;

- A substantial computation which took seven hours on a Sun-4 workstation took twenty minutes on the network of 25 workstations.

We emphasize that we do not claim that the Sobol algorithm is always superior to the Monte Carlo algorithm. We do not even claim that it is always superior for financial derivatives. After all, the test results reported here are only for one particular CMO. However, we do believe it will be advantageous to use the Sobol algorithm for many other types of financial derivatives.

2   The Monte Carlo Algorithm

In this section, we give a brief discussion of the theory underlying the Monte Carlo algorithm for the problem of multivariate integration. For more details, see for example, [11], [12], [15], and [34].

We now present the classical Monte Carlo algorithm. For brevity, we refer to it as the Monte Carlo algorithm. Let $t_1, \ldots, t_n$ be n randomly selected points which are independent and uniformly distributed over $D = [0,1]^d$, the d dimensional unit cube. Consider a function f from the space $L_2(D)$ of L_2-integrable functions. The problem is to approximately compute

$$I(f) = \int_D f(x)dx$$

using function evaluations at randomly chosen points. The classical Monte Carlo algorithm is given by

$$I(f) \approx U_n(f) = U_n(f; t_1, \ldots, t_n) = \frac{1}{n} \sum_{i=1}^n f(t_i). \tag{1}$$

The main idea underlying the Monte Carlo algorithm for multivariate integration is to replace a continuous average by a discrete average over randomly selected points.

The expected value of the estimate $U_n(f; t_1, \ldots, t_n)$ as a function of the random variables $t_1, t_2, \ldots, t_n$ is

$$E(U_n(f; t_1, \ldots, t_n)) = I(f).$$

The expected error of the Monte Carlo algorithm for a function f is defined by

$$E_n(f) = \left(\int_{D^n} (I(f) - U_n(f))^2 \, dt_1 \ldots dt_n \right)^{1/2}.$$

It is well known that

$$E_n(f) = \frac{\sigma(f)}{\sqrt{n}},$$

where the variance $\sigma^2(f)$ of f is defined as

$$\sigma^2(f) = \int_D (f(t) - I(f))^2 \, dt = I(f^2) - I^2(f).$$

Clearly, by reducing the variance of the integrand the expected error would also decrease. In fact, this is the main idea underlying the various variance reduction techniques which are often used in combination with the Monte Carlo algorithm. Examples of variance reduction techniques are importance sampling, control variates, antithetic variables, see for example [15]. Antithetic variables will be discussed in Section 7.

Let $B(L_2(D))$ denote the unit ball of $L_2(D)$. The expected error of the Monte Carlo algorithm with respect to the class $B(L_2(D))$ is measured for the worst function f in $B(L_2(D))$,

$$e_n^{\text{wor-MC}}(B(L_2(D)) = \sup_{f \in B(L_2(D))} E_n(f).$$

It can be easily concluded that

$$e_n^{\text{wor-MC}}(B(L_2(D)) = n^{-1/2}. \tag{2}$$

Since the rate of convergence is independent of d and the Monte Carlo algorithm is applicable to a very broad class of integrands, its advantages for high dimensional integration are clear. Nevertheless, the Monte Carlo algorithm has several serious deficiencies, see for example, [21] and [36]. We mention just three of them here. Even though the rate of convergence is independent of the dimension, it is quite slow. Furthermore, there are fundamental philosophical and practical problems with generating independent random points; instead, pseudorandom numbers are used, see [16] and [41]. Finally, the Monte Carlo algorithm provides only probabilistic error bounds, which is not a desirable guarantee for problems where highly reliable results are needed.

3 Low Discrepancy Deterministic Algorithms

In an attempt to avoid the deficiencies of the Monte Carlo algorithm, many deterministic algorithms have been proposed for computing high dimensional integrals for functions belonging to various subsets of $L_2(D)$, see e.g. [21], [22], [40], [43], [23], and [37]. One class of such deterministic algorithms is based on low discrepancy sequences. First, we define discrepancy, which is a measure of deviation from uniformity of a sequence of points in D. Then, very briefly, we give the theoretical basis of the low discrepancy sequences to be used as sample points for computing multivariate integrals. For more detailed description and treatment of these results, see [21].

For $t = [t_1, \ldots, t_d] \in D$, define $[0, t) = [0, t_1) \times \cdots \times [0, t_d)$. Let $\chi_{[0,t)}$ be the characteristic (indicator) function of $[0, t)$. For $z_1, \ldots, z_n \in D$, define

$$R_n(t; z_1, \ldots, z_n) = \frac{1}{n} \sum_{k=1}^{n} \chi_{[0,t)}(z_k) - t_1 t_2 \cdots t_d.$$

The L_2 (or L_∞) *discrepancy* of $z_1, \ldots, z_n$, is defined as the L_2 (or L_∞) norm of the function $R_n(\cdot; z_1, \ldots, z_n)$, i.e.,

$$\|R_n(\cdot; z_1, \ldots, z_n)\|_2 = \left(\int_D R_n^2(t; z_1, \ldots, z_n) \, dt \right)^{1/2},$$

$$\|R_n(\cdot; z_1, \ldots, z_n)\|_\infty = \sup_{t \in D} |R_n(t; z_1, \ldots, z_n)|.$$

Roth, [29] and [30], proves that

$$\inf_{z_1, \ldots, z_n} \|R_n(\cdot; z_1, \ldots, z_n)\|_2 = \Theta(n^{-1} (\log n)^{(d-1)/2}). \tag{3}$$

Of special interest for numerical integration are infinite low discrepancy sequences $\{z_k\}$ in which the definition of z_k does not depend on the specific value of the number of points n. Examples of such infinite sequences are the Halton [9] and Sobol [33] sequences. For the reader's convenience, we give a short description of both these sequences later in this section. The following bounds hold for the Halton and the Sobol sequences

$$\|R_n(\cdot; z_1, \ldots, z_n)\|_2 \leq \|R_n(\cdot; z_1, \ldots, z_n)\|_\infty = O\left(\frac{(\log n)^d}{n} \right). \tag{4}$$

This is the best upper bound known and it is widely believed that it is sharp for these sequences, see [21].

We stress that the constants in the bounds (3) and (4) depend on the dimension d and good estimates of these constants are not known. Bounds with known constants and n independent of d are studied in [42] and [44].

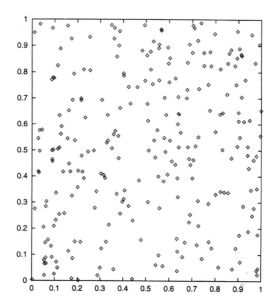

Figure 1: 256 Random points in the unit square

Sequences satisfying the upper bound in (4) are known as *low discrepancy sequences* or *quasi-random sequences*. We will refer to them as *low discrepancy sequences* or *deterministic sequences*.

Clearly, the idea behind the low discrepancy sequences is for any rectangular $[0, t)$ the fraction of the points within $[0, t)$ to be as 'close' as possible to its volume. That way, the low discrepancy sequences cover the unit cube as 'uniformly' as possible by reducing gaps and clustering of points. This idea is illustrated in Figure 1 and Figure 2.

We now state the theoretical bases for the use of low discrepancy sequences as sample points for multivariate integration. Let $V(f) < \infty$ be the variation of f on D in the sense of Hardy and Krause, see [21], and let $\{z_k\}$ be a low discrepancy sequence. The *Koksma-Hlawka inequality* guarantees that

$$\left| I(f) - \frac{1}{n} \sum_{k=1}^{n} f(z_k) \right| \leq V(f) \, \|R_n(\cdot \,; z_1, \ldots, z_n)\|_\infty. \tag{5}$$

Upper bounds in terms of L_2 discrepancy have also been proven, see [21]. Therefore, (4) and (5) provide a worst case assurance for the use of low discrepancy sequences as sample points for numerical integration of functions with bounded variation in the sense of Hardy and Krause. Furthermore, we stress that the deterministic algorithms, based on low discrepancy sequences, have better rates of convergence than the Monte Carlo algorithm.

In addition, low discrepancy sequences also have good properties in the average case setting. Indeed, let $F = C_d$ be the class of real continuous

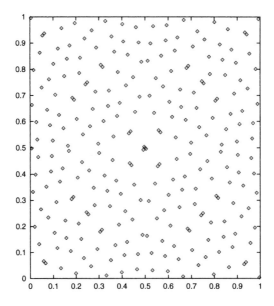

Figure 2: 256 Sobol points in the unit square

functions defined on D and equipped with the classical Wiener sheet measure w. That is, w is Gaussian with mean zero and covariance kernel R_w defined as

$$R_w(t, x) \overset{\text{def}}{=} \int_{C_d} f(t)f(x)w(df) = \min(t, x) \overset{\text{def}}{=} \prod_{j=1}^{d} \min(t_j, x_j),$$

where $t = (t_1, \ldots, t_d)$, $x = (x_1, \ldots, x_d)$ for $t, x \in D$.
    Define

$$x_k = 1 - z_k, \qquad k = 1, 2, \ldots,$$

where $\{z_k\}$ is a low discrepancy sequence. We approximate the integral of f from C_d by the arithmetic mean of its values at x_k,

$$I(f) = \int_D f(t)dt \approx U_n(f) = \frac{1}{n} \sum_{k=1}^{n} f(x_k), \qquad \forall f \in C_d.$$

Woźniakowski [43] relates the average case error of $U_n(f)$ with the L_2 discrepancy,

$$\int_{C_d} (I(f) - U_n(f))^2 w(df) = \int_D R_n^2(t; z_1, \ldots, z_n)dt. \tag{6}$$

Thus, (3), (4), and (6) provide an average case assurance for the use of the sequence $\{x_k\}$ as a set of sample points for numerical integration of the functions in C_d equipped with the classical Wiener sheet measure. Using the identity $\chi_{[0,t)}(z_k) = \chi_{(1-t,1]}(x_k)$ and Proposition 2.4 in [21], we conclude that if $\{z_k\}$ is a low discrepancy sequence then $\{x_k\}$ is also a low discrepancy sequence.

3.1 Halton Points

We give a short description of the Halton low discrepancy sequence.

Let $p_1, p_2, \ldots, p_d$ be the first d prime numbers. Any integer $k \geq 0$ can be uniquely represented as $k = \sum_{i=0}^{\lceil \log k \rceil} a_i p_j^i$ with integers $a_i \in [0, p_j - 1]$. The radical inverse function ϕ_{p_j} is defined as

$$\phi_{p_j}(k) = \sum_{i=0}^{\lceil \log k \rceil} a_i p_j^{-i-1}. \tag{7}$$

The Halton d-dimensional points are defined as

$$z_k = (\phi_{p_1}(k), \phi_{p_2}(k), \ldots, \phi_{p_d}(k)), \qquad k \geq 0.$$

3.2 Sobol Points

We give a short description of the Sobol low discrepancy sequence.

Assume first that $d = 1$. A one-dimensional Sobol sequence is generated as follows. For $i = 1, 2, \ldots, w$, let $v_i = m_i/2^i$ be a sequence of binary fractions with w bits after the binary point, where $0 < m_i < 2^i$ are odd integers. The numbers v_i are called *direction numbers*. Assume for a moment that they are already generated; we will discuss their generation later.

In Sobol's original algorithm, a one-dimensional Sobol sequence is generated by

$$x_k = b_1 v_1 \oplus b_2 v_2 \oplus \cdots \oplus b_w v_w, \qquad k \geq 0$$

where $k = \sum_{i=0}^{\lceil \log k \rceil} b_i 2^i$ is the binary representation of k and $\oplus$ denotes a bit-by-bit exclusive-or operation. For example, $k = 3$ is 11 in base 2. If $v_1 = 0.1$ and $v_2 = 0.11$ then

$$b_1 v_1 \oplus b_2 v_2 = 0.1 \oplus 0.11 = 0.01.$$

Antonov and Saleev [2] suggest a shuffling of the original Sobol sequence which preserves good convergence properties and which can be generated much faster. This version of the Sobol sequence is used here.

The sequence of direction numbers v_i is generated by a primitive polynomial, see [16] and [33], with coefficients in the field Z_2 with elements $\{0, 1\}$. Consider, for example, the primitive polynomial

$$P(x) = x^n + a_1 x^{n-1} + \ldots + a_{n-1} x + 1$$

of degree n in $Z_2[x]$. Then the direction numbers are obtained from the following recurrence formula

$$v_i = a_1 v_{i-1} \oplus a_2 v_{i-2} \oplus \cdots \oplus a_{n-1} v_{i-n+1} \oplus v_{i-n} \oplus (v_{i-n}/2^n), \qquad i > n,$$

where the last term v_{i-n} is shifted right n places. The initial numbers $v_1 = m_1/2, \ldots, v_n = m_n/2^n$ are such that m_i is odd and $0 < m_i < 2^i$ for $i = 1, 2, \ldots, n$.

Consider now an arbitrary $d \geq 1$. Let $P_1, P_2, \ldots, P_d$ be d primitive polynomials in $Z_2[x]$. Denote by $\{x_k^i\}_{k=1}^{\infty}$ the sequence of one-dimensional Sobol points generated by the polynomial P_i. Then the sequence of d-dimensional Sobol points is defined as

$$x_k = (x_k^1, x_k^2, \ldots, x_k^d).$$

4   Software System for Computing High Dimensional Integrals

4.1   Algorithms

The theory presented in the previous sections suggests that the low discrepancy deterministic algorithms provide an interesting alternative to the Monte Carlo algorithm for computing high dimensional integrals. We have developed and tested a distributed software system for computing multivariate integrals on a network of workstations. The deterministic algorithms and the Monte Carlo algorithm are implemented. The software utilizes the following sequences of sample points:

- Halton points;

- Sobol points;

- Uniformly distributed random points.

The user can choose the sequence of sample points from a menu. The software is written in a modular way so other kinds of deterministic and random number generators can be easily added. One or several multivariate functions defined over the unit cube of up to 360 variables can be integrated simultaneously. The number of variables could be extended as well.

A routine for generating Sobol points is given in [28]. However, we have made major improvements and we stress that the results reported in this paper were obtained using FINDER and not the routine in [28]. One of the improvements was developing the table of primitive polynomials and initial direction numbers for dimensions up to 360.

The software uses various kinds of random number generators. More specifically, the random number generators ran1 and ran2 from the first edition of Numerical Recipes [27], and RAN1 and RAN2 from the second edition of Numerical Recipes [28] are used because of their wide availability and popularity.

All of the above random number generators are based on linear congruential generators with some additional features. For more details on these random number generators we refer to [27] and [28].

4.2 Systems

Since workstation clusters and networks provide cost-effective means to perform large-scale computation, we have built and debugged a software system under PVM 3.2 (Parallel Virtual Machine) for computing multivariate integrals. This system runs on a heterogeneous network of machines. PVM is a software package that allows a network of heterogeneous Unix computers to be used as a single large parallel computer. Thus large computational problems can be solved by using the aggregate power of many computers.

A master/slave model is used as the programming paradigm for computing multivariate integrals. In this model, the master program spawns and directs some number of slave processes. Each of the slave processes generates a sequence of sample points specified by the master and evaluates the integrand at those points. A partial sum of integrand values is returned to the master by each of the slave processes. The master combines partial sums as specified by an algorithm, computes an approximate value of the integral, and checks a termination criterion. If the termination criterion is not satisfied, the master spawns a new round of computations. This process continues until the termination criterion is satisfied or some prespecified upper bound of the number of sample points is reached. At the end, the master returns the final result of computation and timing information. In addition, the master process keeps information about each spawned process. If some host dies, the master reallocates its job to some other host.

In a multiuser network environment the load balancing method can be one of the most important factors for improving the performance. A dynamic load balancing method is used. Namely, the Pool of Tasks paradigm is especially suited for a master/slave program. In the Pool of Tasks method the master program creates and holds the 'pool' and sends out tasks to slave programs as they become idle. The fact that no communication is required between slave programs and the only communication is to the master makes our integration problem suitable for the Pool of Tasks paradigm.

5 Finance (CMO) Problem

We now consider a finance problem which is a typical Collateralized Mortgage Obligation (CMO). A CMO consists of several bond classes, commonly referred to as tranches, which derive their cash flows from an underlying pool of mortgages. The cash flows received from the pool of mortgages are

divided and distributed to each of the tranches according to a set of prespecified rules. The cash flows consist of interest and repayment of the principal.
The technique of distributing the cash flows transfers the prepayment risk
among different tranches. This results in financial instruments with varying
characteristics which might be more suitable to the needs and expectations
of investors. For more details on CMOs, we refer to [5]. We stress that the
amount of obtained cash flows will depend upon the future level of interest
rates. Our problem is to estimate the expected value of the sum of present
values of future cash flows for each of the tranches.

We now give some of the details related to the studied CMO and the way it
is reduced to the problem of multivariate integration. The CMO[1] consists of
ten tranches. Denote them by A, B, C, D, E, G, H, J, R, Z. Throughout this
section, we describe results on tranche A in more details. Results for other
tranches are similar unless stated explicitly. The underlying pool of mortgages
has a thirty-year maturity and cash flows are obtained monthly. This implies
360 cash flows. The monthly cash flows are divided and distributed according
to some prespecified rules. The actual rules for distribution are rather complicated and are given in details in the prospectus describing the financial
product.

For $1 \leq k \leq 360$, denote by

$$C \quad - \quad \text{the monthly payment on the underlying pool of mortgages;}$$
$$i_k \quad - \quad \text{the appropriate interest rate in month } k;$$
$$w_k \quad - \quad \text{the percentage prepaying in month } k;$$
$$a_{360-k+1} \quad - \quad \text{the remaining annuity after month } k.$$

Recall that the remaining annuity a_k is given by

$$a_k = 1 + v_0 + \cdots + v_0^{k-1}, \qquad k = 1, 2, \ldots, 360,$$

with $v_0 = 1/(1+i_0)$ and i_0 the current monthly interest rate. In the notation
above, C and $a_{360-k+1}$ are constants; i_k and w_k are stochastic variables to be
determined below.

We now describe the interest rate model. Assume that the interest rate i_k
is of the form

$$i_k = K_0 e^{\xi_k} i_{k-1} = K_0^k i_0 e^{\xi_1 + \cdots + \xi_k},$$

where $\{\xi_k\}_{k=1}^{360}$ are independent normally distributed random variables with
mean 0 and variance σ^2, and K_0 is a given constant. In our case $\sigma^2 = 0.0004$
is chosen.

Suppose that the prepayment model w_k, as a function of i_k, is computed
as

$$\begin{aligned}
w_k &= w_k(\xi_1, \ldots, \xi_k) = K_1 + K_2 \arctan(K_3 i_k + K_4) \\
&= K_1 + K_2 \arctan(K_3 K_0^k i_0 e^{\xi_1 + \cdots + \xi_k} + K_4),
\end{aligned}$$

---

[1] It is labeled as 'Fannie Mae REMIC Trust 1989-23'.

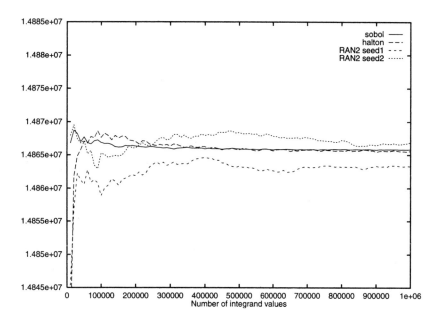

Figure 3: Sobol and Halton runs for tranche A and two Monte Carlo runs using RAN2

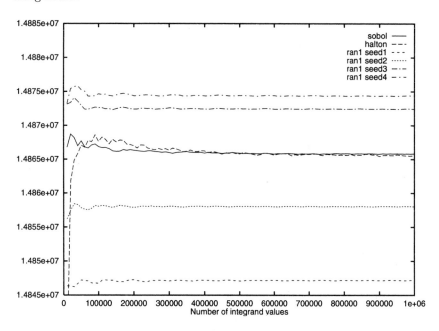

Figure 4: Sobol and Halton runs for tranche A and four Monte Carlo runs using ran1

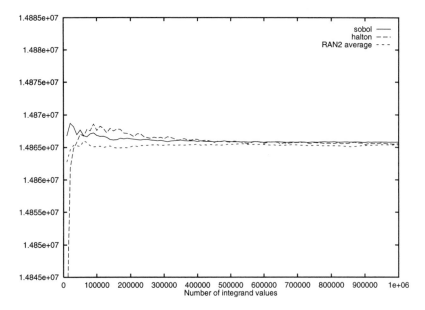

Figure 5: Sobol and Halton runs for tranche A and an average of twenty Monte Carlo runs using RAN2

of sample points on the x-axis is correct only for the deterministic algorithms. The actual number of sample points for the averaged Monte Carlo graph is twenty times the number of sample points on the x-axis. The results of the deterministic algorithms and the averaged Monte Carlo result are approximately the same. We thus conclude that to achieve similar performances, we have to take about 20 times more random than deterministic sample points.

In Figure 6, an automatic termination criterion is applied to Sobol, Halton, and three Monte Carlo runs using RAN2 from [28] as the pseudorandom generator. We choose a standard automatic termination criterion. Namely, when two consecutive differences between consecutive approximations using $10,000\,i$, $i = 1, 2, \ldots, 100$, sample points become less than some threshold value for all of the tranches of the CMO, the computational process is terminated. With the threshold value set at 250, the Sobol run terminates at 160,000 sample points, the Halton run terminates at 700,000 sample points, and the three Monte Carlo runs terminate at 410,000, 430,000, and 780,000 sample points, respectively. Hence, the Sobol run terminates 2 to 5 times faster than the Monte Carlo runs.

We also stress that even though the Sobol algorithm terminates faster, the result is more accurate than two of the results achieved by the Monte Carlo algorithm. In Section 8, we show that the value of the integral for tranche A by using 20,000,000 sample points is about $m = 14,865,801$. Using this value of m, we compute the absolute error for each of the results at the point of termination:

- The error of 160,000 Sobol sample points is 379;

- The error of 700,000 Halton sample points is 69;

- The error of 410,000 random sample points with initial seed 1 is 1,298;

- The error of 430,000 random sample points with initial seed 1147902781 is 2,177;

- The error of 780,000 random sample points with initial seed 1508029952 is 180.

Automatic termination criteria are often used in computational practice. It is of interest to study the relation between the threshold value and the actual error of approximation. See Paskov [25] for the approximation of linear operators in the average case setting assuming that arbitrary linear continuous functionals can be computed. It is proved that standard termination criteria work well. The corresponding problem for approximation of linear operators in the average case setting with information consisting only of function values is still open.

We stress that the conclusions observed in this section for the Monte Carlo and Sobol algorithms hold also for the rest of the tranches. Although, the Halton algorithm performs better than the Monte Carlo algorithm for tranche A and a few other tranches it does not perform well for most of them. Therefore our emphasis for the remainder of this section will be on Sobol rather than Halton points.

7 Antithetic Variables

As already mentioned in Section 2, the expected error of the classical Monte Carlo algorithm depends on the variance of the integrand. Therefore by reducing the variance of the integrand the expected error would also decrease. Various variance reduction techniques such as importance sampling, control variates, antithetic variables and others, see for example [15], are often used with the classical Monte Carlo algorithm.

For the low discrepancy algorithms, the error bound depends on the variation of the integrand, see Section 3. Therefore the error bound will be decreased by reducing of the variation. Reducing the variation of the integrand has not been studied as extensively as reducing the variance. See, however, [4], [17], [20], and [36]. This is a major area for further research.

An important advantage of the classical Monte Carlo algorithm and of the deterministic algorithms studied here, is that they can be utilized very generally. This is important in a number of situations:

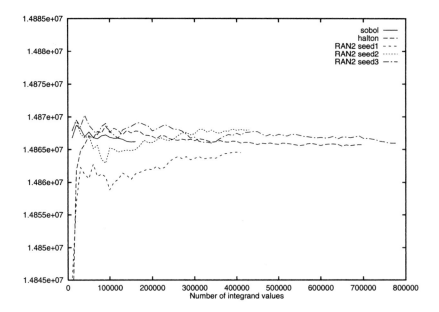

Figure 6: Automatic termination criterion applied to Sobol, Halton, and three Monte Carlo runs using RAN2 for tranche A

- If a financial house has a book with a wide variety of derivatives, it is advantageous to use algorithms which do not need to be tuned to a particular derivative;

- If a new derivative has to be priced, then there is no immediate opportunity to tailor a variance reduction technique to a particular integrand.

Although variance reduction techniques can be very powerful, they can require considerable analysis before being applied. We will therefore limit ourselves here to just one variance reduction technique; antithetic variables. The advantage of antithetic variables is that it can be easily utilized. Tests reveal that it is superior to the classical Monte Carlo algorithm for our CMO problem. We emphasize that antithetic variables is not a palliative; it can be inferior to the classical Monte Carlo algorithm.

For brevity, we refer to the antithetic variables variance reduction technique combined with the Monte Carlo algorithm as the antithetic variables algorithm. This algorithm is based on the identity

$$\int_D f(x)dx = \int_D g(x)dx, \quad \text{where } g(x) = \frac{1}{2}\left(f(x) + f(1-x)\right) \text{ for } f \in L_2(D).$$

Clearly, if the variance $\sigma^2(g)$ of g is much smaller than the variance $\sigma^2(f)$ of f then the Monte Carlo algorithm for g will have much smaller expected error.

All of the above random number generators are based on linear congruential generators with some additional features. For more details on these random number generators we refer to [27] and [28].

4.2 Systems

Since workstation clusters and networks provide cost-effective means to perform large-scale computation, we have built and debugged a software system under PVM 3.2 (Parallel Virtual Machine) for computing multivariate integrals. This system runs on a heterogeneous network of machines. PVM is a software package that allows a network of heterogeneous Unix computers to be used as a single large parallel computer. Thus large computational problems can be solved by using the aggregate power of many computers.

A master/slave model is used as the programming paradigm for computing multivariate integrals. In this model, the master program spawns and directs some number of slave processes. Each of the slave processes generates a sequence of sample points specified by the master and evaluates the integrand at those points. A partial sum of integrand values is returned to the master by each of the slave processes. The master combines partial sums as specified by an algorithm, computes an approximate value of the integral, and checks a termination criterion. If the termination criterion is not satisfied, the master spawns a new round of computations. This process continues until the termination criterion is satisfied or some prespecified upper bound of the number of sample points is reached. At the end, the master returns the final result of computation and timing information. In addition, the master process keeps information about each spawned process. If some host dies, the master reallocates its job to some other host.

In a multiuser network environment the load balancing method can be one of the most important factors for improving the performance. A dynamic load balancing method is used. Namely, the Pool of Tasks paradigm is especially suited for a master/slave program. In the Pool of Tasks method the master program creates and holds the 'pool' and sends out tasks to slave programs as they become idle. The fact that no communication is required between slave programs and the only communication is to the master makes our integration problem suitable for the Pool of Tasks paradigm.

5   Finance (CMO) Problem

We now consider a finance problem which is a typical Collateralized Mortgage Obligation (CMO). A CMO consists of several bond classes, commonly referred to as tranches, which derive their cash flows from an underlying pool of mortgages. The cash flows received from the pool of mortgages are

divided and distributed to each of the tranches according to a set of prespecified rules. The cash flows consist of interest and repayment of the principal.
The technique of distributing the cash flows transfers the prepayment risk
among different tranches. This results in financial instruments with varying
characteristics which might be more suitable to the needs and expectations
of investors. For more details on CMOs, we refer to [5]. We stress that the
amount of obtained cash flows will depend upon the future level of interest
rates. Our problem is to estimate the expected value of the sum of present
values of future cash flows for each of the tranches.

We now give some of the details related to the studied CMO and the way it
is reduced to the problem of multivariate integration. The CMO[1] consists of
ten tranches. Denote them by A, B, C, D, E, G, H, J, R, Z. Throughout this
section, we describe results on tranche A in more details. Results for other
tranches are similar unless stated explicitly. The underlying pool of mortgages
has a thirty-year maturity and cash flows are obtained monthly. This implies
360 cash flows. The monthly cash flows are divided and distributed according
to some prespecified rules. The actual rules for distribution are rather complicated and are given in details in the prospectus describing the financial
product.

For $1 \leq k \leq 360$, denote by

$$C \quad - \quad \text{the monthly payment on the underlying pool of mortgages;}$$
$$i_k \quad - \quad \text{the appropriate interest rate in month } k;$$
$$w_k \quad - \quad \text{the percentage prepaying in month } k;$$
$$a_{360-k+1} \quad - \quad \text{the remaining annuity after month } k.$$

Recall that the remaining annuity a_k is given by

$$a_k = 1 + v_0 + \cdots + v_0^{k-1}, \qquad k = 1, 2, \ldots, 360,$$

with $v_0 = 1/(1 + i_0)$ and i_0 the current monthly interest rate. In the notation
above, C and $a_{360-k+1}$ are constants; i_k and w_k are stochastic variables to be
determined below.

We now describe the interest rate model. Assume that the interest rate i_k
is of the form

$$i_k = K_0 e^{\xi_k} i_{k-1} = K_0^k i_0 e^{\xi_1 + \cdots + \xi_k},$$

where $\{\xi_k\}_{k=1}^{360}$ are independent normally distributed random variables with
mean 0 and variance σ^2, and K_0 is a given constant. In our case $\sigma^2 = 0.0004$
is chosen.

Suppose that the prepayment model w_k, as a function of i_k, is computed
as

$$w_k = w_k(\xi_1, \ldots, \xi_k) = K_1 + K_2 \arctan(K_3 i_k + K_4)$$
$$= K_1 + K_2 \arctan(K_3 K_0^k i_0 e^{\xi_1 + \cdots + \xi_k} + K_4),$$

---

[1]It is labeled as 'Fannie Mae REMIC Trust 1989-23'.

where K_1, K_2, K_3, K_4 are given constants.

The cash flow in month k, $k = 1, 2, \ldots, 360$, is

$$
\begin{aligned}
M_k &= M_k(\xi_1, \ldots, \xi_k) \\
&= C(1 - w_1(\xi_1)) \cdots (1 - w_{k-1}(\xi_1, \ldots, \xi_{k-1})) \times \\
&\quad (1 - w_k(\xi_1, \ldots, \xi_k) + w_k(\xi_1, \ldots, \xi_k)a_{360-k+1}).
\end{aligned}
$$

This cash flow is distributed to the tranches according to the rules of the CMO under consideration. Let $G_{k;T}(\xi_1, \ldots, \xi_k)$ be the portion of the cash flow M_k for month k directed to the tranche T. The form of the function $G_{k;T}$ is very complex. Here, it suffices to say that it is a continuous function which is a composition of min functions and smooth functions. By min function we mean a function which is the minimum of two functions.

To find the present value of the tranche T for month k, $G_{k;T}(\xi_1, \ldots, \xi_k)$ has to be multiplied by the discount factor

$$
u_k(\xi_1, \ldots, \xi_{k-1}) = v_0 v_1(\xi_1) \cdots v_{k-1}(\xi_1, \ldots, \xi_{k-1}),
$$

with

$$
v_j(\xi_1, \ldots, \xi_j) = \frac{1}{1 + i_j(\xi_1, \ldots, \xi_j)} = \frac{1}{1 + K_0^j i_0 e^{\xi_1 + \cdots + \xi_j}}, \qquad j = 1, 2, \ldots, 359.
$$

Summing up the present values for every month k, $k = 1, 2, \ldots, 360$, for tranche T will give us the present value PV_T,

$$
PV_T(\xi_1, \ldots, \xi_{360}) = \sum_{k=1}^{360} G_{k;T}(\xi_1, \ldots, \xi_k) u_k(\xi_1, \ldots, \xi_{k-1}).
$$

We want to compute the expected value $E(PV_T) = E(PV_T(\xi_1, \ldots, \xi_{360}))$. By change of variables, it is easy to see that

$$
E(PV_T) = \int_{[0,1]^{360}} PV_T(y_1(x_1), \ldots, y_{360}(x_{360})) \, dx_1 \cdots dx_{360},
$$

where $y_i = y_i(x_i)$ is implicitly given by

$$
x_i = \frac{1}{\sqrt{2\pi}\sigma} \int_{-\infty}^{y_i} e^{-t^2/(2\sigma)} \, dt. \tag{8}
$$

Therefore, our problem is reduced to a problem of computing ten multivariate integrals over the 360-dimensional unit cube. We stress that after generating a point $(x_1, \ldots, x_{360})$ in the 360-dimensional unit cube, the point $(y_1, \ldots, y_{360})$ has to be computed by finding the value of the inverse normal cumulative distribution function at each $x_k, k = 1, 2, \ldots, 360$. We use software available through NETLIB to compute the points $(y_1, \ldots, y_{360})$. Then the function values $PV_T(y_1, \ldots, y_{360})$ are computed for all tranches T.

We now discuss some of the smoothness properties of the integrand PV_T, where T is one of the tranches. Since the change of variables in (8) is based on smooth functions we may restrict our discussion to PV_T as a function of $y_1, \ldots, y_{360}$. Recall that

$$PV_T(y_1, \ldots, y_{360}) = \sum_{k=1}^{360} G_{k;T}(y_1, \ldots, y_k)u_k(y_1, \ldots, y_{k-1}).$$

Clearly, u_k is a smooth function. As mentioned before, $G_{k;T}$ is a composition of min functions and smooth functions. That is why we believe that the function PV_T has finite variation in the sense of Hardy and Krause and, that is why we decided to use low discrepancy sequences for these type of integrands. Of course, it would be desirable to have a good bound on the variation of PV_T. Unfortunately, due to the complex form of PV_T, this is a very difficult task.

As reported in the Introduction, it is widely believed that the theoretical advantage of low discrepancy sequences degrades with increase of dimension and low discrepancy sequences are effective, in general, only for $d \leq 30$. Although the CMO problem is apparently of dimension 360, some of the tranches are of lower dimension. To understand the performance of the different algorithms for the CMO problem it is important to know the dimensions of the various tranches.

The integrand PV_T of tranche T depends on the cash flows $G_{k;T}$. The cash flow $G_{k;T}$ has an important property. If the tranche T is retired at month k_T, i.e., the whole principal is paid off, then $G_{k;T}$ is zero for month $k > k_T$. More precisely,

$$G_{k;T}(\xi_1, \ldots, \xi_k) = 0 \qquad \forall k > k_T = k_T(\xi_1, \ldots, \xi_{k_T})$$

with $\xi_1, \ldots, \xi_k$ corresponding to a particular interest rate scenario.

We checked computationally that k_T is typically smaller than 360. We generated $n = 3,000,000$ random sample points and determined the maximum k_T for each tranche T. These numbers are reported in Table 1. As we see, the dimension of only one tranche is 360. However, for the other tranches, the dimension is still high.

The concept of the dimension of a tranche can be further relaxed. Intuitively, if the dependence of PV_T is negligible on the variables $x_{k+1}, x_{k+2}, \ldots, x_{360}$ then only the first k variables $x_1, x_2, \ldots, x_k$ contribute substantially to the integral of PV_T. This concept of *effective dimension* can be defined rigorously as follows.

Let f be a function with domain $[0,1]^d$ and let $\varepsilon > 0$. The *effective dimension* $K(\varepsilon)$ of f is the smallest integer $k \in [1, d]$ such that

$$\left| \int_{[0,1]^k} f(x_1, \ldots, x_k, 0, \ldots, 0) \, dx_1 \cdots dx_k - \int_{[0,1]^d} f(x) dx \right| \leq \varepsilon \left| \int_{[0,1]^d} f(x) dx \right|.$$

Tranche	Maximum of k_T
A	189
B	250
C	278
D	298
E	309
G	77
H	91
J	118
R	360
Z	167

Table 1: The maximum of the numbers k_T obtained by generating 3,000,000 random sample points for each tranche of the CMO problem

We apply this definition of effective dimension to the ten tranches of the CMO problem. Clearly, the effective dimension is a function of ε. What value of ε should we choose? In practice, the CMO problem does not have to be solved with high accuracy, see also the discussion in Section 8, and so we choose $\varepsilon = 0.001$. We checked computationally that the values of the integrals of all tranches are about $a10^6$ with $a \in [2, 42]$. Hence by taking $\varepsilon = 0.001$ we introduce an absolute error which is on the order of several thousand dollars.

We estimate $K_T(\varepsilon)$ for $\varepsilon = 0.001$, where $K_T(\varepsilon)$ denotes the effective dimension of tranche T. To compute $K_T(\varepsilon)$ we need a fairly accurate approximation of the integral $I(PV_T) = \int_{[0,1]^{360}} PV_T(x)dx$. We approximate $I(PV_T)$ by using 20,000,000 sample points as explained in Section 8. Using these results we approximately compute the effective dimensions for each of the tranches. The results are summarized in Table 2. As we see, the effective dimension, although smaller, is still high for most of the tranches.

6   Comparison of the Deterministic and Monte Carlo algorithms

We now present the results of extensive testing of the deterministic and Monte Carlo algorithms for the CMO problem. For the reader's convenience, the results are summarized in a number of graphs.

Figure 3 shows the results for tranche A of Sobol, Halton, and Monte Carlo runs with two randomly chosen initial seeds. The pseudorandom generator RAN2 from [28] is used to generate random sample points for the Monte Carlo runs. Figure 3 exhibits behavior that is common in all our comparisons of the

Tranche	$K_T(\varepsilon)$
A	131
B	175
C	212
D	239
E	261
G	42
H	63
J	84
R	338
Z	114

Table 2: The approximate values of $K_T(\varepsilon)$ for $\varepsilon = 0.001$

Sobol algorithm with the Monte Carlo algorithm. For the Halton algorithm see the discussion at the end of this section.

- The Monte Carlo algorithm is sensitive to the initial seed;

- The deterministic algorithms, especially the Sobol algorithm, converge significantly faster than the Monte Carlo algorithm;

- The convergence of the deterministic algorithms, especially of the Sobol algorithm, is smoother than the convergence of the Monte Carlo algorithm. This makes automatic termination easier for the Sobol algorithm; see the discussion below;

- The Sobol algorithm outperforms the Halton algorithm.

Figure 4 shows the same Sobol and Halton runs but a different random number generator, namely ran1 from [27], is used to generate four Monte Carlo runs using four randomly chosen initial seeds. The plot again exhibits sensitivity of the Monte Carlo algorithm to the initial seed. The plot also indicates that there are some problems with this random number generator since the Monte Carlo results seem to lie on horizontal lines when the number of sample points exceeds 300,000 and it seems unlikely that they will converge to the same value. This claim is also supported by the fact that the same effect has been observed for sixteen additional Monte Carlo runs. We assume that is why ran1 has been replaced by a different random number generator in [28]. Figure 4 is included to show that the results can be affected by the poor performance of the random number generator.

Figure 5 plots the same Sobol and Halton runs versus the arithmetic mean of twenty Monte Carlo runs using RAN2 from [28]. We stress that the number

Tranche	r
A	2.82
B	2.78
C	3.09
D	3.61
E	4.70
G	2.80
H	2.81
J	3.14
R	1.55
Z	2.50

Table 3: The ratios r for the ten tranches of the CMO

We must, of course, remember that the cost of one evaluation of g is equal to the cost of two function evaluations of f. Hence, the antithetic variables algorithm is preferable to the Monte Carlo algorithm only if the reduction of the variance is by at least a factor of two. In general, let

$$r = \frac{\sigma(f)}{\sqrt{2}\,\sigma(g)}.$$

Then the expected error of the antithetic variables algorithm is r-times smaller than the expected error of the Monte Carlo algorithm. Furthermore, the cost of both algorithms is about the same. Since r can be smaller than one for some functions, the antithetic variables variance reduction technique, although simple to use, does not work in general and should be used with care, see also [15].

We tested the antithetic variables algorithm for the CMO problem with the ten tranches. Let f_T be the integrand for tranche T. We approximately computed $\sigma(f_T)$, $\sigma(g_T)$, and $r_T = \sigma(f_T)/(\sqrt{2}\,\sigma(g_T))$ for all tranches T. The results for the ratios r_T are given in Table 3. Hence, the antithetic variables algorithm has at least 1.55 and at most 4.70 smaller expected error than the Monte Carlo algorithm for the ten tranches of the CMO. Therefore the antithetic variables algorithm works better than the Monte Carlo algorithm for the CMO problem.

We also tested the antithetic variables technique combined with Sobol points for the CMO problem. Since it did not perform as well as the Sobol algorithm we omit the results.

We now present graphs of the results obtained for the deterministic and antithetic variables algorithms for the CMO problem.

Figure 7 is analogous to Figure 3 in Section 6. It shows the results of Sobol, Halton, and antithetic variables runs with two randomly chosen initial seeds. Again, Figure 7 exhibits the sensitivity of the antithetic variables algorithm

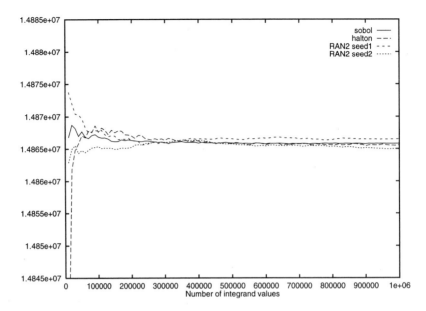

Figure 7: Sobol and Halton runs for tranche A and two antithetic variables runs using RAN2

to the initial seed. However, as it should be expected the antithetic variables algorithm works better than the Monte Carlo algorithm. That is why, the spread and the jaggedness of antithetic variables runs in Figure 7 are smaller than the corresponding ones for Monte Carlo runs in Figure 3.

Figure 8 plots the same Sobol and Halton runs versus the arithmetic mean of twenty antithetic variables runs using RAN2 from [28]. We stress that the number of sample points on the x-axis is correct only for the deterministic algorithms. The actual number of sample points for the averaged antithetic variables graph is twenty times the number of sample points on the x-axis. Clearly, the Sobol and the averaged antithetic variables graphs are very close after 450,000 points.

8   Small Number of Sample Points

In this section we compare the performance of the deterministic algorithms with the Monte Carlo and antithetic variables algorithms for a small number of sample points. Results for a small number of points are sometimes of special importance for people who evaluate CMOs and other derivative products. They need algorithms which can evaluate a derivative in a matter of minutes. Rather low accuracy, on the order of 10^{-2} to 10^{-4}, is often sufficient. The integrands are complicated and computationally expensive. Furthermore, many

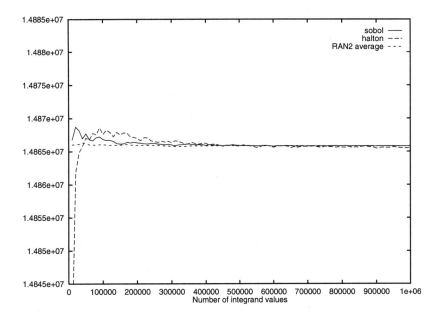

Figure 8: Sobol and Halton runs for tranche A and an average of twenty antithetic variables runs using RAN2

may have to be evaluated on a daily basis with limited computational resources, such as workstations. We must, therefore, limit ourselves to only a small number of sample points.

In this section, by a small number of points we mean 4000 sample points. As we shall see, this still leads to reasonable results and takes less than a couple of minutes of workstation CPU time. We believe that this sample size may yield comparable results for other mortgage-backed securities and interest rate derivatives.

We drop the Halton algorithm from consideration in this section since it is outperformed by both the Monte Carlo and antithetic variables algorithms for 4000 sample points.

¿From now on, let f be the integrand for tranche A and let

$$g(x) = \frac{1}{2} \left(f(x) + f(1 - x) \right), \qquad \text{for } x \in D.$$

Consider the twenty results

$$U_{MC}^{(1)}(f), U_{MC}^{(2)}(f), \ldots, U_{MC}^{(20)}(f) \tag{9}$$

obtained by the Monte Carlo algorithm and the twenty results

$$U_{AV}^{(1)}(g), U_{AV}^{(2)}(g), \ldots, U_{AV}^{(20)}(g) \tag{10}$$

obtained by the antithetic variables algorithm with different initial seeds, each using 4,000 f function evaluations. For 4000 Sobol points, we obtained

$$U_S(f) = 14,868,261. \tag{11}$$

To compute the relative errors of all presented results, we need to know the true or almost true value $I(f)$ of the integral. In Section 7 we showed that the antithetic variables algorithm works better than the Monte Carlo algorithm for the CMO problem. That is why it is reasonable to compute an approximation m to $I(f)$ using 20,000,000 function evaluations with the antithetic variables algorithm. For tranche A, we obtain

$$I(f) \approx m = 14,865,801.$$

We now compute the relative error for each of the results in (9), (10), and (11). In Table 4, we indicate how many times out of 20 the Sobol algorithm has a smaller relative error than the antithetic variables algorithm. As we see, the Sobol algorithm wins for 8 of the tranches, it ties for 1, and it loses for 1. However, for 5 of the tranches the Sobol algorithm wins decisively. In total, the Sobol algorithm wins 139 times and loses 61 times. Therefore, the antithetic variables algorithm sometimes performs better than the Sobol algorithm but in most cases it does not.

In Table 5 we report the smallest and the largest relative errors of the twenty antithetic variables runs in (10). The last column of Table 5 reports the relative errors for the Sobol algorithm.

Note that each of these runs performs very well since the relative error is always at most 0.005. The Sobol algorithm performs even better; it achieves accuracies within range from one part in a thousand to one part in a million, depending on the tranche. We add that it takes approximately 103 seconds to compute the Sobol results and about 113 seconds to compute the antithetic variables results for the ten tranches; each algorithm uses 4000 function evaluations and it is run on a Sun-4/630-M140 workstation.

The results of comparison between the Monte Carlo and the Sobol algorithms are summarized in Table 6 and Table 7. As it can be expected, the performance of the Sobol algorithm versus the Monte Carlo algorithm is even better than the performance of the Sobol algorithm versus the antithetic variables algorithm. In this case, the Sobol algorithm wins decisively for all tranches. In total, the Sobol algorithm wins 177 times and loses 23 times.

9   Statistical Analysis

In this section, we perform statistical analysis of the Monte Carlo and antithetic variables algorithms for a small number of points. We also compare these two randomized algorithms with the Sobol algorithm.

Tranche	Antithetic variables	Sobol
A	9	11
B	1	19
C	6	14
D	10	10
E	11	9
G	2	18
H	3	17
J	2	18
R	8	12
Z	9	11

Table 4: Number of 'wins' of the antithetic variables algorithm and the Sobol algorithm

Tranche	The smallest error	The largest error	Sobol error
A	9.850831e-06	1.078310e-03	1.654482e-04
B	1.684136e-06	8.965597e-04	1.766718e-06
C	2.351182e-06	1.017674e-03	2.154940e-04
D	3.762572e-06	8.231360e-04	2.626482e-04
E	2.111560e-05	5.663271e-04	2.243416e-04
G	1.179616e-06	1.677086e-04	7.305784e-06
H	2.368361e-06	6.108946e-04	3.811441e-05
J	6.517283e-06	7.988188e-04	3.172316e-05
R	3.910856e-05	4.748614e-03	1.179118e-03
Z	1.456121e-05	1.979723e-03	3.784033e-04

Table 5: The smallest and the largest relative error of 20 antithetic variables results and the relative error of the Sobol result each using 4000 sample points

Again, we study tranche A. Consider the one thousand results

$$U_{MC}^{(1)}(f), U_{MC}^{(2)}(f), \ldots, U_{MC}^{(1000)}(f) \tag{12}$$

obtained by the Monte Carlo algorithm and the one thousand results

$$U_{AV}^{(1)}(g), U_{AV}^{(2)}(g), \ldots, U_{AV}^{(1000)}(g) \tag{13}$$

obtained by the antithetic variables algorithm with different initial seeds, each using 4,000 evaluations of f. Let

$$m_{MC} = \frac{1}{1000} \sum_{i=1}^{1000} U_{MC}^{(i)}(f) = 14,864,881,$$

Tranche	Monte Carlo	Sobol
A	3	17
B	0	20
C	3	17
D	3	17
E	2	18
G	0	20
H	0	20
J	0	20
R	8	12
Z	4	16

Table 6: Number of 'wins' of the Monte Carlo algorithm and the Sobol algorithm

Tranche	The smallest error	The largest error	Sobol error
A	3.339877e-05	2.633738e-03	1.654482e-04
B	1.037307e-04	3.040330e-03	1.766718e-06
C	3.329284e-05	3.341837e-03	2.154940e-04
D	1.099033e-04	3.536105e-03	2.626482e-04
E	7.673315e-05	3.665118e-03	2.243416e-04
G	3.094756e-05	5.107036e-04	7.305784e-06
H	9.123779e-05	1.454661e-03	3.811441e-05
J	5.152820e-05	1.768061e-03	3.172316e-05
R	3.564157e-05	9.429284e-03	1.179118e-03
Z	5.896356e-05	3.749949e-03	3.784033e-04

Table 7: The smallest and the largest relative error of 20 Monte Carlo results and the relative error of the Sobol result each using 4000 sample points

$$m_{AV} = \frac{1}{1000} \sum_{i=1}^{1000} U_{AV}^{(i)}(g) = 14,865,749$$

be the sample means, and let

$$s_{MC} = \frac{1}{999} \sum_{i=1}^{1000} (U^{(i)}(f) - m_{MC})^2 = 216,929,148,$$

$$s_{AV} = \frac{1}{999} \sum_{i=1}^{1000} (U^{(i)}(g) - m_{AV})^2 = 37,331,468$$

be the sample variances of these one thousand Monte Carlo and antithetic variables results, respectively.

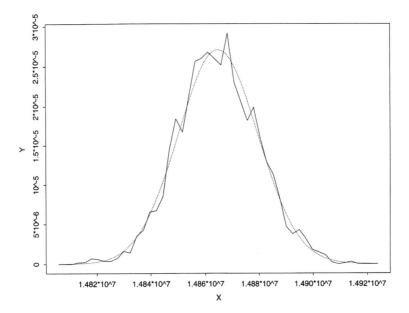

Figure 9: Density function (continuous line) based on one thousand Monte Carlo results each using 4,000 sample points generated by RAN2 and the normal density function (dashed line) with mean m_{MC} and variance s_{MC}

According to the Central Limit Theorem, see e.g. [15], the distributions of $U_n(f)$ and $U_n(g)$ are normal for $n \to \infty$ with the same mean $\int_D f(x)dx = \int_D g(x)dx$ and variances $\sigma^2(f)/n$ and $\sigma^2(g)/n$, respectively. To check how far we differ from the normal distribution we plot the density functions of these one thousand results, see Fig. 9 and Fig. 10, by using the statistical package S-Plus. We conclude that these are very good approximations of the normal distribution.

Further statistical analysis require the knowledge of the true value m of the integral. In Section 8, we showed that it is reasonable to assume that $m \approx 14,865,801$. We seek the number of antithetic variables runs, each using 4,000 sample points, which after averaging would give an error δ with probability η. This procedure is widely used in statistical analysis, see [13].

That is, let $\xi_1, \xi_2, \ldots, \xi_k$ be the results for k antithetic variables runs each using 4,000 f function evaluations. As already discussed, we may assume that they are approximately normally distributed with mean m and variance $s = s_{AV}$. Hence we want to find k such that

$$P\left\{\left|\frac{1}{k}\sum_{i=1}^{k}\xi_i - m\right| \le \delta\right\} \ge 1 - \eta.$$

Note that the sum $\frac{1}{k}\sum_{i=1}^{k}\xi_i$ is normally distributed with mean m and variance

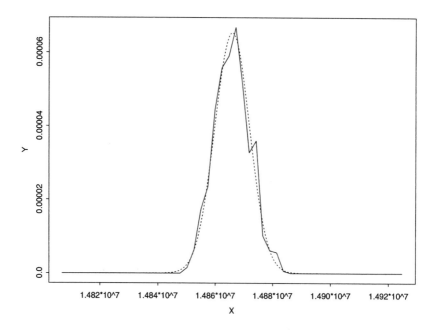

Figure 10: Density function (continuous line) based on one thousand anti-thetic variables results each using 4,000 sample points generated by RAN2 and the normal density function (dashed line) with mean m_{AV} and variance s_{AV}

s/k, i.e., standard deviation $\sqrt{s/k}$. To guarantee an error δ with confidence level 68%, i.e., with $\eta = \sqrt{2/\pi} \int_0^1 e^{-t^2/2} dt \approx 0.68269$ which corresponds to one standard deviation from the mean, we need to determine k_1 such that $\sqrt{s/k_1} = \delta$. In this case,

$$k_1 = s/\delta^2 = 37,331,468/\delta^2. \tag{14}$$

To guarantee an error δ with confidence level 95%, i.e., with

$$\eta = \sqrt{2/\pi} \int_0^2 e^{-t^2/2} dt \approx 0.9545$$

which corresponds to two standard deviations from the mean, we need to determine k_2 such that $2\sqrt{s/k_2} = \delta$. Therefore, in this case,

$$k_2 = 4s/\delta^2 = 149,325,872/\delta^2. \tag{15}$$

We now compare the performance of the Sobol algorithm with the anti-thetic variables algorithm. Since $U_S(f) = 14868261$ is the result for 4,000

Tranche	with confidence 68%	with confidence 95%
A	3	10
B	20	79
C	4	15
D	2	8
E	2	7
G	9	36
H	7	27
J	12	47
R	4	15
Z	2	8

Table 8: Number of antithetic variables runs each using 4,000 random points needed to achieve the worst error of the n Sobol points with $n = 4100 - 20i$, $i = 0, 1, \ldots, 9$

Sobol points, the error in this case is

$$|U_S(f) - m| = 2460.$$

Set $\delta = 2460$. Due to (14), to achieve an error at most δ of 4,000 Sobol points with confidence level 68%, we need to obtain approximately 7 antithetic variables runs each using 4,000 sample points and then to average them. Due to (15), to achieve error at most δ with confidence level 95%, we need to obtain approximately 25 antithetic variables runs.

Clearly, the above analysis depends on the assumption about the true answer of the integral. Furthermore, one may argue that by chance the result from Sobol points might happen to be very close to the true answer. That is why we performed similar analysis for the ten Sobol results that use $n = 4,100 - 20i$, $i = 0, 1, \ldots, 9$, sample points and we set δ as the worst error of the ten Sobol results for tranche A. Then 7 antithetic variables runs obtained before are replaced by 3 runs.

Results for the other tranches are summarized in Table 8. We emphasize that the number of antithetic variables runs are for the worst error of n Sobol points for $n = 4,100 - 20i$, $i = 0, 1, \ldots, 9$. Hence, for a fixed i, the number of antithetic variables runs could be even higher than the corresponding number in Table 8. From Table 8 we conclude that we need from 7 to 79 antithetic variables runs each using 4,000 sample points to achieve an error with confidence level 95% comparable with the worst error of n Sobol points, $n = 4100 - 20i$, $i = 0, 1, \ldots, 9$.

For the Monte Carlo algorithm, we proceed analogously. The results are summarized in Table 9. From Table 9 we conclude that we need from 27 to 607 Monte Carlo runs each using 4,000 sample points to achieve an error

Tranche	with confidence 68%	with confidence 95%
A	14	56
B	153	607
C	44	173
D	35	137
E	53	212
G	64	255
H	38	153
J	67	266
R	13	52
Z	7	27

Table 9: Number of Monte Carlo runs each using 4,000 random points needed to achieve the worst error of the n Sobol points with $n = 4100 - 20i$, $i = 0, 1, \ldots, 9$

with confidence level 95% comparable with the worst error of n Sobol points, $n = 4100 - 20i$, $i = 0, 1, \ldots, 9$.

Remark 1

Similar statistical analysis may also be performed for a large number of sample points. Due to the large computational cost of these simulations, we compare Sobol results with twenty antithetic variables results (instead of 1000 results as for a small number of sample points), each with different initial seed and using 1,000,000 function evaluations. Results for all tranches are summarized in Table 10. We emphasize that the number of antithetic variables runs are for the worst error of n Sobol points for $n = 1,050,000 - 10,000i$, $i = 0, 1, \ldots, 9$.

By comparing Table 8 and Table 10, we conclude that the superiority of the Sobol algorithm over the antithetic variables algorithm for a large number of sample points is greater than for a small number of sample points.

Similar conclusion also holds for the Monte Carlo algorithm. Proceeding analogously, we summarize the results of this comparison in Table 11.

10 Timing Results and Parallel Speedup for the CMO Problem

Since workstation clusters and networks provide a cost-effective means to perform large-scale computation, we have built and debugged a distributed software system under PVM 3.2 for computing multivariate integrals on a network of workstations. PVM is a software package that allows a network of

Tranche	with confidence 68%	with confidence 95%
A	55	219
B	58	230
C	26	103
D	27	108
E	15	60
G	8	32
H	4	16
J	5	18
R	13	50
Z	5	18

Table 10: Number of antithetic variables runs each using 1,000,000 random points needed to achieve the worst error of the n Sobol points with $n = 1,050,000 - 10,000i$, $i = 0, 1, \ldots, 9$

Tranche	with confidence 68%	with confidence 95%
A	434	1735
B	443	1769
C	246	982
D	350	1399
E	331	1324
G	62	245
H	31	122
J	44	176
R	30	120
Z	28	110

Table 11: Number of Monte Carlo runs each using 1,000,000 random points needed to achieve the worst error of the n Sobol points with $n = 1,050,000 - 10,000i$, $i = 0, 1, \ldots, 9$

heterogeneous Unix computers to be used as a single large parallel computer. In this section we report the timing results of the different generators on a single workstation. Then the speedup of the distributed on a network of workstations software system is also measured for the CMO problem.

The CPU time in seconds for simultaneous evaluation of the ten tranches is given in Table 12 for Sobol, Halton, RAN2 from [28], RAN1 from [28], ran1 from [27], and ran2 from [27] generators. The real time in seconds, as should be expected, is slightly higher than the CPU time. It is given in the second column since it is later compared with the real time for the network of work-

Generator	CPU time	Real time
Sobol	25634	25654
Halton	29814	29881
RAN2	31501	31607
RAN1	28092	28126
ran1	30911	30946
ran2	28320	28332

Table 12: Timing results in seconds for evaluation of the CMO using 1,000,000 sample points on a single workstation

stations to derive the parallel speedup. These results have been obtained on a single Sun-4/630-M140 at the Department of Computer Science, Columbia University. All programs have been compiled with the gcc v2.4 compiler with the highest level O2 of optimization. Clearly, the generator of Sobol points is the fastest one. However, we do not regard this as the most significant reason for the use of Sobol points.

Since the speed of the Sun-4/630-M140 is 4.2 MFLOPS, we note from the above table that it takes approximately 107,500 floating point operations to generate one 360-dimensional Sobol point and to evaluate integrands for this problem. Since the cost of generating one 360-dimensional Sobol point is significantly smaller than the evaluation of the integrands, it takes about 100,000 floating point operations to evaluate only integrands. Clearly, this is a rather rough estimate. We stress that some interest rate and prepayment models used in practice are significantly more complicated and more time-consuming to evaluate than the ones considered in this paper.

We now discuss the parallel speedup for the finance problem achieved with the distributed on a network of workstations software system. The software system for computing multivariate integrals was tested on up to 25 SUN workstations and the timing results were measured. The random number generator RAN2 and the Sobol generator were used. The results are given in Table 13. We now compare the real time entry of RAN2 in Table 12 with entries in Table 13. Let N be the number of workstations. In our case N is 5,10,15,20,25. The speedup for the first three entries is about $0.99N$. Then it degrades slightly to $0.92N$ and $0.9N$ for the fourth and the fifth entries, respectively. We stress that this degradation of the performance is partially due to the fact that the last several machines added for the test, were slightly slower than the Sun-4/630-M140, which was used to measure the time on a single workstation.

Similar speedups are measured for the Sobol generator. The speedups for the first three entries are $0.96N, 0.98N, 0.98N$, respectively. The speedup for the last two entries is $0.91N$.

Number of machines	Real time for RAN2	Real time for Sobol
5	6382	5360
10	3170	2625
15	2113	1750
20	1705	1408
25	1398	1125

Table 13: Timing results in seconds for evaluation of the CMO using 1,000,000 sample points generated by RAN2

Therefore, networks of workstations are a cost-effective way to perform these computations. The measured speedup for the parallel/distributed software system for both the Monte Carlo and deterministic algorithms is at least $0.9N$ for $N \leq 25$.

11 Discussion and Future Directions

In this section we present some thoughts about why the Sobol points are so successful. We end with some directions for future research.

We give two reasons which we believe explain, at least in part, the success of Sobol points in solving the CMO problem.

First, let us assume that there are no prepayments. This implies that the cash flow from the underlying pool of mortgages for every month is constant. Due to the time value of the money this means that the present value of the cash flow is a decreasing function of the number of the month. In other words, the present value of the cash flow for earlier months is greater than the present value of the cash flow for later months. The presence of prepayments and rules of distribution to different tranches will distort this monotonicity property. However, this property will be reflected to some extent in the cash flows of the ten tranches. It is a well-known property of the Sobol points[2], see [33], that the low dimensional components are more uniformly distributed than the high dimensional components. As pointed out by Sobol in [35], the Sobol points will be more efficient for numerical integration of functions for which the influence of the x_j-th variable decreases as j increases. Since this property also holds for many other financial derivatives we expect that the Sobol points will provide a powerful alternative to the Monte Carlo and antithetic variables algorithms.

Second, appropriate choices for the initial direction numbers increase the uniformity of the Sobol points in the 360-dimensional unit cube. We believe

---

[2]Other low discrepancy sequences also satisfy this property, see [21]

that our choices of the initial direction numbers contributes to the successful performance of the Sobol points.

The Halton algorithm did not perform as well as the Sobol algorithm. We believe that this behavior is due to the fact that the Halton points are less uniformly distributed than the Sobol points; especially in high dimensions and sample sizes of 1,000,000 or less, see also [18].

We end this section by suggesting some directions for future work:

- Compare the performance of low discrepancy and Monte Carlo algorithms on other financial derivatives;

- Test the performance of other known low discrepancy sequences on various derivatives;

- As mentioned in Section 8, results for a small number of samples are often of special interest in finance. It would be attractive to design new deterministic sequences which are very uniformly distributed for a small number of points;

- Characterize analytic properties of classes of financial derivatives and design new algorithms tuned to these classes;

- Study error reduction techniques for deterministic algorithms;

- There are numerous open theoretical problems concerning high dimensional integration and low discrepancy sequences; see [42] for some of them. We believe that their solution will aid in the design of better algorithms for finance problems.

Acknowledgements

This research was supported in part by the the National Science Foundation and the Air Force Office of Scientific Research. I am grateful to Goldman Sachs for providing the finance problem.

An earlier report on this work was presented by Prof. J. Traub at the Bank of England Conference at the Isaac Newton Institute, Cambridge, U.K. in May 1995.

I express my gratitude to I. Vanderhoof for his invaluable help, numerous discussions and suggestions, and introducing me to Goldman Sachs.

I also express my gratitude to J. Traub and H. Woźniakowski for numerous discussions, suggestions, and guidance on research and on the preparation of this paper.

I would like to thank A. Belur for his help, advice, and cooperation in providing the finance problem.

I am also grateful to T. Boult and S. Baker for their numerous discussions and suggestions during the preparation of this paper.

I appreciate discussions with N. Marinovich, J. Tilley, J. Langsam, P. Karasinski, D. Schutzer, B. Bojanov, S. Tezuka, A. Werschulz, I. M. Sobol, B. Shukhman, V. Temlyakov, S. Heinrich, E. Novak, and T. Warnock.

I would like sincerely to thank the Mortgage Research Group at Goldman Sachs and people associated with it and particularly P. Niculescu, R. Wertz, and J. Davis who gave me a chance to greatly improve my knowledge in mortgage-backed securities and interest rate derivatives.

I also appreciate the help of P. Cheah in building the distributed platform used in the software system for computing integrals.

References

[1] Adler, S.J., *The Geometry of Random Fields*, Wiley, 1981.

[2] Antonov, I.A., & Saleev, V.M., An Economic Method of Computing LP_τ-sequences, *USSR Computational Mathematics and Mathematical Physics* **19**, 252–256, 1979.

[3] Bratley, P. & Fox, B.L., Algorithm 659, implementing Sobol's quasirandom sequence generator, *ACM Trans. Math. Software* **14**, 88–100, 1988.

[4] Chelson, P., *Quasi-random Techniques for Monte Carlo Methods*, PhD Dissertation, The Claremont Graduate School, 1976.

[5] Fabozzi, F.J., *Handbook of Mortgage Backed Securities*, Probus Publishing Co., 1992.

[6] Fabozzi, F.J., *Fixed Income Mathematics*, Probus Publishing Co., 1988.

[7] Fox, B.L., Algorithm 647, implementation and relative efficiency of quasirandom sequence generators, *ACM Trans. Math. Software* **12**, 362–376, 1986.

[8] Geweke, J., Monte Carlo Simulations and Numerical Integration. To appear in *Handbook of Computational Economics* edited by H. Amman, D. Kendrick, & J. Rust, Elsevier, 1996.

[9] Halton, J.H., On the efficiency of certain quasi-random sequences of points in evaluating multi-dimensional integrals, *Num. Math.* **2**, 84–90, 1960.

[10] Halton, J.H. & Smith, G.B., Algorithm 247, radical-inverse quasi-random point sequence, *Commun. ACM* **7**, 701–702, 1964.

[11] Halton, J.H., A retrospective and prospective survey of the Monte Carlo method, *SIAM Review* **12**, 1–63, 1970.

[12] Hammersley, J. & Handscomb, D., *Monte Carlo Methods*, Methuen, 1964.

[13] Hoel, P.G., *Elementary Statistics*, Wiley, 1976.

[14] Janse van Rensburg, E.J. & Torrie, G.M., Estimation of multidimensional integrals: Is Monte Carlo the best method? *J. Phys. A.: Math. Gen.* **26**, 943–953, 1993.

[15] Kalos, M.H. & Whitlock, P.A., *Monte Carlo Methods*, Volume I, Wiley, 1986.

[16] Knuth, D.E., *Seminumerical Algorithms*, vol. 2 of *The Art of Computer Programming*, Addison Wesley, 1981.

[17] Maize, E., *Contributions to the theory of error reduction in quasi-Monte Carlo methods*, PhD Dissertation, The Claremont Graduate School, 1981.

[18] Morokoff, W.J. and Caflisch, R.E., Quasi-random sequences and their discrepancies, *SIAM J. Scientific Computing* **15**, 1251–1279, 1994.

[19] Morokoff, W.J. and Caflisch, R.E., Quasi-Monte Carlo integration, *J. Comp. Physics* **122**, 218–230, 1995.

[20] Moskowitz, B. & Caflisch, R.E., Smoothness and dimension reduction in quasi-Monte Carlo methods, to appear in *Math. Comp. Modeling*.

[21] Niederreiter, H., *Random Number Generation and Quasi-Monte Carlo Methods*, CBMS-NSF, 63, SIAM , 1992.

[22] Novak, E., *Deterministic and Stochastic Error Bounds in Numerical Analysis*, Lecture Notes in Math., Springer Verlag, 1988.

[23] Paskov, S.H., Average case complexity of multivariate integration for smooth functions, *J. Complexity* **9**, 291–312, 1993.

[24] Paskov, S.H., Computing high dimensional integrals with applications to finance, *Joint Summer Research Conference on Continuous Algorithms and Complexity,* Mount Holyoke College, June, 1994.

[25] Paskov, S.H., Termination criteria for linear problems, *J. Complexity* **11**, 105–137, 1995.

[26] Paskov, S.H. & Traub, J.F., Faster valuation of financial derivatives, *The Journal of Portfolio Management* **22**, 113–120, 1995.

[27] Press, W., Teukolsky S., Vetterling, W., & Flannery, B., *Numerical Recipes in C*, First Edition, Cambridge University Press, 1988.

[28] Press, W., Teukolsky S., Vetterling, W., & Flannery, B., *Numerical Recipes in C*, Second Edition, Cambridge University Press, 1992.

[29] Roth, K.F., On irregularities of distribution, *Mathematika* **1**, 73–79, 1954.

[30] Roth, K.F., On irregularities of distribution, IV, *Acta Arith.* **37**, 67–75, 1980.

[31] Rust, J., Using randomization to break the curse of dimensionality, *Social Systems Research Institute Working Paper Series*, No. 9429, 1994.

[32] Sarkar, P.K. & Prasad, M.A., A comparative study of pseudo and quasi random sequences for the solution of integral equations, *J. Computational Physics* **68**, 66–88, 1978.

[33] Sobol, I.M., On the distribution of points in a cube and the approximate evaluation of integrals, *USSR Computational Mathematics and Mathematical Physics* **7**, 86–112, 1967.

[34] Sobol, I.M., *Numerical Monte Carlo Methods* (in Russian), Izdat 'Nauka', Moscow, 1973.

[35] Sobol, I.M., Quadrature formulas for functions of several variables satisfying general Lipschiz condition, *USSR Computational Mathematics and Mathematical Physics* **29**, 935–941, 1989.

[36] Spanier, J. & Maize, E., Quasi-random methods for estimating integrals using relatively small samples, *SIAM Review* **36**, 19–44, 1994.

[37] Tezuka, Shu, A generalization of Faure sequences and its efficient implementation, *Technical Report, IBM Research, Tokyo*, 1994.

[38] Traub, J.F., Average case computational complexity of high-dimensional integration with applications to finance, *NSF Symposium on Simulation and Estimation*, Department of Economics, University of California, Berkeley, August, 1994.

[39] Traub, J.F., Solving hard problems with applications to finance, *Thirteenth World Computer Congress, IFIP 94*, Hamburg, August, 1994.

[40] Traub, J.F., Wasilkowski, G.W., & Woźniakowski, H., *Information-based Complexity*, Academic Press, 1988.

[41] Traub, J.F. & Woźniakowski, H., The Monte Carlo algorithm with a pseudorandom generator, *Mathematics of Computation* **58**, 323–339, 1992.

[42] Wasilkowski, G.W. & Woźniakowski, H., Explicit cost bounds of algorithms for multivariate tensor product problems, *J. Complexity* **11**, 1–56, 1995.

[43] Woźniakowski, H., Average case complexity of multivariate integration, *Bull. AMS (New Series)* **24**, 185–194, 1991.

[44] Woźniakowski, H., Tractability and strong tractability of linear multivariate problems, *J. Complexity* **10**, 96–128, 1994.